Recent Mammals of Alaska

Recent Mammals of Alaska

S. O. MacDonald

and

Joseph A. Cook

University of Alaska Press
Fairbanks, Alaska

University of Alaska Press
P.O. Box 756240
Fairbanks, AK 99775-6240

hardcover edition ISBN 978-1-60223-047-7
paperback edition ISBN 978-1-60223-072-9

The Library of Congress has cataloged the hardcover edition as follows:

    MacDonald, S. O. (Stephen O.)
     Recent mammals of Alaska / by S. O. MacDonald and Joseph A. Cook.
      p. cm.
     Includes bibliographical references and index.
     ISBN 978-1-60223-047-7 (cloth : alk. paper)
     1. Mammals—Alaska. I. Cook, Joseph A. II. Title.
     QL719.A4M33 2008
     599.09798—dc22

                2008045199

Cover design by Dixon Jones, Rasmuson Library Graphics

Text design and layout by Paula Elmes, ImageCraft Publications & Design

Cover illustrations: brown bear, walrus, humpback whale, moose, and caribou by John Schoen, senior scientist, Audubon Alaska. Lynx, marmot, and Dall's sheep courtesy U.S. Fish and Wildlife Service Alaska Image Library. Back cover: snowshoe hare and sparring caribou © William D. Berry, courtesy Berry Studios. Woolly mammoth illustrated by Charles Douglas, reproduced courtesy of the Canadian Museum of Nature, Ottawa, Canada.

This publication was printed on acid-free paper that meets the minimum requirements for ANSI / NISO Z39.48–1992 (R2002) (Permanence of Paper).

To the Memory of

Olaus J. Murie (1889–1963)
and
Margaret "Mardy" E. Murie (1902–2003)

Olaus and Mardy by their home at the foot of the Grand Tetons in Wyoming, 1953 (Elizabeth English photo).

True basic research . . . means living with the animals, trying to think as they do, establishing an intimate relationship with the creatures that reveals their motivation in all they do. Such intimate, on-the-ground contact with animals, for as long as it takes to get the desired information, leads to an understanding of nature which is desperately lacking in this age of human exploitation of the planet.

—Olaus J. Murie (A. Murie 1961:vi)

Mardy on the honeymoon trail, 1924 (Murie Collection).

My prayer is that Alaska will not lose the heart-nourishing friendliness of her youth—that her people will always care for one another, her towns remain friendly and not completely ruled by the dollar—and that her great wild places will remain great, and wild, and free, where wolf and caribou, wolverine and grizzly bear, and all the Arctic blossoms may live on in the delicate balance which supported them long before impetuous man appeared in the North.

This is the great gift Alaska can give to the harassed world.

—Mardy Murie (1978:Preface)

Sparring caribou, *Rangifer tarandus* (W. D. Berry)

Contents

Preface, ix
Acknowledgments, xi

Introduction, 1

A Brief History of Mammalogy and Mammal Collectors in Alaska, 4
Appreciating the Limits of Our Knowledge of Alaska's Mammalian Diversity, 5
Beringian Coevolution Project, 18
Specimen Representation: Time and Space, 20
Study Area, 25
Regional Overviews, 26
Alaska's Island Archipelagos, 37

Materials and Methods, 41

Fieldwork, 41
Collaborations, 43
Additional Museum Studies, 43
Laboratory Studies, 43
Museum Specimens, 44
Accounts of Species and Subspecies, 44

The Recent Mammals of Alaska, 49

Faunal Composition, 49
List of Species, 50

Order PROBOSCIDEA, 53
    Family ELEPHANTIDAE—elephants, 53

Order SIRENIA, 56
    Family DUGONGIDAE—dugong and sea cow, 56

Order PRIMATES, 59
    Family HOMINIDAE—apes and humans, 59

Order RODENTIA, 62
    Family SCIURIDAE—squirrels, 62
    Family CASTORIDAE—beavers, 77
    Family DIPODIDAE—jumping mice, 80
    Family CRICETIDAE—voles, lemmings, deermice, woodrats, 84
    Family MURIDAE—Old World rats and mice, 111
    Family ERETHIZONTIDAE—New World porcupines, 115

Order LAGOMORPHA,   118
   Family OCHOTONIDAE—pikas,   118
   Family LEPORIDAE—hares and rabbits,   121

Order SORICOMORPHA,   127
   Family SORICIDAE—shrews,   127

Order CHIROPTERA,   143
   Family VESPERTILIONIDAE—evening bats,   143

Order CARNIVORA,   151
   Family FELIDAE—cats,   151
   Family CANIDAE—wolf, coyote, and foxes,   156
   Family URSIDAE—bears,   165
   Family OTARIIDAE—eared seals,   174
   Family ODOBENIDAE—walrus,   181
   Family PHOCIDAE—hair seals,   184
   Family MUSTELIDAE—weasels,   194
   Family PROCYONIDAE—raccoons,   212

Order ARTIODACTYLA,   214
   Family CERVIDAE—antlered ungulates,   214
   Family BOVIDAE—horned ungulates,   225

Order CETACEA Brisson,   234
   Family BALAENIDAE—right whales,   234
   Family BALAENOPTERIDAE—rorquals,   239
   Family ESCHRICHTIIDAE—gray whale,   247
   Family DELPHINIDAE—dolphins,   250
   Family MONODONTIDAE—beluga and narwhal,   259
   Family PHOCOENIDAE—porpoises,   263
   Family PHYSETERIDAE—sperm whales,   267
   Family ZIPHIIDAE—beaked whales,   271

Literature Cited,   275

Appendices,   337
   Appendix 1. Locations and Names of Alaska Quadrangle Topographic Maps,   338
   Appendix 2. Regional Occurrence of Alaska Land Mammals,   340
   Appendix 3. Number of Alaska Type Specimens by Institution,   343
   Appendix 4. Specimens of Alaska Mammals by Institution,   344
   Appendix 5. Habitats of Alaska Mammals,   347
   Appendix 6. Conservation Status of Alaska Mammals,   353
   Appendix 7. Introductions and Translocations of Mammals in Alaska,   357
   Appendix 8. Mammal Occurrence on Islands in Southeast Alaska,   366
   Appendix 9. Mammal Occurrence on Islands in Southcentral Alaska,   370
   Appendix 10. Mammal Occurrence on Islands in Southwest Alaska,   371
   Appendix 11. Mammal Occurrence on Islands in Western Alaska,   373
   Appendix 12. Pleistocene Mammals of Alaska and Yukon Territory,   374

Index,   381

Preface

More years than we care to reflect on have slipped by since we began assembling all the elements necessary for this catalog of the Recent mammals of Alaska. Our primary goal was to construct a usable taxonomic and geographic framework from the ever-expanding wealth of information on the mammals that occur or have occurred in Alaska since the end of the last ice age (the Holocene, approximately ten thousand years before present). We include mammals from this time period to emphasize the point that the composition of the mammalian fauna has not been static. Some of the information on these mammals was scattered in scientific publications, unpublished reports, or the experiences of informed individuals, but the foundation of this catalog is the mammal specimens preserved in natural history museums. The most notable collection of those archived materials is at the University of Alaska Museum of the North, where we both worked for many years.

This reference is admittedly incomplete and is best viewed as a work in progress. We hope that the catalog will stimulate others to fill in the many gaps in our knowledge of Alaska mammals, including a reinvigorated appraisal of the taxonomic framework that is currently serving as the basis for primary scientific investigations as well as important management decisions. Lack of basic information on systematics, distribution, and population status is a deficiency of considerable consequence as Alaska experiences rapid change. Change is accelerating in Alaska, due not just to a warming planet, but also to the exponential growth of other human-caused impacts.

This catalog has been a labor of love that is grounded in our lifelong passion for Alaska, its wildlife, and its wildness. Our concern centers on the possibility that this special place and its intact communities of life will fail to flourish. For the sake of future generations, our hope is not only that we will learn to appreciate Alaska's richness, but that, in the end, we will become better stewards of this incomparable land and its inhabitants.

Sketch by William D. Berry

Acknowledgments

A summary monograph like this one results only because of the work and encouragement of many people who ultimately are not credited on the author-line. Those individuals facilitated this research in a wide variety of ways, ranging from University of Alaska colleagues such as Brina Kessel, whose private donation established the faculty position of curator of mammals, and Dean Paul Reichardt, who first supported a graduate assistantship for the mammal collection; to private individuals like Ralph Seekins, who freely provided vehicles for extended field expeditions, and Paul Packard, who repeatedly opened his home on the Kenai to tired, hungry, and dirty field crew members, and Bruce J. Hayward whose generous donations endowed field mammalogy in Alaska; to individuals from state and federal agencies, like Kim Hastings, Tom Hanley, Cole Crocker-Bedford, Rod Flynn, and Sara Wesser who understood what we were up to, navigated their respective bureaucratic mazes, and cheerfully supported our field and laboratory investigations.

We sincerely apologize to those we may have inadvertently failed to name. The following people are specifically thanked for their generous assistance: **Private**—T. Bethel, R. Bishop, K. Bovee, S. Brewington, J. Burns, L. and J. Carson, R. Carstensen, F. Cook, L. Cook, N. Cook, T. Cook, S. Fiscus, J. Florie, C. Garst, S. Geraghty, D. Harbor, B. Hartley, B. J. Hayward, R. Himschoot, the late M. E. Isleib, R. Jahnke, L. Johnson, M. MacDonald, N. MacDonald, O. MacDonald, D. and D. Nelson, J. Owens, R. L. Rausch, D. Rice, L. Robertson, B. Slough, G. Streveler, M. Tapia, T. and K. Wills, and the Friends of the University of Alaska Museum. **Alaska Department of Fish and Game**—N. Barten, K. Blejwas, T. Boudreau, G. Carroll, R. Clarke, E. Crain, J. Dau, S. DuBois, H. Golden, J. Gustafson, J. Hechtel, R. Hunter, M. Kirchhoff, P. Koehl, C. Land, D. Larsen, T. McDonough, M. McNay, D. Parker McNeill, T. Osborne, D. Person, K. Persons, B. Porter, M. Robus, C. T. Seaton, J. Schoen, J. Selinger, M. Sigman, R. Sinnott, K. Titus, L. van Daele, J. Whitman. **U.S. Bureau of Land Management**—J. Denton. **U.S. Fish and Wildlife Service**—S. Brockmann, S. Ebbert, E. Grossman, W. Hanson, N. Holmberg, D. Klein, E. Lance, J. Lindell, J. McClung, L. Saperstein, P. Schempf, S. S. Talbot. **U.S. Geological Service**—K. Sage, S. L. Talbot, M. Bogan, C. Ramotnik. **U.S. Forest Service**—J. Baichtal, S. Blatt, M. Brown, E. Campbell, J. Canterbury, D. Chester, R. Claire, G. DeGayner, T. DeMeo, M. Dillman, J. Falk, M. Goldstein, R. Guhl, C. Iverson, D. Johnson, L. Kvaalen, T. O'Connor, P. Robertson, A. Russell, K. Rutledge, T. Schenck,

C. Sietz-Warmuth, T. Shaw, T. Suminski, S. Wise-Eagle. **National Park Service**—J. Burch, L. Dolle-Molle, N. Guldager, M. Henderson, W. Leacock, H. Lentfer, C. McIntyre, D. Moore, B. Schultz, D. Sharp, S. Swanson, J. Taggart. **Idaho State University**—K. Bell, K. Gamblin, D. Lucid, M. Lucid, E. Tomasik. **Sheldon Jackson College**—M. Siefert, B. Colthrap, M. Ahlgren. **University of New Mexico**—D. Crawford, N. Dawson, J. Frances, D. Goade, A. Hope, A. Koehler, A. Lynch, J. Malaney, R. McCain, C. McLaren, R. Nofchissey, B. Schaaf. **Western New Mexico University**—Interlibrary Loan Department. **Vantaa Research Centre** (Finland)—H. Henttonen, J. Laakkonen, J. Niemimaa. **Institute of Biological Problems of the North** (*Magadan, Russia*)—the late F. Chernyavski, N. Dokuchaev, A. Lazutkin, A. Tsvetkova. **USDA National Parasite Laboratory**—E. Hoberg. **North Slope Borough, Division of Wildlife Management**—C. George, T. Albert, R. Suydam. **Yukon Department of Environment**—T. Jung. **University of Saskatchewan**—S. Kutz, B. Wagner, B. Delehanty, C. Jardine. **Tufts University**—S. Telford. **Curators and collection managers of museums**—M. Carleton, J. Dunnum, R. Fisher, W. Gannon, A. Gardner, S. George, R. S. Hoffmann, G. J. Kenagy, D. Nagorsen, J. Patton, S. Peurach, V. R. Rausch, M. Reich, A. Schwandt-Arbogast, R. Timm, C. G. van Zyll de Jong, and the support of the students and professional staff of the **University of Alaska Museum of the North and the University of Alaska Fairbanks Institute of Arctic Biology**—K. Bagne, L. Barrelli, A. Batten, M. Ben-David, D. Bender, J. Bender II, J. Blake, S. Brunner, J. Bryant, F. S. Chapin, H. Chen, C. Conroy, J. Demboski, T. Dyasuk, F. Fay, A. Fedorov, V. Fedorov, K. Fisher, K. Galbreath, R. Ganglof, A. Goropashnaya, D. D. Gibson, S. Gray, R. D. Guthrie, J. Haddix, A. Jonaitis, K. Kellie, K. Kielland, D. Klein, J. Levino-Chythlook, E. Lessa, S. Lewis, D. Murray, L. Olson, C. Parker, E. Rexstad, A. Runck, S. Runck, K. Stone, E. Waltari, M. Ward, M. Wike, and in particular, B. Jacobsen, D. McDonald, and G. Jarrell. G. Jarrell and D. McDonald worked tirelessly to develop the ARCTOS database, generate specimen lists, and provide preliminary distribution maps. L. Olson and B. Jacobsen generously provided specimen access and information.

Over the past few decades, our field, museum, and laboratory studies on Alaska mammals were supported through grants or logistical assistance from the National Science Foundation (9876837, 9981915, 0196095), Hayward Fund, National Park Service, U.S. Fish and Wildlife Service (0415668), U.S. Forest Service, Bureau of Land Management, U.S. Geological Service, Alaska Sea Otter Commission, Alaska Harbor Seal Commission, Coastal Marine Institute, National Marine Fisheries Service, and the Alaska Department of Fish and Game. G. V. Byrd, S. Ebbert, J. Findley, D. D. Gibson, B. Jacobsen, C. Jones, S. Brunner, L. Olson, R. L. and V. R. Rausch, and J. Whitman provided critical reviews of the manuscript.

Finally, we thank Berry Studios (www.berrystudios.biz) for granting us permission to use some of the fine illustrations of Alaska mammals by the late Fairbanks artist William D. Berry (1926–1979); and thanks to Orien MacDonald for use of several of his mammal illustrations.

All errors are solely the responsibility of the authors.

Introduction

Mammals have long been important to Alaskans. From the earliest indigenous peoples who colonized this vast region tracking wildlife across the Bering Land Bridge, and the Russian colonists who were lured later by vast quantities of sea otter pelts, to the large-scale American whaling operations that followed several decades thereafter, mammals have played a key role in the exploration and colonization of Alaska. More recently, humans have been drawn to Alaska because it represents one of the last places on the planet to still experience wilderness and enjoy plentiful wildlife.

In the future, informed decisions regarding human impact on natural environments will require a more detailed understanding of Alaska's biota. In particular, the impacts of global climate change, development of mineral and petroleum reserves, and deforestation of the boreal and coastal forests on these high-latitude ecosystems must be documented, monitored, and mitigated.

This catalog provides the first comprehensive overview of the 116 species of mammals that have been documented in Alaska and adjacent waters (Fig. 1) during the last ten thousand years. We summarize their taxonomy, distribution, status, habitat affinities, and fossil history. No similar reference has been available for mammals found in Alaska, although numerous research efforts have focused on specific components of the mammalian fauna.

William H. Dall, in 1870, was the first to publish a list of Alaska's mammals in his *Alaska and Its Resources,* followed three-quarters of a century later (in 1946) by Frank Dufresne's *Alaska's Animals and Fishes.*

The first synoptic treatment of mammals that included Alaska was Gerrit Miller's 1924 *List of North American Recent Mammals,* which was updated with coauthor Remington Kellogg in 1955. Not until 1959, however, when E. Raymond Hall and Keith R. Kelson published *The Mammals of North America,* was a single reference available that included all known Alaska taxa. That monograph also provided the first maps that defined the distributional boundaries of these species based on museum specimens. Hall updated that publication in 1981.

A book of regional importance, published by the University of Kansas Museum of Natural History in 1956, was *Mammals of Northern Alaska on the Arctic Slope* by James Bee and E. Raymond Hall. A useful, though limited, publication that appeared in 1965 was *Distribution of Alaskan Mammals* by Richard Manville and Stanley Young, and in 1973 and 1978 the Alaska Department of

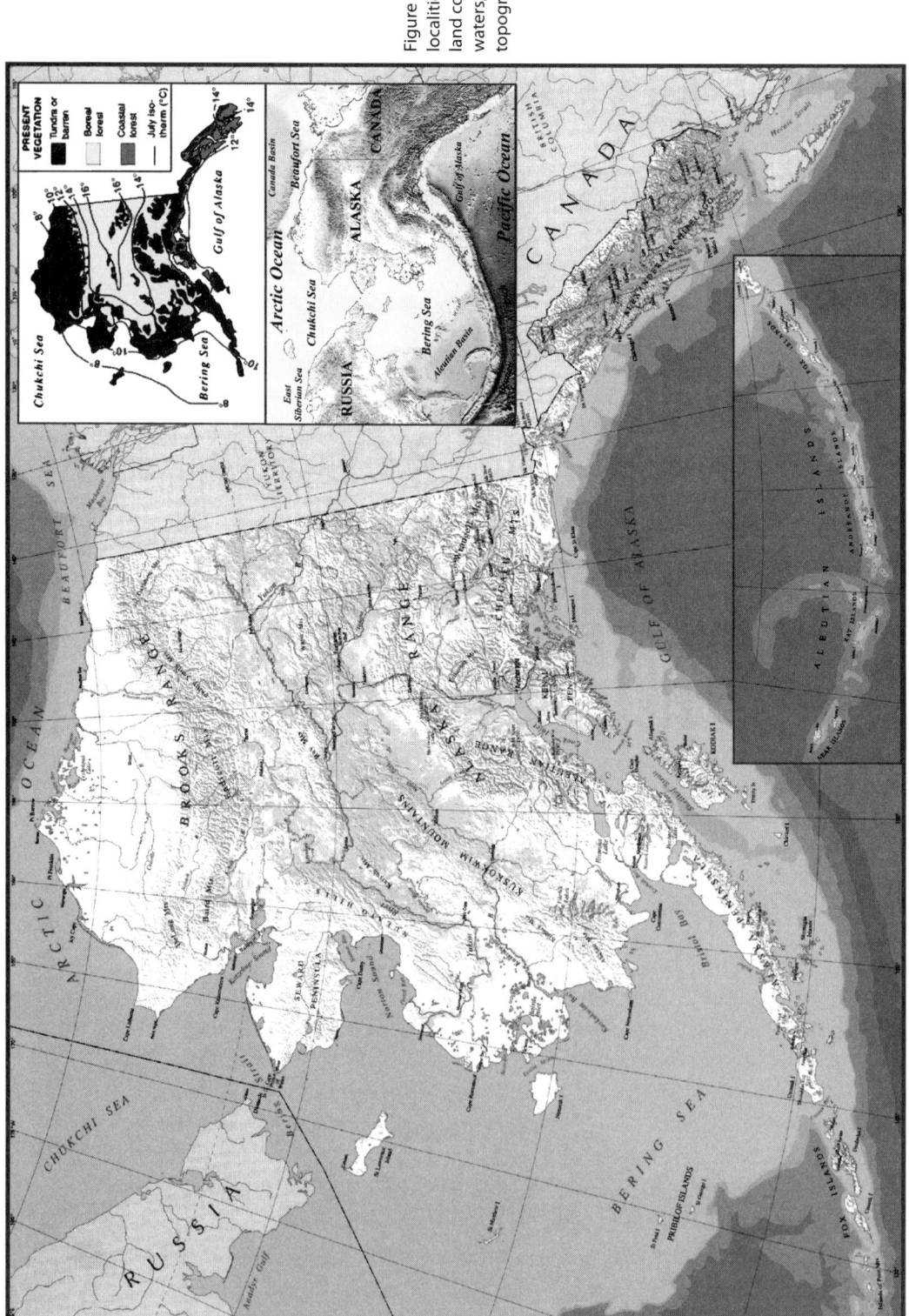

Figure 1. Alaska localities, major land cover, marine waters, and sea floor topography.

Fish and Game published accounts with detailed maps of Alaska game species and their habitat.

Since Hall's 1981 publication, more than thirty taxonomic changes due to various revisions and a dozen new species have significantly altered our understanding of Alaska's mammalian fauna. Collaborations among state and federal agencies, bush communities, and university museums over the past two decades also created a wealth of new distributional and taxonomic information. Those efforts more than doubled the number of specimens and associated data that form the basis of this catalog of Alaska mammals.

These specimens are key to investigations aimed at documenting and exploring historic conditions of Alaska's environments (MacDonald and Cook 2007). By representing discrete slices of space and time across Alaska's biomes, archived specimens provide unique opportunities to apply established (e.g., morphology; Yom-Tov and Yom-Tov 2005) and new methods (e.g., stable isotope ecology; Hirons et al. 2001) to investigations aimed at assessing environmental change. Thus, natural history collections establish baselines and are becoming central to studies aimed at understanding the consequences of climate warming and other perturbations (May 1993; Millien et al. 2006; Suarez and Tsutsui 2004). Similarly, identifying and monitoring contaminants (Chapman 2005) or newly emerging pathogens (Brooks and Hoberg 2006; Kutz et al. 2004; Yates et al. 2002) can be facilitated by specimen collections (Fig. 2). High concentrations of persistent toxic substances in marine mammals include both naturally occurring toxicants, such as heavy metals, as well as the anthropogenic substances, such as polychlorinated biphenyls (PCBs) and chlorinated hydrocarbon pesticides. Marine mammals bioaccumulate contaminants due to their long life span, high position in food webs, and tendency to accumulate body fat. Tissue samples have become critical to efforts to monitor pollutants in real time (Hoekstra et al. 2005). Associated web-accessible electronic databases (http://www.arctos.database.museum) and rigorous curatorial standards provide a significant integrated resource for ecological, evolutionary, epidemiological, and toxicological research on boreal organisms (Graham et al. 2004).

Among the serious challenges now facing the scientific community is the need to effectively communicate and educate the general public about the implications of environmental change. Natural history museums provide an important portal through scholarly publications, exhibits, and the internet into the everyday work of scientists by helping to digest technical studies into terms more accessible by the lay public (Wake 1988). Due to increased specialization, scientists tend to work in ever-narrowing fields of research yet the most significant findings are often at the interface of distinct disciplines. With critical environmental issues now facing human societies, there is an urgent need for scientists such as systematists, toxicologists, molecular geneticists, ecologists, wildlife veterinarians, and emerging pathogen specialists to begin to interact, collaborate, and understand the diverse perspectives of each discipline (Fig. 3). Another challenge, therefore, involves the need to take holistic or multidisciplinary approaches when tackling complex environmental or emerging medical problems. Because a single specimen is often used in many different studies, museums have become key to efforts to bridge across discrete scientific disciplines. Diverse studies thus are tied immediately to the same organism, which represents a discrete data-collecting event in time and space. Interdisciplinary studies are one of the hallmarks of museums, and today numerous studies can be tied together through their uses of different materials archived from the same specimen and hence the same data-collecting event. Finally, by integrating ongoing policy, management, and research initiatives, museum collections are now vital to ongoing environmental investigations and resource management decisions in the North (Fig. 3). Challenges for museums include the need to sample regularly and archive material representative of biotic communities in ways that will be valuable to diverse investigations in the future. One example of an integrative effort is ongoing collaborations underway between mammalogists, parasitologists, and human disease specialists through the Beringian Coevolution Project.

A Brief History of Mammalogy and Mammal Collectors in Alaska

Numerous individuals supported by a diverse set of institutions (Fig. 3) have conducted field expeditions aimed at documenting the distribution and taxonomy of mammals in Alaska. We cover only the early expeditions that resulted in the exploratory efforts to document the diversity and distribution of mammals in Alaska in this brief synopsis. Mammalogy is a broad field and many productive mammalogists have investigated various aspects of mammalian biology in Alaska. Of particular note is work at the University of Alaska Fairbanks by comparative physiologists like Laurence Irving (e.g., Dawson 2007; Irving 1939), P. F. Scholander (Scholander et al. 1942), Brian Barnes (Barnes 1989), and ecologists Dave Klein (Klein 1968), John Bryant (Bryant et al. 1983), and Terry Bowyer (Bowyer et al. 1998). Similarly, agencies such as the Alaska Department of Fish and Game, U.S. Fish and Wildlife Service, USDA Forest

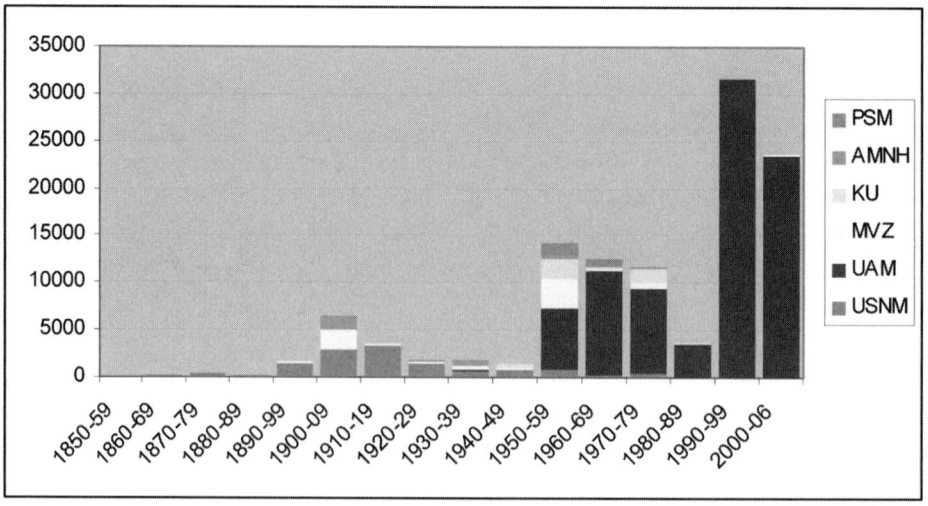

Figure 2. Collecting efforts (number of mammal specimens preserved) over the years by various institutions (acronyms are defined on pages 46–48).

Appreciating the Limits of Our Knowledge of Alaska's Mammalian Diversity

Epic geologic events over the last two million years, combined with fluctuating climates and habitats, led to repeated separation of Alaska's terrestrial fauna followed by association with either Siberian or North American continental elements. Similarly, marine mammals were separated and reunited by events such as the emergence of the Bering Land Bridge. Substantial geographic barriers (e.g., mountain ranges, shallow oceanic straits) have set the evolutionary stage for allopatric divergence and continue to shape biotic diversity across this region. Alaska is a particularly attractive place to study evolutionary diversification and adaptation, but a significant limitation remains the poorly documented biota. Herein, we particularly focus on a few negative consequences that result from our limited understanding of taxonomy (and related geographic variation) in the 116 mammalian species that have occurred in Alaska since the end of the last ice age.

Systematics is the science broadly concerned with biological diversity, while **taxonomy** is more specifically the theory and practice of classifying organisms (Mayr 1969). A reliable taxonomic framework is essential for pure and applied investigations of Alaska's biota. At the start of any study, investigators need to identify the relevant taxon (unit) for analysis or management (e.g., species, subspecies, population). Many investigators (or bureaucrats) might assume that we now have this basic information well in hand, but there is much that we do not know about even a "well studied" group like mammals. Worldwide, the number of known mammal species on earth increased about 5 percent (260 new species) in the ten years between 1993 and 2003 (Wilson and Reeder 2005). The situation is similar in Alaska where new species (e.g., Demboski and Cook 2001; Dokuchaev 1997) or distinct populations (Weckworth et al. 2005) regularly are being discovered or rediscovered (e.g., Cook et al. 2006). Most members of the mammalian fauna of Alaska have not been systematically reviewed since the early investigations by members of the U.S. Biological Survey over a century ago. Taxonomic descriptions that emanated from those early studies relied principally on the assessment of morphologic characters from a limited number of specimens. Pioneering high-latitude mammalogists, such as Merriam, Osgood, Howell, and Miller, were extremely skilled in recognizing and describing geographic variation, especially considering the relatively few specimens available at that time for detailed study. Since those initial taxonomic studies, technological and theoretical advances have greatly enhanced our ability to describe geographic variation, delineate species' boundaries, and identify distinct populations or evolutionarily significant units (ESUs) (Avise 2000). A series of recent papers is applying these techniques in reassessments of many of Alaska's mammals (e.g., Steller's sea lion; Baker et al. 2005; Bickham et al. 1996). A significant limitation to completing this fundamental task (i.e., developing a rigorous taxonomic framework) is the lack of available specimens (especially marine mammals) that represent populations from large regions of the state (e.g., North Slope and Brooks Range, Yukon-Kuskokwim Delta, most islands, and, perhaps surprisingly, many state and federally managed lands). Fundamental taxonomic questions remain unanswered. How distinctive are the Cook Inlet beluga whales (O'Corry-Crowe et al. 2002), Alexander Archipelago wolf (Weckworth et al. 2005), Prince of Wales flying squirrel (Bidlack and Cook 2002), and various populations of harbor seals (Burg et al. 1999; Shaughnessy and Fay 1977; Westlake and O'Corry-Crowe 2002)? Are *Sorex alaskanus, S. ugyanak, Dicrostonyx groenlandicus, Alces americanus, Cervus canadensis,* and *Martes caurina* valid species?

Field research on mammals in Alaska has been hampered due to the incomplete accounting of mammals. An ecological investigation of small mammals on the Kenai Peninsula (Bangs 1979) reported the occurrence of the meadow vole, *Microtus pennsylvanicus,* a species that has not otherwise been documented from the peninsula. Unfortunately, voucher specimens are not available to test the validity of that published species identification. Farther north, a series of seminal studies of "super-cooling" in hibernating arctic ground squirrels (*Spermophilus parryii*) (Barnes 1989; Buck and Barnes 1999) was not aware that more than one species of ground squirrels may occur in Alaska. Recent phylogeographic work (Eddingsaas et al. 2004) suggests that the Brooks and Alaska mountain ranges support distinct species of ground squirrels with different evolutionary histories.

Elsewhere on the planet, our failure to recognize the limitations of existing taxonomies had disastrous conservation consequences (Daugherty et al. 1990). A renewed effort in systematic mammalogy is needed to avoid similar mistakes in Alaska, although our incomplete taxonomic framework already has impacted species such as the Pacific marten, *Martes caurina.* Remnant populations of this marten now occur on only two Alexander Archipelago islands (Kuiu and Admiralty), but previously this medium-sized forest carnivore apparently was widespread (MacDonald and Cook 2007). Human introduction of the other marten species (*M. americana*) to several islands and heavy deforestation in the region (Cook et al. 2006; Koehler, unpubl. data) may have had dire consequences for the viability of the Pacific marten. Maintenance of viable and well-distributed populations in the presence of impending threats will require a more complete knowledge of the distributions and systematics of Alaska's mammalian fauna.

Taxonomy thus provides the classification system for organisms that is the basis for rigorous project design and efficient prioritization of funding. More than just providing a name, though, a taxonomic framework should accurately reflect the evolutionary relationships among the taxa, because all disciplines within biology are dependent on a basic understanding of evolution (Dobzhansky 1973), including genealogical relationships (Baum et al. 2005).

Phylogeography is a new field aimed at analyzing the geographic distribution of lineages based primarily on gene sequence variation. These molecular approaches provide independent tests of previous hypotheses of geographic variation based on other methods, such as morphological comparisons (Avise 2000). Conceptually, phylogeography bridges the more traditional fields of microevolution (population genetics) and macroevolution (systematics). Spatial relationships of gene genealogies are analyzed to assess the evolutionary history of species and even populations (e.g., population expansion related to colonization of new areas). Assessments of DNA variation, along with new analytical methods based on coalescent theory, are now being combined with palaeoclimatic and geologic investigations to reveal new insights into the distribution and evolution of genetic diversity in Alaska (Waltari et al. 2007).

Current diversity patterns in Alaska's mammals are strongly influenced by both historic events such as ice ages and recent effects such as active wildlife management or habitat conversion. Phylogeographic investigations at high latitudes indicate that Pleistocene events played a large role by repeatedly isolating certain populations during glacial advances. Conservation of these divergent populations (postglacial relicts) should be a high priority in management decisions, because relictual populations often suffer more severely from environmental stochasticity and anthropogenic land-use changes (habitat fragmentation) than nonrelict populations. Phylogeography provides a basis for identifying relict species and populations and ultimately for organizing and prioritizing efforts to conserve their corresponding genetic and phenotypic diversity. These new perspectives may inform ongoing debates related to the effects of resource development on wild populations of mammals in Alaska, but only if museum specimens are available that allow site-intensive (large sample size) and geographically extensive studies to be completed.

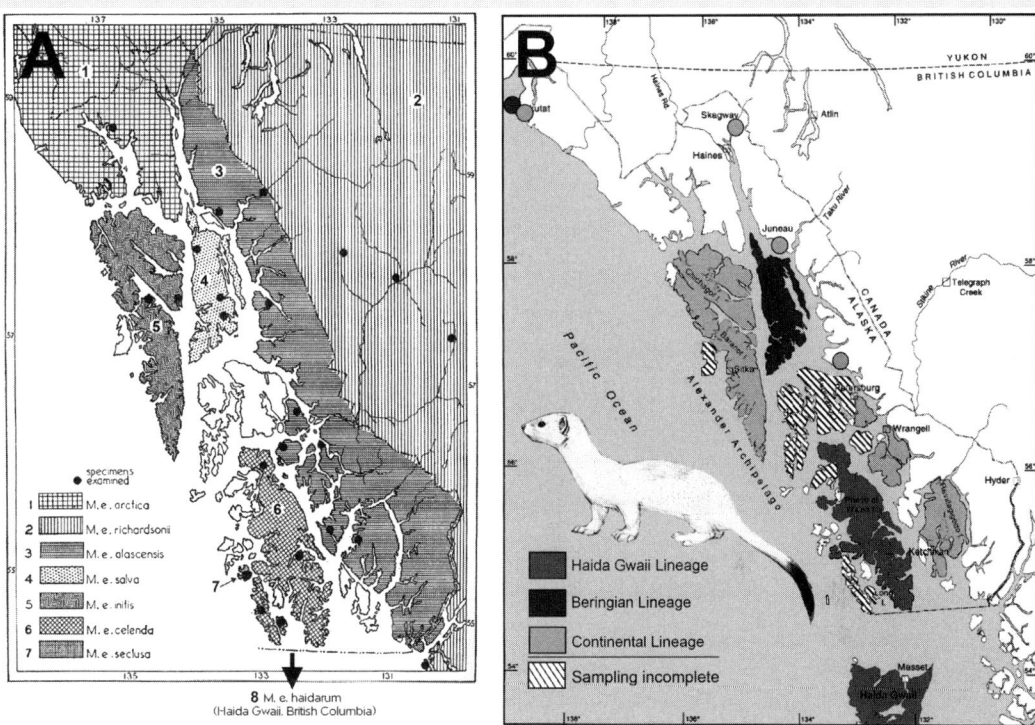

Map showing currently recognized subspecies of ermine, *Mustela erminea*, in southeast Alaska (A) based on skull morphology (after Hall 1951), in contrast to evolutionary lineages (B) based on recent molecular data (after Cook et al. 2006). Are there eight subspecies of ermine in this region or only three? Do the three lineages represent separate species?

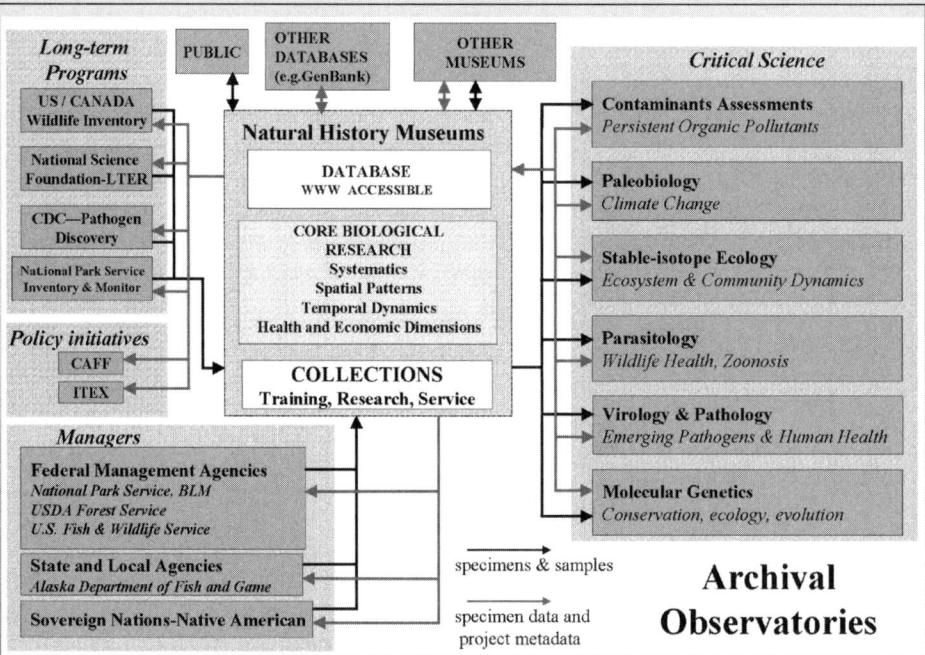

Figure 3. Archival Observatories (aka natural history museums) provide a powerful mechanism for integrated studies of environmental change, emerging pathogens, wildlife conservation and a host of other issues important to society. The developing cyberinfrastructure of web accessible databases such as ARCTOS now links this important specimen-based resource to sample providers, resource managers, major policy initiatives, a diverse array of scientists, other large databases (e.g., GenBank), on-line GIS applications (Berkeley Mapper), and the general public. Because Archival Observatories are often located at major research universities, this model is allowing the next generation of scientists to gain critical exposure to integrated science. Below, the type specimen for one of Alaska's endemic mammals is preserved at the MVZ, University of California Berkeley. That university is a world-leader in training environmental and evolutionary mammalogists.

Holotype (MVZ 51675) of the Alaska marmot, *Marmota broweri*

Service, National Park Service, North Slope Borough Division of Wildlife Management, and, more recently, U.S. Geological Service have supported a number of investigations of mammals by individuals such as Edgar Bailey, Jim Brooks, Kathy Frost, Craig George, Loyal Johnson, Karl Kenyon, Charley Land, Cal Lensink, Jack Lentfer, Lloyd Lowry, Jim Rearden, Vic Scheffer, John Schoen, Sandy Talbot, and Olaf Wallmo, to name only a few. With regard to studies on the distribution and taxonomy of the mammals of Alaska, a primary publication outlet has been the (now defunct) North American Fauna series (Table 1).

Below we briefly summarize a few of the major contributors to our early knowledge on Alaska's mammals.

Early Explorations

The first naturalists of Alaska were the Native peoples who began colonizing the region from Asia at least thirteen thousand years ago. Oral traditions of these cultures have provided key insights into the ecology of high-latitude mammals (e.g., Huntington and Elliott 2002).

The first European contact with Alaska was much later, in 1741, during the Great Northern Expedition led by Vitus J. Bering and including naturalist **Georg W. Steller** (1709–1746). The expedition put briefly ashore on Kayak and Nagai islands, then shipwrecked on one of the Komandorski (Commander) Islands (Ostrov Beringa). A second ship that separated from Bering was commanded by Aleksei Chirikov and it sailed south to discover Alaska's panhandle.

Captain James Cook's (1728–1779) last voyage in 1778 and 1779 for Britain had no naturalist aboard, but his men did secure some specimens. This expedition visited Prince William Sound ("Sandwich Sound"), Unalaska, and St. Matthew Island. William Anderson, surgeon on the *Resolution*, kept a diary and made notes on the birds and mammals seen or collected.

Carl Heinrich Merck visited Kodiak Island in winter of 1789 through July 1790 as naturalist of the Billings Expedition.

Otto von Kotzebue (1787–1846) of the Russian Navy was accompanied by German naturalist Adelbert von Chamisso, who collected (mostly plants) in the Kotzebue Sound area and on St. Lawrence Island in 1816–1817.

Captain F. W. Beechey (1796–1856), commander of the second British ship, *HMS Blossom,* visited Kotzebue Sound and various points north to Barrow between 1825 and 1828. Mammals encountered were reported by John Richardson (1839).

Ilja G. Voznesenskii (1816–1876) collected mammals for the Russian Academy of Sciences (housed at the Zoological Institute in St. Petersburg) during the years 1840–1845 (Alekseev 1987). Specimen localities include Kodiak, Sitka, Saint Michael, Bristol Bay, St. Paul, and Kenai (A. Tsvetkova, pers. comm., 2007).

The U.S. Army Signal Corp and the Smithsonian Institution

American exploration began with naturalists attached to the Russian-American Telegraph Expedition, beginning in 1865. Under the auspices of the Smithsonian

Table 1. Publications of the North American Fauna Series pertinent to Alaska mammals.

No. 4	(1890)	Descriptions of Twenty-six New Species of North American Mammals, by C.H. Merriam. 60 p.
No. 10	(1895)	Synopsis of the American Shrews of the Genus *Sorex*, by C.H. Merriam. Pp. 57–98.
No. 11	(1896)	Synopsis of the weasels of North America, by C.H. Merriam. 44 p.
No. 12	(1896)	Genera and Subgenera of Voles and Lemmings, by G.S. Miller, Jr. 78 p.
No. 13	(1897)	Revision of the North American Bats of the Family Vespertilionidae, by G.S. Miller, Jr. 140 p.
No. 15	(1899)	Revision of the Jumping Mice of the Genus *Zapus*, by E.A. Preble. 42 p.
No. 17	(1900)	Revision of American Voles of the Genus *Microtus*, by V. Bailey. 88 p.
No. 19	(1900)	Results of a Biological Reconnaissance of the Yukon River Region. General Account of the Region. Annotated List of Mammals, by W.H. Osgood. Pp. 21–46.
No. 21	(1901)	Natural History of the Queen Charlotte Islands, British Columbia; and Natural History of the Cook Inlet Region, Alaska, by W.H. Osgood. 87 p.
No. 24	(1904)	A Biological Reconnaissance of the Base of the Alaska Peninsula, by W.H. Osgood. 86 p.
No. 28	(1909)	Revision of the Mice of the American Genus *Peromyscus*, by W.H. Osgood. 285 p.
No. 30	(1909)	Biological Investigations in Alaska and the Yukon Territory: 1. East-central Alaska; 2. Ogilvie Range, Yukon; 3. Macmillan River, Yukon, by W.H. Osgood. 96 p.
No. 32	(1911)	A Systematic Synopsis of the Muskrats, by N. Hollister. 47 p.
No. 37	(1915)	Revision of the American Marmots, by A.H. Howell. 80 p.
No. 41	(1918)	Review of the Grizzly and Big Brown Bears of North America (Genus *Ursus*) with Description of a New Genus, *Vetularctos*, by C.H. Merriam. 136 p.
No. 44	(1918)	Revision of the American Flying Squirrels, by A.H. Howell. 64 p.
No. 46	(1923)	A Biological Survey of the Pribilof Islands, Alaska. Part 1, Birds and Mammals, by E.A. Preble. Pp. 1–128.
No. 47	(1924)	Revision of the American Pikas (Genus *Ochotona*), by A.H. Howell. 57 p.
No. 48	(1926)	Voles of the Genus Phenacomys: I, Revision of the Genus *Phenacomys*. II, Life History of the Red Tree Mouse *(Phenacomys longicaudus)*, by A. Brazier Howell. 64 p.
No. 50	(1927)	Revision of the American Lemming Mice (Genus *Synaptomys*), by A. Brazier Howell. 38 p.
No. 51	(1928)	A Taxonomic Review of the Long-tailed Shrew (Genera *Sorex* and *Microsorex*), by H.H.T. Jackson. 238 p.
No. 54	(1935)	Alaska-Yukon Caribou, by O.J. Murie. 93 p.
No. 56	(1938)	Revision of the North American Ground Squirrels, with a Classification of the North American Sciuridae, by A.H. Howell. 256 p.
No. 61	(1959)	Fauna of the Aleutian Islands and Alaska Peninsula, by O.J. Murie. 364 p.
No. 68	(1969)	The Sea Otter in the Eastern Pacific Ocean, by K.W. Kenyon. 352 p.
No. 74	(1982)	Ecology and Biology of the Pacific Walrus, *Odobenus rosmarus divergens*, Illiger, by F.H. Fay. 279 p.

Institution and the Chicago Academy of Science, **Major Robert Kennicott** (1835–1866), with Charles Pease and H. M. Bannister, landed at St. Michael for the purpose of exploring the Yukon. In July 1866, **William Healey Dall** (1845–1927) traveled north to St. Michael and learned that Kennicott had died in Nulato. Dall was placed in charge of the scientific work and reached Nulato in December 1866. In the spring of 1867, Dall traveled as far east as Fort Yukon. On returning to the mouth of the Yukon River, he learned the enterprise had been abandoned. Instead of leaving with

Robert Kennicott

the others, he remained and finished the work before returning to the States in 1867. The specimens collected were divided between the U.S. National Museum (USNM, now known as the National Museum of Natural History) and the Chicago Academy of Sciences, with most of the latter material destroyed in the great Chicago fire. He published a report on Alaska, its resources, and his adventures (Dall 1870). Dall returned to Alaska for two seasons with the U.S. Coast Survey, 1871–1873, working in southwestern Alaska as far west as Attu.

William Healy Dall

Following Dall, the next significant information on mammals resulted from the activities of the U.S. Army Signal Corps as they established meteorological stations throughout Alaska. Some of these observers gathered specimens for the Smithsonian Institution, and especially noteworthy collectors were **Lucien McShan Turner** (1848–1909) and **Edward William Nelson** (1855–1934). Turner was a meteorologist stationed at St. Michael near Nome, but he was also under direction of Spencer Fullerton Baird, then Secretary of the Smithsonian Institution. Turner joined the Signal Corps and worked in Alaska until 1881, publishing his Contributions to the Natural History of Alaska in 1886. Turner was relieved by Private E. W. Nelson in 1877. Nelson spent over four years in the Arctic and later became an accomplished mammalogist. He was the third Chief of the U.S. Biological Survey. Nelson and **Frederick W. True** (1858–1914), first head

Edward William Nelson

curator of mammals at the USNM and a specialist in whales, published on the mammals of northern Alaska (Nelson and True 1887).

Henry Wood Elliott (1846–1930) studied northern fur seals on the Pribilof Islands in 1874 (Elliott 1896), and sailed north over 370 km to view the polar bears that live year-round on St. Matthew Island (Elliott 1881). Earlier reports "did not cause us to be equal to the sight we saw, for we met bears, yea hundreds of them." Elliott and his party surveyed the island for nine days and were never out of sight of bears. He estimated 250 to 300 bears.

Charles Leslie McKay (1855–1883), also of the Signal Corps, was stationed at Bristol Bay between 1881 and 1883. Frederick W. True provided an annotated list of the mammals collected by McKay from this area in 1886.

Leonhard Stejneger

Leonhard Stejneger (1851–1943), eminent naturalist with a long interest in marine mammals, emigrated to the United States from Norway in 1881 to work for the Smithsonian. The year after his arrival he was sent to Bering Island, which stimulated a passionate interest in Georg Steller's earlier work (and the Steller's sea cow) that culminated in Stejneger's classic biography of Alaska's first naturalist (Stejneger 1936).

John Murdoch and **A. M. Sergeant** coauthored a report (in 1885) on the natural history, including mammals, of the International Polar Expedition to Point Barrow, Alaska (1881–1883), by the U.S. Army Signal Corps.

Edward A. McIlhenny (1872–1949), leader of the McIlhenny Expedition of 1897–1898, spent more than a year at Point Barrow. He also collected at Point Hope, Cape Lisburne, and Wainwright. This expedition was organized for the primary purpose of securing zoological material and was the first in Alaska with this specific objective. Specimens were deposited at the USNM and a summary of this expedition was published (Stone 1900).

The U.S. Biological Survey and the U.S. National Museum

The U.S. Biological Survey (USBS), beginning in the late 1890s, produced many new insights into Alaska's mammalian fauna through surveys conducted

Clinton Hart Merriam

under the direction of **Clinton Hart Merriam** (1855–1942). Those surveys resulted in four important contributions to the North American Fauna series of the USBS based on new specimens from vast sections of the territory. **Clark P. Streator** (1866–1952) participated in the USBS survey in 1895, and his collecting at Yakutat Bay and various points in southeast Alaska provided the first scientific view of the mammals of these coastal sites. A few years later, the Harriman Expedition to Alaska in 1899 included C. Hart Merriam and **Albert Kenrick Fisher** (1856–1948) of the USBS. Their itinerary included visits to Wrangell, Glacier Bay, Sitka, Yakutat Bay, Prince William Sound, Cook Inlet, Kodiak Island, Kukak Bay, Unalaska, the Pribilofs, and St. Matthew and Hall islands.

Principal work of the USBS in Alaska was done under the leadership of **Wilfred H. Osgood** (1875–1947), who was with the USBS from 1897 to 1909 and was later curator at the Field Museum of Chicago. His first trip, in 1899 with Dr. Louis B. Bishop (and Alfred G. Maddren as assistant), covered the territory from Skagway to Whitehorse, then down the Yukon River to St. Michael (Osgood 1900). In 1900, Osgood (1901) visited Cook Inlet with **Edmund Heller** (1875–1939). They started at Seldovia and collected near

Hope, Tyonek, and several other places. In 1902, Osgood's most extensive trip, again with Maddren as assistant and Walter Fleming as camp hand, started from the base of the Alaska Peninsula at Iliamna Bay on 10 July and ended at Cold Bay in late October (Osgood 1904). In 1903, Osgood and **Ned Hollister** (1876–1924) made a second trip into the Yukon, again taking a route from Skagway to Whitehorse, and then floating down to Circle and back to Eagle, where they packed into the Glacier Mountains (Osgood 1909). In 1904, Osgood worked in nearby Yukon Territory.

Wilfred H. Osgood

Olaus J. Murie (1889–1963) was a member of the U.S. Biological Survey and conducted mammalian research similar to that carried out by Osgood. Murie first arrived in the territory in 1920 and remained until 1926. After 1926, he studied Alaska mammals intermittently. Between 1920 and 1922, Murie traveled extensively throughout central Alaska, and in July of 1922 he worked with his brother Adolph in Mt. McKinley National Park. He later traveled north to the Kokrines, Melozitna, Alatna, and the upper Koyukuk, before crossing to the upper Chandalar and on to the Yukon. In 1924, Murie led a Biological Survey Expedition that worked in the Hooper Bay area of the Yukon Delta. This party included **Herbert Brandt, H. B. Conover**, and **Frank Dufresne**. In 1925, he explored the western portion of the Alaska Peninsula. Later that year,

Olaus Murie led an expedition in Hooper Bay in 1924.

he went up the Porcupine River to the Old Crow River. His last extensive assignment was his classic survey of the Aleutian Islands in summers of 1936 and 1937 (Murie 1959).

Frank Dufresne (1896–1966), a member of the 1924 USBS expedition to Hooper Bay, helped shape the first comprehensive Alaska Game Act and subsequently served as the Director of the federal Alaska Game Commission. He published a popular book on selected mammals and fishes of Alaska in 1946.

The Museum of Vertebrate Zoology, University of California, Berkeley

Joseph Grinnell (1877–1939) was in Sitka in the summer 1896. Two years later, in the spring of 1898, he joined prospectors and explored the Kobuk River and Kotzebue region. He wintered at the junction of the Hunt and Kobuk rivers and remained there until June 1899. Those two experiences piqued an interest in Alaska that resulted in his efforts to later organize the Alexander Expeditions financed by **Annie M. Alexander** (1867–1950) and the newly formed Museum of Vertebrate Zoology at Berkeley (Stein 1997), where Grinnell was director.

INTRODUCTION

The **1907 Alexander Expedition** to northern southeast Alaska included Mr. and Mrs. Frank Stephens, Joseph Dixon, Chase Littlejohn, and A. Alexander (Heller 1909).

The **1908 Alexander Expedition** to Prince William Sound consisted of Joseph Dixon, Edmond Heller, **Allen E. Hasselborg** (1876–1956), Louise Kellogg, and A. Alexander (Heller 1910).

Harry S. Swarth, Vancouver Island, 1910 (MVZ photo)

The **1909 Alexander Expedition** to southern and mainland southeast Alaska was conducted by **Harry S. Swarth** (1878–1935) and Allen Hasselborg. The published reports from that fieldwork represented the most intensive surveys yet made in the territory and also the first serious efforts to study faunal areas and relationship of the mammals of specific districts (Swarth 1911a, 1936). Swarth, along with **Joseph S. Dixon** (1884–1952), explored the Stikine River to tidewater near Wrangell in 1919 (Swarth 1922). Swarth also prepared a description of a new lemming (Swarth 1931). Dixon later accompanied a Harvard University group visiting the arctic waters of Siberia and Alaska in 1913–1914. Dixon's notes included Glacier Bay, the base of the Alaska Peninsula, the Aleutian Islands, St. Lawrence Island, Point Hope, Point Barrow, and various points along the Arctic coast as far east as Herschel Island. In 1926, he made extensive studies in Mount McKinley (Denali) National Park (Dixon 1938).

The American Museum of Natural History and Others

The expeditions of **Andrew J. Stone** (1859–1918) from 1901 through 1903 were organized for the American Museum of Natural History to provide exhibits and study specimens from various localities in south-coastal Alaska and the Kenai Peninsula, in particular (Allen 1902, 1903, 1904).

G Dallas Hanna (1887–1970), a well-rounded naturalist and Alaska hand, first went to work for the U.S. Bureau of Fisheries in Alaska, studying wildlife resources at Nushagak on Bristol Bay between 1911 and 1913, followed by teaching school and censusing fur seals in the Pribilofs. Years later, in 1955, he served as Director of the Naval Arctic Research Laboratory at Barrow. His Ph.D. dissertation was on the Alaska fur seals, and other publications on Alaska mammals include the mammals of the St. Matthew Island (1920) and the Pribilofs (1923).

Victor H. Cahalane (1901–1993), former chief biologist for the National Park Service and author of *Mammals of North America* (1947), published a biological survey of Katmai National Monument (Cahalane 1959).

Rudolph Martin Anderson (1876–1961) made a 1908 trip for the American Museum of Natural History along the northern Arctic coast of Canada and eastern Alaska to as far west as Barter Island, then headed inland up the Hulahula River. He later traveled to Flaxman Island for supplies and then returned to the head of the Hulahula River to collect mammals until March 1909. His second major expedition in 1913 was to Nome, Barrow, and along the Arctic coast to Colinson Point. He was the first naturalist to visit the arctic slope of the eastern Brooks Range. He published a catalog of Recent Canadian mammals in 1946.

Hamilton M. Laing (1883–1982) was the naturalist on the Mt. Logan Expedition of the Alpine Club of Canada in 1925. Starting from Cordova, he passed through McCarthy on his way to the upper Chitina River, about 40 km west of the Alaska-Yukon border. His report on mammals from that trip was later published (Laing and Anderson 1929).

Terrence Michael Shortt (1911–1987), ornithologist and artist for the Royal Ontario Museum (ROM), participated in a Canadian expedition in 1936 to collect at Yakutat Bay. Mammal specimens secured during this expedition were deposited in the ROM.

George Willett (1879–1945) studied the birds and mammals of southeast Alaska from 1912 to 1926.

Russell Hendee, Wainwright, 1922

Alfred M. Bailey (1894–1978), former director of the Denver Museum of Natural History, first worked with USBS in southeast Alaska between 1919 and 1921. In 1921–1922, he participated in the Arctic Expedition of the Denver Museum with companion **Russell W. Hendee** (1899–1929). Bailey and Hendee published their notes on the mammals of northwestern Alaska in 1926.

Adolph Murie (1899–1974) was introduced to Alaska beginning in 1922 by his brother Olaus and conducted his landmark study of the wolves in Mt. McKinley National Park (Murie 1944).

James W. Bee and **E. Raymond Hall** (1902–1986) of the University of Kansas Museum of Natural History published *Mammals of Northern Alaska on the Arctic Slope* in 1956. This publication was based on Bee's fieldwork north of the Brooks Range in 1951 and 1952. Assisting him in the field at various times were Hall, **J. Knox Jones, Jr.** (1929–1992), and Edward G. Campbell.

Robert L. Rausch held a research and supervisory position with the U.S. Public Health Service in the Zoonotic Disease Section of the Arctic Health Research Center. This research center was established to investigate health problems of northern indigenous people. Dr. Rausch first began work in Alaska in Barrow in March 1949, and shortly thereafter he made first contact with the Nunamiut at Anaktuvuk Pass. For twenty-five years, he served as chief of the Zoonotic Disease

R. L. Rausch

Section and oversaw a team of as many as twenty-five individuals in four units: Virology, Bacteriology, Entomology, and Helminthic Infections. Investigations were focused on rabies, brucellosis, tularemia, arboviruses, cystic echinococcosis, trichinosis, alveolar echinococcosis, diphyllobothriasis, and other human

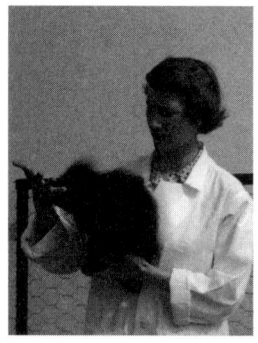

V. R. Rausch

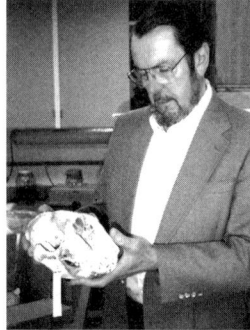

Francis "Bud" Fay
(1927–1994)

pathogens, and included remote field expeditions to gather samples followed by studies in laboratories located on the University of Alaska campus in Fairbanks. **Virginia "Reggie" R. Sacressen** joined the Zoonotic Disease Section in 1950 and she married Robert in 1953. Six years later, **Francis "Bud" H. Fay** (1927–1994) became part of the research team that was then active in field surveys of mammals aimed at isolating and identifying helminths and other parasites. In 1951–1952, Rausch's team first defined the life cycle of the cestodes responsible for alveolar echinococcosis a devastating disease whose source had eluded investigators for more than one hundred years (Eckert 2007). Their surveys of mammals on St. Lawrence Island indicated that arctic foxes (*Vulpes lagopus*) and root voles (*Microtus oeconomus*) were hosts. During the inventories of northern, western, and central Alaska, they prepared a large number of museum specimens that resulted in a series of publications on the mammalian hosts (e.g., Rausch 1953a) and their associated nematode (e.g., *Trichinella*) and cestode (e.g., *Diphyllobothrium*) parasites. After the center closed in 1974, the Rauschs continued work in Alaska, Siberia, and Canada. To date, Robert Rausch has published about three hundred papers in his distinguished career, many with a focus on Alaska's fauna. He and Reggie built arguably the most significant museum collection available anywhere, a collection that integrates across both mammalian hosts and related parasites. Given the temporal span of the collection, it provides ample opportunities for assessing changing environmental conditions at high latitudes.

The Alaska Cooperative Wildlife Unit and University of Alaska Museum of the North

The University of Alaska's first president, Charles E. Bunnell, established a university museum in Fairbanks in the 1920s. Beginning in 1926, he sent **Otto W. Geist** (1888–1962) to collect archaeological and paleontological material across remote regions of the Alaska Territory. Much of that material was sent to the American Museum of Natural History and is known as the Frick Collection after the financial sponsor of Geist's fieldwork. Childs Frick

Otto Geist with some of the many, many boxes of paleontological specimens he collected for the University of Alaska Museum and the Frick Laboratory associated with the American Museum of Natural History (UAF photo).

was the son of steel magnate Henry Clay Frick and sponsored museum expeditions throughout the West in the early half of the twentieth century. Geist, who had learned to prepare mammal and bird specimens from Olaus Murie, collected some of the earliest specimens in the mammal collection.

In 1950, with the establishment of the Alaska Cooperative Wildlife Research Unit on the Fairbanks campus, a collection of mammals was initiated by Cooperative Unit personnel. Started as a teaching collection, it soon incorporated specimens from various research projects in Alaska. It was curated by volunteer faculty until 1961, when the mammal collection (and birds and plants) was placed within the University of Alaska Museum under Director L. J. Rowinski. In 1972, the bird and mammal collections were combined into the museum's Terrestrial Vertebrates Department, and Brina Kessel became the first curator of the vertebrate collection.

Brina Kessel and Olaus Murie preparing museum specimens, Sheenjek Valley, 1956.

The Alaska Department of Fish and Game (ADFG) deposited large series of wolves, wolverines, lynx, sea otters, and phocid seals at the University of Alaska Museum (UAM, now University of Alaska Museum of the North). Scientists of the U.S. Public Health Service also deposited important series of specimens from a number of localities. Other important contributions over the years came from William O. Pruitt Jr., J. L. Buckley, W. L. Libby, Fred C. Dean, D. L. Chesemore, John J. Burns, T. O. Osborne, H. Golden, J. Hechtel, and J. S. Whitman. Burns and Francis Fay were instrumental in building one of the largest collections of pinnipeds worldwide.

In 1979, the UAM hired Stephen O. MacDonald to work with the mammal collection. Gordon H. Jarrell replaced MacDonald in September 1984, and after a short stint in Texas in the early 1990s, he returned to Alaska as Collection Manager for Mammals (and Frozen Tissues) and later as Chief Information Officer for the Museum. Jarrell and later Dusty McDonald were integral to development and implementation of the ARCTOS database. UAM hired Joseph A. Cook as the first Curator of Mammals in 1990. He held that position until 2001. Under his leadership, the mammal collection acquired new staff positions, trained numerous students, and grew from eighteen thousand cataloged specimens to over seventy-five thousand. He also founded the Alaska Frozen Tissue Collection and cryogenically preserved tissues from more than thirty thousand specimens were archived by 2001. During the late 1990s, research activity was high in the mammal collection with a series of undergraduate and graduate student projects on Alaska mammals completed. A large number of loans (more than eighty annually representing thousands of individual specimens) also were made to investigators elsewhere. Cook and MacDonald continued to deposit large series of specimens at UAM from their National Park Service fieldwork through 2004. In 2002, Sylvia Brunner continued collaborations with the various marine mammal subsistence commissions and reinvigorated the marine mammal

Gordon Jarrell salvaging a whale (S. Brunner photo).

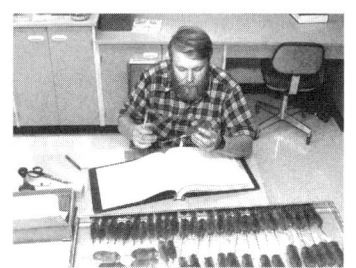

Stephen O. MacDonald

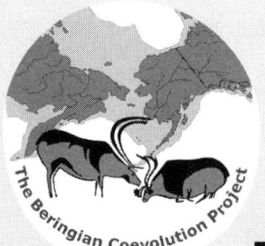

Beringian Coevolution Project

The Beringian Coevolution Project (BCP) is a collaborative effort based at the University of New Mexico (UNM). The USDA National Parasite Collection, Finnish Forest Research Institute, University of Saskatchewan, Russian Academy of Sciences, and a number of other institutions have participated in field expeditions and subsequent research projects (Cook et al. 2005; Hoberg et al. 2003). The primary objectives of the BCP are to (1) provide a detailed and spatially extensive resource of museum specimens from several key high-latitude areas that have not yet been inventoried; (2) develop a comparative framework for the Arctic to examine the history of host-parasite systems that are phylogenetically and ecologically disparate, providing the basis for detailed studies in coevolution, cospeciation, and rates of diversification in associated hosts and parasites; (3) explore forces that have structured high-latitude biomes, including exchange across the northern continents using comparative phylogeographic analyses; (4) build a spatial and temporal foundation for biotic investigations in the Arctic by identifying and further characterizing regions of endemism and contact zones between divergent lineages; and (5) help train the next generation of environmental scientists through integrated field studies.

Successful field expeditions in Siberia, Alaska (including nearly all its national parks and preserves), and Canada have been conducted by parasitologists and mammalogists each year since 1999. Those inventories have visited more than 250 collection sites and preserved over twenty thousand mammalian specimens that represent a diverse set of preparations, including more than eleven thousand lots of endoparasites and five thousand lots of ectoparasites. Survey crews preserve tissues (heart, liver, kidney, spleen, and lung) and embryos in liquid nitrogen as cryogenic archives. These materials are the basis for a growing number of publications (now exceeding one hundred twenty) and completed graduate

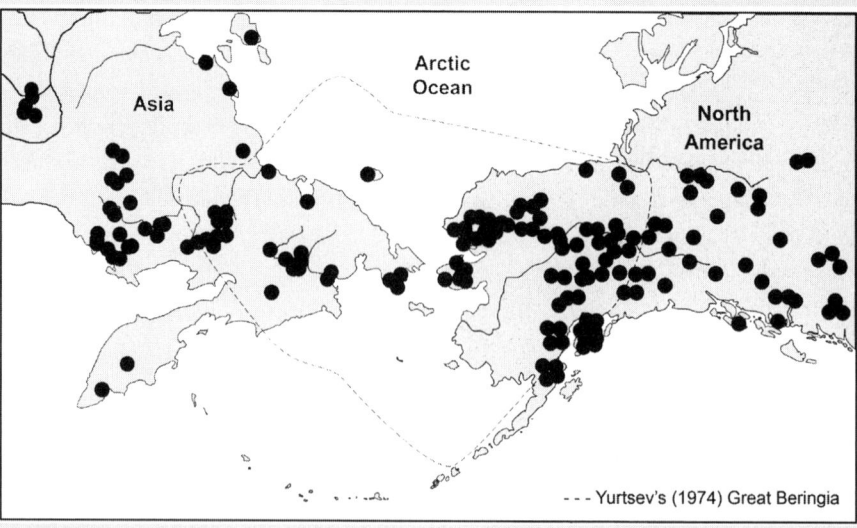

BCP intensive collecting sites, 1999–2006.

student theses and dissertations (sixteen to date) that are tied together through their use of BCP's collection of voucher specimens. This foundation is crucial for conservation efforts in the face of increasing anthropogenic impacts at high latitudes. Several mammal species (e.g., caribou, muskoxen, moose, and some marine mammals) are important for the physical, cultural, and economic well-being of local residents. Ensuring optimal health of these wild populations is critical for Alaskans and the maintenance of intact northern ecosystems. Many factors influence wildlife health in the North, including directional climate change (Kutz et al. 2005). Warming temperatures and changing patterns of precipitation, together with an increase in the frequency and severity of extreme climatic events and other environmental stressors, may adversely affect wildlife that form the basis of rural subsistence economies. Some mammals occasionally serve as sylvatic reservoirs for human pathogens. In rural communities, in particular, close linkage of humans and wild mammals also increases the risk of disease transmission, with the potential for adverse effects on human health. Environmental changes now being experienced at high latitudes may also alter transmission dynamics of parasitic and other infectious diseases (Jenkins et al. 2006), including those zoonotics transmissible to humans.

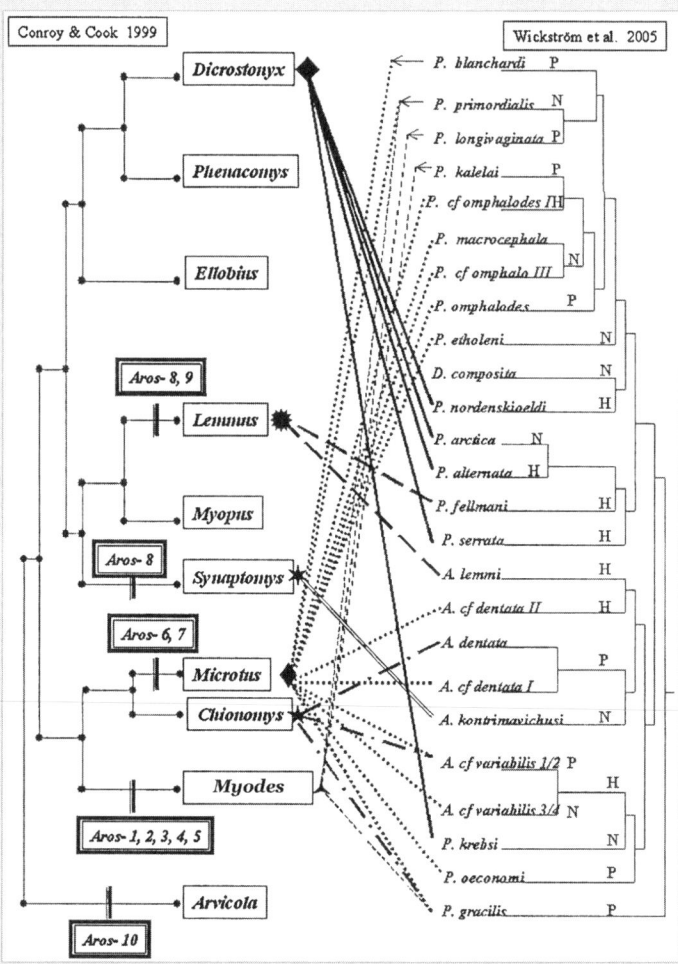

The historical biogeography of Beringia has shaped coevolutionary relationships among arvicoline rodents and their associated cestodes (see Cook et al. 2005 for details).

Joseph A. Cook

stranding network in Alaska. She managed to salvage significant cetacean and pinniped specimens for the museum. Brandy Jacobsen became Collection Manager of Mammals in 2002. Link Olson was appointed Curator of Mammals at UAM in 2003 and has developed an active program that encompasses all aspects of the university's missions related to training, research, and public service.

Specimen Representation: Time and Space

Student collaborators (see Literature Cited), clockwise from top left: Allison Bidlack, Karen Stone, John Demboski, Chris Conroy, Ellen Lance, Kirstin Hunter, and Amy Runck.

Theoretically, collections can be used to help establish baseline environmental conditions or address questions related to impacts on particular mammal species, but only if these museum archives are sufficiently deep (temporally) and broad (spatially). So how well sampled are mammals across Alaska? The answer to this question is driven by the nature of a particular study. While analyses of higher-level systematics may require relatively few samples to represent each taxon, population-level studies often require sample sizes of more than twenty individuals to represent particular localities. Pathogenic surveys of viruses, on the other hand, may require very large sample sizes (more than five hundred) if a particular pathogen exists in low frequencies (less than 5 percent). Studies aimed at documenting avian bird influenza caused by the H5N1 virus, for example, were limited by the lack of samples available for screening. By plotting the location of existing specimens, we can begin to obtain a sense of both spatial and temporal coverage (Fig. 4).

Alaska has not been further subdivided into smaller geopolitical units like counties, so to explore **spatial coverage** the U.S. Geological Service's 1/250,000 scale quadrangle (quad) maps provide finer-scale units of resolution across this vast region. There are 153 of these quadrangle maps that cover the state of Alaska (Appendix 1). Coverage of each quad map varies, but roughly averages about 15,000 km^2, an area more than twice the size of the state of Rhode Island. We use these maps as a basis of finer-scale assessments of the coverage of our museum archives by tallying specimens records from each of these quadrangle maps.

First, we examine one of the most commonly encountered, widespread, and arguably best-sampled small mammal in Alaska, the northern red-backed vole (*Myodes rutilus*). For this exercise (Fig. 5), the museum archive for this species across all years (UAM samples) provides an opportunity for rather detailed studies (more than one hundred specimens available) within thirty quads, with less coverage (thirty to ninety-nine specimens) available for another twenty-five quad areas. Many of these quads occur near population centers or areas easily accessed by roads, although good coverage in several (e.g., Katmai, southern slope of Brooks Range) are the result of federally supported inventory projects. Thus, this species could be studied effectively across about 35 percent of Alaska.

The coverage for other species is considerably bleaker. Beaver (Fig. 6) have been commercially harvested throughout much of Alaska for nearly

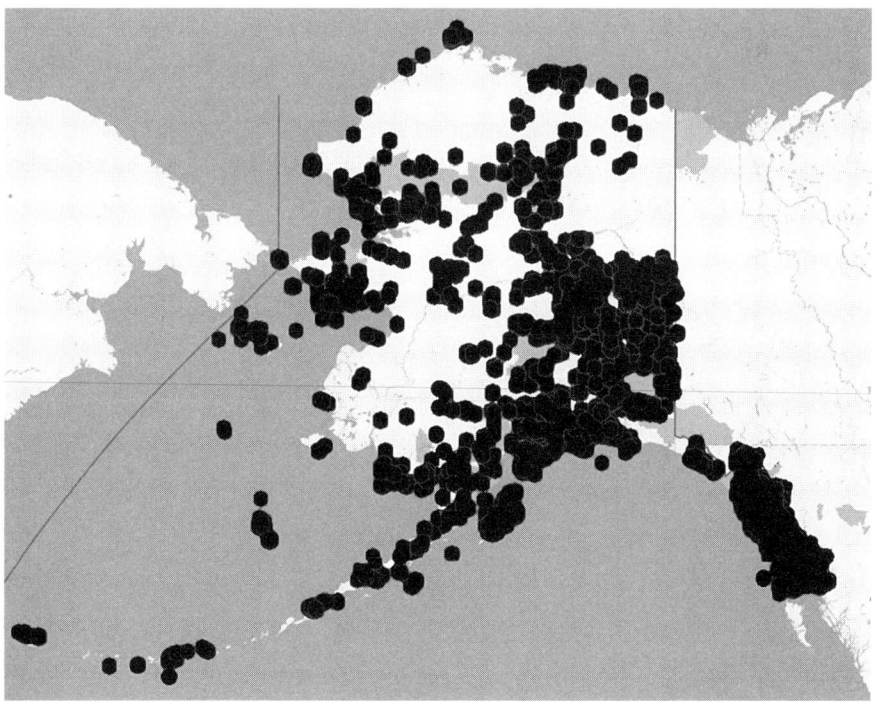

Figure 4. Cumulative collecting localities across Alaska for UAM between 1991 and 2004.

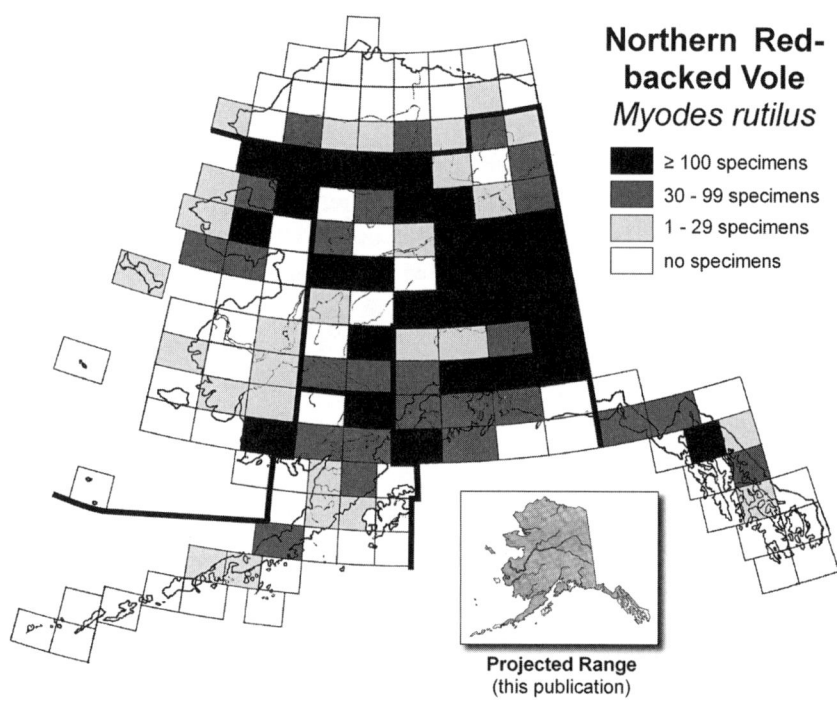

Northern Red-backed Vole
Myodes rutilus

- ⬛ ≥ 100 specimens
- ⬛ 30 - 99 specimens
- ⬜ 1 - 29 specimens
- ☐ no specimens

Projected Range
(this publication)

Number of UAM Specimens	Number of Quads with Specimen(s)	Number of Quads by Specimen Count			Number of Specimens by Region					
		1 - 29	30 - 99	≥ 100	SE	SC	C	SW	W	N
10,370	82 (67%)*	27 (33%)	25 (30%)	30 (37%)	357 (3%)	1291 (13%)	6898 (66%)	445 (4%)	1206 (12%)	173 (2%)

* percentage of total in parentheses

Figure 5. Distribution of UAM specimens of northern red-backed vole by USGS quad.

two centuries and yet relatively few specimens, most from quads close to Fairbanks, have been preserved. In this case, less than 5 percent of the state has meaningful representation. Similarly, for moose (Fig. 7), an intensively studied species, surprisingly limited archival material exists. Marten (Fig. 8) are better represented in some parts of Alaska, primarily because they have been the focus of a series of studies on stable isotope ecology, molecular genetics, and nematode parasites in the last decade. Those investigations resulted in the preservation of large series of voucher specimens, especially from the islands of southeast Alaska.

With regard to **temporal coverage**, if we plot all specimens for a particular year we gain some sense of our available sampling for retrospective studies of environmental conditions. In 1997, a year of high activity for UAM, nearly four thousand mammal specimens were archived from Alaska. Plots of the distribution of these specimens (Fig. 9) indicate that there are regions of the state for which we might be able to investigate environmental conditions using this material, but there are many others where we cannot. We will never be able to go back to 1997 and collect those areas; essentially, our opportunity to explore historic or baseline conditions for these species or their environ-

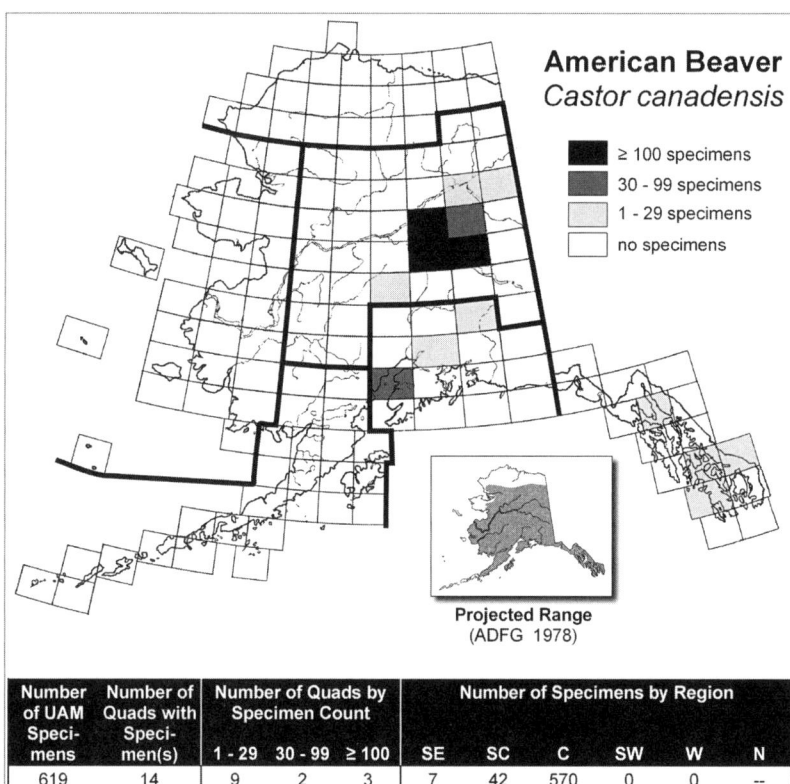

American Beaver
Castor canadensis

	≥ 100 specimens
	30 - 99 specimens
	1 - 29 specimens
	no specimens

Projected Range
(ADFG 1978)

Number of UAM Speci-mens	Number of Quads with Speci-men(s)	Number of Quads by Specimen Count			Number of Specimens by Region					
		1 - 29	30 - 99	≥ 100	SE	SC	C	SW	W	N
619	14 (14%)*	9 (64%)	2 (14%)	3 (22%)	7 (1%)	42 (7%)	570 (92%)	0 (0%)	0 (0%)	-- (--)

* percentage of total in parentheses

Figure 6. Distribution of UAM specimens of American beaver by USGS quad.

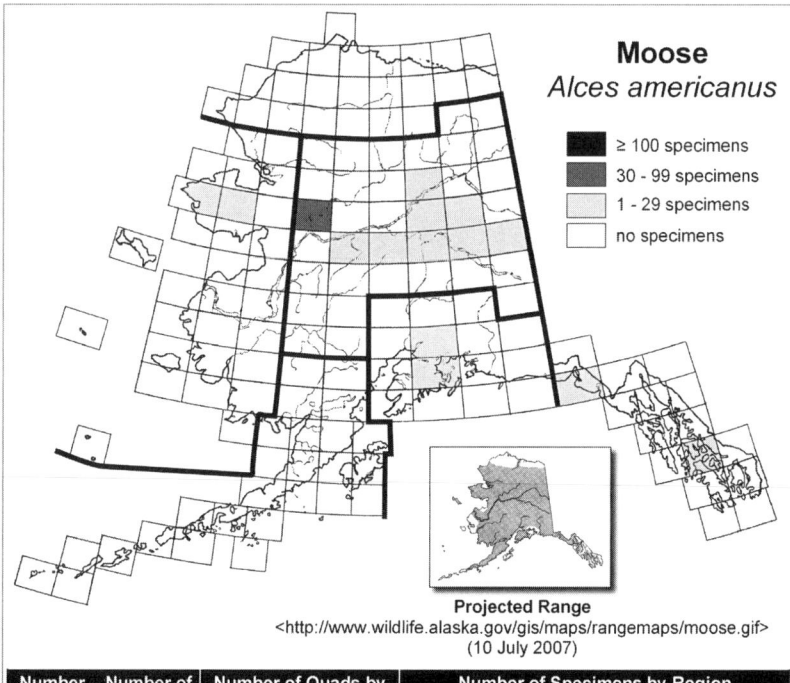

Moose
Alces americanus

	≥ 100 specimens
	30 - 99 specimens
	1 - 29 specimens
	no specimens

Projected Range
<http://www.wildlife.alaska.gov/gis/maps/rangemaps/moose.gif>
(10 July 2007)

Number of UAM Speci-mens	Number of Quads with Speci-men(s)	Number of Quads by Specimen Count			Number of Specimens by Region					
		1 - 29	30 - 99	≥ 100	SE	SC	C	SW	W	N
107	16 (14%)*	15 (94%)	1 (6%)	0 (--)	3 (3%)	2 (2%)	100 (93%)	0 (--)	2 (2%)	0 (--)

* percentage of total in parentheses

Figure 7. Distribution of UAM specimens of moose by USGS quad.

Figure 8. Distribution of UAM specimens of marten by USGS quad.

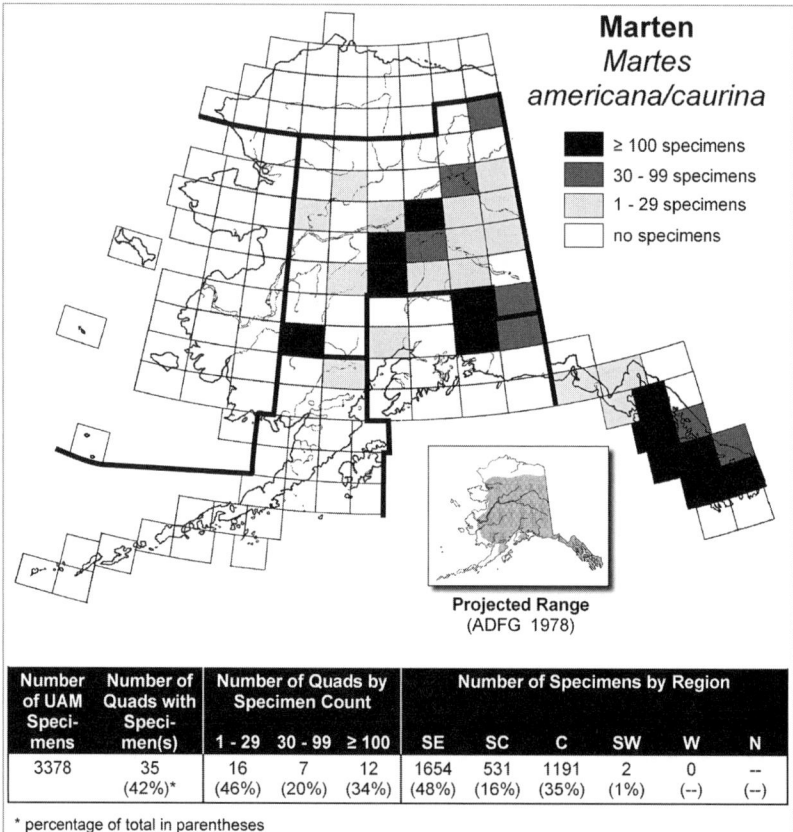

Marten
Martes americana/caurina

- ≥ 100 specimens
- 30 - 99 specimens
- 1 - 29 specimens
- no specimens

Projected Range
(ADFG 1978)

Number of UAM Specimens	Number of Quads with Specimen(s)	Number of Quads by Specimen Count			Number of Specimens by Region					
		1 - 29	30 - 99	≥ 100	SE	SC	C	SW	W	N
3378	35 (42%)*	16 (46%)	7 (20%)	12 (34%)	1654 (48%)	531 (16%)	1191 (35%)	2 (1%)	0 (--)	-- (--)

* percentage of total in parentheses

Figure 9. Distribution of all mammal specimens collected for UAM in 1997 by USGS quad.

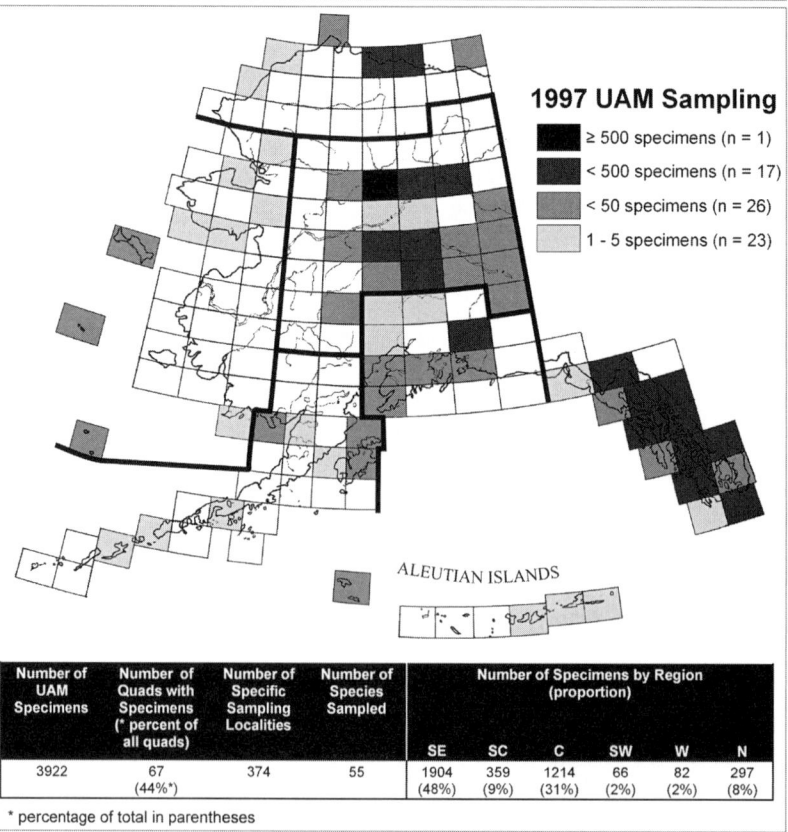

1997 UAM Sampling

- ≥ 500 specimens (n = 1)
- < 500 specimens (n = 17)
- < 50 specimens (n = 26)
- 1 - 5 specimens (n = 23)

ALEUTIAN ISLANDS

Number of UAM Specimens	Number of Quads with Specimens (* percent of all quads)	Number of Specific Sampling Localities	Number of Species Sampled	Number of Specimens by Region (proportion)					
				SE	SC	C	SW	W	N
3922	67 (44%*)	374	55	1904 (48%)	359 (9%)	1214 (31%)	66 (2%)	82 (2%)	297 (8%)

* percentage of total in parentheses

ments has been lost. Now is the time for renewed efforts aimed at building the archival resource.

Study Area

Alaska is immense, spanning a distance greater than 3,000 km east to west, and nearly 1,800 km north to south; the total surface covers more than 170 million ha (Fig. 1). Freshwater is abundant in the many thousands of lakes and twelve major river drainages. Five rivers (Yukon, Porcupine, Stikine, Alsek, and Taku) also drain large regions of adjacent Canada and are important corridors for wildlife movement. Winter temperatures govern the flow of most of Alaska's rivers. Usually beginning in late October and extending into early June for the northernmost streams, thick layers of ice form and may effectively cease flow completely during the coldest months. Permafrost is extensive throughout the northern third of the state with isolated patches over the central portions of Alaska. Climate warming predictions indicate significant changes in the extent of permafrost in the next few decades. No permafrost is found in southcentral, or in the coastal zones of southeast Alaska, the Alaska Peninsula, or the Aleutian Islands. In the last decade, significant impacts to terrestrial and marine biomes have been documented due to warming conditions (ACIA 2005; Chapin et al. 2006; Serreze et al. 2000).

Mountain ranges influence climatic regimes in Alaska and help to define biotic subregions across this vast state. The Brooks Range separates the northern and central regions, while the Alaska Range separates central from southcentral Alaska. Other major mountain ranges include the Chugach Mountains, which form the north boundary of the Gulf of Alaska and together with the Wrangell Mountains merge with the St. Elias Mountains to the east. The latter mountains extend southeastward through Canada and southeast Alaska to

Mouth of Stikine River, mainland southeast Alaska (J. Schoen photo).

become the Coast Range. With the exception of the Brooks Range, many peaks higher than 3,000 m occur in these mountains. The highest peak (6,194 m) in North America, Denali (Mount McKinley), is located in the central Alaska Range.

The marine environment (Fig. 1 insert map) that borders Alaska is similarly expansive, with more than 52,000 km of coastline surrounding the mainland and extensive island archipelagos. Coastal waters off Alaska include the eastern Bering Sea and Gulf of Alaska of the North Pacific Ocean, and the eastern Chukchi and western Beaufort seas of the Arctic Ocean. These waters support some of the most productive fisheries worldwide and are home to the largest marine mammal populations in the United States. Along the coast, mountain ranges influence prevailing weather patterns as they produce up to 5,000 mm of precipitation in southeast Alaska, and up to 3,500 mm along the northern coast of the Gulf of Alaska. Precipitation averages about 1,500 mm along the Alaska Peninsula and Aleutian Islands and then decreases dramatically to the north, with an average of 300 mm in the central region and less than 150 mm in the northern region. Throughout Alaska, the portion of the total annual precipitation produced by snowfall increases with elevation and latitude. Greatest annual precipitation occurred at MacLeod Harbor on Montague Island in the Gulf of Alaska with 8,432 mm in 1976. Snowfall records are from northwest of Valdez at Thompson Pass and include records for season (1952–53: 24,752 mm), month (February 1953: 7,569 mm), and twenty-four-hour period (December 1955: 1,575 mm). In Alaska, mean annual temperatures range from 5° C along the southern coast to –12° C on the north slope of the Brooks Range.

Regional Overviews

Alaska is frequently broken into six distinct subregions (Fig. 10). Thirty-two ecological regions also have been delineated (Fig. 11), and land ownership and management units are complex across the state (Fig. 12).

Southeast Alaska is dominated by maritime ecosystems due to the Alexander Archipelago. The region has over 17,000 km of shoreline and extends from about Icy Bay southward to the British Columbian border at the Dixon Entrance. East to west, southeast Alaska is a mosaic of hundreds of islands strewn along the mainland coast (Fig. 13). The mainland is narrow and bordered to the east by the Coast Mountains and Wrangell–St. Elias ranges. Summers are cool and damp and winters comparatively mild. Precipitation is heavy, ranging from 1,500 to over 3,500 mm annually. At low elevations, winter precipitation may fall as rain or snow. Lowlands are heavily forested with western hemlock (*Tsuga heterophylla*), Sitka spruce (*Picea sitchensis*), and cedars (*Chamaecyparis nootkatensis*, *Thuja plicata*) with a dense understory of huckleberry (*Vaccinium* spp.), devil's club (*Oplopanax horridus*), and alders (*Alnus* spp.). Higher elevations on the mainland support glaciers and some of these reach tidewater and likely form significant barriers to movement of most terrestrial mammals. Six major rivers (Alsek, Chilkat, Taku, Whiting, Stikine, and Unuk)

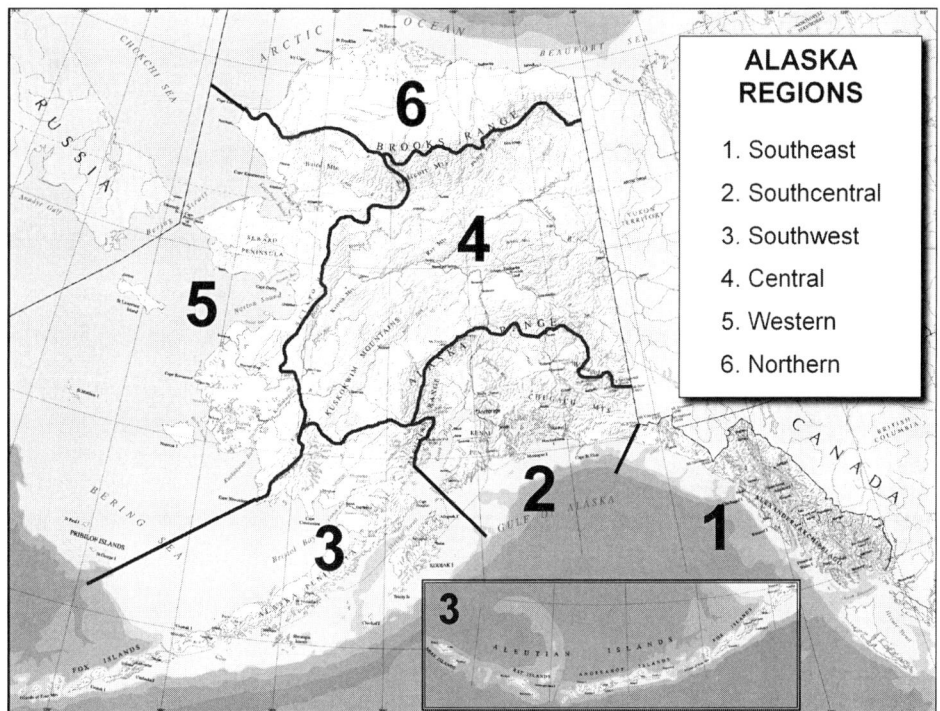

Figure 10. Six primary regions are commonly used to delineate Alaska. For this publication, we generally follow ADFG Game Management Unit boundaries (http://wildlife.alaska.gov).

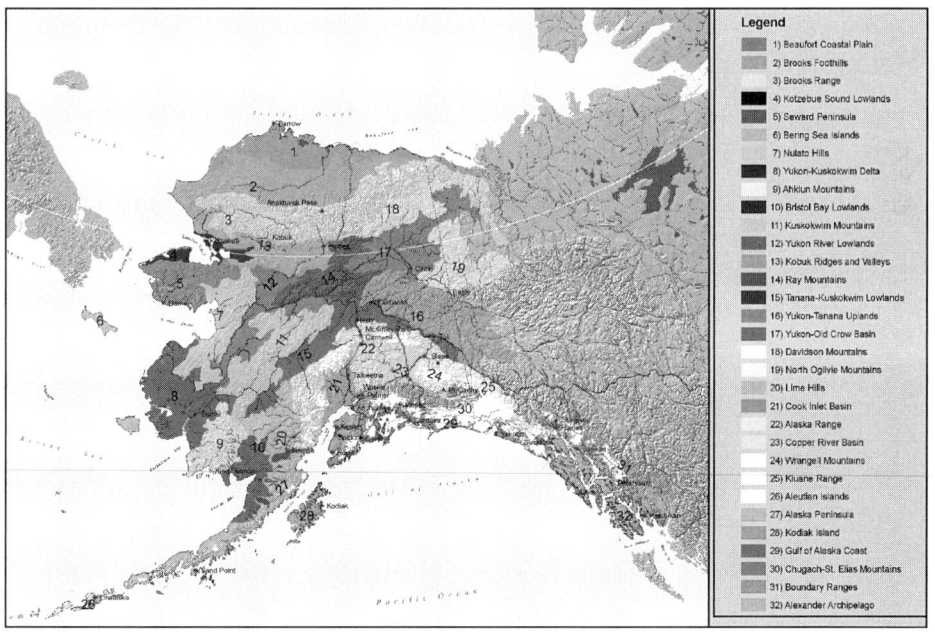

Figure 11. Ecoregions of Alaska (Nowacki et al. 2001) provide additional perspectives on questions related to the distribution and abundance of Alaska's mammals.

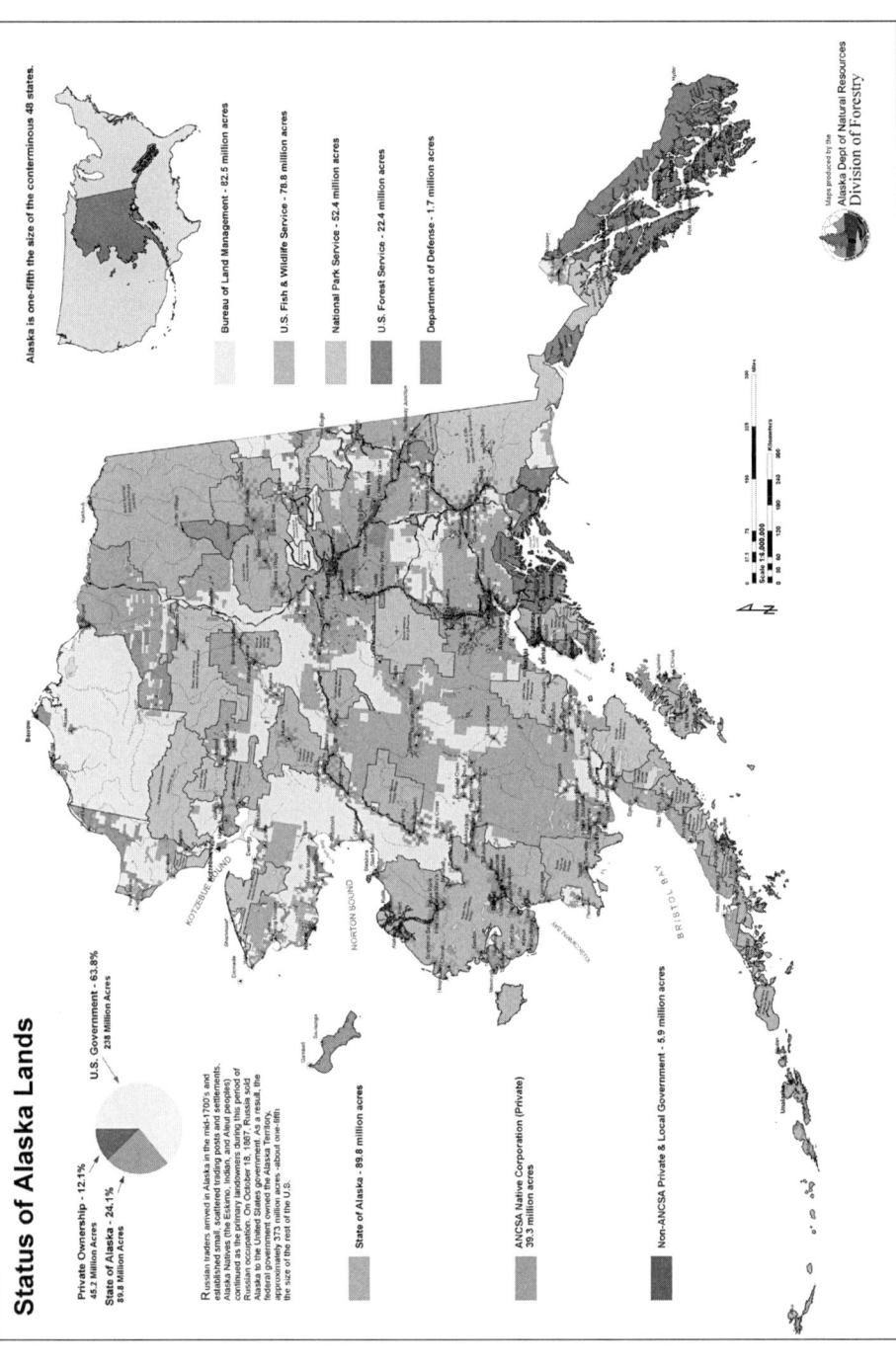

Figure 12. Investigations and conservation efforts related to mammals in Alaska requires a basic appreciation of the complex ownership and management of Alaska lands (modified map of the Alaska Department of Natural Resources, Division of Forestry).

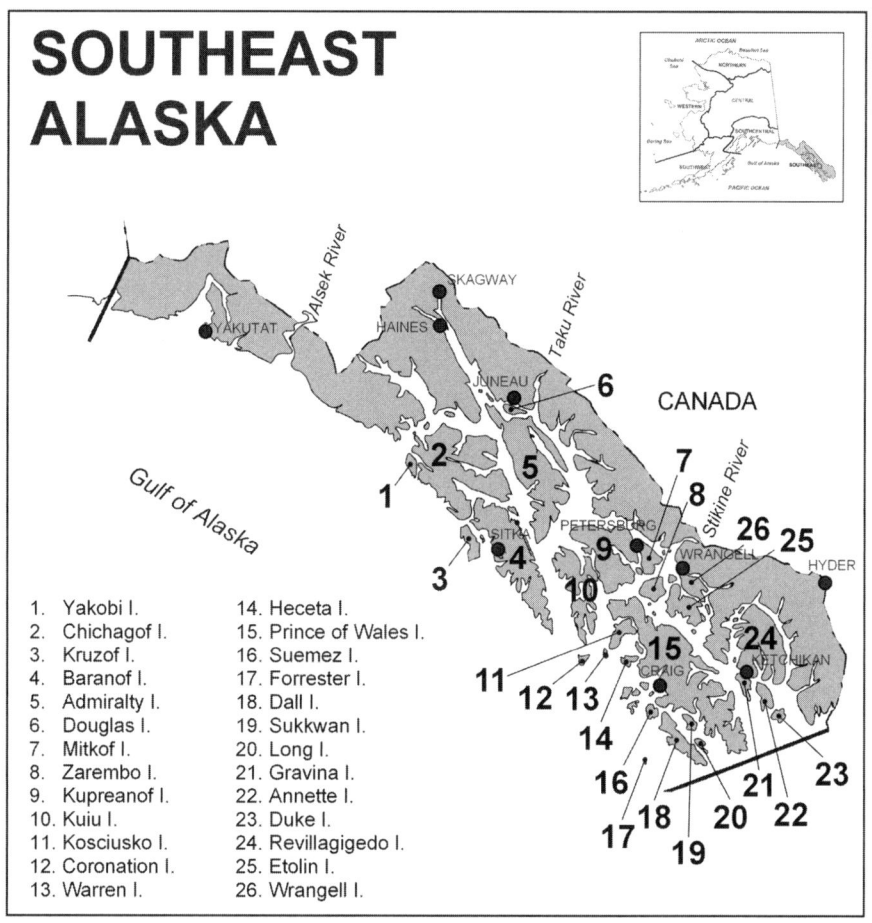

SOUTHEAST ALASKA

SKAGWAY
YAKUTAT
Alsek River
HAINES
Taku River
JUNEAU
6
CANADA
2
5
1
7
Stikine River
8
Gulf of Alaska
SITKA
PETERSBURG
26 25
4
9
WRANGELL
3
10
HYDER
15
24
11
CRAIG
KETCHIKAN
12 13
14
16
21 23
17 18 20 22
19

1. Yakobi I.
2. Chichagof I.
3. Kruzof I.
4. Baranof I.
5. Admiralty I.
6. Douglas I.
7. Mitkof I.
8. Zarembo I.
9. Kupreanof I.
10. Kuiu I.
11. Kosciusko I.
12. Coronation I.
13. Warren I.
14. Heceta I.
15. Prince of Wales I.
16. Suemez I.
17. Forrester I.
18. Dall I.
19. Sukkwan I.
20. Long I.
21. Gravina I.
22. Annette I.
23. Duke I.
24. Revillagigedo I.
25. Etolin I.
26. Wrangell I.

Figure 13. Major islands in southeast Alaska.

pass through the rugged coastal mountains from interior British Columbia, forming narrow corridors to the rest of continental North America.

Alaska's Panhandle supports a rich mammalian fauna (see Appendix 2), with many species found nowhere else in the state such as the Pacific marten (*Martes caurina*), fisher (*Martes pennanti*), four species of bats (*Lasionycteris noctivagans, Myotis* spp.), and five species of rodents (e.g., *Myodes gapperi, Peromyscus keeni*). Nonnative species include brown rat (*Rattus norvegicus*), wapiti (*Cervus canadensis*), and raccoon (*Procyon lotor*). Sitka black-tailed deer (*Odocoileus hemionus sitkensis*), mountain goats (*Oreamnos americanus*), wolves (*Canis lupus*), black and brown bears (*Ursus americanus, U. arctos*), river otters (*Lontra canadensis*), and mink (*Neovison vison*) are characteristic inhabitants of the region. Once extirpated, sea otters (*Enhydra lutris*) were reintroduced and are again numerous along the Southeast's outer coasts. The region's rich marine environment supports many mammals, such as Steller's sea lions (*Eumetopias jubatus*), harbor seals (*Phoca vitulina*), harbor (*Phocoena phocoena*) and Dall's porpoises (*Phocoenoides dalli*), humpback whales (*Megaptera novaeangliae*), killer whales (*Orcinus orca*), and several species of beaked whales. California sea lions (*Zalophus californianus*) and elephant seals (*Mirounga angustirostris*) have become more numerous in recent years.

Open woodland of dwarfed shore pines *(Pinus contorta)*, Baker Island, southeast Alaska (R. Carstensen photo).

Old growth Sitka spruce *(Picea sitchensis)*, Kosciusko Island, southeast Alaska (R. Carstensen photo).

Southcentral Alaska includes the Chugach, Talkeetna, Wrangell, and Kenai mountains and the southern slopes of the Alaska Range. Major river drainages include the Susitna, Beluga, Chakachatna, Kenai, Matanuska, Copper, and Bering rivers. Associated with the region are the marine waters of Cook Inlet, the northern Gulf of Alaska, and Prince William Sound, which is a complex of fiords, bays, coves, and hundreds of islands (Fig. 14) surrounded by mountains with glaciers, rivers, and numerous lakes.

The climate of the region varies from maritime to continental. Temperate rain forest comprised of Sitka spruce and western hemlock dominate lowland areas close to the coast from Prince William Sound to lower Cook Inlet,

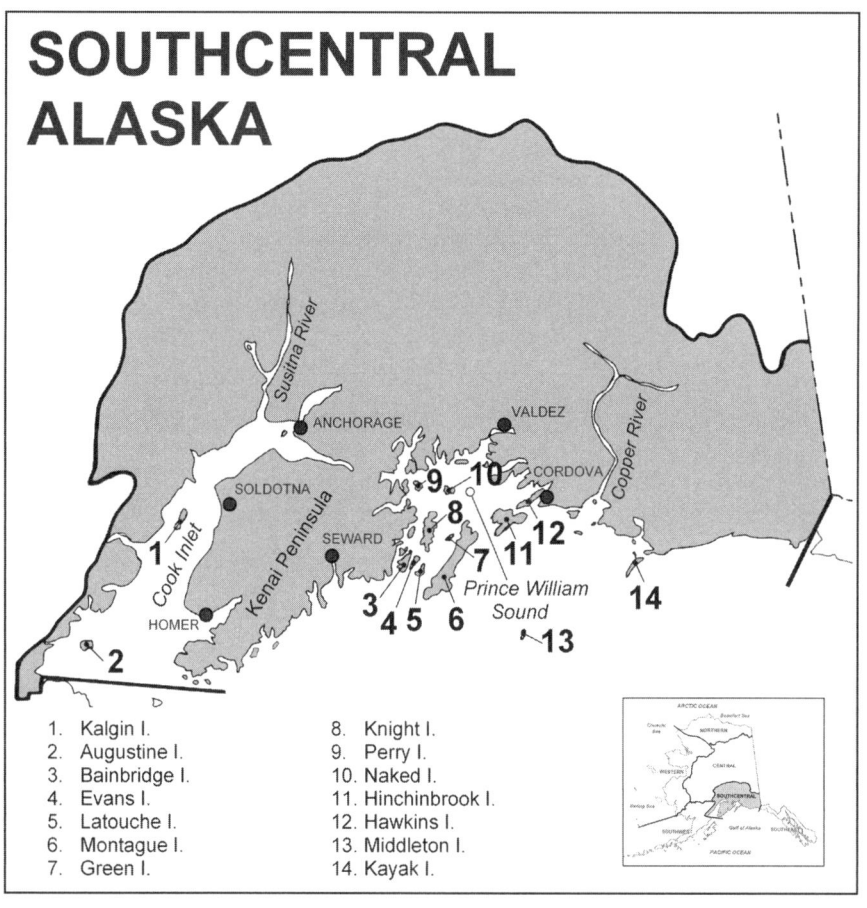

SOUTHCENTRAL ALASKA

1. Kalgin I.
2. Augustine I.
3. Bainbridge I.
4. Evans I.
5. Latouche I.
6. Montague I.
7. Green I.
8. Knight I.
9. Perry I.
10. Naked I.
11. Hinchinbrook I.
12. Hawkins I.
13. Middleton I.
14. Kayak I.

Figure 14. Major islands in southcentral Alaska.

while various associations of white spruce (*Picea glauca*), black spruce (*Picea mariana*), paper birch (*Betula papyrifera*), balsam poplar (*Populus balsamifera*), black cottonwood (*Populus trichocarpa*), and quaking aspen (*Populus tremuloides*) are common farther inland. Low and tall shrub communities composed primarily of willow (*Salix* spp.), alder (*Alnus* spp.), and shrub birch (*Betula* spp.) are common throughout the subalpine zone, while dwarf shrubs and a variety of herbaceous communities are common in alpine areas and lower-elevation wetlands.

Forty-six species of terrestrial mammals inhabit the region, including mountain goats, Dall's sheep (*Ovis dalli*), black bears, coyotes (*Canis latrans*), hoary marmots (*Marmota caligata*), and singing voles (*Microtus miurus*). Sitka black-tailed deer, found throughout Prince William Sound, are the result of translocations during the last

Hidden Lake, Kenai Peninsula, southcentral Alaska (USFWS photo).

Humpback whales off Montague Island, Prince William Sound (N. Moomey photo).

31

A fog-bound Aleutian island (NOAA photo).

century from southeast Alaska stocks. Marine habitats support sea otter, harbor seal, Steller's sea lion, and many species of whales, including an isolated population of beluga (*Delphinapterus leucas*).

Southwest Alaska includes the areas of Bristol Bay, the Gulf of Alaska south of Kennedy Entrance (including Kodiak and other islands), the Alaska Peninsula, the Aleutian Islands, the drainages of the Nushagak River, the Wood River/Tikchik Lakes, and Togiak River (Fig. 15). The region includes maritime, transitional, and continental climatic zones with mean annual temperatures above freezing. Vegetation in the region is varied, with grasslands, shrublands, and wetland meadows comprising most of the vegetation types on the peninsula and in the Aleutians. Coastal forest of mostly Sitka spruce reaches its western limit on Afognak and Kodiak islands and scattered locations on the adjacent Alaska Peninsula, while interior forest composed of white and black spruce, cottonwood, and birch trees extends into the region near Bristol Bay at the base of the Alaska Peninsula. The island terrain is rugged, often fog-bound, and dominated by young and occasionally active volcanoes as this is where the Pacific Plate plunges deeply beneath the North American Plate. Shemya, on the western end of the Aleutian Islands, has experienced winds of at least 207 km/hr.

The marine waters of this region provide important habitats for numerous whales and other marine mammals. Kodiak Island and the Alaska Peninsula host the largest concentrations of big brown bears in the world. Among the forty-six terrestrial mammals known from this region, none occurs naturally west of the Fox Islands in the far eastern Aleutians. Many islands in the region are now home to nonnative species, including brown and roof rats, house mice (*Mus musculus*), North American deermice (*Peromyscus maniculatus*), beavers (*Castor canadensis*), snowshoe hares (*Lepus americanus*), mountain goats, Sitka black-tailed deer, arctic ground squirrels (*Spermophilus parryii*), wapiti, muskrats

SOUTHWEST ALASKA

1. Attu I.	8. Kanaga I.	15. Unimak I.	22. Kodiak I.
2. Agattu I.	9. Adak I.	16. Unga I.	23. Afognak I.
3. Shemya I. (Semichi Is.)	10. Atka I.	17. Nagai I.	24. Barren Is.
4. Kiska I.	11. Amlia I.	18. Simeonof I.	25. Walrus Is.
5. Amchitka I.	12. Umnak I.	19. Big Koniuji I.	26. Hagemeister I.
6. Semisopochnoi I.	13. Unalaska I.	20. Chirikof I.	
7. Tanaga I.	14. Akutan I.	21. Tugidak I.	

Figure 15. Major islands in southwest Alaska.

(*Ondatra zibethicus*), and red squirrels (*Tamiasciurus hudsonicus*). Domestic cattle (*Bos* spp.), horses (*Equus* spp.), sheep (*Ovis aries*), bison (*Bison bison*), and for a time even wild boar (*Sus scrofa*) have been allowed to run feral on a number of islands in this region.

Central Alaska is bounded by the Brooks Range to the north and the Alaska Range (including the highest peak in North America) to the south. This extensive region is dominated by the basins of the Yukon, Kuskokwim, Koyukuk, Tanana, and Porcupine rivers. The region has a continental climate, with the warmest summer and coldest winter temperatures in the state. Average winter minimums in this area range from –25° to –35° C. The coldest temperature ever recorded in Alaska was –62° C at Prospect Creek on 23 January 1971. Precipitation and wind are normally light.

The vegetation of the region is primarily either closed or open canopied forests comprised of various associations of white spruce, black spruce, quaking aspen, paper birch, balsam poplar, and tamarack (*Larix laricina*) trees. Tree line varies from around 300 m along the lower stretches of the Yukon River to 1,100 m near the Alaska-Canada border. Low and tall shrub communities composed primarily of willow, alder, and shrub birch occur on floodplains, lowland

Typical boreal forest habitat in central Alaska.

boggy areas, and mountain slopes. Low shrub and herbaceous communities are prevalent in alpine areas and lower-elevation wetlands.

The interior of Alaska supports healthy populations of moose (*Alces americanus*), caribou (*Rangifer tarandus*), Dall's sheep, black and brown bears, furbearers, snowshoe hares, and other small mammals. Among the forty-four mammals known to occur in the region, only the woodchuck (*Marmota monax*) is unique to this region of the state. The ranges of the pygmy shrew (*Sorex hoyi*) and the taiga vole (*Microtus xanthognathus*) are largely confined to this portion of Alaska.

Western Alaska encompasses a region of maritime and continental influence that extends southward from Cape Lisburne, the Seward Peninsula, and the Yukon-Kuskokwim Delta to Cape Newenham. Besides the lower portions of the Yukon and Kuskokwim rivers, other major drainages include the Noatak, Kobuk, Selawik, Kanektok, and Goodnews rivers (Fig. 16). Associated marine waters are comprised of Kotzebue Sound and the Chukchi Sea and Norton Sound and the Bering Sea with St. Lawrence, St. Matthew, Nunivak,

Figure 16. Major islands in western Alaska.

Nelson, and the Pribilof islands west of the Yukon-Kuskokwim Delta. Sea ice formation in the Chukchi and Bering seas begins in October, and the ice pack can persist through late June. The vegetation of this region is primarily tundra comprised of herbaceous sedges, grasses, and low-growing forbs, lichens, and dwarf shrubs. Boreal forest communities extend into the region along valley bottoms of some major rivers. Coastal areas are treeless likely because winds are frequently high, but

Fur seals near the village of St. Paul, Pribilof Islands (Steve McCutcheon, Anchorage Museum of History and Art, SM01310).

this condition may also reflect the continuing influence of a treeless history during Pleistocene glacial advances. The boreal forest simply may not have had enough time to colonize the region. Summer temperatures are moderated by the open waters of the Bering Sea, but winter temperatures are more severe due to sea ice during the coldest months.

Yukon Delta in autumn (Yukon Delta National Wildlife Refuge photo).

Most of the region's forty terrestrial species occur only on the mainland. Islands in the Bering Sea are home to several endemic mammals, including the Pribilof Island shrew (*Sorex pribilofensis*, found only on St. Paul Island) and the insular vole (*Microtus abbreviatus*, restricted to St. Matthew Island and nearby Hall Island). Muskoxen (*Ovibos moschatus*) were brought to Nunivak Island from Greenland in the 1930s. A variety of marine and sea ice habitats of the Bering and Chukchi seas support populations of walrus (*Odobenus rosmarus*) and spotted (*Phoca largha*), ringed (*Pusa hispida*), and bearded seals (*Erignathus barbatus*), as well as beluga, bowhead (*Balaena mysticetus*), and other whales. Two-thirds of the world's remaining northern fur seals (*Callorhinus ursinus*) breed on the Pribilof Islands in the Bering Sea.

Northern Alaska's climate is cold and arid, with only about 150 to 200 mm of precipitation. Summers are short and cool, while winters are long, dark, windy, and cold. The region is bounded to the north by the Arctic Ocean and topography varies southward from the sandy barrier islands and lowlands on the Arctic coastal plain, through the rolling plateaus and hills of the arctic foothills, to the more rugged Brooks Range and associated mountains. Permafrost is continuous except for local anomalies such as deep lakes and major rivers. The dominant vegetation is wet, moist, and relatively dry alpine tundra comprised of herbaceous sedges, grasses, and low-growing forbs, lichens, and dwarf shrubs. Low and tall shrub communities comprised primarily of willow, alder, and shrub birch occur primarily along floodplains and fairly well-drained foothill slopes

Walrus on an ice floe in the northern Bering Sea, western Alaska (USFWS photo).

Part of the Porcupine caribou herd, Arctic National Wildlife Refuge, northern Alaska (Ken Whitten photo).

and river bottoms. Some of the major northward-flowing rivers draining into the polar seas are the Canning, Sagavanirktok, Colville, Ikpikpuk, Kuk, and Utukok rivers.

Land mammals include caribou, wolves, weasels (*Mustela erminea, M. nivalis*), arctic ground squirrels, and lemmings (*Dicrostonyx groenlandicus, Lemmus trimucronatus*). Muskoxen from Nunivak Island were reestablished in portions of their formal range in northern Alaska in the late 1960s. None of the region's twenty-six terrestrial species is entirely endemic there, although the barren ground shrew (*Sorex ugyunak*) and Alaska marmot (*Marmota broweri*) are found primarily in this region of the state. Arctic waters and associated sea ice are home to a number of marine mammals, including several species of seals and whales. Arctic foxes (*Vulpes lagopus*) and polar bears (*Ursus maritimus*) can be found on the polar ice and along the coast. Reduction in the extent and duration of sea ice due to warming trends has major implications for persistence of several of these ice-associated species.

Polar bear on Arctic Ocean ice, northern Alaska (Ken Whitten photo).

Oil developments on the coastal plain at Prudhoe Bay, northern Alaska (USFWS photo).

Alaska's Island Archipelagos

For the past 150 years, island systems have provided key insights into fundamental biological processes. Studies of island populations were central to the development of evolutionary (Darwin 1859) and ecological (MacArthur and Wilson 1963, 1967) theory. Although sometimes characterized as simplified subsets of nearby mainland biotas, island populations often have their own unique set of characteristics, depending largely on their history, size, and degree of isolation. For example, islands can be classified as either "Darwinian islands" (formed by the appearance of land from the water) or "fragment islands" (formed by rising sea level that isolated particular areas of land from others) (Gillespie and Roderick

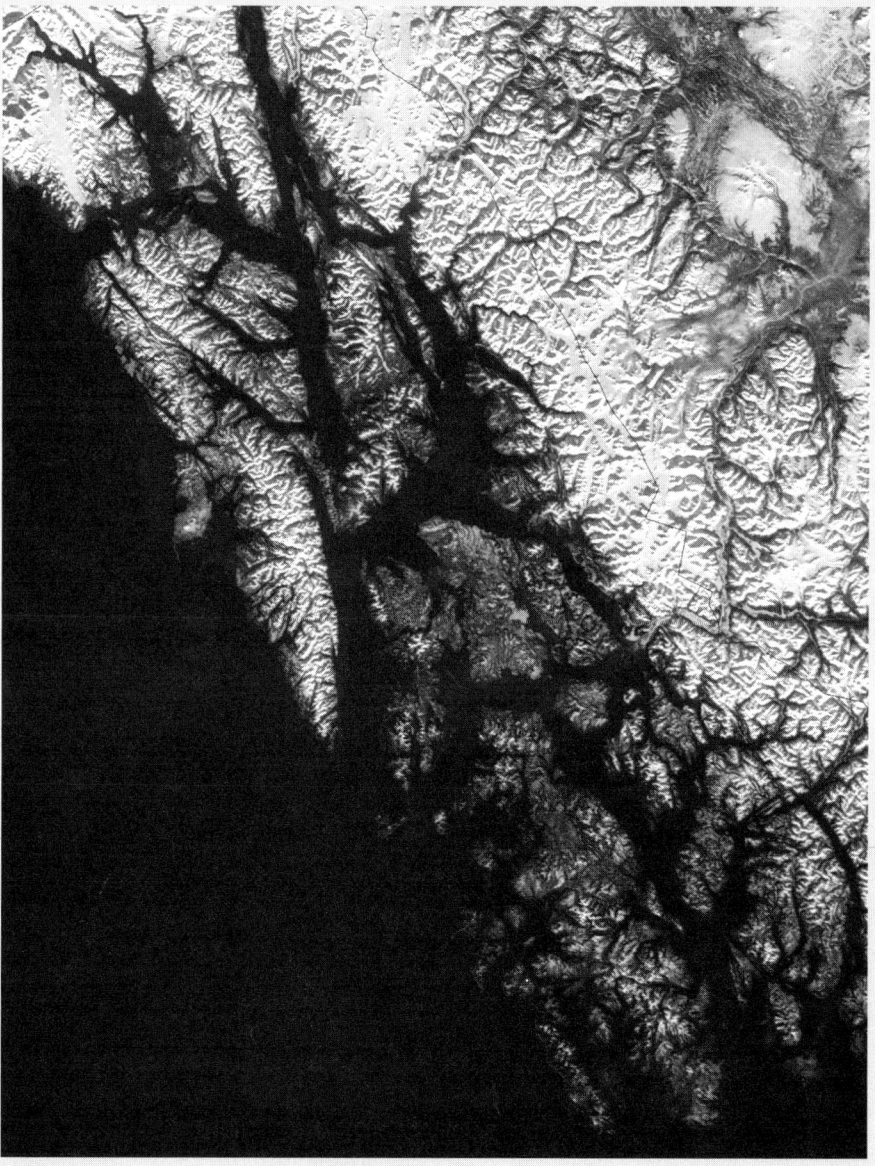

Alexander Archipelago, southeast Alaska (NASA photo)

2002). In general, larger islands support a more diverse fauna than smaller islands. Younger islands that are close to the mainland tend to support less divergent populations due to continuous exchange of individuals. In contrast, older, more isolated islands typically harbor populations that readily can be identified as distinctive or endemic. Large distant islands have higher proportions of endemic species (Johnson 2003). Contrary to popular thinking, islands may be prime generators of biotic diversity on the earth, thus elevating their conservation importance. A substantial portion of existing continental diversity may have originated offshore in island archipelagos (Filardi and Moyle 2005). These kinds of considerations have significant implications for the study of mammals in Alaska because the extraordinary number of islands that ring the North Pacific coast support substantial populations of mammals. In many cases, these populations are distinctive.

Over eighteen hundred named islands are found along the northern coast of the Gulf of Alaska and in the Bering Sea (see Appendices 8–11 for listings of Alaska islands documented with mammal specimens). The two largest archipelagos in U.S. possession are the Alexander Archipelago and Aleutian Islands. Among all U.S. islands, twenty-seven of the thirty-five largest occur in Alaska. Most of these islands are logistically difficult to access, so relatively few studies have been completed on their faunas and floras. In a few cases, however, intensive fieldwork is starting to provide the samples needed to reveal aspects of the ecology, conservation status, biogeography, and evolutionary history of diversification of these insular biomes. Hence, Alaska's islands are natural laboratories for investigations in ecology and evolution (e.g., Foster 1965; Grant 1998; Johnson 2003). We are just beginning to grasp and explore the superb possibilities provided by the numerous insular populations of mammals in Alaska.

The Aleutian Islands are a western continuation of the Aleutian Range on the Alaska Peninsula. These islands arc westward toward Siberia for more than 1,900 km along the edge of the Pacific Plate. Most islands are relatively small and volcanic in origin. Five primary island groups are found in the archipelago: the Fox Islands, Andreanof Islands, Islands of the Four Mountains, Rat Islands, and Near Islands. World commerce is tied to shipping and trade routes through the Aleutian Islands with more than thirty-six hundred tankers, container ships, freighters, and fishing vessels passing through Unimak Pass annually. More than 180 ships have gone aground in the Aleutians since the 1700s (Morkill 2005). Recent groundings in this rugged area increase the possibility of catastrophic oil spills (e.g., vessels like the M/V *Selendang Ayu* in 2004 off Unalaska Island) or introduction of exotic species like brown rats to these relatively pristine island ecosystems. These events have raised concern about impacts to both marine and terrestrial mammals and birds (Gibson and Byrd 2007; Major et al. 2006). Olaus Murie's (1959) pioneering studies on the Aleutians were important to the development of the field of conservation biology due to his cautionary report of the vulnerability of native, insular biotas to the introduction of exotic predators, in this case red fox and associated prey sources such as arctic ground squirrels, to a large number of islands (Bailey 1993; Ebbert and Byrd 2002).

The **Alexander Archipelago** extends nearly 500 km off the southeastern coast of Alaska. This archipelago is less linear in distribution than the Aleutian Chain and consists of more than a thousand islands, including Prince of Wales, Chichagof, Admiralty, Baranof, Revillagigedo, Kupreanof, and Kuiu islands; all larger than 100,000 ha in size. Several of these islands are very rugged due to mountain ranges that rise more than 1,000 m. Deep, wide channels separate most of the islands from other land, although some islands are separated by narrow passages that allow many populations to remain highly connected to each other or to the mainland. Those close connections appear to allow the frequent exchange of more vagile mammal species. Several subregions (MacDonald and Cook 2007) were identified within this large archipelago primarily based on shared species composition. Most islands are largely independent biomes, but considerable testing of this hypothesis needs to be completed in Alaska.

A dynamic geologic history shaped the topography and climate of the region and these, in turn, influenced the persistence and evolution of mammals in the Alexander Archipelago during the Pleistocene. Lower sea levels (up to 100 m less than current conditions) persisted during glacial advances when the region was depressed under the weight of thick ice sheets. These cold episodes were followed by significant upward rebound following deglaciation at a time when sea levels also were rising. Later, the avenues of recolonization available to mammals during the Holocene were also influenced by physiographic features. Geologic events thus played a large role in determining connectivity among sets of islands or between particular islands and the mainland. Some islands (e.g., Douglas, Revillagigedo, and Mitkof islands) are closely connected to the mainland, while others are extremely isolated (e.g., Forrester Island). Mitkof, Kupreanof, and Kuiu islands essentially form a peninsula that extends westward from the mainland and several "mainland" species (e.g., moose, American marten) are now actively colonizing the more distant Kuiu island after island hopping across Mitkof and Kupreanof islands.

Most mammals endemic to the Alexander Archipelago are restricted to southern outer islands, such as Prince of Wales Island (MacDonald and Cook 2007). This pattern of endemism likely reflects the long-term isolation of these organisms from mainland populations (Cook et al. 2006), isolation that perhaps extends into the Pleistocene (Carrara et al. 2007). In a few cases (e.g., Queen Charlotte Islands ermine, *Mustela erminea haidarum*), a genetic footprint of a coastal refugium has been uncovered. These deep molecular signatures suggest that subspecies like this ermine persisted in isolation somewhere along the continental shelf, when sea levels were lower and the rest of the region was covered by glacial ice. An overview of the taxonomic status, biogeographic history, molecular ecology, and conservation status of the mammals of the Alexander Archipelago (MacDonald and Cook 2007) suggested that recognition, investigation, and careful management of endemic organisms on this archipelago were long overdue.

The Kodiak Archipelago includes Kodiak Island (the second-largest U.S. island), Afognak Island, and about twenty smaller islands. Kodiak Island has long intrigued mammalogists due to its potential role as an ice-free refugium (Rausch 1969) and the abundance of large brown bears found there (Troyer and Hensel 1969; van Daele 2003). The presence of brown bears led to protection of about 65 percent of the island in the Kodiak National Wildlife Refuge. Similar to many other islands in Alaska, exotic species also inhabit the island. Sitka black-tailed deer, mountain goat, beaver, muskrat, red squirrel, snowshoe hare, and arctic ground squirrels were translocated to the island. Nearby Afognak Island hosts a healthy population of introduced wapiti and some individuals have now dispersed to smaller nearby islands and recently to Kodiak Island.

Mammals on other island systems in Alaska have been the focus of a variety of investigations. Robert L. Rausch and others conducted a series of studies on small mammals, carnivores, and their associated helminth parasites (e.g., Rausch et al. 1990) on **St. Lawrence Island** in the Bering Sea. This is the fifth-largest U.S. island

(5,135 km²) and it supports a fauna that was heavily influenced by the repeated formation of the Bering Land Bridge during the full glacial events of the Pleistocene. About 350 km southwest of St. Lawrence Island is the **St. Matthew Island** group, which was the focus of an investigation of the population dynamics and impact of an introduced herd of reindeer (Klein 1968, 1987). Hall and St. Matthew islands are home to an endemic vole, *Microtus abbreviatus* (Cook and Klein 1999), a close (perhaps conspecific) relative of the singing vole (*Microtus miurus*) of the mainland. Nearly 400 km farther south are the **Pribilof Islands**, where St. Paul Island supports an endemic shrew (*Sorex pribilofensis*) and intensive efforts to prevent the introduction of exotic rats have been under way for several years.

Archipelagos are being heavily impacted throughout the planet due to habitat conversion, overexploitation, introduction of exotics, and a host of related assaults (Quammen 1996). Those impacts have led to extremely high rates of endangerment and extinction and, ultimately, the disruption of ecosystem function. Unfortunately, our baseline documentation of many archipelagos, and especially the island systems of Alaska, is incomplete (as experienced, for example, in **Prince William Sound** following the devastating *Exxon Valdez* oil spill there in 1989). Lack of comprehensive monitoring programs or even basic inventories will stymie efforts to understand and conserve these unique (endemic) and highly productive biomes (Cook et al. 2006). Nearby, island mammals along the western coast of Canada (e.g., *Rangifer tarandus dawsoni*, *Marmota vancouverensis*), have been listed as threatened or endangered, have been extirpated, or have experienced extinction (COSEWIC 2007). Exotic species are just now gaining a foothold in some of these systems. The expense of managing these invasives will be greatly magnified once they have gained a beachhead. In general, management plans for public lands in Alaska still have not sufficiently addressed these island systems, although a few agencies (e.g., Alaska Maritime National Wildlife Refuge) have established proactive management plans aimed at thwarting further assaults on these incomparable insular faunas and floras.

Materials and Methods

We report taxonomic and distribution information for 116 mammalian species (representing 226 subspecies and monotypic species) in Alaska. Species accounts follow the taxonomic summary of Wilson and Reeder (2005), except as noted. These accounts are based primarily on specimens obtained through fieldwork conducted by the authors and collaborators, specimens archived through several large salvage efforts, a review of literature, and other specimens housed at the University of Alaska Museum of the North (UAM) and thirty-six other natural history museums with substantial holdings from Alaska. This comprehensive review includes specimens cataloged through July 2007.

Fieldwork

Our field investigations began in 1973 when S. O. MacDonald moved to the Chickamin River in southeast Alaska, collected small mammals, and sent those specimens to UAM. J. A. Cook first collected Alaska mammal specimens in 1983 while working out of a wall tent for the U.S. Fish and Wildlife Service on Kodiak Island. Independent field efforts continued sporadically until 1990, when Cook became Curator of Mammals at UAM and we jointly initiated field, museum, and laboratory studies designed to systematically survey the mammals of Alaska. During the next seventeen years, several collaborators and a large number of undergraduate and graduate students at the University of Alaska Fairbanks, and later Idaho State University and University of New Mexico, were involved in a series of field-based inventories of mammals throughout Alaska that increased the depth and breadth of museum holdings (Fig. 17).

Field studies were conducted in collaboration with the Alaska Department of Fish and Game, U.S. Fish and Wildlife Service, U.S. Forest Service, National Park Service, and Bureau of Land Management and often coincided with federally mandated inventory and monitoring efforts (e.g., Cook and MacDonald 2001, 2004a, 2006; Cook et al. 2006). From 1991 to 1999 and 2004 to 2007, the USDA Forest Service and U.S. Fish and Wildlife Service sponsored work in the Chugach National Forest of Prince William Sound (Lance and Cook 1998a, 1998b) and in the Tongass National Forest along the coast of southeastern Alaska (MacDonald and Cook 2007). From 2000 to 2004, we surveyed eleven national parks for the National Park Service (Cook et al. 2004). Since 1991, limited field studies occurred on Bureau of Land Management lands and U.S. Fish and Wildlife Service refuges, often in collaboration with agency biologists (e.g., Cook et al. 2000).

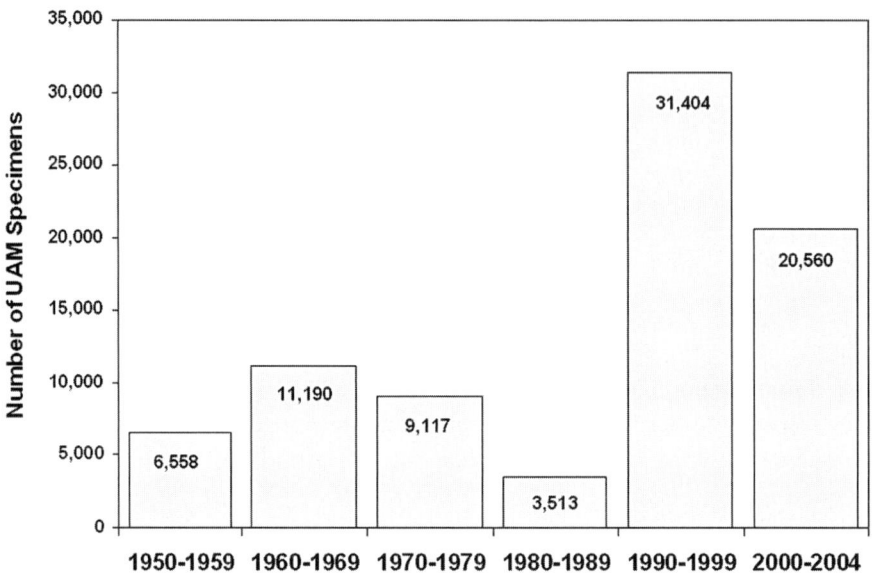

Figure 17. Six decades of collecting effort for UAM.

Our field expeditions were designed to maximize the number and diversity of materials archived by using a variety of sampling methods at each locality we visited. Our field crews consisted of two to six people, depending primarily on the logistics of transportation between remote field sites. For example, small float planes necessitated smaller field crews. Collecting methods followed those outlined in Wilson et al. (1996) and approved by the American Society of Mammalogists (Gannon et al. 2007). These methods included transect lines consisting of a mixture of rat and museum special snap-traps, Sherman live traps, and 1.3-liter pitfall traps set in a wide diversity of habitats. We generally set traps in pairs with each station placed 10 m apart. At most sites, three hundred to five hundred traps were set per night. A mixture of peanut butter and rolled oats was the primary bait. We sampled a variety of habitats at each locality, but focused on ecotonal areas. Elevation transects were used to maximize diversity of species sampled. Field sampling was generally conducted from June to early September when populations were high. Bats were studied using nets or sonogram-based bat detectors (Parker and Cook 1996; Parker et al. 1996, 1997). A shotgun or rifle was used to collect medium-sized mammals (e.g., squirrels). Field notes, individual specimen "AF" or "NK" field catalog pages, and daily trapline data forms are deposited at either UAM or the Museum of Southwestern Biology (MSB).

Materials collected represent geographically extensive and site-intensive collections. Each mammal we collect is assigned a unique field catalog number, and all tissues, parasites, and other subsamples are linked to the original specimen. All mammals sampled are preserved as scientific specimens (vouchers), usually as cleaned skeletons or as fluid preparations in ethanol. In addition to these standard field preparations, a variety of ancillary collections were archived that are now providing a highly integrated view of mammalian communities and associated parasitic diversity. Materials that have been loaned to a large number of independent investigations can be associated easily with a specific (voucher) animal, GPS locality, and date of collection. Embryos were preserved in alcohol or liquid nitrogen. Ectoparasites (e.g., fleas, ticks, and mites) were preserved in ethanol and sent to a number of specialists (e.g., Murrell et al. 2003). Similarly, endoparasites (e.g., helminth worms, protozoans, and viruses) were loaned to various laboratories and form the basis of a growing number of publications (e.g., Hoberg et al. 2003; Laakkonen et al. 2005; Lynch et al. 2007). A variety of tissue samples (e.g., heart, liver, spleen, and kidney) from all specimens were frozen in liquid nitrogen in the field and subsequently archived at UAM or MSB. These specimens have been incorporated into a wide variety of studies related to Alaska's mammals, ranging from zoonotic diseases to molecular conservation genetics (Cook et al. 2005).

Collaborations

In addition to our field expeditions, a number of specimens were provided by state and federal agencies or through collaborative efforts with rural subsistence hunters. The Alaska Department of Fish and Game provided large series of trapper-salvaged carcasses of furbearer species, especially marten, wolverine, lynx, and wolves. Marine mammals were obtained through a series of collaborative efforts with Native communities and subsistence management commissions (e. g., Sea Otter Commission) through the support of the Coastal Marine Institute, National Marine Fisheries Service, and Alaska Department of Fish and Game. Yet another example is more than three hundred sea otters from the Gulf of Alaska that were salvaged after the *Exxon Valdez* oil spill.

Additional Museum Studies

We visited twelve collections with substantial holdings from Alaska (other than UAM and MSB) to review specimens. Electronic databases (when available) were verified by direct reference to voucher specimens. We focused on specimens documenting range extensions, uncommon species, questionable identification or associated data, and difficult to identify species. We also obtained information or loans from an additional twenty-three museums with smaller holdings to confirm problematic distribution or taxonomic records. Overall, a majority of Alaska specimens was originally obtained through our efforts, identified by us, and subsequently archived at either UAM or MSB.

Laboratory Studies

Following fieldwork and subsequent curation of specimens, associated materials were incorporated into a series of investigations, including graduate student projects that were completed in our molecular genetic laboratories, first at the University of Alaska Fairbanks and later Idaho State University and then the University of New Mexico (Bidlack 2000; Conroy 1998; Dawson 2008; Demboski 1999; Durall 1993; Eddingsaas 2001; Frances 2008; Galbreath 2002; Koehler 2006; Lance 1995; Lucid 2003; Parker 1996; Runck 2001,

2006; Stone 2000; Tomasik 2003; Waltari 2005; Weckworth 2003). Materials were also loaned to other institutions (e.g., Wickström et al. 2005) and many of those studies can be tracked through the online database ARCTOS. Those projects were intended largely to further characterize geographic variation, systematics, and ecology of Alaska mammals. Ultimately, these specimens should serve a large number of diverse studies. Overviews of some of these regional, specimen-based studies of Alaska mammals (e.g., Cook et al. 2005, 2006; MacDonald and Cook 2007) are beginning to provide synthetic views of the evolution, conservation, and ecology of these boreal mammals.

Museum Specimens

At least 125,440 specimens, including 169 type specimens (Appendix 3), constitute the collection of Alaska Recent mammals currently preserved in museums (Appendix 4).

Accounts of Species and Subspecies

Species accounts include all currently recognized species and subspecies of mammals that occur, or have recently occurred, in Alaska. For most species, the following topics are discussed:

Scientific Name and Authority

Each scientific name is followed by the name of the author and the year the formal description was published. Parentheses indicate that although the specific name has remained the same, the species has since been assigned to another genus. Nomenclature follows Wilson and Reeder (2005), except when noted otherwise.

Common Name

The common or vernacular name generally follows Wilson and Reeder (2005). Frequently used alternative names are also included.

Systematics

References for taxonomic revisions are provided and taxonomic problems are discussed. Taxa (subspecies, or if monotypic, species) that occur in the state are listed and information for each is presented as follows:
- *Scientific name of taxon* followed parenthetically by author's name, date, and the publication of the original taxonomic description.
- *Type Locality* is the published locality of collection for the type specimen.
- *Type Specimen* is the individual specimen (holotype) that was the basis for the original taxonomic description and is designated in the publication.
- *Range* outlines the general distribution of the taxon, followed by
- *Remarks,* when applicable.

Pertinent information from genetic studies (e.g., DNA variation across the geographic range of particular species) is assessed. Molecular genetic assays provide insight into taxonomic validity and population level differentiation, and are helpful in defining conservation units for management efforts, for screening emerging zoonotics and potential human pathogens, and for reconstructing the historical biogeography of the region.

Global Distribution

A description of the entire range of the species is presented, along with a map showing the range in North America as an insert to the Alaska distribution map.

Alaska Distribution

A general description of distribution in Alaska is followed by more detailed information of occurrence in each of the six regions of the state (as delimited in Fig. 10; see Appendix 2 for regional occurrence by species). A distribution map accompanies each account, with dots representing records from UAM specimens and, when informative, records from other institutions. Open dots are used to denote marginal records that were reexamined and listed (clockwise), starting with the most northern record. Each marginal record includes, when available, latitude and longitude (recorded in decimal degrees rounded to the fifth decimal place), USGS Quadrangle map (see Appendix 1), specific locality, and, parenthetically, number of specimens and institutional acronym. The general range of marine mammals is shaded as illustrated by Angliss and Outlaw (2007).

Habitat

Mammals inhabit a wide range of conditions in both marine and terrestrial environments in Alaska. This section includes a qualitative description of habitat preference for each species based on our field experiences and the published literature. Appendix 5 provides a generalized comparison across all mammals of primary habitat associations as defined by the Alaska Department of Fish and Game (ADFG 2007). Alaska's major land cover (Fig. 1 insert map) and ecoregions are also included (Fig. 11; Nowacki et al. 2001) to provide further perspective on occurrence and habitat association.

Status

Relative abundance of each species in Alaska (common, rare, etc.) is assessed qualitatively based on our fieldwork and the literature review; however, tremendous population fluctuations from year to year characterize many high-latitude species. If a taxon or population has special legal status or has been identified as having special viability concerns, the appropriate citation is noted. The following conservation sources were reviewed: Alaska Department of Fish and Game (ADFG), U.S. Endangered Species Act (ESA), World Conservation Union (IUCN; formerly International Union for Conservation of Nature and Natural Resources), Convention on International Trade in Endangered Species of Wild Fauna and Flora (CITES), and Committee on the Status of Endangered Wildlife in Canada (COSEWIC). See Appendix 6 for a complete listing and source information, and Appendix 7 for a summary of introductions and translocations.

Fossils

Pertinent information on fossils of Alaska mammals is reported.

Specimens

This section lists the total number of specimens preserved in thirty-eight museums through July 2007 (Appendix 4). Specimen representation on islands by region can be found in Appendices 8–11. In several accounts, all known records are listed, following the (telegraphic) style outlined above in "Alaska Distribution." Detailed information on mammal specimens housed at UAM and MSB are available online at http://arctos.database.museum/home.cfm. The mammal collection of the USNM is now available online at http://acsmith.si.edu/, and links to a number of other institutional databases can be accessed through portals of the Mammal Networked Information System (MANIS) at http://manisnet.org/manis.

Abbreviations

Institutions (*visited) and agencies cited in this publication are as follows:

ADFG—Alaska Department of Fish and Game
AMNH—American Museum of Natural History, New York, NY*
AMNWR—Alaska Maritime National Wildlife Refuge
ANSP—Academy of Natural Sciences, Philadelphia, PA
ANWR—Arctic National Wildlife Refuge
BLM—Bureau of Land Management
BMNH—The Natural History Museum, London, England
CAS—California Academy of Sciences, San Francisco, CA
CAS-SU—California Academy of Sciences, Stanford University, CA
CM—Charles R. Conner Museum, Washington State University, Pullman, WA*
CMNH—Carnegie Museum of Natural History, Pittsburgh, PA
CU—Cornell University, Ithaca, NY
DMNS—Denver Museum of Nature and Science, Denver, CO
FMNH—Field Museum of Natural History, Chicago, IL
GZG—Geowissenschaftliches Zentrum der Universität Göttingen, Germany
IWC—International Whaling Commission
KU—University of Kansas, Lawrence, KS*
LACM—Natural History Museum of Los Angeles County, CA*
LSUMZ—Louisiana State University Museum of Natural Science, Baton Rouge, LA

More student collaborators (clockwise from top left): Kurt Galbreath, Doreen Parker, Michael MacDonald, C. Tom Seaton, Grace Leacock, Natalie Dawson, Eric Waltari, Jeff Good, and Darren Zibell.

MCZ—Museum of Comparative Zoology, Harvard University, Cambridge, MA*

MMNH—Minnesota Bell Museum of Natural History, Minneapolis, MN

MNHN—Museum National d'Histoire Naturelle, Paris, France

MNHU—Museum für Naturkunde der Humboldt Universität, Berlin, Germany

MSB—Museum of Southwestern Biology, University of New Mexico, Albuquerque, NM*

MSU—Moscow State University, Moscow, Russia

MSUM—Michigan State University Museum, Lansing, MI

MVZ—Museum of Vertebrate Zoology, Berkeley, CA*

NMC—National Museum of Canada (Canadian Museum of Nature), Aylmer, Quebec

NMFS—United States National Marine Fisheries Service

NPS—United States National Park Service

PSM—Slater Museum of Natural History, University of Puget Sound, Tacoma, WA*

RMNH—Rijksmuseum van Natuurlijke Historie, Leiden (now National Natuurhistorisch Museum, Leiden), The Netherlands

ROM—Royal Ontario Museum, Toronto, Ontario

SDNHM—San Diego Natural History Museum, San Diego, CA

TCWC—Texas Cooperative Wildlife Collection, Texas A&M University, College Station, TX

TTU—Texas Tech University, Lubbock, TX

UAM—University of Alaska Museum of the North, Fairbanks, AK*

UBC—University of British Columbia, Vancouver, British Columbia*

UCB—University of Colorado, Boulder, CO

UCLA—University of California, Los Angeles, CA*

UMMZ—University of Michigan Museum of Zoology, Ann Arbor, MI

UMNH—Utah Museum of Natural History, Salt Lake City, UT

USFS—United States Department of Agriculture Forest Service

USGS—United States Geological Survey

USFWS—United States Fish and Wildlife Service

USNM—United States National Museum of Natural History, Smithsonian Institution, Washington, DC*

UWBM—University of Washington Burke Museum, Seattle, WA*

YPM—Peabody Museum of Natural History, Yale University, New Haven, CT

ZISP—Zoological Institute, St. Petersburg, Russia*

The Recent Mammals of Alaska

Faunal Composition

Including humans, 116 species of mammals, representing 81 genera, 31 families, and 10 orders, have been documented in Alaska and adjacent waters in Recent (post-Pleistocene) times. They include 226 subspecies and monotypic species, of which 88 are endemic to the state (i.e., 39 percent are found only in Alaska) and 43 of these restricted to island localities (i.e., nearly half of Alaska's endemic mammals are island endemics). Carnivores, with thirty-two species, comprise the most speciose group, followed by rodents with twenty-nine species, and whales with twenty-four species. Compared to mammalian faunas at lower latitudes, Alaska has low bat diversity (species richness) and relatively high carnivore diversity. Six species are not native to the state, but have been introduced by humans, and two others (mammoth and sea cow) are now extinct.

To develop a complete appreciation for the contemporary mammalian fauna of Alaska, it is essential to have some perspective on how these communities were assembled prior to Recent times (see Appendix 12 for a complete listing and chronology of the Pleistocene mammalian fauna). First, we recognize that the region has long been shaped by dramatic changes in climate. As climate cooled during the repeated glacial maxima of the Pleistocene, accumulation of ice lowered sea levels and exposed the Bering Land Bridge, the high-latitude crossroads between the northern continents (Hopkins 1967). This terrestrial connection (or barrier for marine mammals) had a profound influence on the composition of Alaska's mammalian communities due to substantial mixing of western North American and eastern Asian faunas. Preliminary data suggest that exchanges were not reciprocal with more Asian species colonizing eastward (Rausch 1994; Waltari et al. 2007).

The Bering Land Bridge is one element of Beringia, the vast ice age refugium (Fig. 18) that once spanned from Siberia to northwestern Canada (O'Neill 2004; Sushkin 1925). Hultén (1937), who studied plants (after a rather fortuitous shipwreck while on his honeymoon) in Kamchatka and then later in Alaska, identified the biogeographical significance of similarities between Siberia and Alaska. The glacial advances that created this large, high-latitude refugium were instrumental in the evolutionary diversification of a number of boreal organisms (Elias et al. 1996; Guthrie and Matthews 1971; Hoberg et al. 2003). Several mammals (Hall 1981; Junge and Hoffmann 1981), and a growing number of associated parasites, are recognized as endemic to the

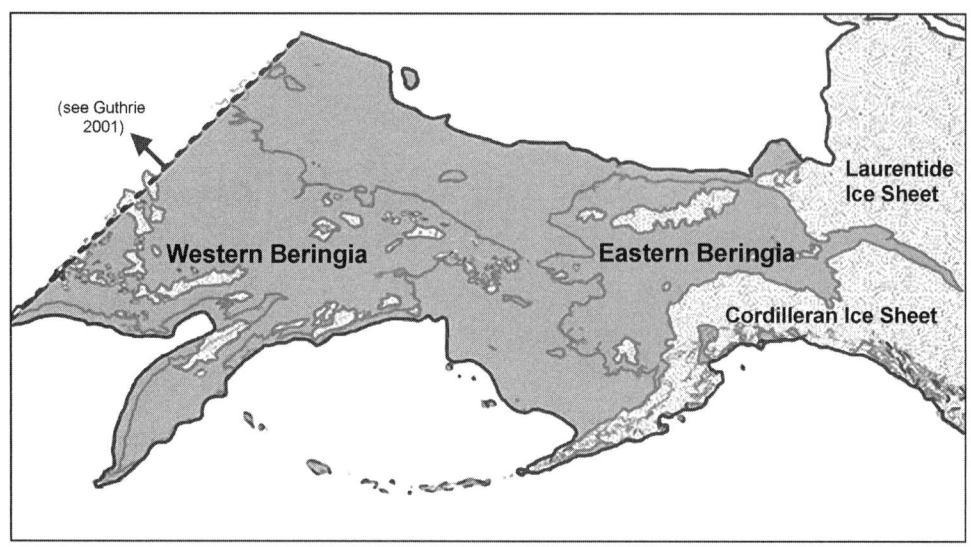

Figure 18. The refugium of Beringia at the height of the latest glaciation, about 18,000 years before present.

region (e.g., Cook et al. 2005; Haukisalmi et al. 2002; Rausch 1994; Wickström et al. 2001, 2003).

The role of Beringia in the historical assembly of high-latitude faunas has been investigated by multiple disciplinary avenues, including paleontology (e.g., Elias et al. 1996; Guthrie 1990a, 2001a, 2003; Guthrie and Matthews 1971; Harington 1978; Kurtén and Anderson 1980; Sher 1971), paleobotany and palynology (e.g., Brubaker et al. 2005; Ritchie and Cwynar 1982), geomorphology, parasitology (Rausch 1994), and more recently, molecular genetic assessments of extant organisms (Abbott and Brochmann 2003; Waltari et al. 2007). Those studies and others (e.g., Hoffmann 1981) demonstrated that the Beringian region supported a diversity of habitats, including steppe-tundra, taiga, dry upland tundra, and wet lowland tundra. This dynamic history sets the stage for understanding the distribution of many of Alaska's contemporary mammals. Interpretation of these changes will be essential to understanding and predicting the influence of current changing conditions on mammals.

List of Species

Scientific and common names used in this publication generally follow Wilson and Reeder (2005) for species and Hall (1981) for subspecies. In the following list of mammals, species introduced to Alaska by humans (summarized in Appendix 7) are identified with an asterisk (*), and those that are now extinct or extirpated are noted with E.

PROBOSCIDEA—elephants

Elephantidae
Mammuthus primigenius, woolly mammoth E

SIRENIA—manatees, dugong, and sea cow

Dugongidae
Hydrodamalis gigas, Steller's sea cow E

PRIMATES—primates

Hominidae
Homo sapiens, human

RODENTIA—rodents

Sciuridae
Glaucomys sabrinus, northern flying squirrel
Marmota broweri, Alaska marmot
Marmota caligata, hoary marmot
Marmota monax, woodchuck
Spermophilus parryii, arctic ground squirrel
Tamiasciurus hudsonicus, red squirrel

Castoridae
Castor canadensis, American beaver

Dipodidae
Zapus hudsonius, meadow jumping mouse
Zapus princeps, western jumping mouse

Cricetidae
Dicrostonyx groenlandicus, collared lemming
Lemmus trimucronatus, brown lemming
Microtus abbreviatus, insular vole
Microtus longicaudus, long-tailed vole
Microtus miurus, singing vole
Microtus oeconomus, root vole
Microtus pennsylvanicus, meadow vole
Microtus xanthognathus, taiga vole
Myodes gapperi, southern red-backed vole
Myodes rutilus, northern red-backed vole
Neotoma cinerea, bushy-tailed woodrat
Ondatra zibethicus, common muskrat
Peromyscus keeni, northwestern deermouse
Peromyscus maniculatus, North American
   deermouse*
Phenacomys intermedius, western heather vole
Synaptomys borealis, northern bog lemming

Muridae
Mus musculus, house mouse*
Rattus norvegicus, brown rat*
Rattus rattus, roof rat*

Erethizontidae
Erethizon dorsatum, North American
   porcupine

LAGOMORPHA—pikas and hares

Ochotonidae
Ochotona collaris, collared pika

Leporidae
Lepus americanus, snowshoe hare
Lepus othus, Alaska hare
Oryctolagus cuniculus, European rabbit*

SORICOMORPHA—shrews, moles, and solenodons

Soricidae
Sorex alaskanus, Glacier Bay water shrew
Sorex cinereus, cinereus shrew
Sorex hoyi, pygmy shrew
Sorex jacksoni, St. Lawrence Island shrew
Sorex monticolus, dusky shrew
Sorex palustris, American water shrew
Sorex pribilofensis, Pribilof Island shrew
Sorex tundrensis, tundra shrew
Sorex ugyunak, barren ground shrew
Sorex yukonicus, Alaska tiny shrew

CHIROPTERA—bats

Vespertilionidae
Eptesicus fuscus, big brown bat
Lasionycteris noctivagans, silver-haired bat
Myotis californicus, California myotis
Myotis keenii, Keen's myotis
Myotis lucifugus, little brown myotis
Myotis volans, long-legged myotis

CARNIVORA—carnivores

Felidae
Lynx canadensis, Canadian lynx
Puma concolor, cougar

Canidae
Canis latrans, coyote
Canis lupus, wolf
Vulpes lagopus, arctic fox
Vulpes vulpes, red fox

Ursidae
Ursus americanus, American black bear
Ursus arctos, brown bear
Ursus maritimus, polar bear

Otariidae
Arctocephalus townsendi, Guadalupe fur seal
Callorhinus ursinus, northern fur seal
Eumetopias jubatus, Steller's sea lion
Zalophus californianus, California sea lion

Odobenidae
Odobenus rosmarus, walrus

Phocidae
Cystophora cristata, hooded seal
Erignathus barbatus, bearded seal

Phocidae (continued)
Histriophoca fasciata, ribbon seal
Mirounga angustirostris, northern elephant seal
Pagophilus groenlandicus, harp seal
Phoca largha, spotted seal
Phoca vitulina, harbor seal
Pusa hispida, ringed seal

Mustelidae
Enhydra lutris, sea otter
Gulo gulo, wolverine
Lontra canadensis, North American river otter
Martes americana, American marten
Martes caurina, Pacific marten
Martes pennanti, fisher
Mustela erminea, ermine
Mustela nivalis, least weasel
Neovison vison, American mink

Procyonidae
Procyon lotor, raccoon*

ARTIODACTYLA——even-toed ungulates

Cervidae
Alces americanus, moose
Cervus canadensis, wapiti or eastern red deer
Odocoileus hemionus, mule deer
Rangifer tarandus, reindeer or caribou

Bovidae
Bison bison, American bison
Oreamnos americanus, mountain goat
Ovibos moschatus, muskox
Ovis dalli, Dall's sheep

CETACEA—whales

Balaenidae
Balaena mysticetus, bowhead
Eubalaena japonica, North Pacific right whale

Balaenopteridae
Balaenoptera acutorostrata, common minke whale
Balaenoptera borealis, sei whale
Balaenoptera musculus, blue whale
Balaenoptera physalus, fin whale
Megaptera novaeangliae, humpback whale

Eschrichtiidae
Eschrichtius robustus, gray whale

Delphinidae
Globicephala macrorhynchus, short-finned pilot whale
Grampus griseus, Risso's dolphin
Lagenorhynchus obliquidens, Pacific white-sided dolphin
Lissodelphis borealis, northern right whale dolphin
Orcinus orca, killer whale
Pseudorca crassidens, false killer whale
Stenella attenuata, pantropical spotted dolphin

Monodontidae
Delphinapterus leucas, beluga
Monodon monoceros, narwhal

Phocoenidae
Phocoena phocoena, harbor porpoise
Phocoenoides dalli, Dall's porpoise

Physeteridae
Kogia breviceps, pygmy sperm whale
Physeter catodon, sperm whale

Ziphiidae
Berardius bairdii, Baird's beaked whale
Mesoplodon stejnegeri, Stejneger's beaked whale
Ziphius cavirostris, Cuvier's beaked whale

Taiga vole, *Microtus xanthognathus* (O. MacDonald)

Order PROBOSCIDEA Illiger, 1811

McKenna and Bell (1997) placed Proboscidea (mastodons, elephants, and kin) in the new rank of "parvorder," in the suborder Tethytheria, Order Uranotheria. One living and six extinct families are usually associated with the Proboscidea (Nowak 1991). These mammals were a dominant group in the Tertiary and Pleistocene (Kurtén and Anderson 1980). Extinct forms were extremely cosmopolitan, occupying a variety of habitats, from low-elevation deserts to mountaintops and on all continents except Australia and Antarctica (Shoshani 1998).

The American mastodon (*Mammut americanum*), family Mammutidae, co-occurred in Eastern Beringia (Alaska and adjacent Canada) with the much more common woolly mammoth during the Upper Pleistocene.

Family ELEPHANTIDAE Gray, 1821—elephants

The earliest fossils of this family are from Late Miocene or Early Pliocene deposits in Africa (Nowak 1991). Along with the three living forms (Shoshani 2005a)—Asian elephant (*Elephas maximus*), African bush elephant (*Loxodonta africana*), and African forest elephant (Loxodonta cyclotis)—the woolly mammoth (*Mammuthus primigenius*) survived well into the Holocene on at least two arctic islands (Guthrie 2004; Vartanyan et al. 1993). Ancient DNA sequences indicate a closer relationship of the woolly mammoth with the Asian, rather than African, elephants (Krause et al. 2006; Yang et al. 1996; but see Thomas et al. 2000).

The long, muscular nose or trunk with nostrils at the end is the most conspicuous external feature of this group. The now-extinct woolly mammoth was about the size of today's Asian elephant (about 3 m high at the shoulder) and with similar teeth (Harington 1995). Its coat consisted of long (up to 90 cm), dark guard hairs and fine underfur. This species had a high, peaked head, a humped, sloping back, a trunk shorter than living forms, and large, curved tusks (smaller on females).

Woolly mammoth, *Mammuthus primigenius* (illustrated by Charles Douglas, reproduced courtesy of the Canadian Museum of Nature, Ottawa, Canada)

Woolly Mammoth

Mammuthus primigenius
(Blumenbach, 1799)

Other Common Names
Tundra mammoth.

Systematics
Mammoths were a group of species in the genus *Mammuthus* that probably originated from *Stegodon* and started to diverge during the Upper Pliocene (some four million years ago) in Africa. Most recent studies suggest *Mammuthus* is a sister group to *Elephas* rather than *Loxodonta* (e.g., Krause et al. 2006). The woolly mammoth, *M. primigenius*, of the Upper Pleistocene was probably derived from the cold-adapted and much larger steppe mammoth, *M. trogontherii*, of the Early Pleistocene (Harington 1995). Mammoths are evolutionarily distinctive from the mastodons (family Mammutidae), although they sometimes shared the same environment.

Mammuthus primigenius
Original Description. 1799. *Elephas primigenius* Blumenbach, Handbuch der Naturgeschichte, 6th ed., p. 697.
Type Locality. Between Dorste and Osterode, Lower Saxony, Germany.
Type Age. Late Pleistocene, Weichselian.
Type Specimen. Lectotype, M1, upper left, designated by Osborn (1942) (GZG.V.010.018).

Remarks. Tooth material used by Blumenbach in his original description long believed lost but recently rediscovered (Reich et al. 2006, 2007, pers. comm.). Osborn (1942) designated also another lectotype from Siberia, but this molar is lost. The absolute age of the lectotype (GZG.V.010.018), based on AMS 14C dating (Groningen University), is 34,340±230/±210 BP (lab reference number GrA-32611).

Global Distribution
Holarctic—Formerly found from northern Europe, including Great Britain, northern Asia, including Japan, and northern North America as far south as Virginia (Kurtén and Anderson 1980).

Alaska Distribution
Woolly mammoths occurred over much of Alaska during the Late Pleistocene. Their remains have been found at numerous sites on the current mainland as well as on Unalaska, King, and St. Lawrence islands. Mammoths on St. Paul Island (Pribilof Islands) persisted to at least fifty-seven hundred radiocarbon years before present (Veltre et al. 2008). Woolly mammoths of dwarfed stature apparently survived until about thirty-seven hundred years before present on the Siberian island of Wrangel (Vartanyan et al. 1993).

Habitat
Woolly mammoths favored cold, dry grasslands, open tundra, and steppe with scattered forest. This habitat was called the "mammoth steppe" by Guthrie (1990a, 2001a) (map 1).

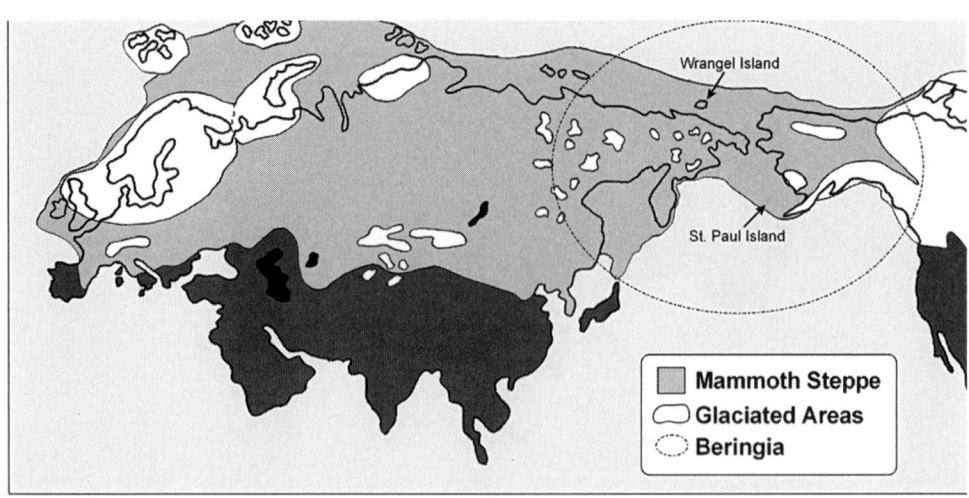

MAP 1. Extent of the Mammoth Steppe during the last glacial maximum (after Matheus 1994).

Mammoths traveled in herds. Stomach contents recovered from frozen animals indicated that they grazed on grass mixed with forbs, shrubs, and parts of trees (Lister and Bahn 1994).

Status

Extinct. Remains of mammoths from the Alaska mainland date to about 11,500 years ago. The population on St. Paul Island in the Bering Sea survived into the Middle Holocene, but appears to have become extinct there before humans reached the island (Guthrie 2004; Veltre et al. 2008).

Fossils

Evidence from fossil bones and frozen carcasses indicates that this species originated in the Middle Pleistocene in Siberia, spreading to North America in the Late Pleistocene (Haynes 1991; Lister and Sher 2001).

Specimen Record

(Guthrie 2004). Pribilof Island, St. Paul Island, Northeast Point, 75 cm below surface (USNM 23455; incomplete upper molar). Additional specimens were reported by Veltre et al. (2008).

Order SIRENIA Illiger, 1811

The Sirenians belong to a clade called Tethytheria, the "subungulates," which includes the elephants, hyraxes, and perhaps the aardvarks and two other orders (Reeves et al. 2002; Rice 1998; Springer et al. 1997). McKenna and Bell (1997) considered Sirenia an infraorder in the suborder Tethytheria, order Uranothera.

Two families comprise this Order: Trichechidae (the manatees) and Dugongidae (dugongs) (Shoshani 2005b).

Family DUGONGIDAE Gray, 1821—dugong and sea cow

Dugongidae is usually divided into two subfamilies: Dugonginae, with the dugong of the Indian Ocean and African and Indo-Malaysian waters as the only surviving member of a once diverse and widespread family; and Hydrodamalinae, the extinct subfamily that included Steller's sea cow, which occurred most recently along the coasts of the Commander and (most likely) westernmost Aleutian islands until being hunted to extinction shortly after its discovery in the eighteenth century.

The Steller's sea cow had a small head, a short neck, stubby foreflippers, and a split tail and was huge, with one adult female measuring over 7 m long (Haley 1978). Finger bones and teeth were totally lacking and they had a wrinkled hide that was hairless, black, and marked with white spots and blotches in some animals (Haley 1978). Sea cows apparently fed primarily on kelp and other algae (Reeves et al. 2002).

Steller's sea cow, *Hydrodamalis gigas* (Michigan Science Art)

Steller's Sea Cow

Hydrodamalis gigas
(Zimmermann, 1780)

Other Common Names

Great northern sea-cow.

Systematics

Domning (1994, 1996) reviewed the taxonomic history of the apparently paraphyletic family Dugongidae and the extinct subfamily Hydrodamalinae. The Late Pleistocene–Recent Steller's sea cow, *Hydrodamalis gigas*, was one of two extinct species in the genus.

The molecular studies of Rainey et al. (1984) and Ozowa et al. (1997) further resolved the systematics of Steller's sea cow. The species was reviewed by Forsten and Youngman (1982).

Hydrodamalis gigas

Original Description. 1780. *Manati gigas* Zimmerman, *Geogr. Gesch. Mensch. Vierf. Thiere*, 2:426.

Type Locality. Bering Island [Ostrov Beringa], Commander Islands, Bering Sea.

Type Specimen. None designated.

Global Distribution

Discoveries of Steller's sea cow remains demonstrate that the distribution of this species once extended around the North Pacific Rim from Baja California, Mexico, to Honshu, Japan (Anderson and Domning 2002). Historically, this species was known only from the Commander Islands (i.e., Bering Island and Copper Island) at the western limit of the Aleutian Islands. During glacial maxima, sea cows, like walruses, evidently ranged much farther south than they did in the Holocene, as indicated by fossils of *H. gigas* from Monterey Bay, California, that dated around nineteen thousand years ago (Harington 1978).

Alaska Distribution

Murie (1959) speculated that this marine mammal may have been only an accidental straggler to the westernmost Aleutian Islands in historical times. Whitemore and Gard (1977) suggested that *H. gigas* once ranged the Aleutian chain, and radiocarbon dates of sea cow bones recently discovered on along western Aleutian Islands demonstrate their presence there at least one thousand years ago (Savinetsky et al. 2004) (map 2). Stejneger (1883:84) quoted reports of Natives that sea cow bones occurred on the Semichi Islands and Agattu Island of the Near Islands, and there are unvalidated reports of a rib found ca. 1842 on Attu Island (Brandt 1861; Domning 1978). The USNM houses various bones of sea cow dating $127,000\pm1,100$ years old unearthed at South Bight on Amchitka Island in the Rat Islands (USNM 170761, 181752, 186807). Gard et al. (1972) reported sea cow remains in Pleistocene interglacial desposits on Amchitka Island. Skeletal elements recovered from Adak Island dated to $1,710\pm70$ years old (Savinetsky et al. 2004). Recently, UAM received a partially hand-worked and charred sea cow rib (UAM 63998) of unclear origin from Adak Island. Another rib, a subfossil dated to approximately 1040 ± 40 years old (corrected age less than 1,000 years old according to Domning et al. 2007), was found in 1998 below an eroding bluff at Dark Cove, Kiska Island, and is housed at UWBM (61309). A human-worked rib bone of a presumed Steller's sea cow, dating $1,611\pm 67$ years old, was discovered by archaeologists working on Buldir Island in 1997 (Savinetsky et al. 2004). After a review of these and other Aleutian records, Domning et al. (2007) concluded that a population of sea cows quite possibly survived at least in the Near Islands into the eighteenth century, perhaps even after 1750.

Habitat

The Steller's sea cow once lived in cold waters near islands along the North Pacific Rim. Domning (1977) suggested that these large toothless animals fed mainly on algae growing in exposed rocky shallows. Information on the natural history of this species is based on the reports of Steller (1751, 1899).

Status

Hydrodamalis gigas is listed as Extinct by IUCN. When first discovered in 1741, Steller's sea cows were fairly numerous at a few areas off the coasts of Copper and Bering islands. Stejneger (1887) estimated the population at the time of discovery at fifteen hundred to two thousand individuals. Russian fur hunting parties had exterminated the last remaining sea cow from Copper Island by 1754, and from Bering Island by 1768 (Forsten and Youngman 1982). Reports of later sightings have been refuted (Heptner 1965). Although "overkill" by Russian hunters is usually blamed for the ultimate demise of Steller's sea cow, Anderson (1995) proposed that their extinction may also have been the result of a combination of predation, competition, and decline in food supplies that occurred when aboriginal human

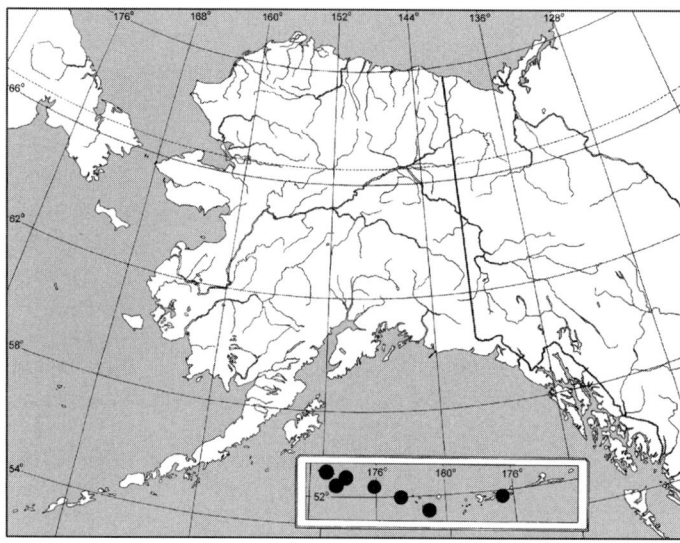

MAP 2. Distribution of Steller's sea cow, *Hydrodamalis gigas*.

● = Specimen Records: Adak Island (Savinetsky et al. 2004; skeletal remains probably housed in Moscow); Adak Island (charred and worked rib sent to UAM but origin unclear); Amchitka Island (3 USNM, subfossils); Buldir Island (Savinetsky et al. 2004; worked piece of rib now probably in Moscow); 51.8778, –177.2263, Kiska Island, Dark Cove (1 UWBM, subfossil); Stejneger (1883) reported nineteenth-century sea cow fossils found on Semichi, Attu, and Agattu islands. A database listing all known Steller's sea cow specimens in museums is available online at <http://www.hansrothauscher.de/steller/seacow.htm> (7 August 2007).

populations colonized mainland coastlines and islands along the North Pacific. Savinetsky et al. (2004) attributed their rapid final extinction to a decline because of unfavorable conditions ("Little Ice Age").

Fossils

The ancestry of this species involves *Metaxytherium* and *Dusisiren* from the Miocene of California, and *Hydrodamalis cuestae* from three-to-eight-million-year-old deposits in California (Berta and Sumich 1999; Domning 1978). Late Pleistocene fossils of *H. gigas* are known from Monterey Bay, California (Jones 1967), and Amchitka Island, Alaska (Whitmore and Gard 1977).

Order PRIMATES Linnaeus, 1758

Order Primates includes humans, apes, monkeys, and prosimians (e.g., lemurs). Groves (2005) divided this order into two suborders with 376 living species in 15 families. Nonhuman primates are found mostly in the tropical and subtropical areas of Central and South America, Africa, and southern Asia. Besides humans (*Homo sapiens*), few species occur as far north in the Americas as southern Mexico, or as far north in Asia as northern Japan. The sole primate in Alaska is *H. sapiens*.

Fossil evidence indicates that primitive ancestors of primates already existed in the latest Cretaceous, although molecular studies suggest that the primate branch is more ancient, originating at least in the mid-Cretaceous (Tavaré et al. 2002).

Family HOMINIDAE Gray, 1825—apes and humans

Under Grove's (2001) scheme that synthesized molecular research, the family of man also includes gorillas, bonobo, chimpanzee, and orangutans. An older scheme, still used by some (e.g., Nowak 1991), placed all nonhuman apes in a separate family, the Pongidae.

The nonhuman members of this family are restricted to equatorial Africa, Sumatra, and Borneo. Hominid fossils date to the Miocene and are known from Africa and Asia (Brunet et al. 2002, 2005).

Hominids are the largest primates, with robust bodies and well-developed forearms. They range in weight from 48 kg to 270 kg. Males are larger than females. Their pollex and hallux are opposable except in humans, who have lost opposability of the big toe. All digits have flattened nails. No hominid has a tail. Numerous skeletal differences between hominids and other primates are related to their upright or semi-upright stance.

Man the hunter (Anchorage Museum at Rasmuson Center b70-28-17)

Human

Homo sapiens
Linnaeus 1758

Other Common Names

Human being, modern man.

Systematics

The genus *Pan* (bonobo, chimpanzee) and the fossil taxon *Australopithecus* were included with *Homo* in subtribe Hominina by McKenna and Bell (1997). Goodman et al. (2001) considered all fossil representatives of the human lineage synonyms of *Homo*.

Homo sapiens is a monotypic species, with no subspecies recognized (Groves 2005; but see Hall 1946, 1981).

Homo sapiens

Original Description. 1758. *Homo sapiens* Linnaeus, *Syst. Nat.*, 10th ed., 1:20.
Type Locality. Uppsala, Sweden.
Type Specimen. None designated.

Global Distribution

Cosmopolitan—The world's population now exceeds 6.5 billion (http://www.census.gov) and is expected to climb to nearly 9 billion by the year 2050 (htttp://www.esa.un.org/unpp).

Alaska Distribution

Humans have been part of Alaska's fauna for at least twelve thousand years, perhaps much longer (see Barton et al. 2004; Madsen 2004; West 1996). The when, where, and how of human colonization of the Americas have long been the subject of considerable debate. Most agree, however, that Alaska must have been the point of entry for some of the earliest pioneers moving eastward into North America from northeastern Asia.

Habitat

Humans' ability to ameliorate their immediate surroundings now allows them to occupy most habitats on the planet. Historically, humans were concentrated along the coast, river corridors, and other areas where they could readily obtain food and shelter. Today the vast majority of humans in Alaska still live below 300 m in elevation.

Status

The human population in Alaska has increased considerably since the 1940s, totaling approximately 670,000 people in 2006, with prediction of an increase to nearly 870,000 by the year 2030 (http://npg.org/states/ak.html). About three-fifths of Alaska's population (in 2000) was concentrated in the southcentral region in the boroughs of Anchorage (42 percent), Matanuska-Susitna (10 percent), and the Kenai Peninsula (8 percent), along with the borough of Fairbanks (13 percent) in the central region. Native Americans indigenous to Alaska comprised about 16 percent of the state's population in 2005.

Fossils

The oldest human skeletal remains from Alaska are from a small karst cave on Prince of Wales Island in southeast Alaska. A human

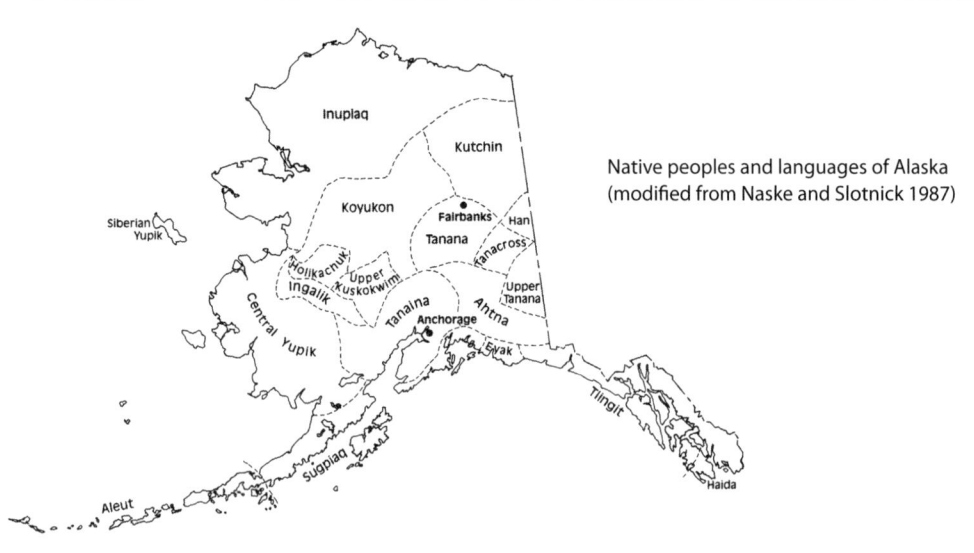

Native peoples and languages of Alaska (modified from Naske and Slotnick 1987)

ALASKA POPULATION

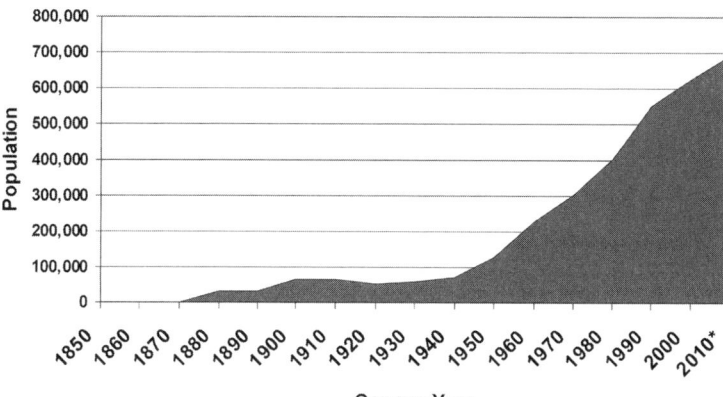

Human population growth in Alaska (*estimated). Source: U.S. Bureau of the Census (http://www.census.gov).

pelvis and mandible had adjusted radiocarbon dates of about 10.3 ka BP (9.2 14C ka BP) (Fedje et al. 2004). Discoveries at this site provide support for humans colonizing the Americas at the end of the last ice age using watercraft along the North Pacific Rim (Dixon 1999).

Specimens

Guthrie (2006: Fig. 1) provided radiocarbon dates of human evidence in Alaska that had been compiled by Yesner (2001).

Order RODENTIA Bowdich, 1821

Rodents comprise the largest order of living mammals, with over 2,200 species and 481 genera in 33 families (Wilson and Reeder 2005). In Alaska, rodents closely follow the carnivores in number of species with twenty-five native and four nonnative species in eighteen genera and six families.

Given the richness and complexity of the Rodentia, formulation of a stable classification has been elusive (see Carleton 1984 and Carleton and Musser 2005 for overviews). Rodents were placed in three suborders: the Sciuromorpha (squirrel-like rodents), the Myomorpha (rat-like rodents), and Hystricomorpha (porcupine-like rodents) (Nowak 1991), but Carleton and Musser (2005) expanded the number of rodent suborders to five, adding Castorimorpha (beavers) and Anomaluromorpha (African scaly-tailed squirrels).

The historical range of the order ranges from Late Paleocene to Recent (Carleton 1984).

Rodents are ecologically diverse and found virtually worldwide, except for Antarctica. All share the distinguishing characteristics of only two pairs of incisors (one upper, one lower) and no canines, leaving a wide gap between incisors and molars called the diastema.

Family SCIURIDAE Fischer von Waldheim, 1817—squirrels and marmots

Worldwide, the family of squirrels comprises at least 278 species in 51 genera (Thorington and Hoffmann 2005). Six species in four genera occur in Alaska.

Harrison et al. (2003) used mitochondrial cytochrome *b* sequences to construct a phylogeny of Marmotinae rodents and found the ground squirrel genus *Spermophilus* paraphyletic with respect to both *Marmota* (marmots) and *Cynomys* (prairie dogs). Thorington and Hoffmann (2005), taking into account advances of molecular studies, recognized five subfamilies of squirrels and provided an overview of sciurid classification.

The evolution of this family dates from at least the Oligocene in Europe and North America, the Miocene in Asia, and the Pleistocene in South America and Africa (McLaughlin 1984).

While the members of this family show a great diversity of form and adaptation, all share common skull characteristics, such as well-developed and round auditory bullae, well-formed postorbital processes, failure of the masseter muscle to penetrate the infraorbital region, ridged and rooted cheek teeth, and a rudimentary thumb. Gliding forms have a patagium connecting the forelegs and hind legs and, unlike other squirrels, are nocturnal rather than primarily diurnal.

Arctic ground squirrel, *Spermophilus parryii* (W. D. Berry)

Northern Flying Squirrel

Glaucomys sabrinus
(Shaw, 1801)

Other Common Names
None.

Systematics
Johnson-Murray (1977) and Thorington et al. (1996) examined morphological relationships between two species of flying squirrels. There has not been a revision of *G. sabrinus* since Howell (1918). A large phylogeographic break found near the British Columbia and Washington border (Arbogast 1999) suggests that the systematic relationships and taxonomic status of this species warrant further investigation (see Arbogast 2007). Hall (1981) recognized three subspecies in Alaska. Manville and Young (1965) included records (source unknown) of a fourth subspecies, *G. s. alpinus,* from six and nine miles north of Juneau. Alaska likely was colonized by this species since the end of Pleistocene (Lessa et al. 2003).

Glaucomys sabrinus griseifrons
Original Description. 1934. *Glaucomys sabrinus griseifrons* A. H. Howell, *J. Mammalogy*, 15:64, February 15.
Type Locality. Lake Bay, Prince of Wales Island, Alaska.
Type Specimen. USNM 256993.
Range. Prince of Wales and adjacent islands, Alexander Archipelago, Southeast Alaska.
Remarks. The description of *G. s. griseifrons* is based on few specimens (Howell 1934). However, analysis of mtDNA sequences of fifty-six specimens revealed fixed nucleotide differences between this subspecies and *G. s. zapheus* (Demboski, Jacobsen, et al. 1998). Expanded studies using larger sample sizes and nuclear and mitochondrial DNA have corroborated those analyses (Bidlack and Cook 2001, 2002), further supporting the validity of this taxon. These diagnostic mutations may indicate a "founder event" on an island within the Prince of Wales archipelago with subsequent spread of squirrels to nearby islands.

Glaucomys sabrinus yukonensis
Original Description. 1900. *Sciuropterus yukonensis* Osgood, *N. Amer. Fauna*, 19:25, October 6.

Type Locality. Camp Davidson, Yukon River, near Alaska-Canada boundary, Yukon Territory.
Type Specimen. USNM 19909/35320.
Range. Yukon Territory through central Alaska.
Remarks. This taxon was named on the basis of two specimens from the Upper Yukon River, 4 km east of the Alaska border in Yukon Territory (Osgood 1900). Youngman (1975) examined the small number of specimens available from Yukon and Alaska and could not distinguish *G. s. yukonensis* or *G. s. alpinus* from *G. s. sabrinus*. He subsequently classified all Yukon and Alaska populations as *G. s. sabrinus*.

Glaucomys sabrinus zaphaeus
Original Description. 1905. *Sciuropterus alpinus zaphaeus* Osgood, *Proc. Biol. Soc. Washington*, 18:133, April 18.
Type Locality. Helm Bay, Cleveland Peninsula, southeast Alaska.
Type Specimen. USNM 136137.
Range. Southeast Alaska and the northwestern coastal region of British Columbia.
Remarks. Fourteen specimens representing *zaphaeus* and *yukonensis* showed little variation in cytochrome *b* sequences across an extensive geographic range (Demboski, Jacobsen, et al. 1998).

Global Distribution
Nearctic—Northern flying squirrels occur in Alaska and Canada, the western contiguous United States, the Black Hills of South Dakota, and the northeastern states to the southern Appalachian Mountains (Thorington and Hoffmann 2005).

Alaska Distribution
This squirrel probably occurs throughout much of forested Alaska, but considerable work remains to adequately document and delimit the distribution of this species within the state (map 3).
Southeast Region. Flying squirrels are known to occur along the coastal mainland of southeast Alaska from Lynn Canal southward, and on the following islands south of Frederick Sound: Barrier (three islands in this group off southwest Prince of Wales Island), Betton, Dall, El Capitan, Etolin, larger unnamed island immediately south of Garden, Gravina, Heceta, Kosciusko, Mitkof, North, Orr, Prince of Wales,

Revillagigedo, Suemez, Thorne, Tuxekan, and Wrangell islands (MacDonald and Cook 2007).

Southcentral Region. The occurrence of flying squirrels in southcentral Alaska needs clarification and further substantiation. Currently the only verifiable records in this region are from the Anchorage area. Two additional but poorly documented specimen records are from the "Cook Inlet District" (USNM 243481) and the "Chuitt River" (USNM 289241), which we assume is synonymous with Chuitna River (Orth 1971) near Tyonek on the west side of Cook Inlet. ADFG (1978) reported this species present throughout the region, including "a small population at Cabin Lake south of Cordova" and "present on the mainland of western and northwestern Prince William Sound," which would include the Kenai Peninsula. No specimens of this species exist in collections from either Prince William Sound or the Kenai Peninsula. Manville and Young (1965) denoted a specimen record in the USNM from apparently the Ninilchik area; however, our search at the national collection for this specimen proved unsuccessful.

The occurrence of flying squirrels in the extensive Wrangell–St. Elias National Park and Preserve is inadequately documented (Cook and MacDonald 2003). In 2002, we were informed by a local resident that trappers along the McCarthy Road occasionally capture flying squirrels

in their marten sets. A trapper from near Twin Lakes on the Nabesna Road said flying squirrels occurred mostly north of the Mentasta Mountains.

Central Region. The distribution of flying squirrels in this region, as elsewhere throughout the state, is poorly documented. Extreme western specimen records are near Tanana (USNM; Dice 1921), while extreme northern specimens are two from about 24 km north of the Yukon River in the general vicinity of Stevens Village (UAM) and one from Charley Creek (Kandik River), Upper Yukon River (MVZ). Osgood (1900) provided secondhand accounts of flying squirrels from as far north as Fort Yukon, eastward toward the Alaska-Yukon border near Chicken, and just inside Yukon at the type locality of *G. s. yukonensis*. ADFG (1978) speculated that this species occurred throughout most of Alaska's boreal forest zone, including the Koyukuk River drainage and the middle and upper Kuskokwim River drainage. A recent report of flying squirrels taken on marten traplines near McGrath in 1990 and 1995 (J. Whitman, pers. comm., 2007) supports ADFG's conjecture.

Southwestern Region. We are unaware of any specimens of flying squirrels from this region; however, ADFG (1978) indicated its occurrence north of the Alaska Peninsula above Iliamna Lake from Cook Inlet west to the Wood River drainage. This and other ADFG reports need to be substantiated.

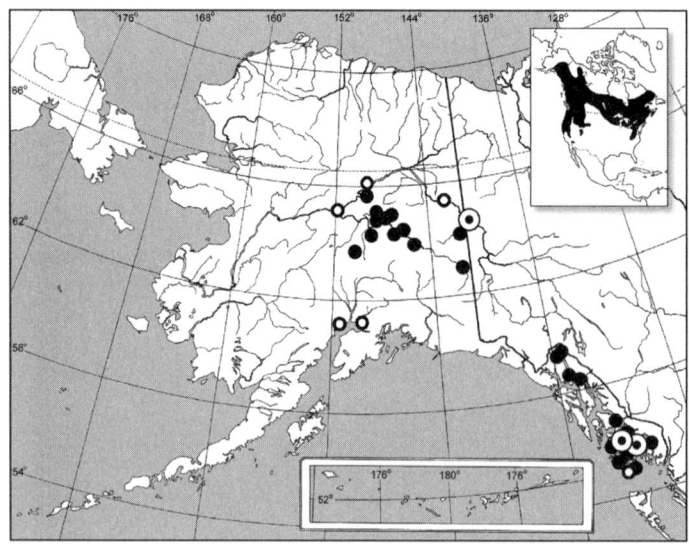

MAP 3. Distribution of northern flying squirrel, *Glaucomys sabrinus*.

⊙ = Type Locality
● = Specimen Record (n = 540)
○ = Marginal Records. 66.28333, –148.75000, Beaver Quad., 20 mi S Lone Mountain, 15 mi N Yukon River (1 UAM); 65.31670, –142.78330, Charley River Quad., Charley Creek [= Kandik River], Yukon River, 80 mi. SE above Circle (1 MVZ); 62.59000, –141.56000, Nabesna Quad., Northway Village (1 UAM); 58.26667, –135.33330, Skagway Quad., Taiya River (1 UAM); 54.80694, –132.76916, Dixon Entrance Quad., Dall Island, Pond Bay (1 UAM); 61.20730, –149.73340, Anchorage Quad., Muldoon Road area (1 UAM); Tyonek Quad., Chuitt [Chuitna] River (1 USNM); 65.16670, –152.06700, Tanana Quad., Tanana (1 USNM).

Habitat

This species inhabits a wide variety of boreal forest and coastal rain forest habitats in Alaska as well as other forested ecosystems across its range. Flying squirrels are assumed to be closely associated with mature and old-growth forests, although specific habitat requirements, particularly in more northerly latitudes, are little known.

In the Alexander Archipelago of southeast Alaska, upland old-growth forests of Sitka spruce and western hemlock were found to be the primary habitat (Smith and Nichols 2003). A study on Prince of Wales Island found the density of large-diameter trees (greater than 74 cm dbh), abundance of *Vaccinium* shrubs, and density of large-diameter (50–74 cm) snags was positively correlated with microhabitat use by *G. s. griseifrons* (Smith et al. 2004). Den sites are usually in natural tree cavities or woodpecker holes (Mowrey and Zasada 1984; Weigl and Osgood 1974). On Mitkof Island in southeast Alaska, squirrels preferentially used cavities in large-diameter snags and diseased trees for denning and shelter (Bakker and Hastings 2002).

Winter habitat of this species in boreal forests of central British Columbia was not limited to old-growth forests. Within younger stands, they selected the largest trees available (Cotton and Parker 2000). In central Alaska in winter, Mowrey and Zasada (1984) found northern flying squirrels nesting, often communally, in witches' brooms, whereas in the central British Columbia study (Cotton and Parker 2000) brooms and snags were used infrequently as nest sites. Pyare et al. (2002) found that northern flying squirrels in southeast Alaska consumed truffles less frequently and consumed a smaller array of truffle genera than described for this species elsewhere in the western United States. The seeds of white spruce (*Picea glauca*) were not a prime dietary item in central Alaska flying squirrels (Brink and Dean 1966), whereas fungi and lichens may be the predominant or only foods eaten at certain times of the year (Mowrey et al. 1981; Pyare et al. 2002).

In our inventories of the Alexander Archipelago, we were most successful capturing flying squirrels close to tidewater in conifer forest with well-developed thickets of huckleberry (*Vaccinium* spp.) and other shrubs. Shrubby forest with occasional large *Thuja* and *Chamaecyparis* cedar trees seemed especially productive. Fewer squirrels were captured at inland, higher elevation sites on islands. Established campgrounds, picnic areas, and forest sites near human dwellings also were productive for capturing flying squirrels.

Status

The status of this nocturnal and elusive squirrel is poorly understood anywhere in the state. It is taken incidentally, sometimes at nuisance levels, in late fall and winter by marten trappers (ADFG 1978).

The possible close association of the island endemic, *G. s. griseifrons*, with old-growth forest slated for harvest in southeast Alaska warrants special concern. This subspecies is listed by IUCN as Endangered (Demboski, Cook, et al. 1998; Hafner et al. 1998), by Alaska Heritage as possibly Imperiled, and as a "species of concern" by West (1991) and Suring et al. (1992). It was listed by ESA as a Candidate until the late 1980s.

Fossils

Fossils of *G. sabrinus* are known from a number of localities, all outside of Alaska, from Late Pleistocene and Holocene strata (Wells-Gosling and Heaney 1984).

Alaska Marmot

Marmota broweri
Hall and Gilmore, 1934

Other Common Names

Arctic marmot, Brooks Range marmot.

Systematics

Marmota broweri was regarded as a synonym of *M. caligata* by Hall and Gilmore (1934) and Hall (1981), but Rausch and Rausch (1965, 1971) and Hoffmann et al. (1979) provided evidence that *broweri* is a distinct species. No subspecies are recognized.

A recent molecular study (Steppan et al. 1999) comparing cytochrome *b* sequences supported the distinctiveness of *M. broweri* from both *M. camtschatica* of eastern Siberia and *M. caligata* of the Nearctic and suggested that the ancestry of the Alaska marmot dates back earlier in the Pleistocene, a view put forward by Rausch and Rausch (1971). These and other studies (e.g., Cardini 2003; Cardini et al. 2005; Polly 2003) have so far failed in their attempts to reject hypotheses of either a Nearctic or Palearctic origin.

Marmota broweri

Original Description. 1934. *Marmota caligata broweri* Hall and Gilmore, *Canadian Field-Nat.*, 48:57, April 2.

Type Locality. "Point Lay, Arctic Coast of Alaska." Restricted by Rausch (1953a:117) to head of Kukpowruk River, Alaska.

Type Specimen. MVZ 51675.

Global Distribution

Nearctic—The Alaska marmot occurs in the mountains north of the Yukon and Porcupine rivers in central and northern Alaska. Reports of marmots, possibly of this species, in the Richardson Mountains in northern Yukon Territory have yet to be substantiated (Youngman 1975; B. Smith, pers. comm., 2005).

Alaska Distribution

The Alaska marmot has been documented from the mountains and foothills of the Brooks Range from Lake Peters westward to Cape Lisburne and Cape Sabine (Childs 1969) (map 4). Dean and Chesemore (1974) reported marmot sign in the northern Baird Mountains near the Nakolik River, and Gardner (1974) sighted a marmot in the vicinity of Mulik Hills on the lower Noatak River. MacDonald and Cook (2002) provided a secondhand report of marmot sign near Copter Peak in the DeLong Mountains. Specimens of this species have been secured by UAM south of the Brooks Range from Spooky Valley in the Ray Mountains northwest of Rampart and in the Kokrines Hills near Horner Hot Springs on the north side of the Yukon River northeast of Ruby.

Habitat

This marmot inhabits boulder fields, rock slides, outcrops, terminal moraines, and talus slopes in alpine tundra with adequate herbaceous forage (Bee and Hall 1956; Hoffmann 1999). Bee and Hall (1956) found them common on the mountain slopes surrounding Lake Peters in the eastern Brooks Range, but noted they were less common or absent away from the lake. Winter dens are relatively permanent for each colony, and some have been used regularly for at least twenty years (Rausch and Rausch 1971).

Status

The status of this Alaska endemic is poorly understood. Alaska marmots apparently are found widely scattered as individuals or in loose colonies across their range (Hoffmann 1999).

Fossils

None is known. The marmot fossils of Late Pleistocene age reported by Yesner (2001; incorrectly listed as *M. flavescens* in Tables 1 and 2) from the Trail Creek caves on the Seward Peninsula are possibly of this species.

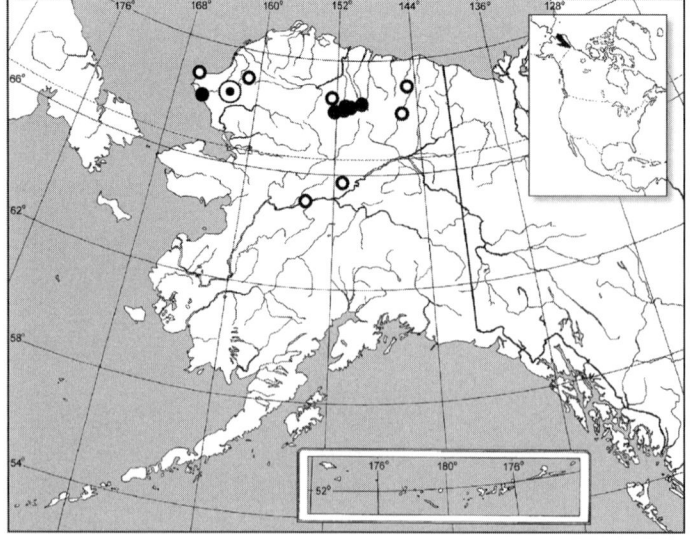

MAP 4. Distribution of Alaska marmot, *Marmota broweri*.

⊙ = Type Locality
● = Specimen Record (n = 81)
○ = Marginal Records. 69.31670, –145.05000, Mt Michelson Quad., Lake Peters (2 MSB); 68.13330, –145.53333, Arctic Quad., Arctic Village (1 PSM); 65.71667, –151.26667, Tanana Quad., Ray Mountains, Spooky Valley (1 UAM); 64.95000, –154.85000, Ruby Quad., Kokrines Hills near Horner Hot Springs (1 UAM); 68.88333, –166.21667, Point Hope Quad., Cape Lisburne (4 UAM); 69.06667, –161.13333, Utukok River Quad., headwaters Utukok River, 8km W Utukok, 12km S Archimedes Ridge (1 UAM); 68.56925, –152.93614, Chandler Lake Quad., Fortress Mountain (2 UAM).

Hoary Marmot

Marmota caligata
(Eschscholtz, 1829)

Common Names
Whistler.

Systematics
The subspecies of *M. caligata* were last revised by Howell (1915). Hoffmann et al. (1979) examined morphometric variation among several subspecies, but not comprehensively. Hall (1981) recognized nine subspecies of *M. caligata*, excluding *M. c. broweri* (see separate species account). Three subspecies occur in Alaska.

Marmota caligata caligata
Original Description. 1829. *Arctomys caligatus* Eschscholtz, *Zool. Atlas*, Part 2, p. 1, pl. 6.
Type Locality. Near Bristol Bay, Alaska.
Type Specimen. Not known to exist.
Range. Central and southern Alaska, Yukon Territory, Mackenzie River Valley in western Northwest Territories, northwestern British Columbia.

Marmota caligata sheldoni
Original Description. 1914. *Marmota caligata sheldoni* A. H. Howell, *Proc. Biol. Soc. Washington*, 27:18, February 2.
Type Locality. Zaikof Bay, Montague Island, Alaska.
Type Specimen. USNM 137319.
Range. Known only from the type locality.

Remarks. Dufresne (1946) mentioned that marmots from Montague Island are frequently melanistic with some almost solid black.

Marmota caligata vigilis
Original Description. 1909. *Marmota vigilis* Heller, *Univ. California Publ. Zool.*, 5:248, February 18.
Type Locality. West shore of Glacier Bay, Alaska.
Type Specimen. MVZ 418.
Range. Known only from the type locality.
Remarks. Most of the specimens from Glacier Bay in MVZ and USNM are distinctly melanistic. Three marmots recently collected by UAM above Excursion Inlet near the eastern boundary of Glacier Bay National Park were also melanistic.

Global Distribution
Nearctic—The hoary marmot has a range extending from Alaska and western Canada south into Washington, Idaho, and northwestern Montana (Thorington and Hoffmann 2005).

Alaska Distribution
Hoary marmots can be found in suitable habitat throughout most of the mountainous regions of Alaska south of the Yukon River (map 5). They are absent from northern and far western Alaska, as well as from islands off the Alaska Peninsula, and all but one island in the Alexander Archipelago of southeast Alaska.
Southeast Region. Marmots have been reported along the mainland from Yakutat to

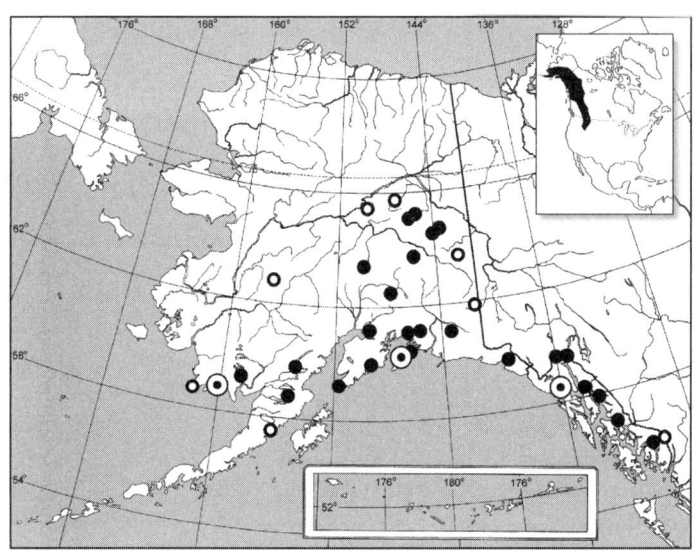

MAP 5. Distribution of hoary marmot, *Marmota caligata*.

⊙ = Type Locality
● = Specimen Record (n = 540)
○ = Marginal Records.. 65.750000, –147.00000, Circle Quad., Victoria Mt, granite rock pile, 3500 ft (1 UAM); 63.66667, –142.21667, Tanacross Quad., Mt Fairplay (2 UAM); 61.88300, –141.20000, McCarthy Quad., Francis Creek, 0.5 mi. N. Ptarmigan Lake (5 AMNH); 56.01667, –130.06694, Bradfield Canal Quad., Salmon River, N of Hyder (2 UAM); 57.56700, –156.03300, Ugashik Quad., Portage Bay (17 USNM); 58.55000, –161.76667, Hagemeister Island Quad., Cape Peirce (2 UAM); 62.91670, –157.00000, Iditarod Quad., Flat, Beaver Mountains (5 USNM); 65.25961, –150.05022, Tanana Quad., Elephant Mountain (1 UAM).

Hyder, and on Douglas Island near Juneau (MacDonald and Cook 2007). The subspecies, *M. c. vigilis*, was described from seven specimens collected at Coppermine Cove, Glacier Bay (Heller 1909).

Southcentral Region. Hoary marmots occur on the mainland of southcentral Alaska, including the Kenai Peninsula, and in Prince William Sound on Hinchinbrook, Montague (as *M. c. sheldoni*), and Knight islands (Heller 1910).

Central Region. This species is found south of the Yukon River in the Yukon-Tanana uplands and throughout the Alaska Range (Buckley and Libby 1957). The hoary marmot's status and distribution between the Porcupine and Yukon rivers in west-central Alaska need clarification as they are known to occur in the Ogilvie Mountains in adjacent Yukon Territory (Osgood 1900; Youngman 1975).

Southwest Region. Marmots are reported from the base of the Alaska Peninsula as far west as Port Moller (Allen 1904; Howell 1915; Murie 1959). They are not known on any island along the Gulf Coast, although USNM (256721) houses an old and discolored partial right mandible of a marmot excavated from an archaeological site on Kodiak Island. Elkins and Nelson (1954) list Sud Island, one of the Barren Islands north of Kodiak, as a site of introduction in about 1930 (source unknown), where Bailey (1976) found them numerous in 1974.

Habitat

Hoary marmots prefer rocky tundra habitats on the precipitous sides of canyons and valleys in the mountains. They occur at sea level where alpine conditions extend to the coast.

Status

Hoary marmots are common in suitable habitat. Two of Alaska's three subspecies, *M. c. sheldoni* and *M. c. vigilis*, are listed by IUCN as Data Deficient (Cook 1998d). The Montague Island marmot, *M. c. sheldoni*, has not been reported since the original collections in 1905 and 1908 (Lance 2002). The current status of *M. c. vigilis* is unknown.

Fossils

Marmota remains of Early Pleistocene age were reported from western Alaska by Guthrie and Matthews (1971). Yesner (2001) reported fossils of marmots (incorrectly as *M. flavescens* in Tables 1 and 2) from central Alaska (Broken Mammoth site, Tanana Valley—Late

Pleistocene and Early Holocene) and western Alaska (Trail Creek caves, Seward Peninsula—Late Pleistocene; interestingly, where neither *M. caligata* nor *M. broweri* currently occur). Marmot subfossils from a cave on Prince of Wales Island in the Alexander Archipelago predate the last glacial maximum (Heaton and Grady 2003). Postglacial remains from sites in central Alaska were reported by Georgina (2001) and Yesner (2001).

Woodchuck

Marmota monax
(Linnaeus, 1758)

Other Common Names

Groundhog.

Systematics

Genetic (Steppan et al. 1999) and craniometric (Cardini et al. 2005) data support dividing the world's marmots into two subgenera: *Petromarmota* (four western North American species, including *caligata*) and *Marmota* (eight Palaearctic species plus the Nearctic species, *monax* and *broweri*). Hall (1981) recognized nine subspecies of *M. monax* across its broad range in North America; one occurs in Alaska.

Marmota monax ochracea

Original Description. 1911. *Marmota ochracea* Swarth, *Univ. California Publ. Zool.*, 7:203, February 18.

Type Locality. Forty-Mile Creek, Alaska.

Type Specimen. MVZ 5872.

Range. East-central Alaska, Yukon Territory, and extreme north-central and northwestern British Columbia.

Remarks. *Marmota ochracea* was described by Swarth (1911b) from two specimens collected in east-central Alaska at the head of Forty-Mile Creek [= Fortymile River?]. Youngman (1975) noted that *M. m. ochracea* is a weakly defined subspecies that intergrades with, and most closely resembles, *M. m. canadensis* to the east, rather than *M. m. petrensis* to the south.

Global Distribution

Nearctic—Woodchucks have the widest range of any marmot in North America, occurring from east central Alaska across central Canada and extending the length of British Columbia south into far northern Idaho and extreme eastern Washington, and then throughout most of the eastern contiguous United States (Hall 1981).

Alaska Distribution

The Alaska distribution of this species appears to be restricted to a relatively narrow band between the Tanana and Yukon rivers from near Nenana southeastward to the Alaska-Yukon border (map 6).

Habitat

Woodchucks in central Alaska are believed to prefer open, well-drained grassy areas and open deciduous forest with an undergrowth of grasses, forbs, and shrubs (ADFG 1978); they are infrequently seen along roadsides and near old buildings. Woodchuck populations reported from outside Alaska have loose social structures with burrow systems that are not clustered together. Individuals are asocial and territorial, and their associations with each other are limited to reproductive efforts (Kwiecinski 1998).

Status

Woodchucks may be rare to uncommon within their limited range.

Fossils

The oldest fossils of *M. monax* are from the mid-Pleistocene (Irvingtonian) in a number of sites in the southeastern United States (Steppan et al. 1999). Morlan (1983) and Harington

(1989) reported postglacial remains from deposits in Yukon Territory.

Arctic Ground Squirrel

Spermophilus parryii
(Richardson, 1825)

Other Common Names

Parry ground squirrel, sik-sik, long-tailed suslik.

Systematics

The taxonomy of *Spermophilus parryii* has been contentious and needs formal revision. *Citellus* has often been employed for this genus (e.g., Rausch 1994), but is not now in general use (Nowak 1991; Wilson and Reeder 2005) and considered invalid by the International Commission on Zoological Nomenclature (ICZN 1956). North American populations have been classified as *S. undulatus* (Bee and Hall 1956; Ellerman and Morrison-Scott 1951; Hall and Kelson 1959; Rausch 1953a), but *S. undulatus* is now applied to a species that occurs in southern Siberia, Mongolia, and northeastern China, while *S. parryii* applies to populations in northeastern Siberia and northwestern North America (Hoffmann et al. 1993; Honacki et al. 1982). Gromov et al.

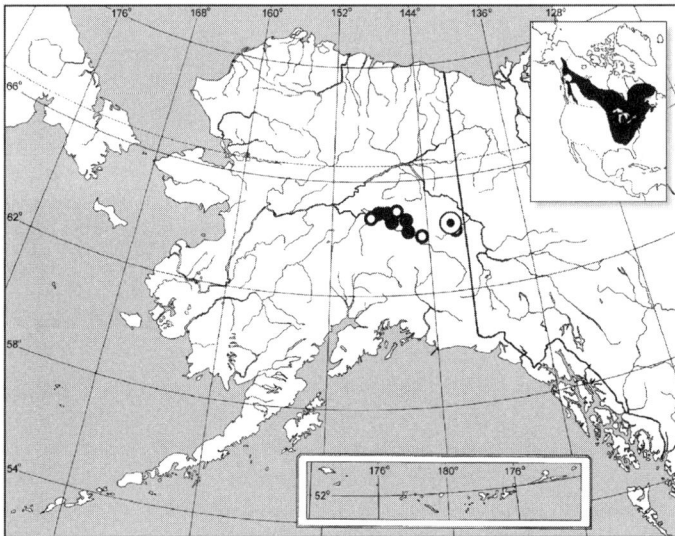

MAP 6. Distribution of woodchuck, *Marmota monax*.

⊙ = Type Locality
● = Specimen Record (n = 50)
○ = Marginal Records. 64.89333, –146.68028, Big Delta Quad., ¼ mile W of Twin Bears Camp, Chena Hot Springs Road (1 UAM); 64.16670, –141.45000, Eagle Quad., Jack Wade Creek (1 MSB); 63.96670, –144.61670, Mt Hayes Quad., Healy River (1 USNM); 64.67361, –148.91333, Fairbanks Quad., Parks Highway (1 UAM).

Woodchuck near Richardson Roadhouse, Tanana Basin, central Alaska, May 1980.

(1965), Nadler et al. (1974), and Eddingsaas (2001) reviewed characters that indicated *S. parryii* is distinctive.

Subspecies of North American populations were last revised by Howell (1938). Of the eight North American subspecies listed by Hall (1981), seven occur in Alaska. Patterns of allozymic (Nadler and Hoffmann 1977) and morphometric (Pearson 1981) variation suggest that perhaps only six subspecies are valid in North America (Nagorsen 1990).

Eddingsaas et al. (2004) conclude that arctic ground squirrels have a long evolutionary history in Alaska, a history that is intricately linked to multiple glacial cycles and, in one case, has a close connection to Siberian populations. The four major clades identified (North, Southwest, West, and Central Alaska) with molecular genetic analyses are geographically distinct and largely match subspecific designations. Squirrels north of the Brooks Range (North clade) are the most distinctive (perhaps at the species level) and suggestive of an early split into arctic and subarctic groups, a pattern previously identified by Nadler and Hoffmann (1977). The introduction of ground squirrels to possibly more than thirty islands in southwest Alaska (Table 2) has complicated interpretation of the evolutionary history of squirrels throughout this region.

Spermophilus parryii ablusus
Original Description. 1903. *Citellus plesius ablusus* Osgood, *Proc. Biol. Soc. Washington*, 16:25, March 19.
Type Locality. Nushagak, Alaska.
Type Specimen. USNM 119815.
Range. Northwestern, southwestern, and southcentral Alaska. Introduced to several locations in the Aleutian Islands.
Remarks. Includes *Citellus stonei* J. A. Allen, 1903, *Bull. Amer. Mus. Nat. Hist.*, 19:537, October 10, type from Stevana Flats, near Port Müller (Allen 1904:278), Alaska Peninsula; not Wrangell, Alaska (see Allen 1903:537).

Spermophilus parryii kennicottii
Original Description. 1861. *A*[*rctomys*] *kennicottii* Ross, *Canadian Nat. and Geol.*, 6:434.
Type Locality. Fort Good Hope, Mackenzie.
Type Specimen. None designated.
Range. Northern Alaska, Yukon Territory, and northwestern Northwest Territories.
Remarks. Includes *Spermophilus barrowensis* Merriam, 1900, *Proc. Washington Acad. Sci.*, 2:20, March 14, type from Point Barrow,

Alaska (USNM 14061/37824), and *Spermophilus beringensis* Merriam, 1900, *Proc. Washington Acad. Sci.*, 2:20, March 14, type from Cape Lisburne, Alaska (USNM 15253).

Spermophilus parryii kodiacensis
Original Description. 1874. *Spermophilus parryi* var. *kodiacensis* J. A. Allen, *Proc. Boston Soc. Nat. Hist.*, 16:292.
Type Locality. Lectotype designated by Howell (1938) from specimen collected on Kodiak Island in 1868.
Type Specimen. USNM 9242/38543.
Range. Restricted to Kodiak Island.
Remarks. Eddingsaas et al. (2004) found a close relationship between mtDNA haplotypes on Kodiak Island and the Shumagin Islands, a finding that questions the native status and taxonomic validity of this subspecies.

Spermophilus parryii lyratus
Original Description. 1932. *Citellus lyratus* Hall and Gilmore, *Univ. California Publ. Zool.*, 38:396, September 17.
Type Locality. Iviktook Lagoon, about 35 miles NW Northeast Cape, St. Lawrence Island, Bering Sea, Alaska.
Type Specimen. MVZ 51172.
Range. Restricted to St. Lawrence Island.

Spermophilus parryii nebulicola
Original Description. 1903. *Citellus nebulicola* Osgood, *Proc. Biol. Soc. Washington*, 16:26, March 19.
Type Locality. Nagai Island, Shumagin Islands, Alaska.
Type Specimen. USNM 59145.
Range. Shumagin Islands.

Spermophilus parryii osgoodi
Original Description. 1900. *Spermophilus osgoodi* Merriam, *Proc. Washington Acad. Sci.*, 2:18, March 14.
Type Locality. Fort Yukon, Alaska.
Type Specimen. USNM 12789/37822.
Range. Yukon Flats area of central Alaska.

Spermophilus parryii plesius
Original Description. 1825. *Spermophilus empetra plesius* Richardson, *in* Parry, *Journal of a second voyage . . .* p. 316, October 6.
Type Locality. Bennett City, head of Lake Bennett, British Columbia.
Type Specimen. USNM 98931.
Range. Southcentral Alaska, southern Yukon Territory, southwestern Northwest Territories, and northern British Columbia,

Table 2. Status of arctic ground squirrels, *Spermophilus parryii*, on islands in southwest Alaska.

Group	Island	Status
Barren Islands	**Ushagat** (58°55.5'N, 152°15'W)	Considered indigenous by Hall (1981). The only one of the Barren Islands with squirrels, where they are abundant (Bailey 1976). Archived specimens at USNM, MSB.
Kodiak Archipelago	**Dark** (58°38.5'N, 152°32.5'W)	Reported present by ADNR (2008). No archived specimens.
	Holiday (57°42.3'N, 152°28.3'W)	Probably introduced from Kodiak Island stock. Archived at MSB, PSM
	Kodiak (57°47'N, 152°24'W)	Probably introduced <1868 (Howell 1938), perhaps by Aleuts (Karlstrom and Ball 1969), from Shumagin Islands (Cook et al., in prep.) or possibly the Semidi Islands (Osgood 1903). Lectotype (USNM 9242/38543) for *kodiacensis* (Allen) designated by Howell (1938) from specimens collected on Kodiak Island by F. Bischoff in 1868. Extant populations reported from Kodiak Airport (UAM, others), Buskin Lake (KU), northeast of Ugak Bay, and possibly near Karluk Lake (Rausch 1953a). Additional specimens at FMNH, MSB, MVZ, PSM, UAM.
	Marmot (58°13'N, 151°50'W)	Possibly indigenous. Reported present as early as 1792 (D.W. Clark, pers. comm.), and abundant on island from 1979 through 1994 (Chumbley et al. 1997). No archived specimens.
	Tugidak (56°30'N, 154°40'W)	Probably introduced. Occurrence reported by Bailey (1993) and ADFG (1995). No archived specimens.
	Woody (57°47'N, 152°20'W)	Probably introduced from Kodiak Island stock. No archived specimens.
Semidi Islands and vicinity	**Aghiyuk** (56°10'N, 156°47'W)	Presence reported by Hatch and Hatch (1983). No archived specimens.
	Aliksemit (56°00'N, 156°40'W)	Presence reported by Hatch and Hatch (1983). No archived specimens.
	Anowik (56°06'N, 156°40'W)	Presence reported by Hatch and Hatch (1983). No archived specimens.
	Chankliut2 (56°08'N, 158°07'W)	Introduced 1918 (Bailey and Faust 1981). No archived specimens.
	Chirikof (55°50'N, 155°37'W)	Possibly indigenous. Excavated skeletal remains radiocarbon dated at 450±50 year-old; island established as a special outpost for ground squirrel fur procurement by Aleksandr Baranov (L. Black, pers. comm.). Archived specimens at MSB.
	Chowiet (56°02'N, 156°42'W)	Presence reported by Hatch and Hatch (1983). Archived specimens at MSB.
	Kak (56°17'N, 157°49'W)	Presence reported by Bailey and Faust (1981). No archived specimens.
	Kateekuk (56°05'N, 156°44'W)	Presence reported by Hatch and Hatch (1983). Archived specimens at UAM, MSB.
	Kiliktagik (56°04'N, 156°39'W)	Presence reported by Hatch and Hatch (1983) as 'Kaliktagik.' Archived specimens at UAM, MSB.
	Nakchamik (56°20'N, 157°49'W)	Presence reported by Bailey and Faust (1981). No archived specimens.
	Unavikshak (56°30'N, 157°43'W)	Presence reported by Bailey and Faust (1981). No archived specimens.

Table 2 (continued)

Group	Island	Status
Shumagin Islands	**Atkins** (55°03.5'N, 159°18.5'W)	Reported present by Byrd et al. (1997).
	Bendel (55°05'N, 159°48'W)	Presence reported by Bailey (1993) and A. Eddingsaas (pers. obs.). No archived specimens.
	Bird (54°49'N, 159°46'W)	Presence reported by Bailey (1993) and Byrd et al. (1997). Specimens archived at UAM, MSB.
	Chernabura (54°47'N, 159°33'W)	Specimens archived at UAM, MSB.
	Heredeen (55°04'N, 159°25'W)	Reported present by Byrd et al. (1997).
	Karpa (55°30.5'N, 160°03'W)	Presence reported by A. Eddingsaas (pers. obs.). No archived specimens.
	Koniuji, Big (55°06'N, 159°33'W)	Probably introduced 1916 (Bailey 1993). Specimens archived at MSB.
	Koniuji, Little (55°00'N, 159°23'W)	Archived specimens at UAM, MSB.
	Nagai (55°05'N, 160°00'W)	Observed on the island by Georg Steller in 1741 (Ebbert and Byrd 2002), which suggested to Bailey (1978) indigenous status on some of the larger Shumagins. Type locality of *nebulicola* Osgood, 1903 (USNM 59145).
	Near (54°56.5'N, 160°03'W)	Presence reported by Bailey (1978, 1993). No archived specimens.
	Simeonof (54°54'N, 159°16'W)	Reported present by Kenyon (1964) and others. Archived specimens at USNM, UAM, MSB.
	Spectacle (55°06.5'N, 159°43.5'W)	Present (Bailey 1993) before 1960 (D.W. Clark, pers. comm.). No archived specimens.
	Turner (55°02'N, 159°49'W)	Presence reported by Bailey (1993). No archived specimens.
Sandman Reefs	**Cherni** (54°38'N, 162°22'W)	Presence reported by A. Eddingsaas (pers. obs.). No archived specimens.
Aleutian Islands	**Amaknak** (53°54.5'N, 166°32'W)	Introduced ca 1895, source unknown (Peterson 1967). No archived specimens.
	Kavalga (51°33'N, 178°48'W)	Translocated 1920 from Unalaska (Murie 1959). Archived specimens at USNM, UAM, MSB.
	Umnak (53°15'N, 168°20'W)	Probably introduced (Eyerdam 1936; Howell 1938). No archived specimens.
	Unalaska (53°35'N, 166°50'W)	Introduced 1896 (Burris and McKnight 1973) or earlier (USNM houses specimens from 1890-1895) from Nushagak stocks (Osgood 1904). Additional specimens at MSB, MVZ.
	Unimak (54°45'N, 165°00'W)	Considered indigenous by Murie (1959), Hall (1981) and others. Archived specimens at USNM, MVZ.

including southeast Alaska in the mountains near Haines and Skagway.

Global Distribution

Holarctic—Arctic ground squirrels range across northeastern Siberia (northeastern Yakutia, Anadyrsk Krai, and Chukotka), Alaska, and northwestern Canada (Yukon Territory, northwestern British Columbia, Northwest Territories, Nunavut, and northern Manitoba) (Hall 1981; Nagorsen 2005; Thorington and Hoffmann 2005).

Alaska Distribution

This species is widely distributed across northern, eastern, and southwestern Alaska (map 7). Indigenous populations are absent from the Aleutian Islands west of Unimak Island.

Southeast Region. The distribution of this squirrel in southeast Alaska is restricted to the mountains north of Haines and Skagway (MacDonald and Cook 2007). Additional sampling along the rugged coastal mountains of the mainland is needed to determine the distribution of this species in the region.

Southcentral Region. There are no recent reports of this species along the Gulf Coast from the Kenai Peninsula and Prince William Sound eastward. A number of MVZ specimens reportedly taken near the turn of the century from several locations on the Kenai Peninsula appear to be in error as no ground squirrels occur there today. The only specimens of this species in Prince William Sound are one taken at Valdez Narrows in 1908 (MVZ; Heller 1910) and several from Worthington Glacier and Thompson Pass in the Chugach Mountains north of Valdez (MVZ, PSM). M. E. Isleib (pers. comm., 1972) reported arctic ground squirrels along the railroad tracks between Portage and Whittier. Ground squirrels are widely distributed throughout Wrangell–St. Elias National Park and Preserve (Cook and MacDonald 2003).

Central Region. The distribution of ground squirrels in central Alaska is poorly documented. Squirrels occur in the Ogilvie Mountains near the Alaska-Canada border, but are absent from the Yukon-Tanana Uplands south of the Yukon River. They inhabit lowland areas of the Yukon Flats (as *S. p. osgoodi*), with an estimated 10 to 20 percent of these Yukon Flats squirrels exhibiting melanism (Guthrie 1967; Howell 1938). Areas in need of inventory include the Kokrines Hills, Kaiyuh Mountains, and the Kuskokwim Mountains. Jack Whitman (pers. comm., 2007) found no sign of ground squirrels in the Beaver Mountains (part of the Kuskokwim Mountains), Sunshine Mountain, or Bitzshtini Mountain.

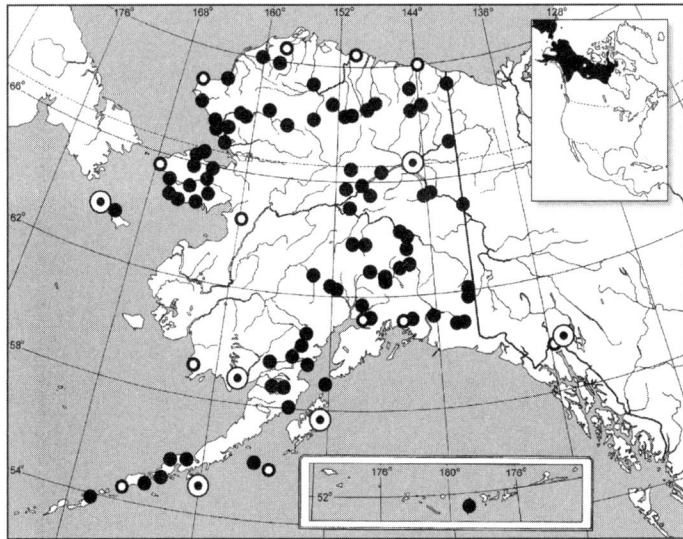

MAP 7. Distribution of arctic ground squirrel, *Spermophilus parryii*.

⊙ = Type Locality
● = Specimen Record (n = 2154)
○ = Marginal Records. 70.48056, –157.41667, Meade River Quad., Meade River Village (=Atkasuk) (13 UAM); 70.43333, –150.40000, Harrison Bay Quad., Colville River Delta (5 UAM); 70.13333, –143.63333, Barter Island Quad., Barter Island, Kaktovik (1 UAM); 59.63742, –136.12914, Skagway Quad., Mount Ashmun (1 UAM); 61.06778, –146.66972, Valdez Quad., Valdez Narrows (1 MVZ); 61.19944, –149.63056, Anchorage Quad., Fort Richardson (2 UAM); 55.83333, –155.61667, Trinity Islands Quad., Chirikof Island (1 MSB); 54.40420, –164.79170, Unimak Quad., Unimak Island, Scotch Cap (1 MVZ); 59.08600, –161.89742, Goodnews Bay Quad., North of North spit (1 UAM); 64.37, –159.67, Norton Bay Quad., Nulato Hills, 11.5 mi. and 147.8 degrees from Debauch Mountain (7 UAM); 65.55000, –167.91667, Teller Quad., Seward Peninsula, Tin City (1 UAM); 68.61667, –166.00000, Point Hope Quad., Kipalog Creek, Cape Dyer (1 UAM).

Southwest Region. Native populations of arctic ground squirrels inhabit the length of the Alaska Peninsula as far west as Unimak Island. The status and distribution of ground squirrels farther west on the Aleutian Chain and on Gulf Coast islands south of the Alaska Peninsula, including Kodiak Island and nearby Ushagat, Marmot, and Chirikof islands, have been clouded by past introductions and translocations (see Table 2).

Western Region. *Spermophilus parryii* is common throughout the northwestern parts of this region, including the Seward Peninsula. Their status south of the peninsula remains poorly documented. In 2005, UAM collected arctic ground squirrels in the Nulato Hills northeast of Unalakleet, and Seavoy (2004a) noted that arctic ground squirrels are found farther south in the Andreafsky Hills and in the Kilbuck Mountains on the south side of the Yukon-Kuskokwim Delta. Among the offshore islands in the Bering Sea, ground squirrels occur only on St. Lawrence Island. There was an unsuccessful attempt to introduce ground squirrels to St. George Island in 1914 (Howell 1938).

Northern Region. Ground squirrels are found in suitable habitat throughout this region, but are most numerous in the mountains and foothills of the Brooks Range (Bee and Hall 1956).

Habitat

Arctic ground squirrels occur in tundra, meadow, riverbank, and lakeshore habitats with loose, friable soils. These well-drained, permafrost-free sites provide good vantage points and adequate supplies of low, early successional vegetation. This species is found from sea level to well above tree line in the mountains. *Spermophilus parryii* is the northernmost hibernator in North America, spending up to nine months sequestered underground (Buck and Barnes 1999).

Status

This species is colonial and locally abundant over much of its range, but there is a paucity of information for western Alaska and many islands in southwestern Alaska. The insular subspecies *kodiacensis*, *lyratus*, and *nebulicola* are listed by IUCN as Data Deficient (Cook 1998a; Appendix 6).

Fossils

Fossil discoveries indicate the presence of arctic ground squirrels in both western and eastern Beringia since the middle Pleistocene (Illinoian) (Guthrie 1984; Jopling et al. 1981; Repenning 1967), with genetic data (Eddingsaas et al. 2004) suggestive of an even earlier presence. The bones, nests (some containing mummies), and seed caches of arctic ground squirrels from the late Pleistocene are common in frozen muck deposits around Fairbanks, an area where they are no longer present (Kurtén and Anderson 1980).

Red Squirrel

Tamiasciurus hudsonicus
(Erxleben, 1777)

Other Common Names

Pine squirrel.

Systematics

This species has not had a significant taxonomic revision since Allen (1898). Study of geographic variation throughout its entire range is needed. Hall (1981) recognized twenty-five subspecies; four occur in Alaska.

Tamiasciurus hudsonicus kenaiensis

Original Description. 1936. *Tamiasciurus hudsonicus kenaiensis* A. H. Howell, *Proc. Biol. Soc. Wash.*, 49:136, August 22.
Type Locality. Hope, Cook Inlet, Alaska.
Type Specimen. USNM 107603.
Range. Kenai Peninsula, Alaska.

Tamiasciurus hudsonicus petulans

Original Description. 1900. *Sciurus hudsonicus petulans* Osgood, *N. Amer. Fauna*, 19:27, October 6.
Type Locality. Glacier, 1870 ft., White Pass, southern Alaska.
Type Specimen. USNM 97457.
Range. Southern Yukon Territory, extreme northwestern British Columbia, and northern southeast Alaska.

Tamiasciurus hudsonicus picatus

Original Description. 1921. *Sciurus hudsonicus picatus* Swarth, *J. Mammalogy*, 2:92, May 2.
Type Locality. Kupreanof Island, 25 mi. S Kake Village, at southern end of Keku Straits, SE Alaska.
Type Specimen. MVZ 8767.
Range. Southeast Alaska, northwestern coast of British Columbia.

Tamiasciurus hudsonicus preblei

Original Description. 1936. *Tamiasciurus hudsonicus preblei* A. H. Howell, *Proc. Biol. Soc. Wash.*, 49:133, August 22.

Type Locality. Fort Simpson, Mackenzie District, Northwest Territories.

Type Specimen. USNM 133862.

Range. Central and southwest Alaska across northern Canada to Saskatchewan.

Global Distribution

Nearctic—The red squirrel ranges widely across Alaska, Canada, and the contiguous United States, with southern extensions into the Rocky Mountains and the Appalachians (Hall 1981).

Alaska Distribution

Red squirrels occur throughout most of forested Alaska, from near the crest of the Brooks Range southward to the base of the Alaska Peninsula, and southeastward to islands in the Alexander Archipelago of southeast Alaska (map 8).

Southeast Region. Red squirrels are native inhabitants of forested habitats throughout the coastal mainland of this region. They were also apparently native to those islands in the Alexander Archipelago south of Frederick Sound and east of Clarence Strait, specifically Betton, Deer, Douglas, Etolin, Gedney, Grant, Hassler, Horseshoe, Kuiu, Kupreanof, Mitkof, Read, Revillagigedo, Sullivan, Tatoosh,

Tongass, Vank, Woronkofski, Wrangell, and Zarembo (MacDonald and Cook 2007). Present-day populations on Gravina and Annette islands may be the result of unreported translocations (see Swarth 1911a: Fig. 1).

The status of the red squirrel as a native resident of Admiralty Island is unclear. Most evidence suggests that this species was introduced to the north end of the island sometime in the late 1940s or early 1950s; they now are believed to occur throughout the island (MacDonald and Cook 1996).

Red squirrels from the Juneau area were successfully transplanted to Baranof and Chichagof islands in 1930 and 1931 (Burris and McKnight 1973). Since then, red squirrels have been found on Hill, Inian, Kruzof, Moser, Partofshikof, and Yakobi islands, and on some of the small islands in Sitka Sound (MacDonald and Cook 1996). An apparently unsuccessful introduction occurred on Prince of Wales Island (Fay and Sease 1985).

Southcentral Region. This species occurs throughout the mainland of southcentral Alaska, but is absent from the islands of Prince William Sound (ADFG 1978; Heller 1910).

Central Region. Red squirrels are widespread, but generally limited in the north to spruce along the southern drainages of the Brooks Range (Bee and Hall 1956).

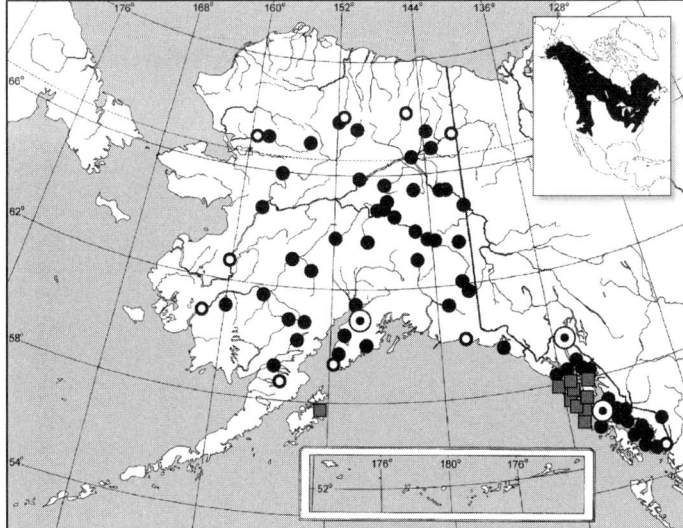

MAP 8. Distribution of red squirrel, *Tamiasciurus hudsonicus*.

⊙ = Type Locality
● = Native
▣ = Specimen Record (n = 1631)
○ = Marginal Records. 68.13330, –145.53330, Arctic Quad., Arctic Village (1 PSM); 67.16700, –141.66700, Coleen Quad., Porcupine River, Rampart House (1 USNM); 55.94361, –130.33361, Prince Rupert Quad., Gwent Cove, Hidden Inlet, Pearse Canal (1 UAM); 60.06700, –142.43330, Bering Glacier Quad., Cape Yakataga (4 CAS); 59.20000, –151.75000, Seldovia Quad., near Cape Elizabeth (12 USNM); 58.67010, –155.42900, Mt Katmai Quad., NE Naknek Lake, Fures Cabin (1 UAM); 60.79170, –161.75000, Bethel Quad., Bethel (1 USNM); 62.65420, –160.20830, Holy Cross Quad., Anvik (1 USNM); 67.08972, –159.78083, Baird Mts Quad., confluence of Kallarichuk River and Kobuk River (1 UAM); 68.13330, –151.75000, Chandler Lake Quad., Summit Anaktuvuk Pass (1 USNM).

Southwest Region. Red squirrels are uncommon in the timbered regions north of the Aleutian Range and Naknek Lake (ADFG 1978; Schiller and Rausch 1956). They were successfully transplanted to Afognak and Kodiak islands in 1952 from Anchorage-area populations. Burris and McKnight (1973) claimed the transplant to Kodiak Island was unsuccessful; however, red squirrels were collected on the island in 1991 (UAM) and populations there are considered healthy (van Daele 2004).

Western Region. The range of this species extends minimally into western Alaska. Specimens have been taken along the Kuskokwim River to near Bethel, northward to the Kobuk River Valley. Woodford (2005a) reported red squirrels occurring on the Seward Peninsula as far west as Council, but this has not been substantiated with specimens.

Habitat

The red squirrel is a characteristic inhabitant of coniferous forest but also occurs in mixed hardwood forests. In northern Alaska, red squirrels occasionally occur beyond tree line in riparian shrub thickets (Banfield 1974; Bee and Hall 1956; Wolff and Zasada 1975).

Status

Red squirrels are common within their core range. Loss of spruce habitat over large areas of southcentral and southwestern Alaska in recent years due to insect infestation may have long-term consequences for this and other conifer-dependent species.

Fossils

The fossil record of the red squirrel extends back to the Irvingtonian. Specimens have been identified from more than thirty Rancholabrean faunas from the central and eastern United States; none is known from Beringia (Kurtén and Anderson 1980; Steele 1998).

Chattering red squirrel (W.D. Berry)

Family CASTORIDAE Hemprich, 1820—beavers

The evolutionary relationship of the Castoridae to other extant families of rodents is unclear (Wahlert 1977; Wood 1959). Helgen (2005) concluded that beavers are most closely allied to geomyoid rodents (gophers and pocket mice). The single extant genus, *Castor*, contains two species, *C. fiber* of Eurasia and *C. canadensis* of North America.

Fossils of this family range from early Oligocene to Recent in North America and Europe, and middle Miocene to Recent in Asia. Fossil castorids include twenty-two extinct genera, including the black bear–sized *Castoroides* from the Pleistocene of North America, including Alaska (Kurtén and Anderson 1980; McLaughlin 1984).

Beavers are large rodents, second only to the South American capybara (*Hydrochoerus hydrochaeris*) in size. Most of their unique characteristics, such as the flat tail, webbed hind feet, protective membranes and valvular flaps, are adaptations to an aquatic existence.

American beaver, *Castor canadensis* (W. D. Berry)

American Beaver

Castor canadensis
Kuhl, 1820

Other Common Names

North American beaver, Canadian beaver.

Systematics

Lavrov and Orlov (1973) considered the Nearctic species, *Castor canadensis*, distinct from the Palearctic species, *C. fiber*. Hall (1981) recognized twenty-four subspecies across North America; two occur in Alaska. A revision of beaver with adequate specimens and modern techniques is needed (MacDonald and Cook 1998).

Castor canadensis belugae

Original Description. 1916. *Castor canadensis belugae* Taylor, *Univ. California Publ. Zool.*, 12:429, March 20.

Type Locality. Beluga River, Cook Inlet region, Alaska.

Type Specimen. MVZ 4224.

Range. Alaska, Yukon Territory, and coastal British Columbia.

Castor canadensis phaeus

Original Description. 1909. *Castor canadensis phaeus* Heller, *Univ. California Publ. Zool.*, 5:250, February 18.

Type Locality. Pleasant Bay, Admiralty Island, Alaska.

Type Specimen. MVZ 209.

Range. Admiralty Island and perhaps Chichagof and Baranof islands, Alexander Archipelago (Heller 1910).

Global Distribution

Nearctic—The American beaver is widely distributed across North America from Alaska to extreme northern Mexico (Hall 1981; Jenkins and Busher 1979) and has been introduced in Eurasia and South America (Helgen 2005).

Alaska Distribution

Beavers are distributed over most of mainland Alaska, although they are poorly documented in museum collections. They are found from near the crest of the Brooks Range south to the middle of the Alaska Peninsula and through southeast Alaska (map 9).

Southeast Region. Beavers occupy suitable habitat along the entire mainland and on most of the islands of the Alexander Archipelago, except Coronation, Warren, Forrester, and perhaps other small, more remote islands in the archipelago (MacDonald and Cook 2007). They are particularly abundant on the Yakutat forelands and in all of the major river valleys of the mainland.

Ten animals from Prince of Wales Island were successfully (re)introduced to Baranof Island in 1927, and possibly Kruzof Island in 1925 (Burris and McKnight 1973).

Southcentral Region. Beavers are found in suitable habitat throughout the mainland of this region. They are generally rare in Prince William Sound, and found only in low numbers on Hawkins and Hinchinbrook islands and from Gravina River to Rude River (ADFG 1978; Crowley 2004a). They are common to abundant along the Copper River, throughout the lowlands near Cape Yakataga, and on Kalgin Island in upper Cook Inlet (ADFG 1978).

Central Region. Beavers occur extensively throughout central Alaska.

Southwest Region. This species is abundant throughout the northern portion of this region and may occur in limited numbers as far south as Port Moller (ADFG 1978). Beavers were introduced to islands in the Kodiak area in 1925 from Cordova stock (Burris and McKnight 1973) and are now well established in suitable habitat on Kodiak, Afognak, Raspberry, and several other islands (ADFG 1978). Kellogg (1936) found beaver bones in Native middens from Kodiak Island; however, these remains might have been transported from the mainland by Native hunters (Murie 1959).

Western Region. Beavers continue to expand their range in the region, occurring as far north as the upper Kugururuk River and in the vicinity of Point Hope near the Chukchi Sea coast (Dau 2004a). They have become numerous in the Selawik and Kobuk River drainages, and continue to expand westward across the Seward Peninsula, with increasing numbers now being reported in the Serpentine River drainage (Gorn 2004). They also have extended their range into areas southeast of Norton Sound, with a small population now residing in the Stebbins-Unalakleet area (ADFG 1978). Beavers are abundant throughout the Yukon-Kuskokwim Delta region (Seavoy 2004a).

Northern Region. Beavers are found along drainages south of the crest of the Brooks Range. Reports of their occasional occurrence farther north (e.g., Bee and Hall 1956) remain unsubstantiated.

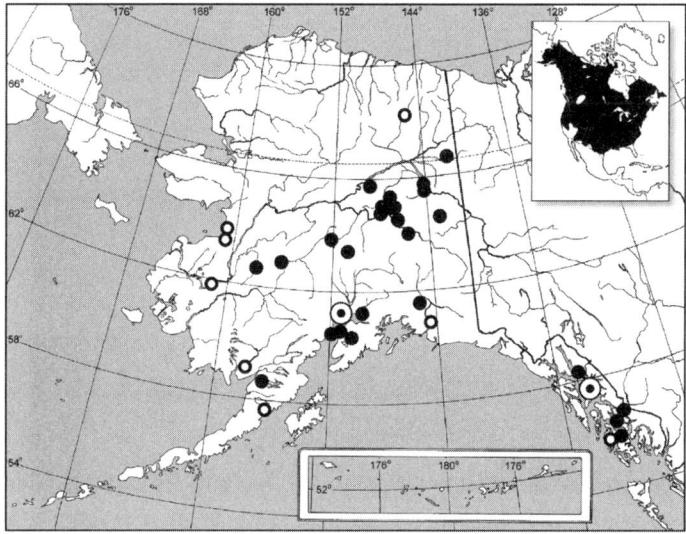

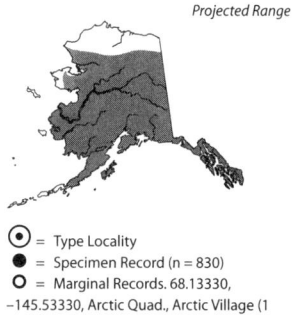

MAP 9. Distribution of American beaver, *Castor canadensis*.

Projected Range

⊙ = Type Locality
● = Specimen Record (n = 830)
○ = Marginal Records. 68.13330, −145.53330, Arctic Quad., Arctic Village (1 PSM); 55.76667, −133.45000, Craig Quad., Heceta Island, Chuck Lake (1 UAM); 60.84580, −144.51700, Cordova Quad., Bremner River (2 USNM); 57.75000, −156.25000, Ugashik Quad., Becharof Lake, Paulik Mt. (1 USNM); 58.95000, −158.48300, Dillingham Quad., lower Nushagak River (2 USNM); 61.78300, −161.31670, Russian Mission Quad., Russian Mission (2 USNM); 63.33330, −161.00000, Unalakleet Quad., Golsovia River (1 USNM); 63.86700, −160.78300, Unalakleet Quad., Unalakleet River (3 CAS).

Habitat

Beavers inhabit lakes, ponds, marshes, rivers, and streams. They are considered a keystone species due to their creation of wetland habitats for many other species (Noss and Cooperrider 1994).

Status

Alaska beavers are common to abundant in suitable habitat. The insular subspecies *C. c. phaeus* is considered Data Deficient by IUCN (MacDonald and Cook 1998).

Fossils

Three species of *Castor* have been reported from the North American Pleistocene (Kurtén and Anderson 1980), although speciation within the genus remains poorly understood (Fichter 1972). The black bear–sized giant beaver, *Castoroides ohioensis*, coexisted with *Castor canadensis* in eastern Beringia during the Late Pleistocene (Harington 1978; Kurtén and Anderson 1980). A small fragment of an incisor tooth collected from Cripple Creek near Fairbanks in 1942 was the only evidence Harington (1996c) found of giant beaver from Alaska.

Beaver lodge (W. D. Berry)

Family DIPODIDAE Fischer von Waldheim, 1817—jumping mice

This family of birch mice (genus *Sicista*), jumping mice (*Zapus, Napaeozapus, Eozapus*), and jerboas (e.g., genus *Jerboa*) comprises fifty-one species in sixteen genera in Eurasia and North America (Wilson and Reeder 2005). There is ongoing debate whether all should be included in a single family (Holden and Musser 2005; Klingener 1984) or two families (Nowak 1991; Stein 1990).

Three distinct genera of zapodines (one Old World, two New World) are generally recognized (but see Holden and Musser 2005 for a review of disparate views). Two species in the genus *Zapus* occur in Alaska.

Fossil zapodines are known from Lower Pliocene to Recent in Asia, Early Miocene to Recent in North America, and Oligocene to Recent in Europe (Nowak 1991).

Jumping mice are small, slender nocturnal rodents that hibernate. They have very long tails, large hind feet, and deeply grooved orange incisors. The two species that occur in Alaska are very similar in appearance and are best separated by cranial characters.

Meadow jumping mice, *Zapus hudsonius* (B. Hines)

Meadow Jumping Mouse

Zapus hudsonius
(Zimmermann, 1780)

Other Common Names
None.

Systematics
Krutzsch (1954) recognized eleven subspecies of meadow jumping mice across the wide range of this species. An additional subspecies represents isolated populations in New Mexico and Arizona (Hafner et al. 1981). Two subspecies occur in Alaska. Jones (1981) concluded that there was insufficient evidence to warrant subspecific distinctions. Nagorsen (1990) suggested that a detailed study of geographic variation utilizing morphological and genetic data is needed to resolve the taxonomy of this species.

Minimal variation in the cytochrome *b* gene was found in a study comparing individuals from localities throughout Alaska and British Columbia, suggesting a relatively recent colonization (Malaney and Cook unpubl.).

Zapus hudsonius alascensis
Original Description. 1897. *Zapus hudsonius alascensis* Merriam, *Proc. Biol. Soc.Wash.*, 2:223, July 15.
Type Locality. Yakutat Bay, Alaska.
Type Specimen. USNM 73584.

Range. Southern Alaska, extreme southern Yukon, and northwestern British Columbia.

Zapus hudsonius hudsonius
Original Description. 1780. *Dipus hudsonius* Zimmermann, *Geogr. Gesch. Mensch. Theire*, 2:358.
Type Locality. Hudson Bay (Fort Severn), Ontario.
Type Specimen. Not known to exist.
Range. Central Alaska across northern Canada to Ontario.

Global Distribution
Nearctic—The meadow jumping mouse is found across Alaska, Canada, and central and eastern contiguous United States as far south as New Mexico and Alabama (Hafner et al. 1981; Hall 1981; Whitaker 1972).

Alaska Distribution
There are considerable gaps in our knowledge of the distribution of this species in Alaska. Specimens are from the Yukon River southward to the end of the Alaska Peninsula and into southeast Alaska, with a major break in distribution in east-central and most of south-central Alaska (map 10).

Southeast Region. Meadow jumping mice have been found near Haines and Yakutat on the northern mainland, and, curiously, some 400 km farther south at two separate localities on Revillagigedo Island (MacDonald and Cook 2007).

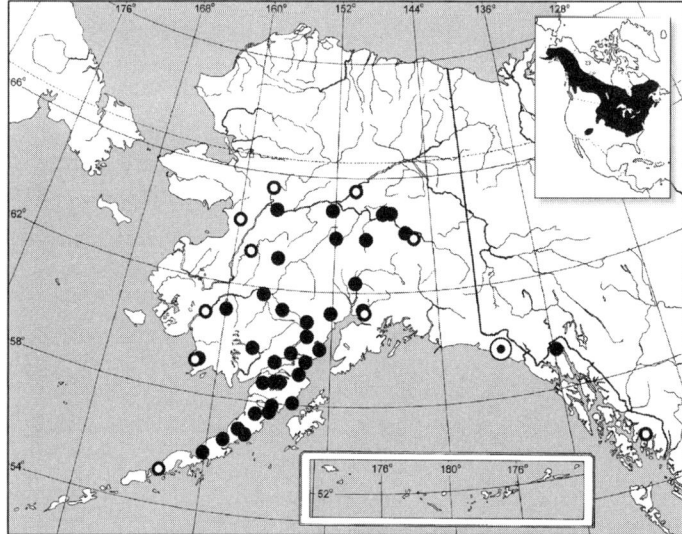

MAP 10. Distribution of meadow jumping mouse, *Zapus hudsonius*.

⊙ = Type Locality
● = Specimen Record (n = 398)
○ = Marginal Records. 65.50000, −150.25000, Tanana Quad., Rampart (2 UAM); 64.03333, −145.55000, Big Delta Quad., Delta Junction, Remington Rd. (20 UAM); 55.76667, −131.08333, Ketchikan Quad., Revillagigedo Island, Portage Cove (7 UAM); 55.81667, −131.36667, Ketchikan Quad., Revillagigedo Island, Orchard Lake (1 UAM); 61.28997, −149.74262, Anchorage Quad., Ft. Richardson (3 UAM); 55.26667, −162.84167, Cold Bay Quad., 6 km NW Cold Bay (2 UAM); 59.08053, −161.89119, Goodnews Bay Quad., north spit of Goodnews Bay (5 UAM); 60.98333, −160.06667, Bethel Quad., Tuluksak River, vicinity of Slate Creek (2 UAM); 63.18333, −158.26667, Ophir Quad., junction Iditarod and Yetna Rivers (1 UAM); 64.3727, −159.5463, Norton Bay Quad., Nulato Hills (1 UAM); 65.48333, −157.76667, Kateel River Quad., Upper Pitka River (3 UAM).

RODENTIA

81

Southcentral Region. This species has yet to be recorded anywhere west of the Anchorage area or the middle Susitna River. No specimens document its occurrence on the Kenai Peninsula, although Osgood (1901) found credible the sighting of a jumping mouse at Hope prior to his visit there in 1900.

Central Region. Specimens have been collected as far north along the Yukon River as Rampart and in the upper Pitka River, which is a tributary of the Koyukuk River southwest of Huslia.

Southwest Region. Meadow jumping mice are found westward to Cape Peirce (Peirce and Peirce 2000b) and along the entire length of the Alaska Peninsula to Unimak Island (unconfirmed, see Murie 1959:326).

Western Region. Meadow jumping mice have been found as far west as the Nulato Hills, Bethel, and the Goodnews River (UAM; Peirce and Peirce 2000b).

Habitat

Meadow jumping mice can be widespread in a variety of herbaceous meadow and shrub habitats, but are usually most abundant in the thick mosaic of vegetation along ponds, streams, and marshes. Jumping mice may require standing water nearby as they are known to drink heavily. During the winter months, they hibernate underground in well-drained soils.

Status

Meadow jumping mice are localized and sometimes abundant.

Fossils

Fossils of this species have not been reported from Alaska or Yukon Territory.

Western Jumping Mouse

Zapus princeps
J. A. Allen, 1893

Other Common Names

None.

Systematics

Krutzsch (1954) recognized eleven subspecies of *Zapus princeps*. Hall (1981) recognized twelve, but Hafner et al. (1981) presented convincing evidence that *luteus* in New Mexico and Arizona is actually a race of *Z. hudsonius*. One subspecies of *Z. princeps* inhabits southeast Alaska.

Cook et al. (unpubl.) found minimal levels of sequence divergence in the mitochondrial cytochrome *b* gene between individuals of *Z. princeps* from southeast Alaska and individuals from southern Canada.

Zapus princeps saltator

Original Description. 1899. *Zapus saltator* Allen, *Bull. Amer. Mus. Nat. Hist.*, 12:3, March 4.

Type Locality. Telegraph Creek, British Columbia.

Type Specimen. AMNH 14408.

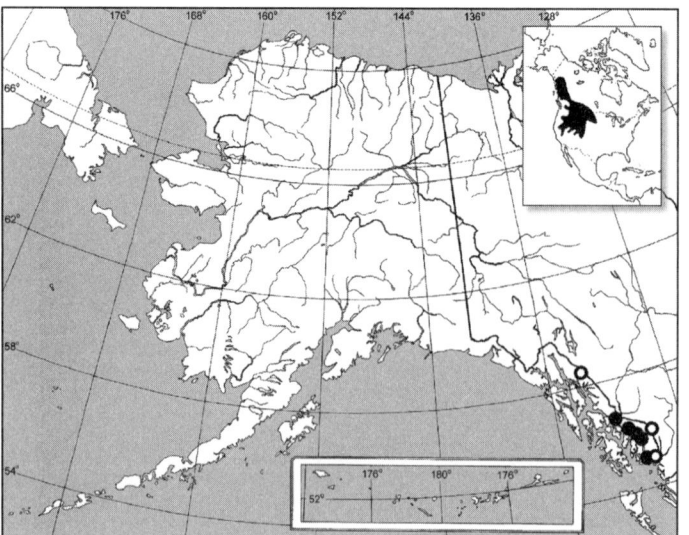

MAP 11. Distribution of western jumping mouse, *Zapus princeps*.

● = Specimen Record (n = 63)
○ = Marginal Records. 58.55000, –133.68333, Taku River Quad., Taku River, 0.8 km N Fish Creek (2 UAM); 55.96667, –130.06667, Ketchikan Quad., Hyder (2 UAM); 54.94361, –130.33361, Prince Rupert Quad., Gwent Cove, Hidden Inlet, Pearse Canal (15 UAM).

Range. Southern Yukon Territory, southeast Alaska, British Columbia, and western Alberta.

Remarks. Jones (1981) considered this taxon indistinct from *Z. p. princeps*.

Global Distribution

Nearctic—Western jumping mice occur in western United States and Canada.

Alaska Distribution

Zapus princeps, which superficially looks very similar to *Z. hudsonius*, has been found in Alaska only along the mainland coast and river systems of the southeast region (map 11). Its occurrence in southeast Alaska was based on one adult male captured by Swarth (1911a) on the Taku River. Our fieldwork documented this species on the mainland from the Taku River southward (MacDonald and Cook 2007).

Habitat

Like *Z. hudsonius*, *Z. princeps* occurs from high mountain meadows to riparian streamsides, ponds, and marshes in the lowlands where moist soils support a dense canopy of grasses, forbs, and shrubs. They hibernate underground in a dry nest chamber during the winter months (Cranford 1999). During active months, they build globular nests in tall grass and at the base of tall willow clumps.

Status

Western jumping mice are localized and often relatively abundant within their restricted Alaska distribution.

Fossils.

Fossils presumed to be ancestral to *Z. princeps* were known from the Upper Pliocene into the Pleistocene of Kansas and Oklahoma (Cranford 1999). Fossil remains of *Zapus* have yet to be reported from eastern Beringia.

Family CRICETIDAE Fischer von Waldheim, 1817—voles, lemmings, deermice, woodrats

The higher-level classification of rodents of the superfamily Muroidea is complicated and at best provisional. A long-standing dilemma has been whether to place the cricetid and murid rodents into one or two separate families. Most recent authors have adopted the single family hypothesis; however, Musser and Carleton (2005) provided sound reasons for two, which is adopted here.

Cricetidae is the second largest family of rodents, comprising nearly a third of the world's species. This diverse group of voles, lemmings, hamsters, and New World rats and mice includes about 681 species and 130 genera in 6 subfamilies (Musser and Carleton 2005). Cricetid rodents constitute the single largest family of mammals in Alaska, with most (thirteen species of voles and lemmings in seven genera) listed as arvicolines (subfamily Arvicolinae). Three species of the New World neotomines (subfamily Neotominae) occur in Alaska and are in the genera *Peromyscus* and *Neotoma*. The geologic range of North American cricetid rodents is late Eocene to Recent (McKenna and Bell 1997).

Cricetids are relatively small, short-lived animals that range up to the muskrat in size (about 1.8 kg). Members of this group are found in many terrestrial habitats, and some are arboreal or aquatic. Voles and lemmings (subfamily Arvicolinae) are restricted to northern regions and tend to undergo frequent, sometimes cyclic, fluctuations in population density. Most live in burrows and along runways through grassy areas. Their flat-crowned cheek teeth are ever-growing and well adapted to a diet of fibrous stems and leaves. New World mice and rats (subfamily Neotominae) generally have longer tails and larger, more conspicuous eyes and ears. They are also apt to be more omnivorous and nocturnal, as well as good climbers.

Long-tailed vole, *Microtus longicaudus* (W. D. Berry)

Collared Lemming

Dicrostonyx groenlandicus
(Traill, 1823)

Other Common Names

Nearctic collared lemming, Northern collared lemming, varying lemming, hoofed lemming.

Systematics

The taxonomy and distribution of *Dicrostonyx* have been problematic for many years. The number of species in recent revisions ranged from a single Holarctic species, *D. torquatus* (Ognev 1963; Rausch 1953a, 1963), to six species in North America alone (Musser and Carleton 2005). Much of the instability and controversy stems from the discovery of extraordinary chromosomal variation that produces subfertile hybrids as well as the occurrence of populations that occupy different tundra biotopes (Youngman 1975). Several of these distinctive populations are more interfertile than originally suspected and these chromosomal races may not represent separate species (Engstrom 1999).

We follow the application of three species of collared lemmings in North America and only a single species for Alaska (and the resulting modification of subspecies listed by Hall 1981) as proposed by Jarrell and Fredga (1993) and supported by Fedorov, Fredga, et al. (1999), Fedorov and Goropashnaya (1999), and Fedorov and Stenseth (2002). We recognize, however, that this may not be the last word on this subject. As noted by Engstrom (1999), the number of species from Alaska, in particular, is still in doubt. Further studies are needed to document the diversity of collared lemmings and illuminate the evolutionary history that may date to the earliest radiation of arvicoline rodents (Musser and Carleton 1993).

Dicrostonyx groenlandicus exsul

Original Description. 1919. *Dicrostonyx exsul* G. M. Allen, *Bull. Mus. Comp. Zool.*, 62:532, February.

Type Locality. St. Lawrence Island, Bering Sea, Alaska.

Type Specimen. MCZ 11885.

Range. Known only from type locality.

Remarks. Rausch and Rausch (1972) reported fertile progeny from crosses of St. Lawrence Island lemmings with lemmings from the Seward Peninsula.

Dicrostonyx groenlandicus nelsoni

Original Description. 1900. *Dicrostonyx nelsoni* Merriam, *Proc. Washington Acad. Sci.*, 2:25, March 14.

Type Locality. St. Michaels, Norton Sound, Alaska.

Type Specimen. USNM 186500.

Range. Western Alaska.

Dicrostonyx groenlandicus peninsulae

Original Description. 1953. *Dicrostonyx unalascensis peninsulae* Handley, *J. Washington Acad. Sci.*, 43:199, June 24.

Type Locality. Near sea level at Urilia Bay, Unimak Island, Alaska.

Type Specimen. USNM 246377.

Range. Southwestern Alaska.

Dicrostonyx groenlandicus rubricatus

Original Description. 1839. *Arvicola rubricatus* Richardson, *The zoology of Captain Beechey's voyage . . .* , p. 7.

Type Locality. American side of Bering Strait, Alaska.

Type Specimen. Not specified.

Range. Northern Alaska.

Dicrostonyx groenlandicus stevensoni

Original Description. 1929. *Dicrostonyx unalascensis stevensoni* Nelson, *Proc. Biol. Soc. Washington*, 30:145, March 30.

Type Locality. Umnak Island, Alaska.

Type Specimen. USNM 235552.

Range. Known only from type locality.

Dicrostonyx groenlandicus unalascensis

Original Description. 1900. *Dicrostonyx unalascensis* Merriam, *Proc. Washington Acad. Sci.*, 2:25, March 14.

Type Locality. Unalaska, Alaska.

Type Specimen. USNM 99622.

Range. Known only from type locality.

Remarks. In comparison to other collared lemmings, this insular race (along with *D. g. stevensoni* on Umnak Island) is large, pale in coloration, and does not turn white or develop digging claws in the winter (Gilmore 1933; Rausch and Rausch 1972).

Global Distribution

Nearctic (but includes Wrangel Island, Russia)—Collared lemmings occur from Alaska to Hudson Bay and throughout the high arctic islands of northern Canada and in northern Greenland (Hall 1981; Nagy and Gower 1999). Fedorov, Fredga, et al. (1999) and Smirnov and Fedorov (2003) included the collared lemmings of Russia's Wrangel Island in this species.

Alaska Distribution

Collared lemming populations are found along the treeless tundra zones of northern, western, and southwestern Alaska, extending into the eastern Aleutian Islands as well as on St. Lawrence Island in the Bering Sea (map 12).

Central Region. Rausch (1951) reported seldom seeing this lemming south of the crest of the Brooks Range; however, UAM houses a specimen (UAM 6812) taken east of Coldfoot at the divide between the Koyukuk and Chandalar rivers (approximately 100 km south of the Brooks Range crest). Specimens at the USNM purportedly from Fort Yukon and Ruby near the Yukon River were probably acquired from other localities.

Small numbers of collared lemmings, described as a distinct and isolated subspecies *nunatakensis* by Youngman (1975), have been found in rocky alpine tundra in the Ogilvie Mountains north of Dawson in Yukon Territory, suggesting their possible occurrence inside Alaska in the mountains north of Eagle.

Southwest Region. Collared lemmings occur along the length of Alaska Peninsula and into the Aleutian Islands as far west as Umnak Island (Peterson 1967). The recent discovery of collared lemmings on the west side of Cook Inlet near The Cone (UAM 64499) and Kirschner Lake (UAM 64451-64455), suggest the occurrence of this species farther north in the Chigmit

Mountains and beyond (D. McDonald, pers. comm.).

Western Region. St. Lawrence Island and Sledge Island near Nome (USNM) are the only Bering Sea islands known to harbor collared lemmings.

Northern Region. Collared lemmings are found throughout northern Alaska, from the low coastal plain to high in the mountains of the Brooks Range (Bee and Hall 1956; Rausch 1951).

Habitat

The collared lemming is well adapted to the extremes of cold, snow, and ice of arctic conditions. They are generally found occupying higher, drier, rockier tundra than *Lemmus trimucronatus*, the brown lemming (Banfield 1974). Bee and Hall (1956) found them closely associated with cotton-grass sedges (*Eriophorum* spp.).

Status

Collared lemmings can be scarce to abundant, with population size varying considerably between years. The status of insular populations is not known. Island populations under the names *D. exsul*, *D. unalascensis stevensoni*, and *D. u. unalascensis* were listed by IUCN as Data Deficient (Cook 1998b).

Fossils

The first occurrence of *Dicrostonyx* fossil records in North America was about 1.2 million years ago (Repenning 2001). Fossils are known from Fairbanks and a number of other

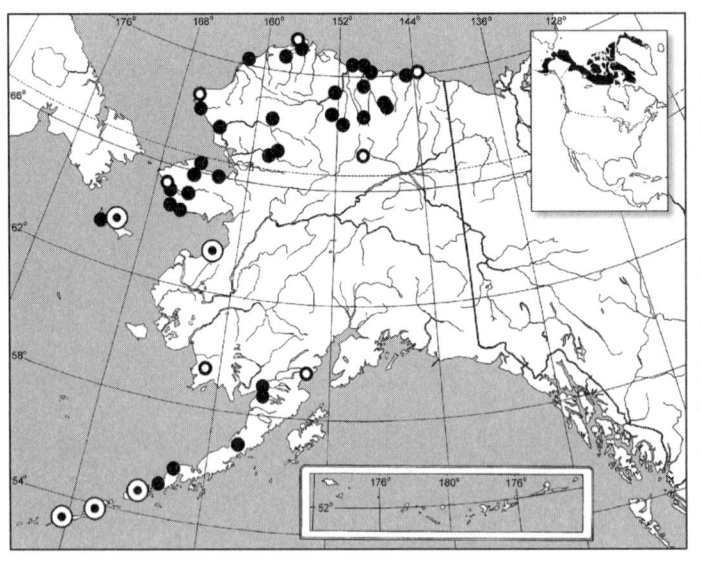

MAP 12. Distribution of collared lemming, *Dicrostonyx groenlandicus*.

⊙ = Type Locality
● = Specimen Record (n = 398)
○ = Marginal Records. 71.28333, –156.78333, Barrow Quad., Barrow (6 UAM); 70.13333, –143.63333, Barter Island Quad., Kaktovik (2 UAM); 67.33333, –149.50000, Chandalar Quad., Wiseman, Chandalar-Koyukuk divide (1 UAM); 59.52361, –153.92806, Iliamna Quad., 4 miles W of The Cone (1 UAM); 59.32667, –161.12861, Goodnews Bay Quad., Goodnews River, mi. 28 (1 UAM); 65.41667, –167.16667, Teller Quad., Lost River area (3 UAM); 68.65222, –166.23056, Point Hope Quad., Cape Dyer region (1 UAM).

central Alaska localities (Dixon 1984; Guilday 1963; Guthrie 1968a; Kurtén and Anderson 1980; Porter 1988).

Brown Lemming

Lemmus trimucronatus
(Richardson, 1825)

Other Common Names
Nearctic brown lemming, Siberian lemming, black-footed lemming.

Systematics
New World and Old World lemmings form a closely related complex. Their unstable taxonomic history was summarized by Rausch and Rausch (1975b). Rausch (1953a) synonymized Nearctic lemmings under Palearctic *L. sibiricus*, a taxonomic arrangement elaborated by Rausch and Rausch (1975b) and maintained by Hall (1981) but not Musser and Carleton (2005). Jarrell and Fredga (1993) proposed the existence of at least four circumpolar species of *Lemmus*, and that the populations in North America and those in northern Chukotka (as *L. sibiricus chryosogaster* by some) together constitute one species, *L. trimucronatus*. That view was supported in mtDNA studies by Fedorov, Goropashnaya, et al. (1999).

We tentatively follow Jarrell and Fredga (1993) in recognizing *L. trimucronatus* as the valid name for brown lemmings in Alaska while acknowledging that this matter requires further investigation. A modification of Hall (1981) of the seven subspecies found in Alaska is as follows:

Lemmus trimucronatus alascensis
Original Description. 1900. *Lemmus alascensis* Merriam, *Proc. Washington Acad. Sci.*, 2:26, March 14.
Type Locality. Point Barrow, Alaska.
Type Specimen. USNM 186499.
Range. Extreme northern Alaska.

Lemmus trimucronatus harroldi
Original Description. 1931. *Lemmus harroldi* Swarth, *Proc. Biol. Soc. Washington*, 44:101, October 17.
Type Locality. Nunivak Island, Alaska.
Type Specimen. CAS 6294.
Range. Known only from type locality.

Lemmus trimucronatus helvolus
Original Description. 1828. *Arvicola (Lemmus) helvolus* Richardson, *Zool. Jour.*, 3:517.

Type Locality. Near the headwaters of one of the southern tributaries of Peace River, or between there and the Jasper House region, Alberta (Preble, *N. Amer. Fauna*, 27:182, October 26, 1908). See Smith and Edmonds (1985).
Type Specimen. Not known to exist.
Range. Mountains of southern Yukon Territory, the extreme southwestern Northwest Territories, Alberta, British Columbia, and, presumably, the coastal mountains of southeast Alaska near Haines.

Lemmus trimucronatus minusculus
Original Description. 1904. *Lemmus minusculus* Osgood, *N. Amer. Fauna*, 24:36, November 23.
Type Locality. Kakhtul River, near junction with Malchatna River, Alaska.
Type Specimen. USNM 119612.
Range. Southwestern Alaska mainland.

Lemmus trimucronatus nigripes
Original Description. 1894. *Myodes nigripes* True, *Diagnoses of new North American mammals*, p. 2, April 24 (preprint of *Proc. U.S. Nat. Mus.*, 17:242, November 15, 1894).
Type Locality. St. George Island, Pribilof Islands, Alaska.
Type Specimen. USNM 59152.
Range. Known only from type locality.

Lemmus trimucronatus subarcticus
Original Description. 1956. *Lemmus trimucronatus subarcticus* Bee and Hall, *Univ. Kansas Mus. Nat. Hist., Miscl. Publ.*, 8:109, March 10.
Type Locality. Lake Schrader, long. 145°09' 50" W, lat. 69°24' 28" N, 2900 ft, Brooks Range, Alaska.
Type Specimen. USNM 303204 (originally catalogued as KU 50876).
Range. Brooks Range, northern Alaska.

Lemmus trimucronatus yukonensis
Original Description. 1900. *Lemmus yukonensis* Merriam, *Proc. Washington Acad. Sci.*, 2:27, March 14.
Type Locality. Charlie Creek [= Kandik River], Yukon River, Alaska.
Type Specimen. USNM 98849.
Range. Central and western Alaska.
Remarks. The Canadian subspecies *trimucronatus* (Richardson 1825) has been collected at the Yukon-Alaska border at Rampart House on the Porcupine River (Hall 1981). DNA sequences (cytochrome *b*) by Fedorov et al. (2003) from specimens collected in east-central Alaska revealed two divergent (6 percent) haplotypes, one Beringian (Alaska,

eastern Siberia), the other Canadian arctic (east of the Mackenzie Delta). These divergent haplotypes contact in the vicinities of the upper Yukon River, far-eastern Alaska Range, and the upper Susitna River in Southcentral Alaska.

Global Distribution

Holarctic—The modern range of this species comprises northeastern Siberia, Alaska, and northern Canada as well as southward into central British Columbia (Hall 1981; Jarrell and Fredga 1993; Nagorsen 2005).

Alaska Distribution

Brown lemmings inhabit most of mainland Alaska as well as two Bering Sea islands (map 13).

Southeast Region. In 2004, brown lemmings were discovered (by UAM) at two localities in the mainland mountains northwest of Haines (MacDonald and Cook 2007).

Southcentral Region. This species has not been found on the Kenai Peninsula, in the Talkeetna Mountains, in the Chugach Mountains, or on any islands in the region.

Central Region. Records occur throughout this region.

Southwest Region. Brown lemmings are found throughout the mainland of the Alaska Peninsula. An old specimen (USNM 59150) listed from Nagai Island in the Shumagin Islands needs to be verified by further survey work.

Western Region. Brown lemmings occur on the mainland as well as on Nunivak Island and

St. George Island, Pribilof islands (as *L. t. nigripes*). Lemmings from St. George Island were twice introduced to St. Paul Island (Hall 1981; Preble 1923; True 1899), but apparently without success (Byrd and Norvell 1993).

Habitat

This species occurs in a variety of arctic, alpine tundra, and taiga habitats (Banfield 1974; Pitelka and Batzli 1993). Above tree line they are usually associated with wet sedge-grass tundra, at lower elevations in spruce bogs and wet meadows (Buckley and Libby 1957; Kessel et al. 1982).

Status

Population levels of brown lemmings can vary considerably between years. In northern Alaska between 1946 (at least) and the early 1970s, their numbers fluctuated about every three years (Pitelka 1967; Pitelka and Batzli 2007).

Fossils

The fossil history of this species was reviewed by Koenigswald and Martin (1984). The earliest fossil records of the brown lemming in North America are the Middle Pleistocene (Irvingtonian) Cape Deceit Fauna of northwestern Alaska (Guthrie and Matthews 1971). In central Alaska, Guthrie (1968a) and Kurtén and Anderson (1980) reported Late Pleistocene fossils from the Fairbanks area, and Georgina (2001) found them present in Lime Hills strata ranging in age from

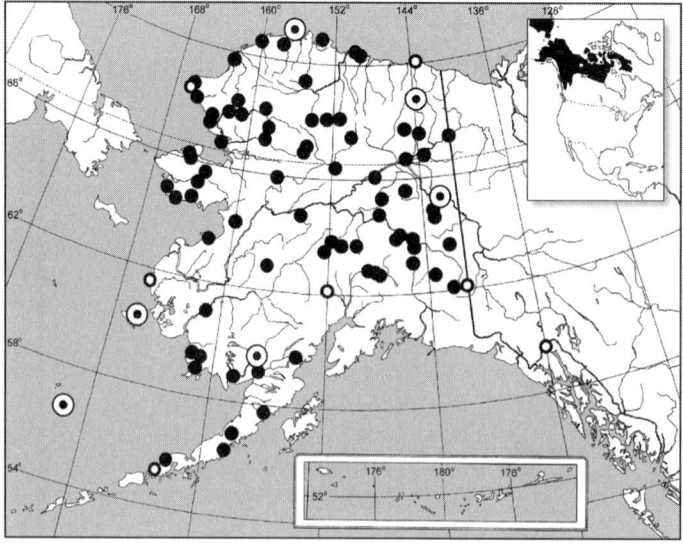

MAP 13. Distribution of brown lemming, *Lemmus trimucronatus*.

⊙ = Type Locality
● = Specimen Record (n = 4505)
○ = Marginal Records. 70.13333, –143.63333, Barter Island Quad., Barter Island, Kaktovik (4 UAM); 62.03118, –141.12948, Nabesna Quad., Braye Lakes (1 UAM); 59.62533, –136.08933, Skagway Quad., Tohitkah Mountain (1 UAM); 59.63742, –136.12914, Skagway Quad., Mount Ashmun (2 UAM); 62.29500, –152.05139, Talkeetna Quad., 8 km W of mouth of Fourth of July Creek (1 UAM); 55.33333, –162.80000, Cold Bay Quad., Izembek (1 UAM); 61.53190, –166.09170, Hooper Bay Quad., Hooper Bay (16 USNM); 68.35000, –166.78333, Point Hope Quad., Point Hope (4 UAM).

Late Pleistocene through Late Holocene. Pre-last glacial remains have been found in a cave on Prince of Wales Island, southeast Alaska (Heaton and Grady 2003).

Insular Vole

Microtus abbreviatus
Miller, 1899

Other Common Names
St. Matthew Island vole.

Systematics
Rausch and Rausch (1968) reviewed the taxonomy and systematic relationships of the insular vole, noting its close association with *M. miurus* of the Alaska mainland. An analysis of DNA sequence data (Conroy and Cook 2000b) found *M. abbreviatus* to be a sister (perhaps conspecific) taxon of *M. miurus*.

M. abbreviatus was first separated into two subspecies by Merriam (1900), an arrangement followed by Hall and Cockrum (1952), Hall (1981), and Musser and Carleton (2005).

Microtus abbreviatus abbreviatus
Original Description. 1899. *Microtus abbreviatus* Miller, *Proc. Biol. Soc. Washington*, 13:13, January 31.
Type Locality. Hall Island, Bering Sea, Alaska.
Type Specimen. USNM 15540/22429.
Range. Known only from type locality.

Microtus abbreviatus fisheri
Original Description. 1900. *Microtus abbreviatus fisheri* Merriam, *Proc. Washington Acad. Sci.*, 2:23, March 14.
Type Locality. St. Matthew Island, Bering Sea, Alaska.
Type Specimen. USNM 97976.
Range. Known only from type locality.

Global Distribution
Beringian—The insular vole is restricted to the St. Matthew Islands group, in the Bering Sea of Alaska.

Alaska Distribution
Microtus abbreviatus occurs on two of the three islands in the St. Matthew Islands (Cook and Klein 1999) (map 14).

Habitat
Colonies of this insular species are found most commonly in the vegetation of the moist, relatively well-drained lowlands and on the lower slopes of the islands. Their burrow systems are often dug near rocky outcroppings and small streams (Cook and Klein 1999; Rausch and Rausch 1968).

Status
Insular voles have a restricted island distribution. Their population densities fluctuate widely, with the last high reported in 1993 (Cook and Klein 1999). Both subspecies are listed as Data Deficient by IUCN (Cook 1998c).

Fossils
No fossils are known for this species.

MAP 14. Distribution of insular vole, *Microtus abbreviatus*.

⊙ = Type Locality
Specimen Record (n = 129)

Long-Tailed Vole

Microtus longicaudus
(Merriam, 1888)

Other Common Names

Coronation Island vole.

Systematics

Hall (1981) recognized fifteen subspecies of *M. longicaudus*, of which three occur in Alaska. The extensive karyotypic (Judd and Cross 1980) and mtDNA (Conroy and Cook 2000a) variation reported for this species, and the lack of a comprehensive study of morphologic variation throughout the range (Nagorsen 1990), suggest the need for a taxonomic revision.

Conroy and Cook (2000a) investigated sequence variation in the cytochrome *b* gene to recontruct the biogeographic history of this species across western North America. They identified two well-supported clades in Alaska. One included the islands and north mainland coast of southeast Alaska and extended marginally into the Wrangell Mountains. The other was found south along the mainland, and then on the eastern slope of the coast ranges from central Alaska through British Columbia south to Washington. Individuals representing these two clades were in contact near Haines. They suggested that the minimal haplotype divergence within these northern populations indicated recent (post-Pleistocene) coloniza-

tion of Alaska by ancestors of both clades (but see "Fossils," below).

Microtus longicaudus coronarius

Original Description. 1911. *Microtus coronarius* Swarth, *Univ. California Publ. Zool.*, 7:131, January 12.

Type Locality. Egg Harbor, Coronation Island, Alaska.

Type Specimen. MVZ 8721.

Range. Coronation, Warren, and Forrester islands, Alexander Archipelago.

Remarks. This subspecies was described as a distinct species, *M. coronarius*, by Swarth (1911a) on the basis of larger body size in a small number of specimens collected on Coronation, Warren, and Forrester islands (Swarth 1933). While some authors (e.g., Hall 1981, but not Map 457) have continued to retain *coronarius* as a distinct species, most (e.g., Baker et al. 2003; Conroy and Cook 1998; Musser and Carleton 2005) have relegated it to subspecific status.

Karyotypes of two voles from Coronation Island are indistinguishable from voles from Mitkof Island and the Haines area (UAM unpublished data). Furthermore, recent studies (Conroy 1998; Cook et al. 2001) using DNA sequences failed to detect divergence of populations of *coronarius* from others in the Alexander Archipelago.

Microtus longicaudus littoralis

Original Description. 1933. *Microtus mordax littoralis* Swarth, *Proc. Biol. Soc. Washington*, 46:209, October 26.

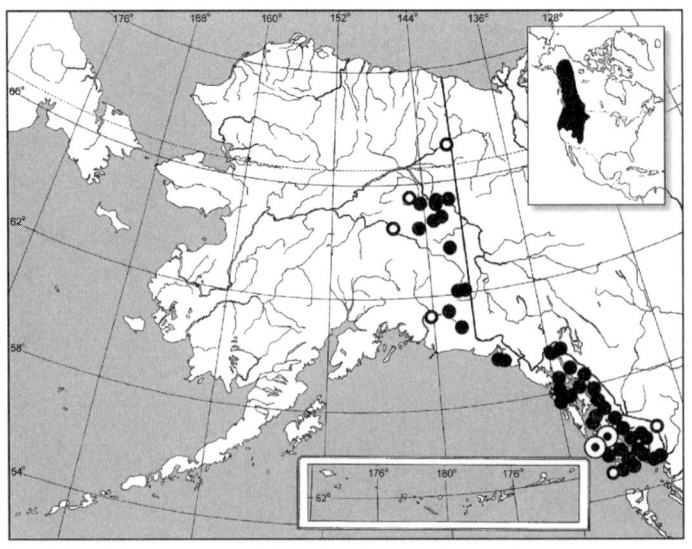

MAP 15. Distribution of long-tailed vole, *Microtus longicaudus*.

⊙ = Type Locality
● = Specimen Record (n = 1829)
○ = Marginal Records. 67.16700, –141.66700, Coleen Quad., Porcupine River, 28 mi. above Coleen River (9 USNM); 55.91667, –130.01667, Ketchikan Quad., Hyder (2 UAM); 54.82139, –133.52806, Dixon Entrance Quad., Forrester Island, Eagle Harbor (4 UAM); 61.31793, –144.23532, Valdez Quad., SW of Summit Lake (15 UAM); 64.46700, –146.93300, Big Delta Quad., N side Salcha River (1 KU); 65.48333, –145.41667, Circle Quad., Steese Hwy, Eagle Summit, mile 111 (1 UAM).

Type Locality. Shakan, Prince of Wales Island, Alaska [locality unclear as Shakan is on Kosciusko Island].
Type Specimen. MVZ 8642.
Range. Southeast Alaska and adjacent coastal British Columbia.

Microtus longicaudus vellerosus
Original Description. 1899. *Microtus vellerosus* J. A. Allen, *Bull. Amer. Mus. Nat. Hist.*, 12:7, March 4.
Type Locality. Upper Liard River, British Columbia.
Type Specimen. AMNH 14403.
Range. East-central Alaska southeastward through Yukon Territory, central British Columbia, to extreme western Alberta.

Global Distribution
Nearctic—Long-tailed voles range widely throughout western North America (Hall 1981; Smolen and Keller 1987).

Alaska Distribution
Long-tailed voles inhabit the far eastern portions of the state (map 15).

Southeast Region. The long-tailed vole occurs extensively throughout this region. They have been found in suitable habitat throughout the coastal mainland and on islands of the Alexander Archipelago that include Admiralty, Anguilla, Annette, Baker (sign), Chichagof, Coronation, Dall, Dog, Douglas, Etolin, Forrester, Hoot, Kosciusko, Kuiu, Kupreanof, Lester, Marble, Mary, Mitkof, Moser, Noyes, Orr, Owl, Prince of Wales, Revillagigedo, Santa Rita, Shelikof, Stevenson, Suemez, Sukkwan, Sullivan, Thorne, Tuxekan, Warren, Woewodski, Wrangell, and Zarembo islands (MacDonald and Cook 2007). Long-tailed voles have not been documented on Baranof Island, despite its close proximity to populations on Chichagof Island.

Southcentral Region. This species is known from a number of sites east of the Copper River (Cook and MacDonald 2003).

Central Region. Long-tailed voles are found as far west as the Salcha River east of Fairbanks, and north to the upper Porcupine River.

Habitat
Long-tailed voles are found in a wide variety of habitats, including grassy openings in forests, meadows, shrub thickets, rocky mountain slopes, and around streams, lakes, and marshes (Banfield 1974).

Status
Microtus longicaudus is common and periodically abundant in suitable habitat. It is listed as Data Deficient by IUCN as *M. coronarius* (Conroy and Cook 1998).

Fossils
Pre-glacial and Early Holocene remains of this species have been recovered in southeast Alaska from cave deposits on Prince of Wales Island (Fedje et al. 2004). Remains from caves on the mainland near Wrangell are thought to be from Late Pleistocene to Early Holocene in age (Heaton 2001; Heaton and Grady 2003).

Singing Vole

Microtus miurus
Osgood, 1901

Other Common Names
Alaska haymouse, Alaska vole, Toklat vole.

Systematics
Microtus miurus and *M. abbreviatus* of St. Matthew Island are thought to be sister taxa or even conspecific based on mtDNA sequence data (Conroy and Cook 2000b). These species were considered members of the subgenus *Stenocranius* by Rausch (1964), and closely related to *M. gregalis* from the Palearctic (Ognev 1950; Rausch 1964; Rausch and Rausch 1968). Other data clearly support the distinctiveness of the New and Old World forms (Anderson 1960; Fedyk 1970; Vorontsov and Lyapunova 1986). Zagorodnyuk (1990) emphasized the distant kinship of *M. gregalis* and *M. miurus* by placing them in different subgenera. Analysis of mtDNA sequences (Conroy and Cook 2000b) also found *M. gregalis* distantly related to *M. miurus*.

Hall (1981) recognized five subspecies of *M. miurus*, with four found in Alaska. The relationships of these subspecies are not well understood and need further study.

Microtus miurus cantator
Original Description. 1947. *Microtus cantator* Anderson, *Bull. Nat. Mus. Canada*, 102:161, January 24.
Type Locality. Mountain top near Tepee Lake, lat. 61°35' N, long. 140°22' W, North slope St. Elias Range, Yukon.
Type Specimen. NMC 17236.
Range. Southwestern Yukon Territory and adjacent Alaska.

Microtus miurus miurus

Original Description. 1901. *Microtus miurus* Osgood, *N. Amer. Fauna*, 21:64, September 26.

Type Locality. Head Bear Creek, in mountains near Hope City, Turnagain Arm, Cook Inlet, Alaska.

Type Specimen. USNM 107175.

Range. Southcentral Alaska.

Microtus miurus muriei

Original Description. 1931. *Microtus muriei* Nelson, *J. Mammalogy*, 12:311, August 24.

Type Locality. Kutuk River (tributary of Alatna River), Endicott Mountains, Alaska.

Type Specimen. USNM 243482.

Range. Northern Alaska and Yukon Territory.

Remarks. Includes *Microtus miurus paneaki* Rausch, 1950, *J. Washington Acad. Sci.*, 40:135, April 21, type from Tolugak Lake (lat. 68°24' N, long. 152°10' W), Brooks Range, Alaska (USNM 290296).

Microtus miurus oreas

Original Description. 1907. *Microtus miurus oreas* Osgood, *Proc. Biol. Soc. Washington*, 20:61, April 18.

Type Locality. Toklat River, Alaskan Range, Alaska.

Type Specimen. USNM 148596.

Range. Alaska Range.

Global Distribution

Nearctic—Alaska, Yukon Territory, and adjacent Northwest Territories.

Alaska Distribution

Singing voles inhabit many of the mountainous areas of Alaska (map 16).

Southcentral Region. Singing voles occur throughout the Alaska Range (including the far eastern Nutzotin Mountains) and in the Aleutian Range east of Iliamna Lake (UAM). They are also found in the Talkeetna Mountains, the far western Chugach Mountains near Anchorage, and on the Kenai Peninsula (Allen 1902; Fuller 1981; Osgood 1901; Rausch 1953a). The occurrence of singing voles eastward along the Chugach Mountains and into the Wrangell and St. Elias mountain ranges is poorly documented with only three specimens from Chisana (Cook and MacDonald 2003).

Central Region. Singing voles are found along the south slope of the Brooks Range, with recent records (UAM) from the Ray Mountains extending their known range considerably southward toward the center of the state. This species is absent from the Yukon-Tanana uplands, despite Clough (1976) claiming to have collected a *M. miurus* on Mount Kathyrn and another on nearby Twin Mountains in the middle Charley River region in 1975. One of these specimens (USNM 512822) was recently examined by the authors and identified as *Microtus oeconomus*, not *M. miurus*. Singing voles do, however, occur farther east and north of the Yukon River in the

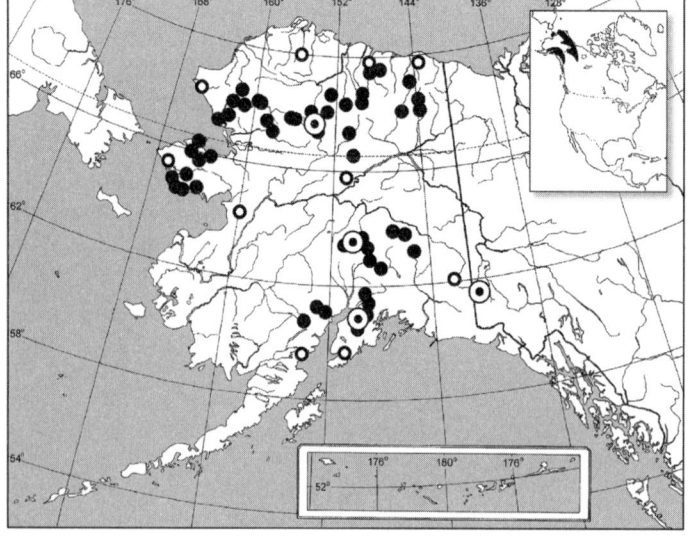

MAP 16. Distribution of singing vole, *Microtus miurus*.

⊙ = Type Locality
● = Specimen Record (n = 2874)
○ = Marginal Records. 69.96667, −155.76667, Ikpikpuk River Quad., Colville River region (1 UAM); 69.80000, −148.68333, Sagavanirktok Quad., Dalton Hwy, Franklin Bluffs (1 UAM); 69.71111, −143.61583, Demarcation Point Quad., 6 miles NE VABM Bitty on the Jago River (2 UAM); 65.63333, −150.85000, Tanana Quad., Ray Mountains, Crash Creek, 5 mi SW of Mt Tozi (2 UAM); 62.06485, −142.04648, Nabesna Quad., Chisana (2 UAM); 59.83300, −150.75000, Seldovia Quad., Sheep Creek, Kenai Peninsula (64 AMNH); 59.52361, −153.92806, Iliamna Quad., 4 miles W of The Cone (2 UAM); 64.3718, −159.5572, Norton Bay Quad., North Fork Unalakleet River (1 UAM); 65.40000, −167.10000, Teller Quad., Lost River area (3 UAM); 68.30000, −165.53333, Point Hope Quad., Cape Thompson, Ogotoruk Creek area (86 UAM).

Ogilvie Mountains of Yukon Territory (Youngman 1975).

Western Region. Singing voles occur extensively throughout the mountains north of the Kobuk River, on the Seward Peninsula, and in the Nulato Hills northeast of Unalakleet (UAM).

Northern Region. This species is found throughout the mountains and foothills of the Brooks Range.

Habitat

Microtus miurus inhabits a variety of tundra and taiga-tundra habitats on well-drained sites, sometimes along streambanks, near or above tree line. The burrow systems of this semicolonial species are often associated with willows (*Salix*) and rocky areas. Toward the end of summer, this species becomes more vocal and caches piles of vegetation for winter use (Batzli and Henttonen 1993; Bee and Hall 1956; Douglass 1984; Galindo and Krebs 1985; Murie 1961; Quay 1951; Rausch 1953a; Youngman 1975).

Status

Singing voles can be moderately abundant in suitable habitat, but population densities vary considerably between years (Pruitt 1968).

Fossils

Fossils ranging in age from Pleistocene (Rancholabrean) through Recent are known from a variety of sites in eastern Beringia (Georgina 2001; Harington 1978; Jopling et al. 1981). Near Fairbanks, where the species no longer occurs, singing vole remains have been found in every major stratigraphic unit, from Illinoian to terminal Wisconsinan (Guthrie 1968a; Kurtén and Anderson 1980).

Root Vole

Microtus oeconomus
(Pallas, 1776)

Other Common Names

Tundra vole, northern vole.

Systematics

The conspecific status of Palearctic and Nearctic populations was confirmed by Zimmermann (1942) and supported by subsequent studies (Galbreath and Cook 2004; Nadler et al. 1976, 1978; Ognev 1964; Rausch 1953a).

Hall (1981), following a revision by Paradiso and Manville (1961), recognized ten

subspecies in North America. All ten occur in Alaska. Study of allozymic variation of five Alaska populations revealed little difference among subspecies, with the exception of *M. o. operarius* from Hinchinbrook Island (Lance and Cook 1998a). Mitochondrial sequences of the cytochrome *b* gene indicated that the Montague Island vole, *M. o. elymocetes*, was the most genetically divergent of all of the North American subspecies (Galbreath 2002; Galbreath and Cook 2004) and otherwise largely corroborated previous work based on allozymes.

It remains unresolved if the correct specific name for the Holarctic root vole is *oeconomus* rather than possibly *ratticeps* or, less likely, *kamtschaticus* (Hall 1981; Musser and Carleton 2005).

Microtus oeconomus amakensis

Original Description. 1930. *Microtus amakensis* O. J. Murie, *J. Mammalogy*, 11:74, February 11.

Type Locality. Amak Island, Bering Sea, Alaska.

Type Specimen. USNM 246449.

Range. Known only from type locality.

Microtus oeconomus elymocetes

Original Description. 1906. *Microtus elymocetes* Osgood, *Proc. Biol. Soc. Washington*, 19:71, May 1.

Type Locality. East side of Montague Island, Prince William Sound, Alaska.

Type Specimen. USNM 137323.

Range. Known only from type locality.

Microtus oeconomus innuitus

Original Description. 1900. *Microtus innuitus* Merriam, *Proc. Washington Acad. Sci.*, 2:21, March 14.

Type Locality. Northeast Cape, St. Lawrence Island, Bering Sea, Alaska.

Type Specimen. USNM 99373.

Range. Known only from type locality.

Microtus oeconomus macfarlani

Original Description. 1900. *Microtus macfarlani* Merriam, *Proc. Washington Acad. Sci.*, 2:24, March 14.

Type Locality. Fort Anderson, Anderson River, Mackenzie.

Type Specimen. USNM 9155/37347.

Range. Northern and central Alaska, Yukon Territory, northwestern Northwest Territories, and extreme northwestern British Columbia.

Remarks. Includes *M. operarius endoecus* Osgood, 1909, *N. Amer. Fauna*, 30:23, June 21, type from the mouth of Charlie Creek [Kandik

River], Yukon River, about 50 mi. above Circle, Alaska (USNM 128327).

Antell (1987) concluded that populations of root voles (previously included under *M. o. macfarlani*) from Haines Junction, Yukon Territory, and in the vicinity of Skagway (previously as *M. o. yakutatensis*) warrant recognition as a new, and as yet undescribed, subspecies.

Microtus oeconomus operarius

Original Description. 1893. *Arvicola operarius* Nelson, *Proc. Biol. Soc. Washington*, 8:139, December 28.

Type Locality. St. Michael, Norton Sound, Alaska.

Type Specimen. USNM 14379/22225.

Range. Western and southwestern Alaska.

Remarks. Includes *M. kadiacensis* Merriam, 1897, *Proc. Biol. Soc. Washington*, 11:222, July 15, type from Kodiak Island, Alaska (USNM 65827), and *M. o. gilmorei* Setzer, 1952, *Proc. Biol. Soc. Washington*, 65:75, April 25, type from Point Lay, lat. 69°46′ N, long. 163° 04′ W, Alaska (USNM 293109).

Microtus oeconomus popofensis

Original Description. 1900. *Microtus unalascensis popofensis* Merriam, *Proc. Washington Acad. Sci.*, 2:22, March 14.

Type Locality. Popof Island, Shumagin Islands, Alaska.

Type Specimen. USNM 97956.

Range. Shumagin Islands.

Microtus oeconomus punukensis

Original Description. 1932. *Microtus innuitus punukensis* Hall and Gilmore, *Univ. California Publ. Zool.*, 38:399, September 17.

Type Locality. Big [= north] Punuk Island, near east end of St. Lawrence Island, Alaska.

Type Specimen. MVZ 51392.

Range. Known only from the Punuk Islands.

Microtus oeconomus sitkensis

Original Description. 1897. *Microtus sitkensis* Merriam, *Proc. Biol. Soc. Washington*, 11:221, July 15.

Type Locality. Sitka, Alaska.

Type Specimen. USNM 73839.

Range. Northern outer islands of the Alexander Archipelago, southeast Alaska.

Microtus oeconomus unalascensis

Original Description. 1897. *Microtus unalascensis* Merriam, *Proc. Biol. Soc. Washington*, 11:222, July 15.

Type Locality. Unalaska, Alaska.

Type Specimen. USNM 30772/42672.

Range. Unalaska Island and vicinity.

Microtus oeconomus yakutatensis

Original Description. 1900. *Microtus yakutatensis* Merriam, *Proc. Washington Acad. Sci.*, 2:22, March 14.

Type Locality. North shore Yakutat Bay, Alaska.

Type Specimen. USNM 98005.

Range. Northern coast of southeast Alaska.

Remarks. Antell (1987) suggested the designation of a new subspecies, *M. o. littoralis* (type from Yakutat, Alaska, No. 85-315 in Charles R. Conner Museum), for populations of root voles from Bartlett Cove northward along the coast but not including the area immediately north of Yakutat Bay, which he retained as *M. o. yakutatensis*. He also concluded that populations of root voles in the vicinity of Skagway (previously as *M. o. yakutatensis*) and from Haines Junction, Yukon Territory (previously included under *M. o. macfarlani*), warrant recognition as a new, and as yet undescribed, subspecies.

Global Distribution

Holarctic—Root voles have a vast range across northern Eurasia but are restricted in North America to Alaska and adjacent northwestern Canada (Musser and Carleton 2005).

Alaska Distribution

Root voles have an exceptionally broad distribution across Alaska, occurring throughout nearly all of the mainland as well as on many islands (map 17).

Southeast Region. Root voles inhabit the northern mainland of this region as far south as the upper east side of Lynn Canal. They also occur on Baranof (including Catherine "Island," which is actually part of Baranof), Chichagof, Haenke, Inian, Lemesurier, Seal, and Yakobi islands (MacDonald and Cook 2007).

Southcentral Region. This vole has been documented (Appendix 9) on Chenega, Disc, Egg, Eleanor, Evans, Crafton, Hawkins, Hinchinbrook, Kayak, Knight, Montague, and Naked islands in Prince William Sound, and on Kalgin Island in Cook Inlet (MVZ; UAM; Heller 1910).

Central Region. Root voles are found throughout central Alaska.

Southwest Region. The root vole occurs extensively throughout the mainland and on the following islands (Appendix 10): Afognak, Aiktak (root vole found dead but probably transported there from Ugamak; S.

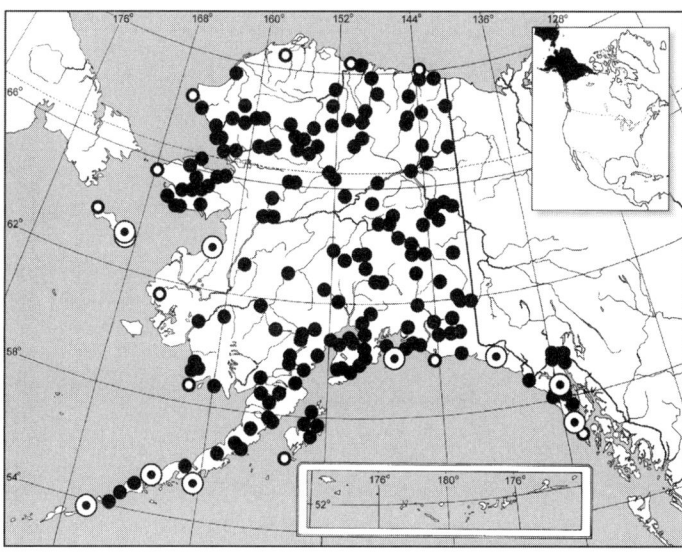

MAP 17. Distribution of root vole, *Microtus oeconomus.*

⊙ = Type Locality
● = Specimen Record (n = 10,633)
○ = Marginal Records. 70.48056, –157.41667, Meade River Quad., Meade River village (10 UAM); 70.42500, –150.40500, Harrison Bay Quad., Colville Village (3 UAM); 70.16667, –143.58333, Barter Island Quad., Barter Island (1 UAM); 56.59028, –134.86028, Port Alexander Quad., Baranof Island, Plotnikof Lake (34 UAM); 59.80000, –144.58333, Middleton Island Quad., Kayak Island (3 UAM); 56.41667, –154.75000, Trinity Islands Quad., Tugidak Island, S end of Tugidak Island (1 UAM); 58.55000, –161.76667, Hagemeister Island Quad., Cape Peirce (5 UAM); 61.50000, –165.00000, Hooper Bay Quad., Old Chevak (5 UAM); 63.78333, –171.75000, St. Lawrence Island, Gambell old village site (6 UAM); 65.61667, –168.08333, Teller Quad., Wales (6 UAM); 68.35000, –166.78333, Point Hope Quad., Point Hope (21 UAM).

Ebbert, pers. comm., 2008), Akutan, Amak, Avatanak, Barrier, Glen, Kodiak, Round, Sanak, Shumagins (specimens from Popof and Unga islands), Tigalda, Tugidak, Ugamak, Ukolnoi, Unalaska, Unalga, and Unimak islands (UAM; USNM; AMNH; Byrd and Williams 2002; Murie 1959; Paradiso and Manville 1961; Peterson 1967).

Western Region. Root voles are widely distributed throughout the mainland of western Alaska and on St. Lawrence Island and all three of the Punuk Islands (Fay and Sease 1985).

Northern Region. The root vole occurs extensively throughout northern Alaska.

Habitat

Root voles inhabit a variety of open herbaceous-dominant habitats at various elevations (Banfield 1974; Lance and Cook 1998b), occurring most abundantly in wet sedge and grass-forb meadows and bogs.

Status

This species is widespread and common, with densities fluctuating considerably between years (Pruitt 1968). It is listed as Data Deficient by IUCN as *M. o. amakensis, M. o. elymocetes, M. o. innuitus, M. o. popofensis, M. o. punukensis,* and *M. o. sitkensis* (Lance and Cook 1998b).

Fossils

Some authors have hypothesized that *M. oeconomus* is a relatively recent colonizer of North America (Galbreath 2002; Lance and Cook 1998a; Rausch 1994). Reports of fossils from northern Yukon Territory that date to late Illinoian (Jopling et al. 1981; Zakrzewski 1985) suggest a deeper history. Fossils that date from the Early Holocene have been recovered from caves on Prince of Wales Island, southeast Alaska (Heaton and Grady 2003), an island south of this species' current distribution.

Meadow Vole

Microtus pennsylvanicus
(Ord, 1815)

Other Common Names

None.

Systematics

Hall (1981) recognized twenty-six subspecies across the broad range of this species; four occur in Alaska. A morphological analysis by Snell and Cunnison (1983) found no conspicuous subdivision, suggesting that subspecific designations may be inappropriate, but a comprehensive revision is long overdue.

Microtus pennsylvanicus admiraltiae

Original Description. 1909. *Microtus admiraltiae* Heller, *Univ. California Publ. Zool.*, 5:256, February 18.

Type Locality. Windfall Harbor, Admiralty Island, Alaska.

Type Specimen. MVZ 118.

Range. Known only from type locality.

Microtus pennsylvanicus alcorni

Original Description. 1951. *Microtus pennsylvanicus alcorni* Baker, *Univ. Kansas Publ., Mus. Nat. Hist.*, 5:105, November 28.

Type Locality. 6 mi. SW Kluane, 2550 ft., Yukon.

Type Specimen. KU 21552.

Range. Southern Alaska, extreme northwestern British Columbia, and southern Yukon Territory.

Microtus pennsylvanicus rubidus

Original Description. 1940. *Microtus pennsylanicus rubidus* Dale, *J. Mammalogy*, 21:339, August 13.

Type Locality. Sawmill Lake, near Telegraph Creek, British Columbia.

Type Specimen. MVZ 30738.

Range. Southeast Alaska, northwestern British Columbia.

Remarks. This taxon was synonymized with *M. p. drummondii* by Anderson (1946) and Cowan and Guiguet (1965).

Microtus pennsylvanicus tananaensis

Original Description. 1951. *Microtus pennsylvanicus tananaensis* Baker, *Univ. Kansas Publ., Mus. Nat. Hist.*, 5:107, November 28.

Type Locality. Yerrick Creek, 21 mi. W, 4 mi. N Tok Junction, Alaska.

Type Specimen. KU 21509.

Range. Central Alaska.

Global Distribution

Nearctic—This vole has a broad range across North America, occurring from Alaska and Canada south to New Mexico and Georgia (Hall 1981).

Alaska Distribution

Meadow voles occur throughout the boreal zone of central and southern Alaska (map 18).

Southeast Region. Meadow voles have a restricted distribution in southeast Alaska, occurring along the mainland valleys of the Chilkat, Taku, and Stikine rivers, and in the Alexander Archipelago on Admiralty, Mitkof, and Wrangell islands (as perhaps an accidental), and near the Stikine River, including Stikine delta islands of Kadin, Sergief, and Vank (MacDonald and Cook 2007).

Southcentral Region. Meadow voles are absent from the Kenai Peninsula (contra Bangs 1979), Prince William Sound, and eastward along the Gulf Coast.

Central Region. Northernmost records are from Old John Lake near Arctic Village and Bettles on the upper Koyukuk River.

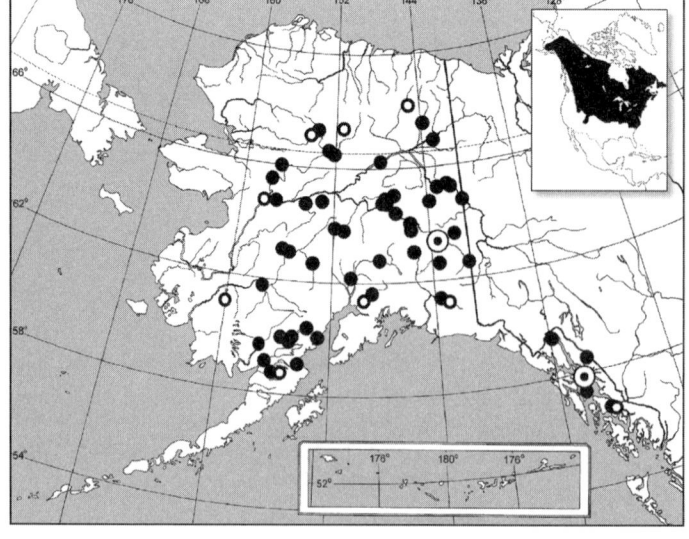

MAP 18. Distribution of meadow vole, *Microtus pennsylvanicus*.

⊙ = Type Locality
● = Specimen Record (n = 3351)
○ = Marginal Records. 68.05000, –145.00000, Arctic Quad., Old John Lake (6 UAM); 56.45278, –132.60278, Petersburg Quad., Vank Island, Mud Bay (2 UAM); 61.36556, –143.44250, McCarthy Quad., Ruby Lake (28 UAM); 61.17139, –149.71306, Anchorage Quad., Fort Richardson (1 UAM); 58.67010, –155.42900, Mt. Katmai Quad., NE Naknek Lake, Fures Cabin (2 UAM); 60.98333, –160.06667, Bethel Quad., Tuluksak River, vicinity of Slate Creek (17 UAM); 64.55000, –158.10000, Nulato Quad., Nulato (2 USNM); 67.10289, –154.27153, Survey Pass Quad., SE side of Walker Lake (2 UAM); 67.45311, –150.85003, Wiseman Quad., unnamed lake near North Fork Koyukuk River (16 UAM).

Southwest Region. This vole was first recorded in the southwest region by Osgood (1904). Schiller and Rausch (1956) and Cook and MacDonald (2006) documented this species on the Alaska Peninsula as far west as Naknek Lake.

Western Region. Meadow voles extend into this region from central Alaska near the headwaters of the Kobuk River at Walker Lake, and down the Kuskokwim River to at least the Tuluksak River.

Habitat

As their name implies, meadow voles inhabit open herbaceous habitats such as grassy meadows, fields, and marshes, often in riparian situations.

Status

This species is common and sometimes abundant in suitable habitat, but is listed as Data Deficient by IUCN as *M. p. admiraltiae* (MacDonald et al. 1998).

Fossils

Youngman (1975) suggested that *M. pennsylvanicus* is a probable postglacial immigrant to extreme northwestern North America. Jopling et al. (1981), however, reported fossils of this species of early Wisconsin age from Old Crow Basin deposits in Yukon Territory.

Taiga Vole

Microtus xanthognathus
(Leach, 1815)

Other Common Names

Chestnut-cheeked vole, yellow-cheeked vole, yellow-nosed vole.

Systematics

Microtus xanthognathus was placed in the subgenus *Microtus* by Miller (1896) and Bailey (1900), but may have affinities with the subgenus *Aulacomys* (Zagorodnyuk 1990). Although conventionally viewed as closely related to (Anderson 1960), or possibly conspecific with (Hall and Kelson 1959), *M. chrotorrhinus*, morphological, chromosomal, ecological, and ethological traits indicate a more distant kinship (Musser and Carleton 2005; Youngman 1975). Phallic morphology (Lidicker and Yang 1986) and DNA sequences (Conroy and Cook 2000b) suggest a possible sister relationship with *M. miurus*.

No subspecies are currently recognized.

Microtus xanthognathus

Original Description. 1815. *Arvicola xanthognatha* Leach, *Zool. Miscl.*, 1:60.
Type Locality. Hudson Bay.
Type Specimen. Not known.

Global Distribution

Nearctic—This species occurs from central Alaska across northern boreal Canada to Hudson Bay (Hall 1981).

Alaska Distribution

The distribution of taiga voles in Alaska (map 19) is not well understood, with specimens indicating a patchy distribution north of the Alaska Range across the central region of the state (Conroy and Cook 1999; Lensink 1954). They have been documented (UAM) as far west as the mouth of the Innoko River, and down the Kobuk River (in the western region) to the Kallarichuk River. The northeastern-most records (Youngman 1975) are from the Alaska-Canada border near the Firth River (in the northern region) and the upper Black River.

Habitat

Taiga voles inhabit fire-successional and riparian, boreal, sphagnum forest habitats near streams and other moist areas (Conroy and Cook 1999; Wolff 1999; Youngman 1975). Like singing voles (*M. miurus*), taiga voles are semicolonial and vocal (Youngman 1975).

Status

Microtus xanthognathus is known from scattered localities within their relatively limited distribution. Their preferred habitat may be short-lived due to successional change, so populations can be ephemeral, patchy, and unpredictable (Wolff 1999). *Microtus xanthognathus* may maintain high densities (Lensink 1954) or fluctuate considerably between years (Rand 1945, 1948; Wolff and Lidicker 1980; Youngman 1975).

Fossils

The few fossils known for this species were reviewed by Conroy and Cook (1999). *Microtus xanthognathus* occurred south of the Laurentide ice sheet during the Late Pleistocene (Guilday et al. 1977), and may have been present in Beringia as well. Fossil teeth from late Illinoian to Holocene are known from Alaska (Guilday et al. 1964; Zakrzewski 1985); however, given the difficulty of distinguishing species of *Microtus* from small series of teeth (Guilday 1982), these records may not predate the Wisconsinan (Guilday et al. 1977;

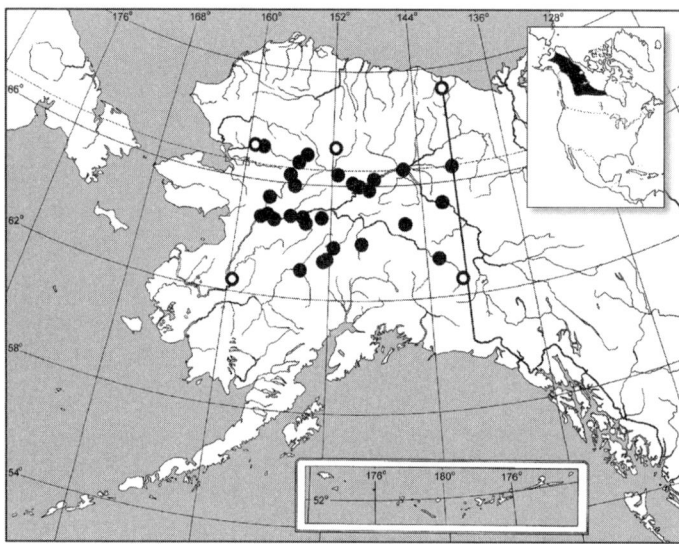

MAP 19. Distribution of taiga vole, *Microtus xanthognathus*.

⊙ = Type Locality
● = Specimen Record (n = 2606)
○ = Marginal Records. 69.33300, –141.00000, Demarcation Point Quad., Alaska–Yukon Boundary (1 USNM); 62.52500, –141.26670, Nabesna Quad., Wellesley Lake (83 UAM); 62.20000, –159.71667, Holy Cross, confluence of Innoko River and Red Wing Slough (1 UAM); 67.08972, –159.78083, Baird Mts Quad., confluence of Kallarichuk River and Kobuk River (10 UAM); 67.20017, –151.74217, Wiseman Quad., ¾ mi N of confluence of Suckik Creek and Timber Creek, 20 mi N of Bettles (1 UAM).

Richards 1988). A mummified Pleistocene *M. xanthognathus* (AMNH 180252) was collected in Chicken, Alaska (Guilday and Bender 1960; Youngman 1975).

Southern Red-Backed Vole

Myodes gapperi
(Vigors, 1830)

Other Common Names

Boreal red-backed vole, Gapper's red-backed vole, southern red-backed mouse.

Systematics

Musser and Carleton (2005), after a review of Pallas (1811) and other authorities, resurrected *Myodes* as the long-overlooked and valid genus for red-backed voles, replacing the more familiar *Clethrionomys*.

Interspecific and intraspecific taxonomic relationships of *Myodes* have been problematic (Cook et al. 2004; Runck 2001). Bee and Hall (1956) and Youngman (1975) considered *gapperi* and *rutilus* conspecific, but did not systematically assess their status.

Runck (2001, 2006) examined phylogeographic variation in this species, with a particular focus on a zone of intergradation with *M. rutilus* on the mainland of southeast Alaska

between Juneau and the southern Behm Canal. In that region, these species are difficult to distinguish morphologically and limited introgression may occur (based on the mitochondrial cytochrome *b* gene). Minimal (but potentially diagnostic) molecular variation distinguishes the island forms from the mainland populations.

Cook et al. (2004) and Runck and Cook (2005) not only supported the specific distinctiveness of *gapperi* and *rutilus*, but also indicated that *gapperi* as currently recognized likely consists of at least three cryptic species in North America. Southeast Alaska populations are the result of western postglacial expansion via trans-mountain river corridors from Pleistocene refugia in the midwestern United States.

Hall (1981) recognized twenty-nine subspecies of southern red-backed vole; four, possibly five, occur in southeast Alaska.

Myodes gapperi phaeus

Original Description. 1911. *Evotomys phaeus* Swarth, *Univ. California Publ. Zool.*, 7:127, January 12.

Type Locality. Marten Arm, Boca de Quadra, Alaska.

Type Specimen. MVZ 8742.

Range. Extreme southern mainland of southeast Alaska and adjacent British Columbia.

Myodes gapperi saturatus

Original Description. 1894. *Evotomys gapperi satu-ratus* Rhoads, *Proc.Acad.Nat.Sci.Philadelphia*, 46:284, October 23.

Type Locality. Nelson, on the banks of a small stream flowing into Kootenai [sic] Lake, British Columbia.

Type Specimen. ANSP 7483.

Range. Central British Columbia southward into northeastern Washington, northern Idaho, and extreme northwestern Montana.

Remarks. This taxon may occur in the upper reaches of Portland Canal based on the range map in Hall (1981:782).

Myodes gapperi solus

Original Description. 1952. *Clethrionomys gapperi solus* Hall and Cockrum, *Univ. Kansas Publ., Mus. Nat. Hist.*, 5:304, November 17.

Type Locality. Loring, Revillagigedo Island, Alaska.

Type Specimen. USNM 74939.

Range. Restricted to Revillagigedo Island.

Myodes gapperi stikinensis

Original Description. 1952. *Clethrionomys gapperi stikinensis* Hall and Cockrum, *Univ. Kansas Publ., Mus. Nat. Hist.*, 5:305, November 17.

Type Locality. Stikine River at Great Glacier, British Columbia.

Type Specimen. MVZ 30735.

Range. Stikine River south from the Flood Glacier, British Columbia (but not including Sergief Island), south to Cleveland Peninsula, Alaska.

Myodes gapperi wrangeli

Original Description. 1897. *Evotomys wrangeli* V. Bailey, *Proc. Biol. Soc. Washington*, 11:130, May 13.

Type Locality. Wrangell, Wrangell Island, Alaska.

Type Specimen. USNM 74724.

Range. Etolin Island (Runck 2001), Wrangell Island, and presumably nearby Sergief Island at the mouth of the Stikine River (Swarth 1922).

Global Distribution

Nearctic—The southern red-backed vole occurs broadly from southeast Alaska across boreal Canada and the contiguous United States.

Alaska Distribution

The southern red-backed vole is restricted to southeast Alaska (map 20). It occurs on the mainland of southeast Alaska from about the Stikine River south, and on a few islands south of Stikine Strait and east of Clarence Strait, namely Bell, Black, Deer, Etolin, Hassler, Misery, Revillagigedo, and Wrangell islands (MacDonald and Cook 2007).

Habitat

Red-backed voles are habitat generalists, occupying a wide variety of habitats but especially forest, woodland, and shrub habitats.

Status

Myodes gapperi is common, occasionally abundant. The subspecies *M. g. solus* is listed as

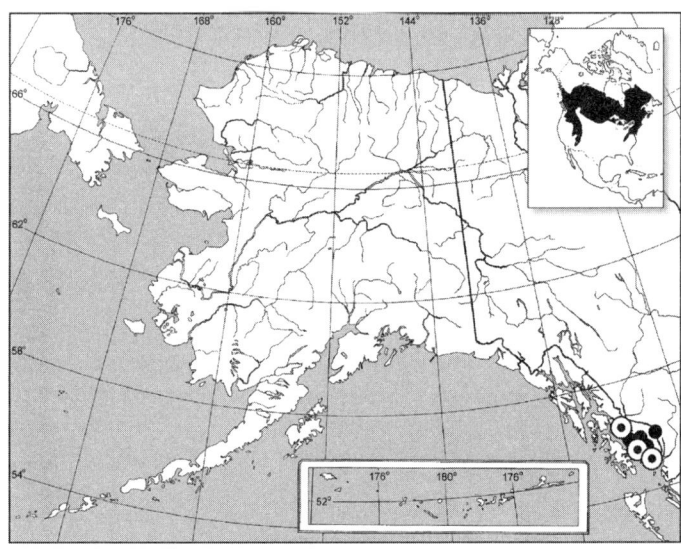

MAP 20. Distribution of southern red-backed vole, *Myodes gapperi*.

⊙ = Type Locality
● = Specimen Record (n = 1358)

Data Deficient by IUCN (Cook and Kirkland 1998).

Fossils

Fossils of *M. gapperi* are known from Irvingtonian (late Kansan) age to early Holocene from a wide variety of sites south of Alaska (Merritt 1981). Heaton and Grady (2003) recovered postglacial remains believed to be of this species from limestone deposits on Wrangell Island.

Northern Red-Backed Vole

Myodes rutilus
(Pallas, 1779)

Other Common Names

Dawson red-backed mouse, northern red-backed mouse, tundra redback vole.

Systematics

See southern red-backed vole account on the use of *Myodes* as the valid genus for red-backed voles. The Holarctic distribution of *M. rutilus* was first advanced by Rausch (1950, 1953a) and later corroborated by Nadler et al. (1976, 1978), Rausch and Rausch (1975a), and Cook et al. (2004).

Hall (1981) recognized eight subspecies in North America; six occur in Alaska. Rausch and Rausch (1975a) noted that the continental subspecies in North America are weakly differentiated.

Myodes rutilus albiventer

Original Description. 1932. *Clethrionomys albiventer* Hall and Gilmore, *Univ. California Publ. Zool.*, 38:398, September 17.
Type Locality. Sevoonga [= Savoonga], 2 mi. E of North Cape, St. Lawrence Island, Bering Sea, Alaska.
Type Specimen. MVZ 51221.
Range. Restricted to St. Lawrence Island.

Myodes rutilus dawsoni

Original Description. 1888. *Evotomys dawsoni* Merriam, *Amer. Nat.*, 22:650, July.
Type Locality. Finlayson River, 3000 ft., a northern source of Liard River, lat. 61°30' N, long. 129°30' W, Yukon.
Type Specimen. NMC 92.
Range. Most of Alaska, northwestern Canada.
Remarks. Includes *Evotomys alascensis* Miller, 1898, *Proc. Acad. Nat. Sci. Philadelphia*, 50:364, October 15, type from St.

Michael, Norton Sound, Alaska (USNM 14359/22226) (see Osgood 1904:34).

Myodes rutilus glacialis

Original Description. 1945. *Clethrionomys dawsoni glacialis* Orr, *J. Mammalogy*, 26:71, February 27.
Type Locality. Glacier Bay, Alaska.
Type Specimen. MVZ 388.
Range. Glacier Bay area of southeast Alaska.
Remarks. An unpublished revision of *Clethrionomys rutilus* by Antell (1987) extended the distribution of the subspecies *C. r. glacialis* to include all populations from Yakutat south to Bartlett Cove.

Myodes rutilus insularis

Original Description. 1910. *Evotomys dawsoni insularis* Heller, *Univ. California Publ. Zool.*, 5:339, March 5.
Type Locality. W side Canoe Passage, Hawkins Island, Prince William Sound, Alaska.
Type Specimen. MVZ 557.
Range. Islands in Prince William Sound.

Myodes rutilus orca

Original Description. 1900. *Evotomys orca* Merriam, *Proc. Washington Acad. Sci.*, 2:24, March 14.
Type Locality. Orca, Prince William Sound, Alaska.
Type Specimen. USNM 98028.
Range. Prince William Sound area.

Myodes rutilus watsoni

Original Description. 1945. *Clethrionomys dawsoni watsoni* Orr, *J. Mammalogy*, 26:73, February 27.
Type Locality. Cape Yakataga, Alaska.
Type Specimen. CAS 8968.
Range. Known only from type locality.

Global Distribution

Holarctic—Northern red-backed voles occur broadly across northern Eurasia (including Sakhalin and Hokkaido islands), Alaska, and northwestern Canada east to Hudson Bay (Henttonen and Peiponen 1982; Musser and Carleton 2005).

Alaska Distribution

This vole is one of Alaska's most ubiquitous and common species, occurring extensively throughout most of the state and at most elevations (map 21).

Southeast Region. The northern red-backed vole occurs along the northern mainland of the region southward to LeConte Bay near the Stikine River (MacDonald and

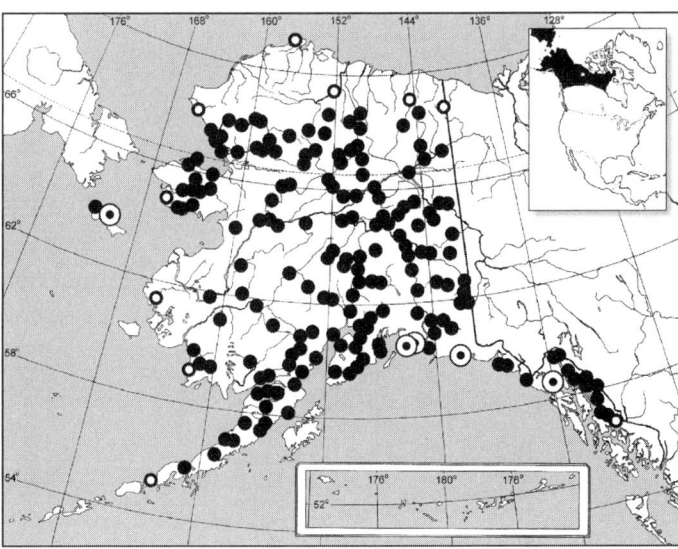

MAP 21. Distribution of northern red-backed vole, *Myodes rutilus*.

⊙ = Type Locality
● = Specimen Record (n = 14,690)
○ = Marginal Records. 71.30420, –156.59330, Barrow Quad., Pt. Barrow (1 KU); 69.37170, –152.13610, Umiat Quad., Umiat (8 KU); 69.08333, –144.61667, Mt Michelson Quad., upper Hulahula River (3 UAM); 68.66667, –141.25000, Table Mtn Quad., Mancha Creek (6 UAM); 56.77278, –132.60750, Petersburg Quad., Jap Creek, 4 mi. NW Le Conte Bay (2 UAM); 55.15000, –162.81250, Cold Bay Quad., Izembek (5 UAM); 59.08053, –161.89119, Goodnews Bay Quad., north spit of Goodnews Bay (1 UAM); 61.33333, –165.16667, Hooper Bay Quad., Old Chevak (4 UAM); 64.80000, –166.46667, Nome Quad., Cape Wooley (4 UAM); 68.10000, –165.75000, Point Hope Quad., Cape Thompson (22 UAM).

Cook 2007; Runck 2006). The only insular populations documented in the region are from Douglas Island and Young Island (one of the Beardslee Islands in Yakutat Bay).

Southcentral Region. *Myodes rutilus* is found throughout the mainland and on Chenega, Esther, Evans, Hawkins, Hinchinbrook, Knight, and Latouche islands in Prince William Sound; Hesketh Island in Kachemak Bay; and Chisik Island near Tuxedni Bay on the west side of Cook Inlet (Appendix 9).

Central Region. Northern red-backed voles occur extensively throughout central Alaska.

Southwest Region. This species is found throughout the region, except on islands.

Western Region. This vole occurs throughout the mainland and on St. Lawrence Island in the Bering Sea.

Northern Region. Northern red-backed voles are found throughout northern Alaska but are most numerous in the Brooks Range and along the major river valleys (Bee and Hall 1956).

Habitat

Red-backed voles inhabit a wide range of habitats but are generally most abundant in forest, woodland, and shrub habitats (Guthrie 1968a; Kessel et al. 1982; MacDonald 1980; West 1974, 1979).

Status

Northern red-backed voles are very common, periodically abundant across their broad range.

Fossils

Fossils of *Myodes*, not assigned to species, have been reported from a number of localities in central Alaska and Yukon (Georgina 2001; Harington 1990; Kurtén and Anderson 1980).

Bushy-Tailed Woodrat

Neotoma cinerea
(Ord, 1815)

Other Common Names

Packrat.

Systematics

Carleton (1980) and Hall (1981) recognized thirteen subspecies of *Neotoma cinerea* across its range; one occurs in Alaska.

Neotoma cinerea occidentalis

Original Description. 1855. *Neotoma occidentalis* Baird, *Proc. Acad. Nat. Sci. Philadelphia*, 7:335, April.

Type Locality. Shoalwater [= Willapa] Bay, Pacific County, Washington.

Type Specimen. USNM 572.

Range. Mackenzie, Northwest Territories, southward along coastal southeast Alaska, coastal and central British Columbia to central Oregon.

Remarks. Includes *Neotoma saxamans* Osgood, 1900, *N. Amer. Fauna*, 19:33, October 6, type from Bennett City, head Lake Bennett, British Columbia (USNM 98923). This taxon was regarded as inseparable from *N. c. occidentalis* by Cowan and Guiguet (1965) and Youngman (1975).

Global Distribution

Nearctic—Bushy-tailed woodrats occur in western North America, from westernmost Northwest Territories and southern Yukon Territory southward to California, Arizona, and New Mexico.

Alaska Distribution

Neotoma cinerea is known only from along the coastal mainland in the southeast region of the state (map 22), but the Wrangell–St. Elias Mountains in southcentral Alaska need to be more thoroughly surveyed.

The only published specimens are reported by Shaw (1962) from the Taku River (collected in 1940) and the Unuk River (1925). In 1963 and 1969, specimens of this woodrat were taken on nunataks in the Juneau Ice Fields (PSM).

A bushy-tailed woodrat collected at the head of Lake Bennett, British Columbia, and one seen at Glacier on the White Pass and Yukon Railroad just inside Alaska (Osgood 1900) indicate a possible mainland distribution at least this far north. Dufresne (1946:138–139) stated that woodrats were "fairly common near the head of Portland Canal and along the Unuk River. It has also been reported from the Stikine River and Taku River watersheds."

Habitat

Bushy-tailed woodrats are found in rocky substrates and occasionally in deserted buildings and mine shafts, from sea level to the summit of mountains (Banfield 1974; Smith 1997).

Status

The status of this species in Alaska is unknown.

Fossils

Fossils of this species are known from late Rancholabrean and Holocene deposits in Wyoming, Idaho, Colorado, New Mexico, and California (Kurtén and Anderson 1980; Smith 1997).

Common Muskrat

Ondatra zibethicus
(Linnaeus, 1766)

Other Common Names

Muskrat.

Systematics

Pietsch (1970) and Boyce (1978) examined patterns of skull variation throughout North America. Hall (1981) recognized sixteen subspecies, of which two occur in Alaska. Pietsch (1970) validated only twelve of these subspecies. A revision using modern techniques is needed.

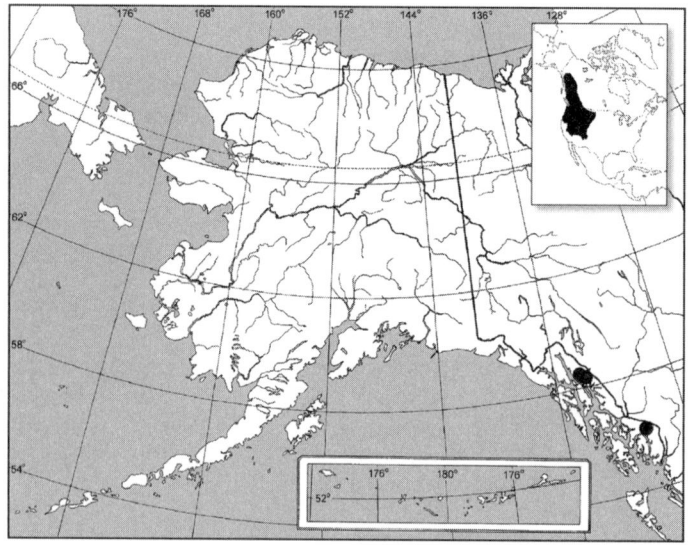

MAP 22. Distribution of bushy-tailed woodrat, *Neotoma cinerea*.

● = Specimen Records (n = 6): 58.50000, –134.25000, Juneau Quad., Juneau Ice Fields (4 PSM); 58.48300, –133.96700, Taku River Quad., mouth lower Taku River (1 USNM); 56.25000, –130.56000, Bradfield Canal Quad., Unuk River (1 USNM).

Ondatra zibethicus spatulatus

Original Description. 1900. *Fiber spatulatus* Osgood, *N. Amer. Fauna*, 19:36, October 6.

Type Locality. Lake Marsh, Yukon Territory.

Type Specimen. USNM 98567.

Range. Western Alaska and northwestern Canada.

Ondatra zibethicus zalophus

Original Description. 1910. *Fiber zibethicus zalophus* Hollister, *Proc. Biol. Soc. Washington*, 23:1, February 2.

Type Locality. Becharof Lake, Alaska Peninsula, Alaska.

Type Specimen. USNM 131488.

Range. Southern Alaska.

Global Distribution

Nearctic—Muskrats inhabit most of North America, ranging from Alaska southward to the Mexican border (Hall 1981). Muskrats have been introduced to numerous locations throughout the Palearctic region, including the Russian Far East (Chernyavskyi 1984).

Alaska Distribution

Muskrats are found throughout most of the Alaska mainland south of the Brooks Range (map 23).

Southeast Region. This species has a limited distribution in southeast Alaska. Documented reports include Yakutat Bay, Haines area, Taku River, Stikine River (including the delta islands of Farm and Sergief), Admiralty Island, and Revillagigedo Island (MacDonald and Cook 1996). Muskrats have also been reported but no specimens secured from Thomas Bay and Kuiu, Kupreanof, Mitkof, Woewodski, and Wrangell islands (MacDonald and Cook 2007).

There were unsuccessful attempts in 1929 to transplant muskrats from Haines to Klawock Lake on Prince of Wales Island (Burris and McKnight 1973).

Southcentral Region. Muskrats are scattered but locally common on the Copper River Delta (Crowley 2004a). They are not documented on any of the islands or mainland of Prince William Sound (Crowley 2004a; Heller 1910). Muskrats have been collected on the Kenai Peninsula as far south as Seldovia (Allen 1904).

Central Region. Yukon Flats, Minto Flats, and Tetlin Lakes are considered some of the most productive areas for muskrats in the state (ADFG 1978). The northern limit of this species in the region is unknown,

but they have been reported as common around Arctic Village (Rausch 1953a), and apparently occur in the John River as far upstream as Hunt Fork (Bee and Hall 1956). Muskrats occur in the upper reaches of the Old Crow River in the far northeastern part of the region (Youngman 1975).

Southwest Region. Muskrats are not known south of Ugashik Lakes on the Alaska Peninsula (ADFG 1978). Murie (1959) was told by residents of False Pass that muskrats were found west to Port Moller, a claim not supported by specimens. Murie (1959) also reported that muskrats were unsuccessfully released on Unalaska Island sometime before 1920.

The Alaska Game Commission transplanted muskrats from the Copper River area to Kodiak, Afognak, Whale, and Spruce islands in 1925; animals were relocated from Long Island to Kodiak Island in 1929 (Burris and McKnight 1973). By 1943 the status of these populations was described as "excellent; abundant and spreading" (Burris and McKnight 1973). Van Daele (2004) indicted that viable populations still occurred in the archipelago.

Western Region. Muskrats have been reported from Kilikmak Creek (UAM), Noatak River (Gardner 1974), Kobuk River (Dean and Chesemore 1974), and from the Seward Peninsula (Quay 1951; UAM). The Yukon-Kuskokwim Delta is considered one of the most productive areas for muskrats in Alaska (ADFG 1978).

An early attempt to translocate seven muskrats from the Nushagak area to the Pribilof Islands in 1913 was a complete failure (Burris and McKnight 1973; Preble 1923).

Northern Region. A muskrat was collected in a lagoon near the mouth of Kissimilouk Creek east of Cape Lisburne (UAM). Bee and Hall (1956) mentioned an unsubstantiated report of muskrats on the Arctic Slope north of the Brooks Range.

Habitat

Muskrats inhabit fresh, brackish, and saltwater marshes, ponds, lakes, rivers, and streams (Banfield 1974; Willner et al. 1980).

Status

This species is widespread and common, sometimes abundant, in suitable habitat.

Fossils

Fossils of muskrats that date back to the latest Illinoian have been reported from sites in

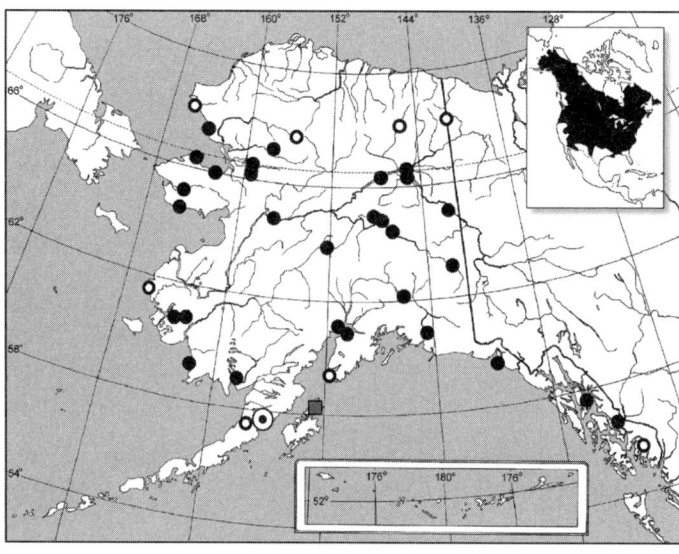

MAP 23. Distribution of common muskrat, *Ondatra zibethicus*.

⊙ = Type Locality
● = Native, ■ = Non-native Specimen Record (n = 706)
○ = Marginal Records. 68.58333, –165.49025, Point Hope Quad., Ogotoruk Creek region, Kisimilok (sic) Creek, pond in lagoon at mouth (2 UAM); 67.23972, –157.61778, Ambler River Quad., BSMS camp 3, Redstone Lake (1 UAM); 68.13300, –145.53300, Arctic Quad., Arctic Village (4 PSM); 68.20000, –141.00000, Table Mtn Quad., Old Crow Flats, Int. Boundary (1 USNM); 55.77360, –131.04170, Ketchikan Quad., Revillagigedo Island, Portage Cove (5 MVZ); 59.43750, –151.70830, Seldovia Quad., Seldovia (1 AMNH); 58.25000, –152.50000, Afognak Quad., Afognak Island [from introduced population] (1 USNM; 2 CMNH); 57.51250, –157.39580, Ugashik Quad., Ugashik, Alaska Peninsula (1 USNM); 61.50000, –166.00000, Hooper Bay Quad., Hooper Bay (1 UAM).

northern Yukon Territory (Jopling et al. 1981), and from the Early Holocene at Lime Hills in central Alaska (Georgina 2001).

Northwestern Deermouse

Peromyscus keeni
(Rhoads, 1894)

Other Common Names

Keen's mouse, white-footed mouse, deer mouse, forest deer mouse, northwestern deer mouse, Sitka mouse, Sitka deer mouse, Sitka white-footed mouse.

Systematics

Two species and five subspecies of *Peromyscus* were once recognized in the southeast region of Alaska (Hall 1981). Hogan et al. (1993) found chromosome, allozyme, and mitochondrial DNA variation supporting the hypothesis that these north-coastal subspecies of *P. maniculatus* and *P. sitkensis* are conspecific. They included them under *P. keeni*, which was recognized as the senior synonym.

Currently described subspecies for the region (as modified from Hall 1981) are listed below. A molecular study that analyzed cytochrome *b* sequences from northwestern deermice from twenty-three islands and six mainland localities (Lucid and Cook 2004) found most island populations to be geneti-

cally distinctive. Areas with divergent populations were largely inconsistent with three of five currently recognized subspecies. Cryptic variation was detected in eight areas not previously identified by morphologic analyses.

Peromyscus keeni algidus

Original Description. 1909. *Peromyscus maniculatus algidus* Osgood, *N. Amer. Fauna*, 28:56, April 17.
Type Locality. Head of Lake Bennett (site of old Bennett City), British Columbia.
Type Specimen. USNM 130013.
Range. Between upper Lynn Canal, Alaska, and southcentral Yukon Territory.

Peromyscus keeni hylaeus

Original Description. 1908. *Peromyscus hylaeus* Osgood, *Proc. Biol. Soc. Washington*, 21:141, June 9.
Type Locality. Hollis, Kasaan Bay, Prince of Wales Island, Alaska.
Type Specimen. USNM 127038.
Range. Prince of Wales Island northward through the central Alexander Archipelago to Glacier Bay.

Peromyscus keeni macrorhinus

Original Description. 1894. *Sitomys macrorhinus* Rhoads, *Proc. Acad. Nat. Sci. Philadelphia*, 46:259, October.
Type Locality. Mouth of Skeena River, British Columbia.
Type Specimen. ANSP 8381.

Range. Mainland coast and adjacent islands of southeast Alaska and British Columbia.

Peromyscus keeni oceanicus
Original Description. 1935. *Peromyscus sitkensis oceanicus* Cowan, *Univ. California Publ. Zool.*, 40:432, November 14.
Type Locality. Forrester Island, Alaska.
Type Specimen. MVZ 20890.
Range. Known only from type locality.

Peromyscus keeni sitkensis
Original Description. 1897. *Peromyscus sitkensis* Merriam, *Proc. Biol. Soc.Washington*, 11:223, July 15.
Type Locality. Sitka, Alaska.
Type Specimen. USNM 73809.
Range. Baranof, Chichagof, Warren, Coronation, and Duke islands of the Alexander Archipelago.

Global Distribution
Nearctic—Peromyscus keeni ranges from southeast Alaska southward along the Pacific Coast to Washington (Hogan et al. 1993).

Alaska Distribution
The Northwestern deermouse is widely distributed and common throughout most of southeast Alaska (map 24), occurring on the mainland as far north as the upper Lynn Canal and Glacier Bay near Tlingit Point and Muir Inlet. In the Alexander Archipelago, it is documented on Admiralty, Anguilla, Annette, Baker, Baranof, Betton, the Brothers (East and West),

Bushy, Cat, Chichagof, Coronation, Dall, Deer, Dog, Douglas, Duke, Esquibel, Etolin, Forrester, Goat, Gravina, Heceta, Inian, Kadin, Kosciusko, Kruzof, Kuiu, Kupreanof, Lincoln, Long, Lowrie, Lulu, Marble, Mary, Mitkof, Moser, Noyes, Orr, Partofshikof, Pow, Prince of Wales, Revillagigedo, St. Ignace, San Fernando, San Juan Bautista, Santa Rita, Sergief, Shelter, Shrubby, Spanish, Suemez, Sukkwan, Swan, Thorne, Tuxekan, Vank, Warren, Woewodski, Woronkofski, Wrangell, and Zarembo islands (MacDonald and Cook 2007).

Habitat
Peromyscus keeni inhabits an extremely wide variety of habitats at various elevations. In southeast Alaska, they appear to favor the forest-beach edge and also have been observed high in large spruce and hemlock trees. They are infamous for inhabiting human dwellings.

Status
Northwestern deermice are common and ubiquitous throughout their range.

Fossils
Storer (2003b) reported *Peromyscus* represented in pre-Wisconsinan deposits from Yukon Territory, and undated skeletal remains have been recovered from limestone cave deposits in southeast Alaska (Heaton and Grady 2003).

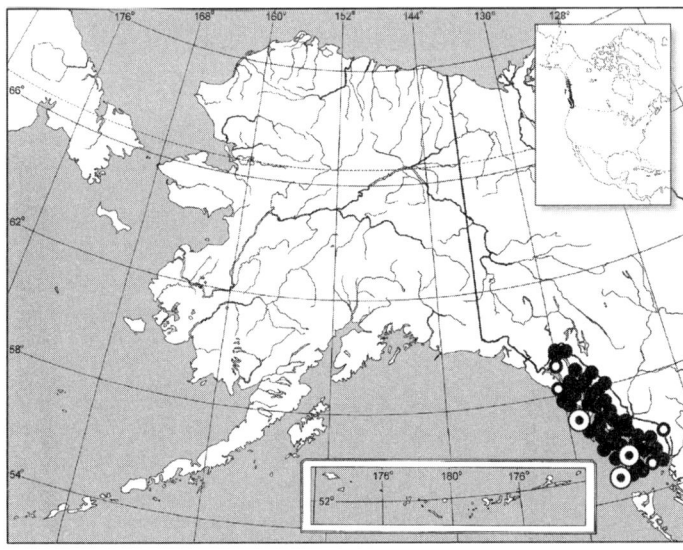

MAP 24. Distribution of northwestern deermouse, *Peromyscus keeni*.

◉ = Type Locality
● = Specimen Record (n = 6769)
○ = Marginal Records. 59.04000, −136.21670, Skagway Quad., Glacier Bay, Muir Glacier moraine (1 CAS); 56.01667, −130.06667, Bradfield Canal Quad., Salmon River Valley, Hyder area (17 UAM); 54.97500, −131.33333, Prince Rupert Quad., Duke Island, Pond Bay (46 UAM); 58.25833, −136.29861, Mt. Fairweather Quad., Inian Island (2 UAM).

North American Deermouse

Peromyscus maniculatus
(Wagner, 1845)

Other Common Names

White-footed mouse.

Systematics

The subspecific status of Alaska populations is undetermined. *Peromyscus maniculatus borealis* has been reported to occur near the Alaska border in Yukon Territory (Slough and Jung 2007; Youngman 1975). *Peromyscus maniculatus algidus* of southwestern Yukon and northwestern British Columbia (Hall 1981) were found to be genetically distinct from both Northwestern deermice, *P. keeni*, and other populations of *P. maniculatus* based on analysis of cytochrome *b* sequence variation (Wike 1998; Lucid and Cook 2004; Lucid and Cook 2007). These far-northwestern populations in Yukon Territory and extreme northern British Columbia may constitute a distinct species, but additional work is needed to delineate their distribution and status.

Global Distribution

Peromyscus maniculatus as currently recognized (Hall 1981) occurs naturally from northwestern Canada to Central America.

Alaska Distribution

Deermice are found close to the Alaska border in central Yukon Territory and far northwestern British Columbia (Nagorsen 2005; Slough and Jung 2007; Youngman 1975); however, as discussed above, the taxonomic affinities of these populations remain unresolved.

Southcentral Region. On 13 June 2008, four *Peromyscus maniculatus* were captured in the Copper River Basin at two sites approximately 1.5 km apart in the vicinity of Milepost 51 of the Richardson Highway near the Tiekel River and Tiekel (map 25). The full extent of this population's occurrence in the area remains to be determined (MacDonald et. al., in press). Previous trapping efforts (Cook and MacDonald 2003; Laing and Anderson 1929; McDonough and Rexstad 2005; UAM unpublished) have not documented deermice elsewhere in the region.

Central Region. Deermice are found close to the Alaska-Yukon border in the vicinities of Dawson, along the Alaska Highway at the Donjek River crossing (Youngman 1975), and most recently in the Ogilvie Mountains (Slough and Jung 2007). Dixon (1938) reported seeing (but did not collect) a deermouse in a cabin at the Kantishna Ranger Station, Denali National Park, in August 1932. From this one sighting, he concluded that this mouse was a rare inhabitant of the park; however, this lone central Alaska record has long been considered questionable and probably the result of human transport from elsewhere.

Southwestern Region. An introduced population of *P. maniculatus* was first discovered in the Aleutian Islands on Shemya Island in 1978. Since then, specimens have been sampled there in 1995, 1999, and 2001 (all UAM).

Habitat

The North American deermouse exhibits a broad tolerance for different habitats as long as they are dry (Banfield 1974). The newly discovered population in the lower Copper River Basin near the foothills of the Chugach Mountains was found at 400 m in elevation in a small, well-drained meadow of emerging cow parsnip (*Heracleum lanatum*) with scattered thickets of high bush cranberry (*Viburnum edule*) and along the dry, grassy edge of the cleared easement for the trans-Alaska pipeline.

Status

The status and distribution of the newly discovered population of deermice in the Copper River Basin have not been fully resolved; however, MacDonald et al. (in press) provided genetic evidence supportive of their specific status as *P. maniculatus* more closely aligned with unambiguous *P. maniculatus* samples 600 km to the east near Atlin, British Columbia, than to samples referable to either *P. keeni* or the unnamed cryptic species from 400 km away in western Yukon Territory and extreme northwestern British Columbia. This finding, in conjunction with the current lack of deermice records elsewhere in far-eastern central or southcentral Alaska, suggested that this population was originally introduced by human agency, perhaps as stowaways in cargo used in the construction of the trans-Alaska pipeline in the mid-1970s.

The introduced population of deermice on Shemya Island in southwestern Alaska are from an unknown source and may be expanding their range beyond the immediate vicinity of the Shemya Airforce Base (D.D. Gibson, pers. comm., 2001).

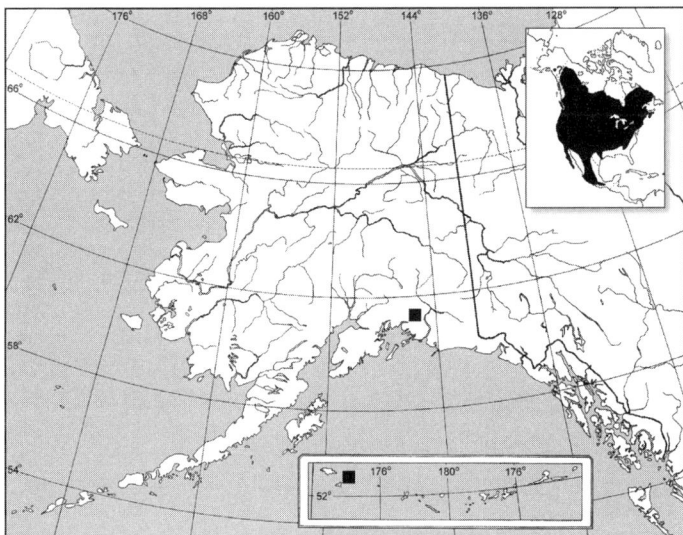

MAP 25. Distribution of North American deermouse, *Peromyscus maniculatus*.

■ = Non-native Specimen Record (n = 36): 61.34785, −145.30388, Copper River Basin, Tiekel River near Tiekel (3 MSB); 61.34062, −145.30719, Copper River Basin, Tiekel River near Tiekel (1 MSB); 52.71667, −174.11667, Aleutian Islands, Shemya Island (1 UAM); 52.72222, −174.11667, Shemya Island (23 UAM); 52.72417, −174.11889, Shemya Island (8 UAM).

Fossils

Fossils of Late Pleistocene age are lacking from Alaska in eastern Beringia (Appendix 12). Considered a post-glacial immigrant from a southern refugia by Youngman (1975), fossil remains of *Peromyscus* sp. in pre-Wisconsinan (last interglacial) deposits near Thistle Creek in Yukon Territory were reported by Storer (2003a).

Western Heather Vole

Phenacomys intermedius
Merriam, 1889

Other Common Names

Heather vole.

Systematics

The taxonomic relationship among eastern and western North American populations of *Phenacomys* remains problematic. Most authors have regarded *ungava* as a subspecies of *intermedius*, whereas Anderson (1946), Peterson (1966), Cowan and Guiguet (1965), Musser and Carleton (1993, 2005), and George (1999b) recognized the specific distinctiveness of the two forms. Further efforts are needed to help resolve this issue.

The taxonomic status of Alaska populations has not yet been determined. An adequate series of specimens from the newly discovered populations in southeast Alaska needs to be

secured and then compared with series from nearby British Columbia, Yukon Territory, and elsewhere.

By inference from Hall (1981), one subspecies may occur in the state.

Phenacomys intermedius intermedius

Original Description. 1889. *Phenacomys intermedius* Merriam, *N. Amer. Fauna*, 2:32, October 30.

Type Locality. A basaltic plateau, 5500 ft., about 20 mi. NNW Kamloops, British Columbia.

Type Specimen. NMC 780.

Range. Central British Columbia south to New Mexico.

Global Distribution

Nearctic—Heather voles occur widely across boreal Canada (*P. ungava*) and the mountainous regions of the western United States and throughout most of British Columbia's mainland (*P. intermedius*) (Hall 1981; Nagorsen 2005).

Alaska Distribution

The occurrence of this species in Alaska is currently known from four specimens recently collected at two mainland localities in southeast Alaska (MacDonald et al. 2004) (map 26).

In 1995, this species was found by UAM in subalpine habitat 2 km inside Canada near Hyder, Alaska. In 1996, a heather vole was trapped above tree line in the Chilkat Range near Excursion Inlet, Glacier Bay National Park. In 1999, three specimens were collected (UAM) from the far southern end of the

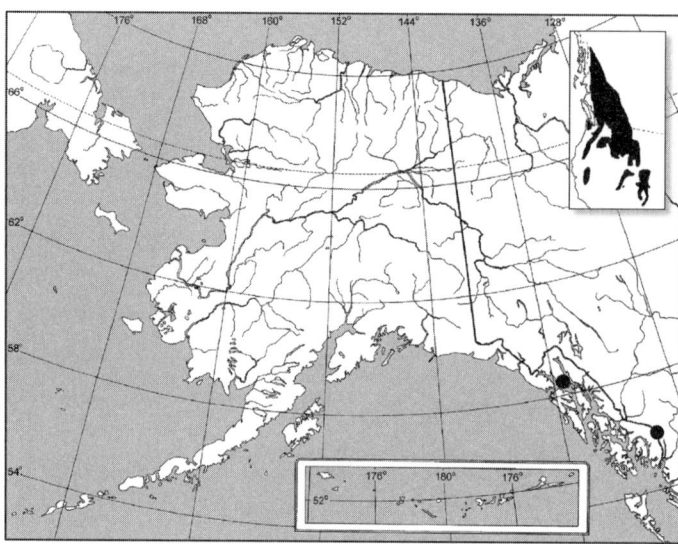

MAP 26. Distribution of western heather vole, *Phenacomys intermedius.*

● = Specimen Record (n = 4):58.41667, –135.43333, Juneau Quad., Excursion Inlet, E side in Chilkat Range (1 UAM); 55.83, –130.05, Ketchikan Quad., Reverdy Mts., Titan Trail N of Hyder, T68S R100E, Sec6 NW ¼ of NW ¼ (3 UAM).

region along the Titan Trail at elevations between 1,128 and 1,220 m near Hyder on 23 September (MacDonald et al. 2004). Further trapping along the coastal mountains of southeast Alaska, and perhaps also in the eastern Wrangell–St. Elias area of southcentral Alaska, should expand our understanding of the status and distribution of this vole within the state.

Habitat

Heather voles usually occur in the mountains near or above tree line in a variety of habitats, including dry coniferous forest and woodlands, meadows near water, and dwarf-low shrub tundra (Banfield 1974; McAllister and Hoffmann 1988; Nagorsen 2005; Youngman 1975).

Status

The status of this species in Alaska is unknown.

Fossils

The fossil record of *Phenacomys* extends to the Middle Pleistocene (Irvingtonian) in most of North America (Martin et al. 1986). Specimens of extinct species of *Phenacomys* from sites in the East Beringian refugium were first found along the Old Crow River and are of Olduvai age (1.67–1.89 million years old; in McAllister and Hoffmann 1988). Guthrie and Matthews (1971) described an extinct species from the Seward Peninsula that later was included under this genus and dated about 2.1 million years ago. Repenning et al. (1987) described a new species of extinct *Phenacomys*

from the Fish Creek Fauna on the Arctic coast of northern Alaska that dated to about 2.4 million years ago. They agreed with Martin et al. (1986) that the members of this genus may be of Asian origin.

Fossils of heather voles have been reported from limestone caves on Prince of Wales Island, Alexander Archipelago, southeast Alaska. All dated material is older than the last glacial maximum (Heaton and Grady 2003).

Northern Bog Lemming

Synaptomys borealis
(Richardson, 1828)

Other Common Names

Lemming mouse.

Systematics

Jarrell and Fredga (1993), following Koenigswald and Martin (1984) and Repenning and Grady (1988), consider *Mictomys* the appropriate generic name for this North American species.

Hall (1981) recognized nine subspecies; two occur in Alaska.

Synaptomys borealis dalli

Original Description. 1896. *Synaptomys (Mictomys) dalli* Merriam, *Proc. Biol. Soc. Washington*, 10:62, March 19.

Type Locality. Nulato, Alaska.

Type Specimen. USNM 49373.

Range. Central Alaska, Yukon Territory and British Columbia.

Synaptomys borealis truei

Original Description. 1896. *Synaptomys (Mictomys) truei* Merriam, *Proc. Biol. Soc. Washington*, 10:63, March 19.

Type Locality. Skagit Valley, Skagit County, Washington.

Type Specimen. USNM 3798/12101.

Range. Southeast Alaska, coastal British Columbia, and northwestern Washington.

Remarks. Includes *Synaptomys (Mictomys) wrangeli* Merriam, 1896, *Proc. Biol. Soc. Washington*, 10:63, March 19, type from Wrangell, Alexander Archipelago, Alaska (USNM 74720).

Global Distribution

Nearctic—This species occurs from Alaska across boreal Canada into the northern contiguous United States (Hall 1981).

Alaska Distribution

The distribution of the northern bog lemming across forested Alaska (map 27) remains poorly documented.

Southeast Region. Mainland records range from Portland Canal northward to White Pass and the upper Lynn Canal. Until recently the only island records were from Wrangell Island (MacDonald and Cook 1996). In 1995, a year of notable bog lemming abundance, animals were seen and collected from estuarine meadows on Betton, Back, and Gravina islands (UAM). In 1999, a bog lemming was secured from Revillagigedo Island (UAM), and in 2004–2005 individuals were taken on Kupreanof and Kuiu islands (MacDonald and Cook 2007).

Southcentral Region. *Synaptomys* is found throughout the mainland of this region, including the Kenai Peninsula, and on Hinchinbrook Island in Prince William Sound.

Central Region. This species occurs throughout the region, probably to the limits of forest.

Southwest Region. Specimens of bog lemmings have been taken as far southwest as Cape Ugyak and Brooks River at the base of the Alaska Peninsula (Cook and MacDonald 2006; Schiller and Rausch 1956).

Western Region. Swanson (1996) reported the capture of bog lemmings (UAM) from seral spruce forest in the upper Kobuk River valley.

Habitat

The northern bog lemming is generally restricted to open habitats including damp meadows, marshes, bogs, and fens that have an abundance of grasses, sedges, mosses, and low shrubs (Banfield 1974).

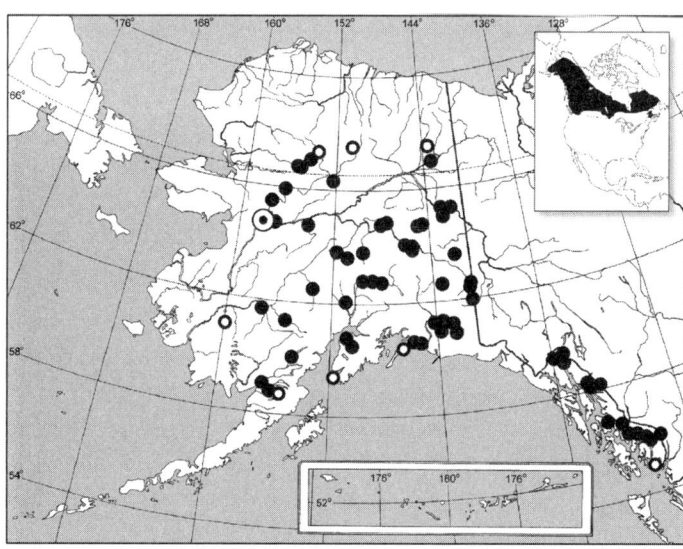

MAP 27. Distribution of northern bog lemming, *Synaptomys borealis.*

⊙ = Type Locality
● = Specimen Record (n = 722)
○ = Marginal Records. 67.45810, –150.86347, Wiseman Quad., unnamed lake near North Fork Koyukuk River (13 UAM); 67.36667, –143.80000, Coleen Quad., Small Lake (2 UAM); 54.98333, –131.00000, Prince Rupert Quad., Foggy Bay, Kirk Point (1 UAM); 60.40000, –146.66667, Cordova Quad., Hinchinbrook Island (1 UAM); 59.43750, –151.70830, Seldovia Quad., Seldovia (7 AMNH); 58.67010, –155.42900, Mt Katmai Quad., NE Naknek Lake, Fures Cabin (2 UAM); 60.98333, –159.98333, Bethel Quad., Tuluksak River, vicinity of Slate Creek (1 UAM); 67.35333, –153.67844, Survey Pass Quad., W side of Takahula Lake (1 UAM).

Status

The distribution of bog lemmings is localized. They are usually uncommon to rare, but can become numerous some years.

Fossils

Fossils of bog lemmings (*Synaptomys* sp.) have been reported from eastern Beringia that date back to the Early Pleistocene (Harington 1978; Irving et al. 1989; Storer 2003b). Georgina (2001) identified remains of *S. borealis* present in Late Holocene deposits from the Lime Hills in central Alaska. Guthrie (1968a) and Youngman (1975) speculated that this species is a postglacial immigrant to the region.

Family MURIDAE Illiger, 1811—Old World rats and mice

Muridae is the largest family of rodents, comprising 730 species and 150 genera in five subfamilies (Musser and Carleton 2005). Three murids, the house mouse (*Mus*) and brown and roof rats (*Rattus*), have been accidentally introduced into Alaska. All have proportionally long, sparsely haired, scaly tails. Their molars have small, rounded cusps arranged in three longitudinal rows.

The geologic range of murines in northern Asia is Early Pliocene to Recent (Carleton and Musser 1984).

House mouse, *Mus musculus,* and brown rat, *Rattus norvegicus* (O. MacDonald)

House Mouse

Mus musculus
Linnaeus, 1758

Other Common Names
None.

Systematics
Musser and Carleton (2005) provided a detailed discussion of *Mus* systematics.

Populations inhabiting northern North America, including Alaska, were believed to be derived from the commensal race *Mus musculus domesticus* (Schwarz and Schwarz 1943). Variation among North American populations has not been studied. This subspecies is considered a distinct species by Marshall and Sage (1981).

Mus musculus domesticus
Original Description. 1772. *Mus domesticus* Rutty, *Essay Nat. Hist. County Dublin*, 1:281.
Type Locality. Dublin, Ireland.
Type Specimen. Not known to exist.
Range. Worldwide.

Global Distribution
This species has spread throughout most of the world through its close association with humans.

Alaska Distribution
Information on the distribution of the house mouse in Alaska is nearly nonexistent, and preserved specimens are few (map 28).

Clark P. Streator of the U.S. Biological Survey considered *Mus* common in forests near Juneau in August 1895. In July 2006, a house mouse was captured in the Mendenhall Wetlands (MSB 149435). Four specimens, dating from 1891 to 1946, are preserved from Wrangell and Sitka (MacDonald and Cook 2007).

Elsewhere in the state, there are recent records from Kasilof, Anchorage, Eagle River, Chugiak, Palmer, Fairbanks, Kodiak Island (and on nearby Hog Island according to Bailey 1993), Unalaska Island, and Kiska Island (UAM; USNM; Bailey 1993; Murie 1959; Peterson 1967). First recorded on St. Paul Island in 1872 (Manville and Young 1965), house mice are currently restricted to the residential areas and the refuse dump (Ebbert and Byrd 2002).

Status
There is little known on the status of this exotic species anywhere in the state. Their presence on St. Paul Island is currently restricted to developed areas (Ebbert and Byrd 2002). The detrimental effect of this invasive species on island populations of seabirds has recently been documented (Wanless et al. 2007).

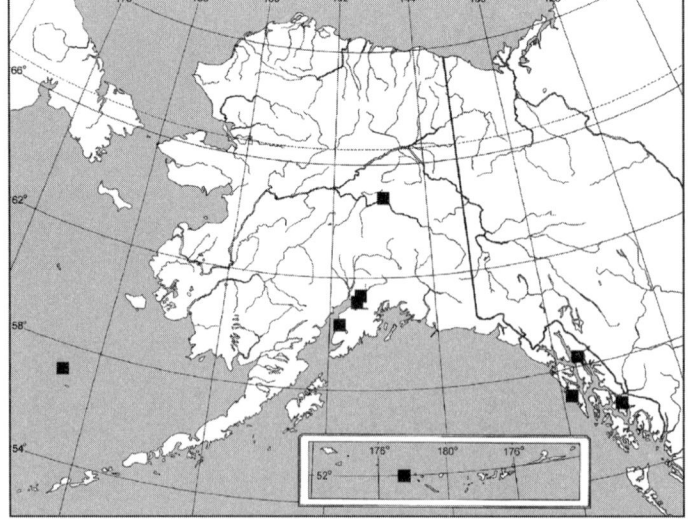

MAP 28. Distribution of house mouse, *Mus musculus*.

■ = Specimen Record (n = 110): 61.18965, –149.70595, Anchorage Quad., Ft. Richardson (1 UAM); 61.6, –149.1, Anchorage Quad., Palmer (1 AMNH, 1 UAM); 61.2, –149.9, Anchorage Quad., Anchorage (34 AMNH, 2 KU, 4 UAM); 61.32222, –149.56667, Anchorage Quad., Eagle River (1 UAM); 61.38333, –149.46667, Anchorage Quad., 2 mi. N, 0.5 mi. E Chugiak (1 UAM); 60.38929, –151.29558, Kenai Quad., mouth Kasilof River (3 MSB); 64.86667, –147.75000, Fairbanks Quad., Fairbanks (3 UAM); 51.97500, –177.50000, Kiska Quad., Kiska Island (1 USNM); 56.46700, –132.37800, Petersburg Quad., Wrangell Island, Wrangell (1 CAS); 58.36204, –134.60478, Juneau Quad., Mendenhall Wetlands (1 MSB); 57.16670, –170.25000, Pribilof Islands Quad., St. Paul Island (32 USNM, 16 CAS, 1 MVZ); 57.05000, –135.33300, Sitka Quad., Baranof Island, Sitka (1 CAS).

Brown Rat

Rattus norvegicus
(Berkenhout, 1769)

Other Common Names
Norway rat, barn rat.

Systematics
Brown rats are not native to North America. Milyutin (1990) and Musser and Carleton (2005) provided an overall review of systematics.

Hall (1981) listed one subspecies for North America; however, the taxonomy of North American populations has been obscured by multiple introductions (Nagorsen 1990).

Global Distribution
Assumed to originally be an inhabitant of southeastern Siberia and northern China (Musser and Carleton 1993), this Old World species has been widely introduced throughout the world including North America (Nagorsen 1990). The brown rat may be more common in colder climates of high latitudes (Kucheruk 1990).

Alaska Distribution
The current status and distribution of the nonnative *R. norvegicus* in Alaska remains poorly understood (map 29). Introduced rats have been reported from the communities of College, Cordova, Craig, Douglas, Fairbanks, Homer, Juneau, Kenai, Ketchikan, King Cove, Kotzebue, Nome, Petersburg, Sitka, St. Michael, Tanana, Valdez, Wasilla, and Wrangell. Island records and reports include Adak, Akutan, Amaknak, Amchitka, Atka, Attu, Baranof, Bat, Bird Rock, Bolshoei Islets, Douglas, Great Sitkin, Kagalaska, Kiska, Kodiak, Little Kiska, Makarius, Mitkof, Ogangen, Prince of Wales, Rat, Revillagigedo, Sanak, Seal Rocks, Sedanka, Shemya, Unalaska, and Wrangell islands (Bailey 1993; Brechbill 1977; Fritts 2006; http://www.stoprats.org/index.htm).

Habitat
Rattus norvegicus is usually a commensal, nonnative rat associated with human-created habitats such as buildings, sewers, and wharves. Feral populations inhabit a variety of habitats on islands and along beaches.

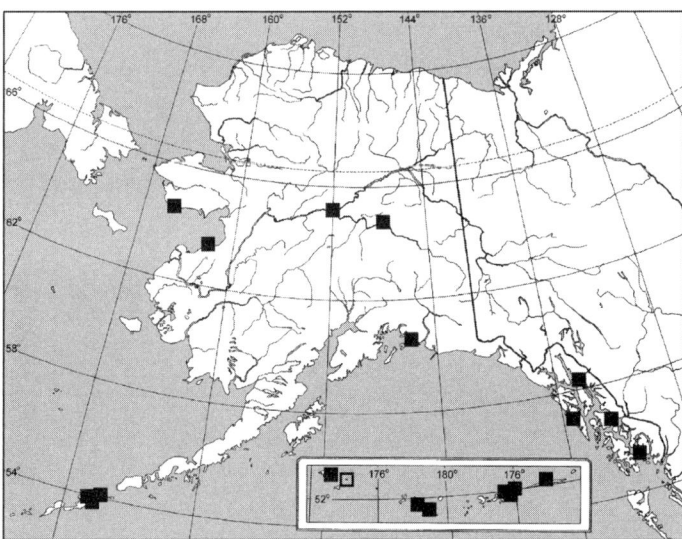

MAP 29. Distribution of brown rat, *Rattus norvegicus* (solid square), and roof rat, *Rattus rattus* (hollow square).

□ = Specimen Record (Taylor and Brooks 1995): Attu quad., Shemya Island (none preserved).

■ = Specimen Record (n = 108): 64.50000, –165.41667, Nome Quad., Nome dump (3 UAM); 65.16667, –152.06700, Tanana Quad., Tanana (4 USNM); 64.90000, –147.80000, Fairbanks Quad., Fairbanks vicinity (13 UAM); 58.30830, –134.40830, Juneau Quad., Juneau (1 USNM); 58.27778, –134.39306, Juneau Quad., Douglas Island, Douglas (1 UAM); 56.80290, –133.17500, Petersburg Quad., Mitkof Island, Petersburg (2 USNM); 55.00000, –131.00000, Ketchikan Quad., Revillagigedo Island, Ketchikan vicinity (2 UAM); 57.05000, –135.33300, Sitka Quad., Baranof Island, Sitka (1 CAS); 60.55000, –145.75000, Cordova Quad., Cordova (1 MVZ); 54.13470, –165.72220, Unalaska Quad., Akutan Island, Akutan (1 MVZ); 53.78417, –166.19344, Unalaska Quad., Sedanka Island (1 UAM); 53.87500, –166.53330, Unalaska Quad., Unalaska Island, Unalaska (1 USNM); 52.20000, –174.20000, Atka Quad., Atka Island, Atka (1 USNM); 52.05761, –176.11092, Adak Quad., Great Sitkin (7 UAM); 51.82917, –176.41111, Adak Quad., Kagalaska Island, Laska Cove (1 UAM); 51.71667, –176.71667, Adak Quad., Adak Island, Adak vicinity (22 UAM); 51.53300, –179.00000, Rat Islands Quad., Amchitka Island (1 USNM); 51.80000, –178.31670, Rat Islands Quad., Rat Island (5 USNM); 52.90250, 173.00000, Attu Quad., Attu Island, Attu vicinity (7 UAM); 63.48333, –162.03330, St Michael Quad., St. Michael (2 USNM).

Status

Introductions of this species to islands worldwide had disastrous effects. The first recorded accidental introduction of *R. norvegicus* to Alaska was sometime before 1780 on Rat Island in the western Aleutians (Ebbert and Byrd 2002). Since then, documented records of this destructive species (especially to burrow-nesting seabirds) include at least twenty-two other islands across the state. Increased transcontinental shipments of freight by ocean freighters elevates the risk of introduction of *Rattus* spp. that should not be underestimated. Coordinated efforts to eradicate established rat populations in Alaska, especially detrimental to Alaska's island faunas, and to prevent their further spread are currently underway (Woodford 2005b).

Roof Rat

Rattus rattus
(Linnaeus, 1758)

Other Common Names

Black rat, Alexandrine rat, ship rat.

Systematics

Based on biochemical and morphological data, two species groups (perhaps incipient species) have been identified, one Oceanian or European, the other Asian. Recent studies (Musser and Carleton 2005) indicate that because of multiple founder events, the widespread "introduced" populations are genetically very similar.

Global Distribution

The roof rat is native to the Indian Peninsula, but has been introduced worldwide, including many isolated islands (Musser and Carleton 2005). Generally less common than the brown rat in many temperate regions, roof rats appear to decline after the arrival of the larger, more aggressive brown rat (Nagorsen 2005).

Alaska Distribution

The first report of this alien rodent in Alaska was the discovery, in 1995, of *R. rattus* remains in military buildings on the Aleutian island of Shemya (Taylor and Brooks 1995) (map 29). This species is now established on the island and has been trapped there every year since 2004 (S. Ebbert, pers. comm., 2008).

Habitat

Rattus rattus is usually associated with urban and agriculture areas. Roof rats introduced on Haida Gwaii (Queen Charlotte Islands) are now found on isolated islands with seabird colonies (Nagorsen 2005).

Status

Roof rats on Shemya Island apparently have become established, persisting in colonies outside buildings during winter (S. Ebbert, pers. comm., 2008).

Family ERETHIZONTIDAE Bonaparte, 1845— New World porcupines

This family of New World porcupines includes sixteen species in five genera (Woods and Kilpatrick 2005). Only the monotypic species, *Erethizon dorsatum*, inhabits North America.

Fossil records of *Erethizon* date back to the Late Pliocene in North America (Woods 1984).

Like the Old World porcupines (Hystricidae), the porcupines of the Americas are relatively large, heavyset rodents that are formidable prey due to a thick outer covering of quills (modified hairs imbedded into skin musculature). All are more or less arboreal and herbivorous. The North American species tends to be solitary except during autumn when males seek out females to breed. In winter, this species is often found high in trees feeding on the inner cambium layers of bark. During severe weather they den up, sometimes communally, in caves, root wads, and other natural cavities.

North American porcupine, *Erethizon dorsatum* (W. D. Berry)

North American Porcupine

Erethizon dorsatum
(Linnaeus, 1758)

Other Common Names
Porcupine.

Systematics
Six subspecies were listed by Hall (1981); two occur in Alaska.

Erethizon dorsatum myops
Original Description. 1900. *Erethizon epixanthus myops* Merriam, *Proc. Washington Acad. Sci.*, 2:27, March 14.
Type Locality. Portage Bay, Alaska Peninsula, Alaska.
Type Specimen. USNM 59140.
Range. Southwestern and central Alaska eastward to Alberta.

Erethizon dorsatum nigrescens
Original Description. 1903. *Erethizon epizanthus* [sic] *nigrescens* J. A. Allen, *Bull. Amer. Mus. Nat. Hist.*, 19:558, October 10.
Type Locality. Shesley River, British Columbia.
Type Specimen. AMNH 20772.
Range. Southeast Alaska, British Columbia, and Washington.

Global Distribution
Nearctic—*Erethizon dorsatum* occurs broadly across North America from Alaska to Labrador, south to Tennessee, Iowa, and Texas, northern Mexico, and California (Hall 1981; Woods 1973).

Alaska Distribution
Porcupines are found throughout most of mainland Alaska, and on several islands along the southeastern coast (map 30).

Southeast Region. Porcupines are found along the coastal mainland and on some islands in the Alexander Archipelago, including Douglas, Etolin, Hassler, Kupreanof, Mitkof, Revillagigedo, and Wrangell islands (MacDonald and Cook 2007).

Southcentral Region. This species is found throughout the mainland, but not on any islands.

Central Region. Porcupines occur throughout central Alaska.

Southwest Region. This species is found throughout the mainland.

Western Region. Porcupines are generally absent from the Yukon-Kuskokwim Delta as well as from northern and western Seward Peninsula.

Northern Region. Porcupines are found in the Brooks Range and along the major wooded streams in the foothills on the Arctic Slope. There are reports of porcupines to the Arctic coast at Point Lay (UAM unpubl.), Barrow (J. C. George, pers. comm.), and Icy Reef (Bee and Hall 1956).

Habitat
Porcupines are a versatile animal, occurring in a wide variety of habitats from closed forest to

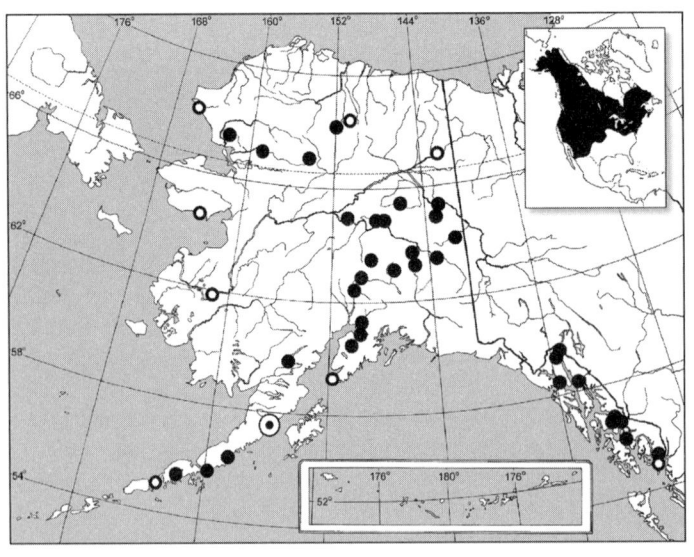

MAP 30. Distribution of North American porcupine, *Erethizon dorsatum*.

⊙ = Type Locality
● = Specimen Record (n = 245)
○ = Marginal Records. 68.27531, −150.65675, Chandler Lake Quad., Nanushuk River (1 UAM); 67.00000, −142.75000, Coleen Quad., Porcupine River, 10 mi. below Coleen River (1 USNM); 54.77500, −130.24170, Prince Rupert Quad., Tongass Island, Fort Tongass (1 USNM); 59.43750, −151.70830, Seldovia Quad., Seldovia (2 AMNH); 55.00000, −163.00000, Cold Bay Quad., Morzhovi Bay, Alaska Peninsula (1 USNM); 61.78300, −161.31670, Russian Mission Quad., Russian Mission (1 USNM); 64.56667, −163.66667, Solomon Quad., Koyana Creek, bluff region (1 UAM); 68.25, −166.0, Point Hope Quad., Cape Thompson area (3 UAM).

open shrub tundra, and from sea level to high elevations in the mountains.

Status

Porcupines are relatively common and widespread.

Fossils

Erethizon fossils have been found in Late Irvingtonian, Wisconsinan, and Holocene deposits in Alaska (Georgina 2001; Harington 1978; Kurtén and Anderson 1980).

Order LAGOMORPHA Brandt, 1855—pikas, hares, and rabbits

This order of two living families of "hare shaped" mammals comprises ninety-two species in thirteen genera worldwide (Hoffmann and Smith 2005). The affinities of the Lagomorpha, once considered only a suborder of Rodentia, remains contentious, although there is a growing body of evidence that supports uniting the Order Lagomorpha with Rodentia in a superorder called Glires (Huchon et al. 2002; Szalay 1985).

The geographical range of the Lagomorpha is Late Paleocene to Recent (Diersing 1984). Early Eocene (ca. fifty-three million years ago) fossil specimens were recently discovered in western India, demostrating that the lagomorphs were already distinct from other mammals by that time (Rose et al. 2008).

Family OCHOTONIDAE Thomas, 1897—pikas

The systematics of this family is unstable and in need of further study (Nowak 1991). The family of pikas (also called mouse hares or conies) includes about thirty living species in the genus *Ochotona* (Hoffmann and Smith 2005) that are further split into five groups based on molecular sequence variation (Niu et al. 2004). Almost exclusively Palearctic in distribution, only one of the two New World species occurs in Alaska.

Twenty-five extinct genera of ochotonids, whose fossil remains date back to the Early Eocene in Eurasia and Early Miocene and Pleistocene in North America, have been recognized (Diersing 1984; McKenna and Bell 1997). Erbajeva et al. (2003) suggested that the extant Eurasian species *O. pusilla* and the two Nearctic species *O. princeps* and *O. collaris* shared a common ancestor that migrated from Asia to North America at the beginning of the Pleistocene.

Pikas have short legs, stocky bodies, rounded ears, and no visible tail. At a distance they look more like a soft-furred rodent than they do their closest relatives, the rabbits and hares. Unlike other lagomorphs, pikas are highly vocal and colonial. As in all other lagomorphs, however, adult pikas possess two pairs of upper incisors, the second pair behind the first and, in males, the scrotum is in front of the penis rather than posterior. Most species are found in close association with talus slopes or piles of broken rock (Nowak 1991).

Collared pika, *Ochotona collaris* (W. D. Berry)

Collared Pika

Ochotona collaris
(Nelson, 1893)

Other Common Names
None.

Systematics

Niu et al. (2004) inferred from mtDNA cytochrome *b* sequences that five major species groups of pikas diverged during the Early Pleistocene. Their northern group comprised *Ochotona alpina, O. hyperborea*, and *O. pallasi* in a North Palearctic subgroup, and *O. princeps* and *O. collaris* in a Nearctic subgroup. Broadbooks (1965) and Youngman (1975) considered *O. collaris* conspecific with *O. princeps*. Gureev (1964) and Corbet (1978) considered *collaris, princeps*, and *hyperborea* conspecific with *alpina*. Reevaluation of morphological (Weston 1981), behavioral (Kawamichi 1981), chromosomal (Vorontsov and Ivanitskaya 1973), and mtDNA (Niu et al. 2004) data indicate that all four are separate species. *O. collaris* is monotypic (Hall 1981; Hoffmann and Smith 2005).

Ochotona collaris

Original Description. 1893. *Lagomys collaris* Nelson, *Proc. Biol. Soc. Washington*, 8:117, December 21.

Type Locality. Near the head Tanana River, about 200 mi. S Fort Yukon, Alaska.

Type Specimen. USNM 14384/36297.

Global Distribution

Nearctic—The range of this species is restricted to Alaska and northwestern Canada (MacDonald and Jones 1987).

Alaska Distribution

Collared pikas occur in mountains of east-central and southern Alaska (map 31).

Southeast Region. *Ochotona collaris* has been documented in southeast Alaska near White Pass (MacDonald and Cook 2007) and along Chilkat Pass in nearby British Columbia (Nagorsen 2005).

Southcentral Region. Collared pikas are found in the Talkeetna, Chugach, and Wrangell mountain ranges (Cook and MacDonald 2003; Rausch 1962), but are absent from the Kenai Peninsula (Cook and MacDonald 2004c). They are present in the mountains north of Valdez (USNM). Manville and Young (1965) noted a specimen from near Cordova (specimen not located) and former Cordova resident M. E. Isleib reported seeing pikas above Eyak Lake in 1976 and 1978.

Central Region. Pikas occur on higher peaks in the Yukon-Tanana highlands (Libby 1958) and along the length of the Alaska Range (Cook and MacDonald 2002, 2004b; Larson 2002; Rausch 1962). They occur in the Ogilvie Mountains of west-central Yukon Territory (Youngman 1975), suggesting their possible presence in adjacent alpine areas north of the upper Yukon River in Alaska. A report of pikas in the Kuskokwim Mountains (USFWS 2005) lacks verification and is contradicted by

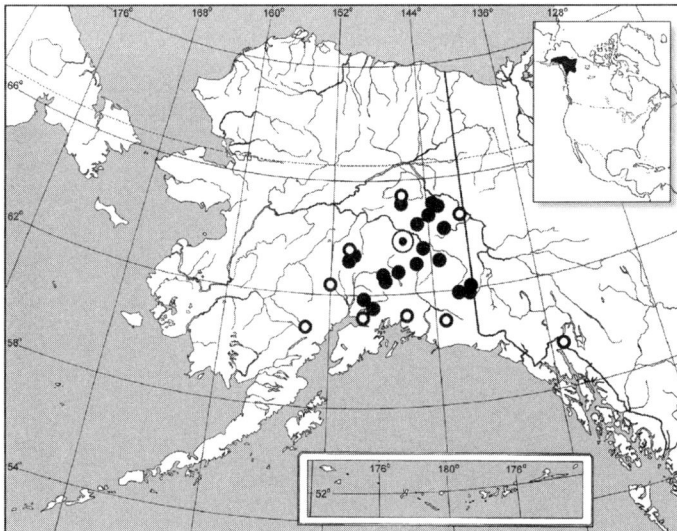

MAP 31. Distribution of collared pika, *Ochotona collaris*.

⊙ = Type Locality
● = Specimen Record (n = 346)
○ = Marginal Records. 65.48333, −145.41667, Circle Quad., Eagle Summit (10 UAM); 64.73300, −141.33300, Eagle Quad., mountains near Eagle (85 USNM); 59.61611, −135.13833, Skagway Quad., White Pass (2 UAM); 61.05845, −143.36338, McCarthy Quad., Pocket Creek (1 UAM); 61.13300, −145.73300, Valdez Quad., Chugiak Mountains, pass N of Valdez (1 USNM); 61.21667, −149.58333, Anchorage Quad., Fort Richardson (2 UAM); 60.76667, −153.85000, Lake Clark Quad., 3.2 km SE of NE corner of Turquoise Lake (1 UAM); 62.29500, −152.05139, Talkeetna Quad., 8 km W of mouth of Fourth of July Creek (1 UAM); 63.71000, −150.59333, Mt McKinley Quad., 2 km SW of Pilgrim Peak (4 UAM).

a failure to find evidence of their occurrence in either the Beaver, Sunshine, or Bitzshtine mountains despite considerable time spent on the ground in these areas between 1987 and 2007 (Jack Whitman, pers. comm., 2007).

Southwest Region. Pikas were recently collected near Turquoise Lake in Lake Clark National Park and Preserve (Cook and MacDonald 2004b), and Osgood (1904) mentioned pika sightings on the mountain near Keejik on Lake Clark. Records of this species farther south in the Aleutian Range is limited to two specimens collected by McKay in 1882 (USNM; True 1886) from the "Chigmit Mountains," but the exact locality is unknown.

Northern Region. MacDonald and Jones (1987) reported sightings of pikas in the eastern Brooks Range by several observers in the Shubelik, Romanzof, and Philip Smith mountains. Dufresne (1946) also mentioned reports of pikas north of the Arctic Circle in parts of the Brooks Range. None of these records, however, has been substantiated by specimen or photograph.

Habitat

Collared pikas form colonies in mountainous terrain. They inhabit rock slides, talus slopes, and large boulders near meadows and patches of vegetation (MacDonald and Jones 1987). Near Lake Louise in southcentral Alaska, Rausch (1962) encountered pikas in a forested valley (white spruce-birch-willow) more than 180 m from the nearest talus. Their burrows were beneath rocks scattered among the trees.

Status

The collared pika is locally common. The loss of habitat due to global warming may pose a long-term threat to this and other montane species (Grayson 2005).

Fossils

MacDonald and Jones (1987) reviewed the fossil history of this species. Guthrie (1973) reported the mummified remains and preserved dung pellets of *O. collaris* from Pleistocene (Wisconsinan) deposits in central Alaska. Fossils of Late Pleistocene and Early Holocene age were reported from the central Tanana Valley by Yesner (2001). Fossils of this species are also known from Yukon Territory (Harington 1977, 1978). Related forms have been described from fossil deposits in Alaska (Guthrie and Matthews 1971) and Yukon (Harington 1978; Kurtén and Anderson 1980).

Family LEPORIDAE Fischer von Waldheim, 1817—hares and rabbits

This familiar family of hares (genus *Lepus*) and rabbits is made up of sixty-one species in eleven genera (Hoffmann and Smith 2005). No subfamilies are usually recognized. Leporids occur naturally on most of the world's major landmasses, and have been introduced (from both wild and domestic stocks) into many areas where they hadn't previously occurred. Two species of hare (*Lepus*) and a nonnative rabbit (*Oryctolagus*) are found in Alaska.

The geological range of Leporidae is Late Paleocene to Recent in Asia, and Middle Eocene to Recent in North America (Diersing 1984).

Rabbits and hares generally have long ears, long hind legs and hind feet, bulging eyes, and a short tuft of a tail. In many species, females are larger than males. In contrast to the pikas, members of this family are usually nocturnal and mostly nonvocal. In addition to a number of morphological and behavioral differences, hares are precocial, born in the open, fully furred, eyes open, and ready to run, while rabbits are altricial, born blind, without fur, and essentially helpless in underground nests.

Snowshoe hare, *Lepus americanus* (W. D. Berry)

Snowshoe Hare

Lepus americanus
Erxleben, 1777

Other Common Names

Varying hare, snowshoe rabbit.

Systematics

A phylogenetic study of North American hares based on cytochrome *b* DNA sequences (Halanych et al. 1999) found *L. americanus* more closely related to species from the southwest United States and Mexico (western American clade) than to other northern latitudes species (arctic clade).

Hall (1981) recognized fifteen subspecies, of which one occurs in Alaska. Nagorsen (1985) found no basis for recognizing subspecies because patterns of cranial variation were largely clinal.

Lepus americanus dalli

Original Description. 1900. *Lepus americanus dalli* Merriam, *Proc. Washington Acad. Sci.*, 2:29, March 14.
Type Locality. Nulato, Alaska.
Type Specimen. USNM 8996/7579.
Range. Alaska and northwestern Canada.
Remarks. Includes *L. a. macfarlani* Merriam as synonymized by Youngman (1975).

Global Distribution

Nearctic—The range of *L. americanus* extends from Alaska, across most of Canada and the northern contiguous states, and down through the Rocky Mountains into Utah and New Mexico (Hall 1981).

Alaska Distribution

The snowshoe hare occurs throughout the taiga of Alaska (map 32).

Southeast Region. Snowshoe hares are limited to the northern mainland of this region. They regularly occur in the Chilkat Valley near Haines and at Dyea in the vicinity of Skagway. Snowshoe hares have been reported from the Taku River, Glacier Bay, Alsek River, and Yakutat (MacDonald and Cook 2007). The report of hares at the mouth of the Stikine River by Manville and Young (1965) has not been substantiated.

Snowshoe hares now present on Douglas Island near Juneau were probably introduced there from Haines stock "a few years previous" (Bailey 1920; also Wenrich 1922). The extant population of hares found on the mainland near Juneau may be derived from those introduced animals.

In 1923 and 1924, the Alaska Game Commission released snowshoe hares from Washington stock to Point Retreat, Admiralty Island; Otstoia Island, Peril Strait; and Smeaton Island, Behm Canal. Stock from the Anchorage area were also released in 1924 on Cape Island, Prince of Wales Island, and Village Island. All these transplant attempts were considered failures (Burris and McKnight 1973).

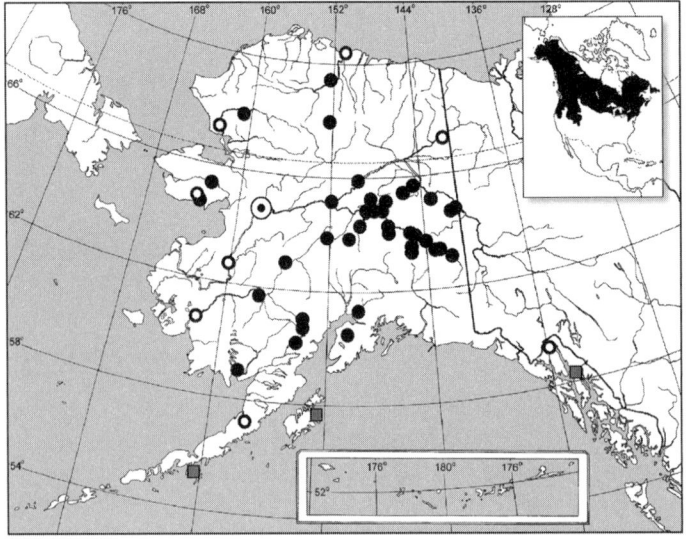

MAP 32. Distribution of snowshoe hare, *Lepus americanus*.

⊙ = Type Locality
● = Native, ■ = Non-native Specimen Record (n = 1190)
○ = Marginal Records. 70.34167, −150.96667, Harrison Bay Quad., Colville River delta (1 UAM); 67.16670, −141.66670, Coleen Quad., Rampart House, Porcupine River (1 USNM); 59.41670, −135.92920, Skagway Quad., Wells, Chilkat River Valley (1 MVZ); 57.18333, −157.25833, Ugashik Quad., Mother Goose Lake Field Station (1 UAM); 60.79170, −161.75000, Bethel Quad., Bethel (58 USNM); 62.65420, −160.20830, Holy Cross Quad., Anvik (3 USNM); 64.90000, −163.66667, Solomon Quad., Council (2 UAM); 67.41670, −163.00000, Noatak Quad., Noatak River, Eli River (1 USNM).

Southcentral Region. Hares occur throughout the mainland of this region including the Kenai Peninsula (Selinger 2004). Along the coast, hares are found east of Cordova to Icy Bay. They are rare along the mainland and absent from the islands of Prince William Sound (ADFG 1978).

Central Region. Snowshoe hares are found throughout the region, but are less numerous and localized in the north.

Southwest Region. Hares occur throughout suitable habitat in the northern portion of the region but do not range far beyond the limits of trees on the Alaska Peninsula (ADFG 1978; Murie 1959); two snowshoe hare specimens from Mother Goose Lake (UAM) south of Ugashik represent the most westerly record.

According to Burris and McKnight (1973), over five hundred snowshoe hares from near Anchorage were released on Kodiak and Afognak islands in 1934. Manville and Young (1965) also mention introductions to Raspberry Island. These transplants were successful, and in 1952 hares from Kodiak Island were moved to adjacent Woody and Long islands. In 1955, fifteen hares from Kodiak Island were introduced to Popof Island (Shumagin Islands), off the Alaska Peninsula, resulting in a substantial population there as early as 1960.

Snowshoe hares were considered common on Marmot Island, east of Afognak Island, by Chumbley et al. (1997); the status of this island population lacks documentation.

Western Region. During population lows these animals rarely occur in tundra areas of the Seward Peninsula and Yukon-Kuskokwim Delta (ADFG 1978). In the Kobuk and lower Noatak valleys they are found most frequently in riparian willow stands along the rivers and streams.

Northern Region. Snowshoe hares occur occasionally in the southern portion of this region and generally are restricted to willows along major water courses (ADFG 1978). Snowshoe hares may have first arrived north of the Brooks Range by colonizing along the Colville River and some of the other major western drainages in the early 1900s (Klein 1995). Snowshoe hares were numerous in the Colville River drainage by the 1990s (Carroll 2004a) and their spread into new areas should be monitored in relation to climate warming.

Habitat

Hares inhabit forests, shrubby woodlands, and riparian shrub thickets (Banfield 1974; Wolff 1980).

Status

Hares are generally common and periodically very abundant. Coastal populations fluctuate erratically and are less cyclic than others (ADFG 1978).

Fossils

Youngman (1975) considered *L. americanus* a postglacial immigrant to northwestern Canada and Alaska from refugia south of the main glacial systems of the Late Pleistocene. Jopling et al. (1981), however, reported fossils of this species dating to the late Illinoian from deposits in the Old Crow Basin, Yukon Territory.

Alaska Hare

Lepus othus
Merriam, 1900

Other Common Names

Alaska arctic hare, Alaska tundra hare, tundra hare.

Systematics

The taxonomic status of *L. othus* in relation to *L. timidus* of Siberia and *L. arcticus* of Canada has been a long-standing controversy (Baker et al. 1983). Some authors have considered the three forms comprising a single holarctic species that should be recognized as *L. timidus* (Chapman et al. 1983; Dixon et al. 1983; Flux 1983; Honacki et al. 1982). Others think that *L. othus* may be conspecific only with *L. timidus* (Bee and Hall 1956; Corbet and Hill 1980; Rausch 1953a, 1963). Rausch (reported by Anderson 1974) has suggested that *L. othus* is a valid species and is not conspecific with *L. timidus* or *L. arcticus*. An analysis of DNA sequences of the cytochrome *b* gene examined evolutionary relationships for eleven species of hares (Halanych et al. 1999) and found that *L. othus*, *L. timidus*, and *L. arcticus* displayed minimal genetic divergence, suggesting a single circumpolar species. In a more recent phylogenetic study of circumpolar arctic hares, Waltari and Cook (2005) tentatively supported the delineation of three species while noting additional sampling in the Chukotka Peninsula at the eastern extreme of far east Russia and

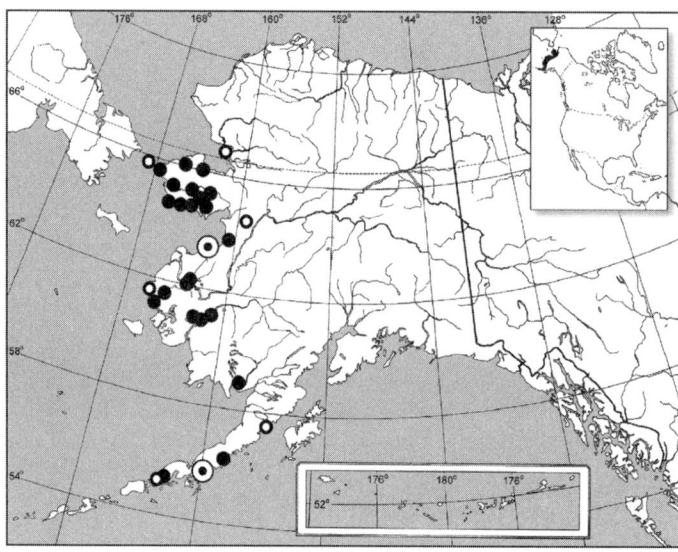

MAP 33. Distribution of Alaska hare, *Lepus othus*.

⊙ = Type Locality
● = Specimen Record (n = 260)
○ = Marginal Records. 66.90000, –162.58300, Kotzebue Quad., Kotzebue (1 UAM); 64.50000, –159.33300, Norton Bay Quad., Nulato River, mountains NW of (2 USNM); 57.56700, –156.03300, Ugashik Quad., Portage Bay and Becharof, between (6 USNM); 55.20000, –162.70000, Cold Bay Quad., Frosty Road, between first and second bridge (1 UAM); 61.66667, –166.00000, Hooper Bay Quad., Near Kokechik Bay, S 18–T18N–R91W (3 UAM); 65.75000, –168.91670, Teller Quad., Little Diomede Island (1 USNM).

independent molecular markers are needed to resolve the relationship between *othus* and *timidus* populations.

Hall (1981) recognized *L. othus* as a distinct species comprised of two subspecies.

Lepus othus othus

Original Description. 1900. *Lepus othus* Merriam, *Proc. Washington Acad. Sci.*, 2:28, March 14.

Type Locality. St. Michael, Norton Sound, Alaska.

Type Specimen. USNM 15883.

Range. Yukon-Kuskokwim Delta northward to Norton Sound.

Lepus othus poadromus

Original Description. 1900. *Lepus poadromus* Merriam, *Proc. Washington Acad. Sci.*, 2:29, March 14.

Type Locality. Stepovak Bay, Alaska Peninsula, Alaska.

Type Specimen. USNM 98068.

Range. Bristol Bay through the Alaska Peninsula.

Global Distribution

Nearctic—The range of *L. othus* is restricted to western and southwestern Alaska.

Alaska Distribution

The Alaska hare is endemic to the coastal tundra areas of western Alaska, and has been documented by a small number of widely scattered specimens from the tip of the Alaska Peninsula north to Kotzebue Sound (Anderson 1978; Klein 1995; Rausch 1963) (map 33).

Southwest Region. *Lepus othus* is found throughout the mainland of the Alaska Peninsula and Bristol Bay region (Murie 1959). There are no island records except for the report by Bailey (1993) of this species being introduced on Chirikof Island in 1891 but disappearing afterwards, perhaps because of foxes.

Western Region. *Lepus othus* is absent or extremely rare north of Kotzebue Sound, with scattered pockets of abundance centering in the western Seward Peninsula and the Yukon-Kuskokwim Delta region (Klein 1995). A curious record is a specimen (USNM 260900) from Little Diomede Island.

Northern Region. Historically, *L. othus* (presumably this species and not *L. arcticus*) was present north of the Brooks Range along the North Slope from the vicinity of the Colville River westward (Bee and Hall 1956; Klein 1995).

Habitat

Alaska hares are found in a variety of coastal tundra habitats and also in shrub communities along streams (Klein 1995). The two subspecies have been associated with distinct habitat types, with *L. o. othus* preferring tundra or alluvial plain, and *L. o. poadromus* living primarily in coastal lowland areas (Best 1999; Best and Henry 1994).

Status

The Alaska hare was formerly present on the western North Slope of northern Alaska but there have been no new reports of these large hares in that region since 1951 (Klein 1995). Circumstantial evidence suggests that the Alaska hare may have declined there after the arrival of the snowshoe hare (*L. americanus*) early in the twentieth century (Klein 1995).

Populations of Alaska hares from Kotzebue to the Yukon-Kuskokwim Delta have remained low since population highs in the 1970s, whereas hare densities on the Alaska Peninsula have been reported low since the early 1950s (Klein 1995). Climate warming may adversely affect this species, perhaps by reducing the extent of its habitat.

Fossils

Fossils of *L. othus* were reported in two Late Rancholabrean sites in central Alaska: Canyon Creek roadcut (Weber et al. 1981) and Porcupine River Cave 1 (Dixon 1984). Both are outside the current range of *L. othus*. It was suggested that the Canyon Creek remains may be *L. arcticus* (Weber et al. 1981), which today is found farther east, from the lower McKenzie River eastward across Arctic Canada.

No *L. timidus* fossils dated prior to the Holocene have been found in eastern Russia (Averianov 1998; Hopkins et al. 1982).

European Rabbit

Oryctolagus cuniculus
(Linnaeus, 1758)

Other Common Names
Old World rabbit, domestic rabbit.

Global Distribution
Wild populations of *O. cuniculus* apparently were confined to southern France, the Iberian Peninsula, and possibly northwestern Africa following the end of the Pleistocene. Introduced and feral populations now occur across much of Europe and in many other parts of the world because the species has been extensively domesticated (Nowak 1991).

Alaska Distribution
Domestic rabbits have been introduced to a number of islands in southern Alaska at various times (map 34).

Southeast Region. Burris and McKnight (1973) indicated that several transplants of rabbits have been attempted in this region but they did not list specific localities. Several residents of Ketchikan reported to us (in 1995) the presence of feral rabbits on Betton Island near Clover Pass, but the current status of that population is unknown.

Southcentral Region. According to O'Farrell (1965) and Burris and McKnight (1973), one male and three female domestic rabbits were released on Middleton Island,

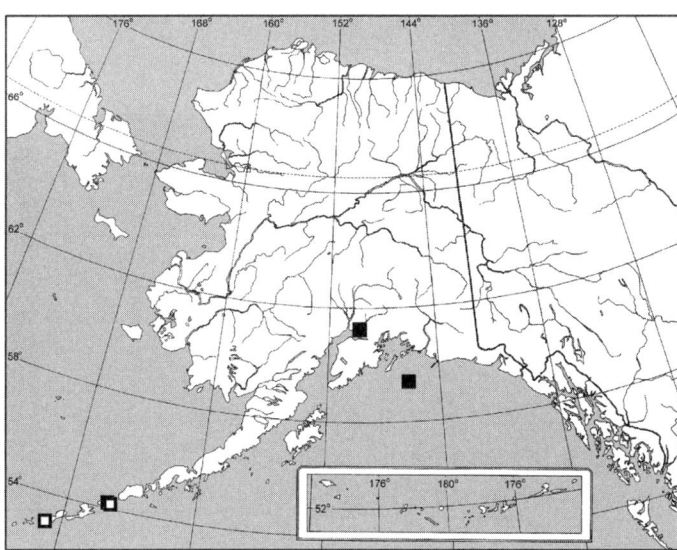

MAP 34. Distribution of European rabbit, *Oryctolagus cuniculus*.

■ = Specimen Record (n = 10). 59.43333, –146.33333, Gulf of Alaska, Middleton Island (9 UAM); Anchorage (1 USNM).
□ = Locality Record (no specimens archived)

Prince William Sound, in 1954. Starting out as semidomestic pets under the houses of island residents, their population increased to fifty by the fall of 1955 and to approximately two hundred by the summer of 1956. Estimates made in the summer of 1962 placed the population on the island at about five thousand rabbits, dropping to about three thousand during the winter of 1962–1963. ADFG collected several rabbits on the island for UAM in 1975.

Southwest Region. The locations of feral populations of domestic rabbits in the Aleutian Islands have been summarized by Burris and McKnight (1973) and Bailey (1993; erroneously as *Lepus capenus*) as follows: Ananiuliak Island (date of introduction ca. 1917, perhaps ca. 1940); Hog Island (in Unalaska Bay; before 1940); Kanaga Island (before 1936); Poa Island (date unknown); Rabbit Island (near Umnak Island; ca. 1940); Tangik Island (date of release unknown); and Umnak Island (1930). Rabbits have also been reported on Anangula Island, and are still present on Ananiuliak, Poa, and Tangik islands (AMNWF 2006). The current status of other populations is unknown.

European rabbit, *Oryctolagus cuniculus* (K. H. McInnes)

Order SORICOMORPHA Gregory, 1910

Small, primitive mammals of uncertain affinities were historically relegated to the Order Insectivora (Yates 1984), but molecular DNA analyses of these organisms have begun to produce phylogenies and classifications that likely more accurately reflect their evolutionary relationships (reviewed by Nowak 1991; see also McKenna and Bell 1997). Because of accumulating evidence for the paraphyletic nature of the former Insectivora, Hutterer (2005) provisionally treated Soricomorpha as a separate order inclusive of the moles, shrews, and solenodons. Thus reconstituted, it comprises 428 species and 45 genera in 4 families, with Soricidae containing 88 percent of all living species.

The geological range of this group dates back to the Late Cretaceous (Yates 1984).

Family SORICIDAE Fischer von Waldheim, 1817—shrews

This monophyletic family consists of 26 genera and 376 species in 3 subfamilies (Hutterer 2005). All Recent North American shrews are soricines. The genus *Sorex* is generally divided into two subgenera, *Otisorex* and *Sorex* (George 1988), and includes seventy-seven species. Ten of these occur in Alaska, making *Sorex* the most speciose genus in Alaska.

Soricids have been present in North America at least since the Eocene, and since Early Oligocene in Europe and Asia (Repenning 1967).

Shrews are among the world's smallest and most active mammals. The Alaska tiny shrew, *Sorex yukonicus*, a newly discovered species that is endemic to Alaska, is the smallest, with an average total length in twenty-nine individuals of 70 mm and weight of just 1.8 grams. Shrews have tiny eyes, pointy snouts, and a metabolic rate that requires them to feed frequently throughout the day and year. Invertebrates are their primary food. Some species use echolocation to navigate and find prey. Other species (outside Alaska) have poisonous salivary glands. Most shrews prefer moist microhabitats and a few species are so well adapted to an aquatic lifestyle (e.g., *Sorex palustris*) that they are occasionally caught in minnow traps.

American water shrew, *Sorex palustris* (O. MacDonald)

Cinereus Shrew

Sorex cinereus
Kerr, 1792

Other Common Names

Common shrew, masked shrew.

Systematics

Hall (1981) recognized five subspecies of *Sorex cinereus* in Alaska, but three of these (*pribilofensis, jacksoni, ugyunak*) have been problematic and are currently considered separate species (see separate species accounts). Populations of *S. cinereus* (sensu stricto) from Alaska that were included in an assessment of mtDNA sequences (cytochrome *b* and ND4 genes) exhibited minimal levels of variation (Demboski 1999; Demboski and Cook 2003). Waltari (2005) identified a zone of secondary contact between *S. cinereus* and *S. ugyunak* that roughly followed the tundra-taiga interface along the southern edge of the Brooks Range.

Sorex cinereus cinereus

Original Description. 1792. *Sorex arcticus cinereus* Kerr, *Animal Kingdom*, p. 206.
Type Locality. Severn Settlement (= Fort Severn), Ontario.
Type Specimen. USNM 84556.
Range. Northern Alaska through most of Canada and western United States.

Sorex cinereus hollisteri

Original Description. 1925. *Sorex cinereus hollisteri* Jackson, *J. Mammalogy*, 6:55, February 9.
Type Locality. St. Michael, Alaska.
Type Specimen. USNM 99305.
Range. Western and southwestern Alaska.
Remarks. A renaming of *Sorex personatus arcticus* Merriam 1900.

Sorex cinereus streatori

Original Description. 1895. *Sorex personatus streatori* Merriam, *N. Amer. Fauna*, 10:62, December 31.
Type Locality. Yakutat, Alaska.
Type Specimen. USNM 73537.
Range. Pacific Coast from Prince William Sound, Alaska, to Washington.

Global Distribution

Nearctic—The cinereus shrew occurs throughout Alaska and Canada and southward along the Rocky and Appalachian mountains (Hutterer 2005). It does not occur in northeast Siberia, Palearctic region, as previously suggested (van Zyll de Jong 1982).

Alaska Distribution

The cinereus shrew is found throughout most of mainland Alaska as well as on a number of islands (map 35).

Southeast Region. *Sorex cinereus* is found along the entire coastal mainland and on islands of the Alexander Archipelago that include Baranof, Bell, Black, Chichagof, Deer, Douglas, Emmons (J. Whitman, pers. comm.), Etolin, Gedney, Grant,

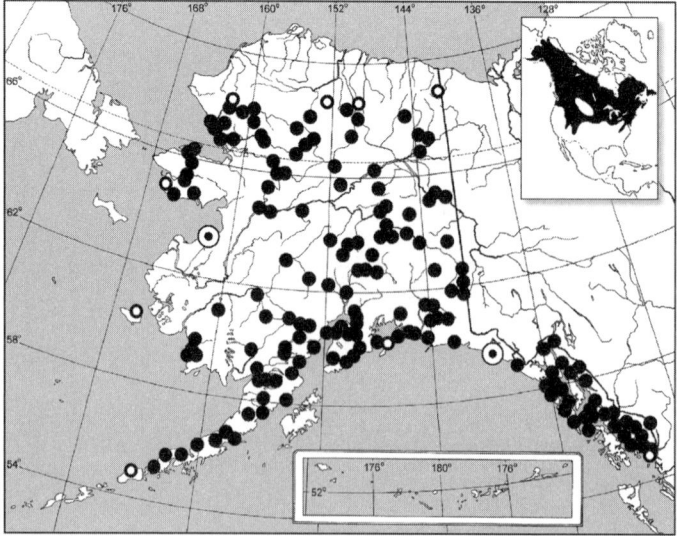

MAP 35. Distribution of cinereus shrew, *Sorex cinereus*.

⊙ = Type Locality
● = Specimen Record (n = 10,884)
○ = Marginal Records. 68.66667, −141.25000, Table Mtn Quad., Mancha Creek (1 UAM); 54.81667, −130.65000, Prince Rupert Quad., 2 km NW of Willard Inlet mouth (3 UAM); 60.16667, −147.25000, Seward Quad., Montague Island (6 UAM); 54.91670, −164.30000, Unimak Quad., Unimak Island, Urilia Bay (2 USNM); 60.38333, −166.18333, Nunivak Island Quad., Nunivak Island, Mekoryuk (2 UAM); 64.87444, −166.21361, Nome Quad., Woolley Lagoon (3 UAM); 68.20972, −159.83139, Misheguk Mtn Quad., Aniralik Lake (1 UAM); 68.57719, −152.95381, Chandler Lake Quad., Fortress Mountain (11 UAM); 68.06667, −149.58333, Philip Smith Mts Quad., Chandalar Shelf, mile 239.4 Dalton Hwy (7 UAM).

Gravina, Halleck, Hassler, Herbert Graves, Kadin, Krestof, Kruzof, Kuiu, Kupreanof, Lemesurier, Lester (Beardslee Island), Mitkof, Moser, Partofshikof, Read, Revillagigedo, Wrangell, and Yakobi islands (MacDonald and Cook 2007). This shrew is not present on Admiralty Island or any of the islands that comprise the Prince of Wales Archipelago, or nearby Zarembo Island.

Southcentral Region. This shrew occurs region-wide and on the following islands: Kalgin Island in Cook Inlet (UAM); Chenega Island, Crafton Island, Elrington Island, Evans Island, Hawkins Island, Hinchinbrook Island, Hoodoo Island, Latouche Island, and Montague Island in Prince William Sound (MVZ, UAM).

Central Region. The cinereus shrew occurs extensively throughout central Alaska.

Southwest Region. This shrew inhabits the mainland and Unimak Island (USNM). Worthy of closer scrutiny are two small, tricolored shrew specimens (USNM 246483, 288092) from west of Unimak Pass on Tigalda Island, one of the Fox Islands group of which Unalaska Island is a member. These specimens, identified as *S. cinereus,* may represent *S. ugyunak* or possibly *S. pribilofensis*.

Western Region. *Sorex cinereus* occurs throughout the western mainland and on Nunivak Island (USNM, UAM).

Northern Region. The status and distribution of northernmost populations of *S. cinereus* in relation to *S. ugyunak* is unclear and in need of further investigation.

Habitat

The cinereus shrew is a dominant and periodically abundant species in a wide variety of habitats throughout its broad range (Nagorsen 1996), and may be especially abundant in riparian areas with dense ground cover (Banfield 1974; MacDonald 1980).

Status

Cinereus shrews are common and widespread throughout the state, but their abundance can fluctuate considerably from year to year.

Fossils

Fossils of this species have been identified from over thirty Pleistocene faunal sites in North America; none is from Alaska (Kurtén and Anderson 1980). Storer (2004a) tentatively identified *S. cinereus* from Middle Pleistocene (Late Irvingtonian) fossil deposits in Yukon Territory.

Pygmy Shrew

Sorex hoyi
Baird, 1857

Other Common Names

None.

Systematics

The pygmy shrew was formerly placed in a separate genus, *Microsorex* (Long 1974), until Diersing (1980) and Junge and Hoffmann (1981) subsumed it as a subgenus of *Sorex*. Allozyme data by George (1988) supported its inclusion in the genus *Sorex* but considered the subgeneric assignment to *Microsorex* invalid because of its close relationship with other species of subgenus *Otisorex*.

One weakly defined subspecies (Youngman 1975) is recognized in Alaska (Diersing 1980; Hall 1981).

Sorex hoyi eximius

Original Description. 1901. *Sorex (Microsorex) eximius* Osgood, *N. Amer. Fauna,* 21:71, September 26.

Type Locality. Tyonek, Cook Inlet, Alaska.

Type Specimen. USNM 107126.

Range. Alaska and Yukon Territory by various authors (e.g., Hall 1981; Long 1999).

Global Distribution

Nearctic—The pygmy shrew is found across the boreal regions of Alaska, Canada, and the northern contiguous United States, with isolated populations in the Appalachian Mountains and the central Rocky Mountains (Nagorsen 1996).

Alaska Distribution

The distribution of this species in Alaska is poorly understood with specimens largely confined to the forested regions of the state (map 36).

Southcentral Region. There is only limited information on the occurrence of this species in the southcentral region of the state. Pygmy shrews have been collected northeast of Anchorage on Fort Richardson (Peirce 2003), near Talkeetna (UAM), at 18 mile Nabesna Road (Rausch 1967), and at an expanded number of sites across the western slope of the Kenai Peninsula (UAM, MSB).

Central Region. This species probably occurs throughout the region.

Southwest Region. Murie (1959) reported a pygmy shrew from west of Lake Clark and

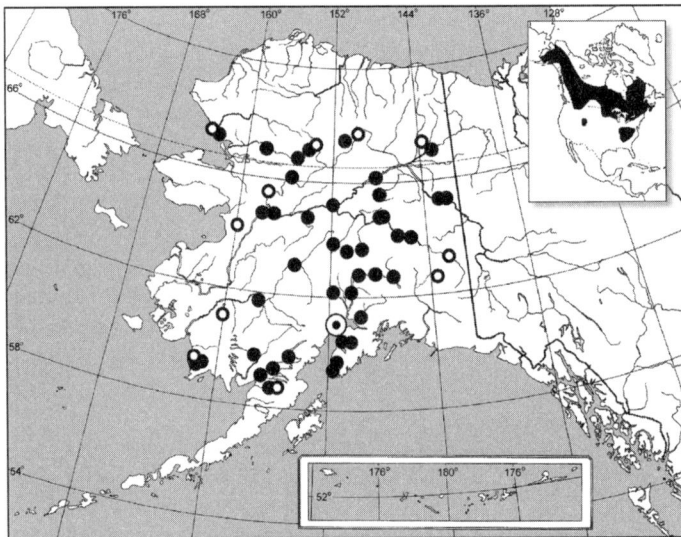

MAP 36. Distribution of pygmy shrew, *Sorex hoyi.*

⊙ = Type Locality
● = Specimen Record (n = 437)
○ = Marginal Records. 67.61667, –149.78333, Chandalar Quad., Dietrich River, mile 207 Dalton Hwy (1 UAM); 67.36667, –143.80000, Coleen Quad., Small Lake (1 UAM); 63.30000, –142.60000, Tanacross Quad., junction Taylor Hwy and Alaska Hwy (1 USNM); 62.56700, –143.53300, Nabesna Quad., Mi. 18 Nabesna Road, (1 Rausch 1967); 58.67590, –155.42030, Mt Katmai Quad., Portage Lake (1 UAM); 59.12583, –161.57250, Goodnews Bay Quad., 0.7 miles NE of Goodnews Bay (1 UAM); 60.98333, –160.06667, Bethel Quad., Nyac area, Tuluksak River, 2 km upstream from confluence with Granite Creek (2 UAM); 64.3745, –159.5580, Norton Bay Quad., North Fork Unalakleet River (1 UAM); 65.48333, –157.76667, Kateel River Quad., Upper Pitka River (8 UAM); 67.26375, –163.67714, Noatak Quad., Kakagrak Hills (1 UAM); 67.35333, –153.67844, Survey Pass Quad., W side of Takahula Lake (1 UAM).

another 129 km up the Kakwok River. This species has been found near Goodnews Bay (Peirce and Peirce 2000a), Turquoise Lake (MU; UAM; Cook and MacDonald 2004b), and near Idavain and Naknek lakes (Cook and MacDonald 2006), the latter constituting first records for Katmai National Park and an extension of range southwestward from Nondalton (Rausch 1967).

Western Region. Records of pygmy shrew from western Alaska are limited to near Nyac at the eastern boundary of the region (UAM), Goodnews River in Togiak National Wildlife Refuge (Peirce and Peirce 2000a; UAM), the Nulato Hills (UAM), the upper Kobuk River valley (MacDonald and Cook 2003; Swanson 1996), the Kakagrak Hills and Igisukruk Mountain in Cape Krusenstern National Monument (UAM; Cook and MacDonald 2004a), and the Waring Mountains (UAM; Cook and MacDonald 2004a).

Northern Region. A skin-only specimen (CAS 24511) from Kaolak River at 70° N latitude is cataloged as a pygmy shrew, but lack of an associated skull has prevented verification.

Habitat

Pygmy shrews can be found in a variety of boreal habitats ranging from various forests, disturbed sites, shrub thickets, meadows, marshes, bogs, and especially riparian situations with dense ground cover near open water (Long 1974; MacDonald 1980).

Status

This shrew is apparently rare to uncommon throughout its range; however, surveys using pitfall traps, the most efficient method to collect this shrew (MacDonald 1980), need to be more extensive.

Fossils

Storer (2003a) included, without elaboration, *Microsorex* sp. in an Early Pleistocene faunal assemblage along the Yukon River in Yukon Territory. Late Pleistocene–Early Holocene remains of this species are common south of its current range (Kurtén and Anderson 1980; Long 1974).

St. Lawrence Island Shrew

Sorex jacksoni
Hall and Gilmore, 1932

Other Common Names
None.

Systematics

The St. Lawrence Island shrew was placed in the *arcticus* species complex by Hall and Gilmore (1932), and in the *cinereus* species complex by Hoffmann and Peterson (1967).

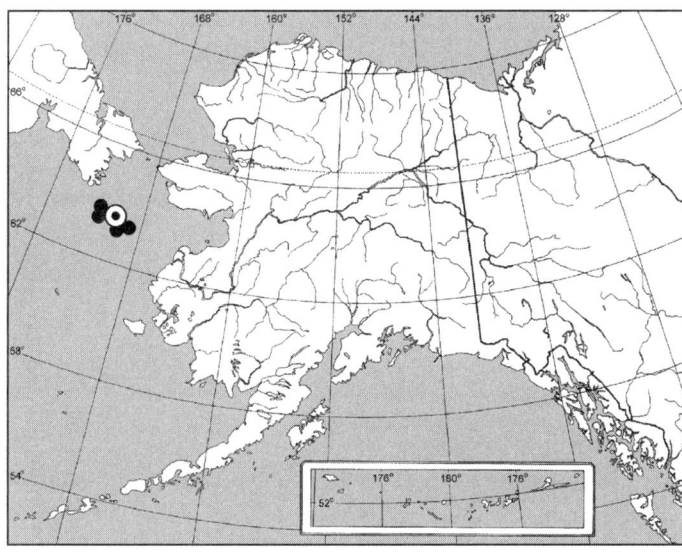

MAP 37. Distribution of St. Lawrence Island shrew, *Sorex jacksoni.*

⊙ = Type Locality
● = Specimen Record (n = 88)

It was then separated from the *cinereus* complex by Junge and Hoffmann (1981). Van Zyll de Jong (1982) included the Beringian shrew taxa *ugyunak, leucogaster,* and *portenkoi* within *S. jacksoni,* but later retained all as distinct species (1991). More recently, Rausch and Rausch (1995) compared shrews from St. Lawrence Island and *S. cinereus* from the Alaska mainland. They could discern no differences in karyotypes, in structure of the glans penis, or in relationships of the medial tines of the incisors, and concluded that the shrews on St. Lawrence Island should be regarded a subspecies of *S. cinereus.*

Results of a recent molecular genetic study by Demboski and Cook (2003) indicated a relatively distant relationship between *jacksoni* and *cinereus* but close unresolved relationship of *jacksoni* with four other Beringian shrews—*camtschatica, portenkoi, pribilofensis,* and *ugyunak.*

Sorex jacksoni

Original Description. 1932. *Sorex jacksoni* Hall and Gilmore, *Univ. California Publ. Zool.,* 38:392, September 17.

Type Locality. Sevoonga [= Savoonga], 2 miles east of North Cape, St. Lawrence Island, Bering Sea, Alaska.

Type Specimen. MVZ 51142.

Distribution. *Beringian*—This shrew is restricted to St. Lawrence Island in the Bering Sea (map 37) but may not be distinct from other Beringian shrews (van Zyll de Jong 1999a).

Habitat

According to Fay and Sease (1985), when abundant, this shrew inhabits old village sites and three major habitat types on the island: bog/wet tundra, alpine/fell-field tundra, and mesic tundra. When scarce, it is found only in fell-field habitats and boulder scree, especially within auklet nesting colonies. In winter, it often invades human dwellings.

Status

St. Lawrence Island shrews are highly variable in numbers from year to year in their limited range (Fay and Sease 1985). This island endemic is listed as Endangered by IUCN (see Appendix 6).

Fossils

None has been reported.

Dusky Shrew

Sorex monticolus
Merriam, 1890

Other Common Names

Montane shrew.

Systematics

Sorex monticolus has long been included in the *Sorex vagrans-obscurus* group (Smith and Belk 1996). *Sorex obscurus* was considered a subspecies of *S. vagrans* (Findley 1955), but studies by Hennings and Hoffmann (1977), Hawes

(1976, 1977), and Junge and Hoffmann (1981) presented evidence that the two are distinct species. The older name *monticolus* has priority over *obscurus*. Allozyme data by George (1988) lent additional support to this classification. A study of sequence variation of the cytochrome *b* gene by Demboski and Cook (2001) and Cook et al. (2001) further refined *S. monticolus* by identifying two distinctive clades, Pacific Coastal and Continental (average uncorrected sequence divergence was 5.3 percent). Contact between members of these clades may occur along the west side of the upper Lynn Canal, northern southeast Alaska. Further studies of contact zones that use additional markers may help clarify the evolutionary dynamics and taxonomic status of these populations.

Hennings and Hoffmann (1977) split *S. monticolus* into eighteen subspecies. Carraway (1990) and Alexander (1996) reduced this number to fourteen subspecies, of which six occur in Alaska. As pointed out by Alexander (1996), the shrews from the coastal islands of southeast Alaska and British Columbia need further analysis to clarify their status.

Sorex monticolus alascensis

Original Description. 1895. *Sorex obscurus alascensis* Merriam, *N. Amer. Fauna*, 10:76, December 31.

Type Locality. Yakutat, Alaska.

Type Specimen. USNM 73539.

Range. Coastal Alaska from Prince William Sound to the Taku Inlet and Admiralty Island (Alexander 1996).

Remarks. Includes *Sorex glacialis* Merriam, 1900, *Proc. Washington Acad. Sci.*, 2:16, March 14, type from Point Gustavus, east side of entrance to Glacier Bay, Alaska (USNM 97709).

Sorex monticolus elassodon

Original Description. 1901. *Sorex longicauda elassodon* Osgood, *N. Amer. Fauna*, 21:35, September 26.

Type Locality. Cumshewa Inlet near old Indian village of Clew, Moresby Island, Queen Charlotte Islands, British Columbia.

Type Specimen. USNM 100597.

Range. Central and southern outer islands of the Alexander Archipelago of southeast Alaska (excluding Coronation and Warren islands) to Haida Gwaii (Queen Charlotte Islands), British Columbia.

Sorex monticolus longicauda

Original Description. 1895. *Sorex obscurus longicauda* Merriam, *N. Amer. Fauna*, 10:74, December 31.

Type Locality. Wrangell, Alaska.

Type Specimen. USNM 74711.

Range. Taku River, Alaska, to River Inlet, British Columbia.

Remarks. Alexander (1996:29) indicated that the correct spelling of this subspecies should be "*longicaudus*" to agree in gender with the masculine *Sorex* and *monticolus*.

Sorex monticolus malitiosus

Original Description. 1919. *Sorex obscurus malitiosus* Jackson, *Proc. Biol. Soc. Washington*, 32:23, April 11.

Type Locality. East side of Warren Island, Alaska.

Type Specimen. MVZ 8401.

Range. Warren and Coronation islands, Alexander Archipelago, and southeast Alaska.

Sorex monticolus obscurus

Original Description. 1891. *Sorex vagrans similis* Merriam, *N. Amer. Fauna*, 5:34, July 30. Renamed *Sorex obscurus* in 1895 by Merriam (*N. Amer. Fauna*, 10:72, December 31).

Type Locality. Timber Creek, Lemhi Mountain (= Salmon River Mts.), 8200 ft. (2,440 m), Lemhi Co., Idaho.

Type Specimen. USNM 23525/30943.

Range. North- and east-central Alaska through the Rocky Mountain states.

Remarks. Shrews from the upper Lynn Canal area of southeast Alaska were included in this taxon by Alexander (1996).

Sorex monticolus shumaginensis

Original Description. 1900. *Sorex alascensis shumaginensis* Merriam, *Proc. Washington Acad. Sci.*, 2:18, March 14.

Type Locality. Popof Island, Shumagin Islands, Alaska.

Type Specimen. USNM 97993.

Range. Southcentral (Kenai Peninsula westward), southwestern, and western Alaska.

Global Distribution

Nearctic—The dusky or montane shrew inhabits western North America from Alaska to northern Mexico (George 1999a).

Alaska Distribution

Dusky shrews are widely distributed across Alaska south of the Brooks Range. This shrew also occurs on many of the islands along Alaska's southern coast (map 38).

Southeast Region. *Sorex monticolus* is found throughout the mainland and on islands in the Alexander Archipelago east of Chatham Strait, including Admiralty, Anguilla,

Annette, Baker, Barrier, Beardslee (including Lester and Young), Bell, Betton, Black, Cap, Coronation, Dall, Deer, Dog, Douglas, Duke, Eagle, Etolin, Forrester, Gedney, Gravina, Hassler, Heceta, Hoot, Hotspur, Inian, Kadin, Kosciusko, Kuiu, Kupreanof, Lemesurier, Long, Lowrie, Lulu, Marble, Mary, Mitkof, Noyes, Owl, Percy, Pleasant, Prince of Wales, Revillagigedo, San, San Fernando, San Juan Bautista, Sangao, Santa Rita, Shelikof, Shelter, Shrubby, Spanish, Stone, Suemez, Sullivan, Tuxekan, Warren, Woewodski, Woronkofski, Wrangell, and Zarembo (MacDonald and Cook 2007).

Southcentral Region. Dusky shrews occur extensively throughout this region, including many islands in Prince William Sound, specifically Disc, Eleanor, Elrington, Esther, Evans, Green, Hawkins, Hinchinbrook, Knight, Latouche, and Montague islands (MVZ; UAM; USNM; Heller 1910).

Central Region. This species has been documented throughout central Alaska.

Southwest Region. *Sorex monticolus* occurs throughout southwestern Alaska and the Alaska Peninsula to Unimak Island (USNM 246467; Murie 1959). It is also found on the Shumagin Islands (USNM and UAM specimens from Atkin, Herendeen, Nagai, Popof, and Unga islands), Sanak Island (USNM), and Ugaiushak Island (UAM). There are also two UAM specimens from Round Island in Bristol Bay, but there are

no records from any island in the Kodiak Archipelago.

Western Region. Dusky shrews occur throughout most of the mainland of western Alaska, including the Seward Peninsula (UAM; MacDonald and Cook 2002), but are not known from any islands in the Bering Sea.

Northern Region. This species is marginally distributed along the northern foothills of the Brooks Range.

Habitat

The dusky shrew is found in many different habitats, from coastal and boreal forests to riparian shrub thickets in the mountains and in the subarctic tundra-taiga transition at higher latitudes (Nagorsen 1996; van Zyll de Jong 1983). Dense understory ground cover in moist or wet situations is a main component of suitable microhabitat (Belk et al. 1990; Doyle 1990; Hawes 1977; MacDonald 1980).

Status

Dusky shrews are common and sometimes abundant.

Fossils

All fossils assigned to this species are from the Late Pleistocene and Holocene (Bonnichsen et al. 1986; Kurtén and Anderson 1980; Miller 1971; Mullican and Carraway 1990). Fossil remains of dusky shrews from Alaska are limited to those recovered with remains of *Peromyscus* (cf *P. keeni*) and *Marmota caligata* (dated beyond the radiocarbon limit of 40,000

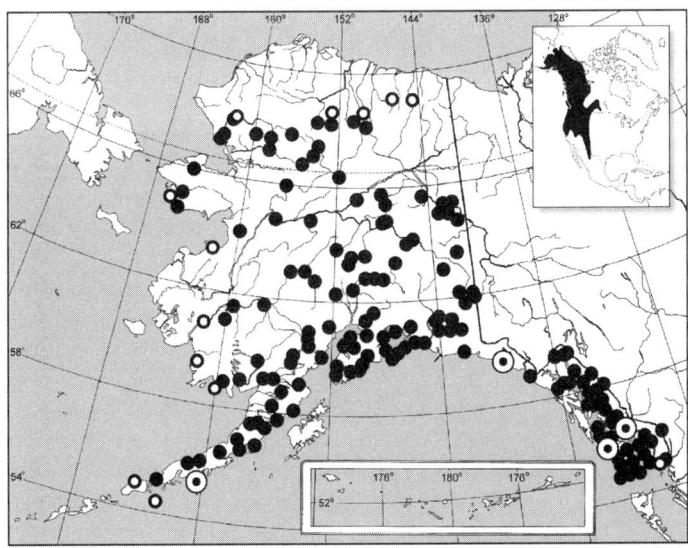

MAP 38. Distribution of dusky shrew, *Sorex monticolus.*

⊙ = Type Locality
● = Specimen Record (n = 5315)
○ = Marginal Records. 69.13330, −146.96670, Mt Michelson Quad., Wahoo Lake (2 KU); 69.08333, −144.61667, Mt Michelson Quad., upper Hulahula River (1 UAM); 54.98333, −131.00000, Prince Rupert Quad., Foggy Bay, Kirk Point (14 UAM); 58.60000, −159.96667, Nushagak Bay Quad., Round Island (2 UAM); 59.37775, −161.70589, Goodnews Bay Quad., Cripple creek between Twin and Cone mountains (1 UAM); 60.71967, −161.75000, Bethel Quad., Bethel (7 USNM); 63.48333, −162.03333, St Michael Quad, St. Michael (1 USNM); 64.80000, −166.46667, Nome Quad., Cape Wooley (2 UAM); 68.13333, −161.63333, Misheguk Mtn Quad., Brooks Range, Kugururok R, T32N R13W Sec 21 (2 UAM); 68.57719, −152.95381, Chandler Lake Quad., Fortress Mountain (2 UAM); 68.62833, −149.59333, Philip Smith Mts Quad., Toolik Lake Research Station (1 UAM).

SORICOMORPHA

133

ybp) from silt deposits in a limestone cave on northern Prince of Wales Island, southeast Alaska (Heaton and Grady 2003).

American Water Shrew

Sorex palustris
Richardson, 1828

Other Common Names

Water shrew, northern water shrew, navigator shrew.

Systematics

Of the nine subspecies listed by Hall (1981), only one occurs in Alaska. *Sorex alaskanus* is restricted to Glacier Bay and is most likely a subspecies of *S. palustris* (see *S. alaskanus* account).

A recent study examining variation in cytochrome *b* sequence data (O'Neill et al. 2005) suggested that *S. palustris* may consist of two species: a boreal eastern form (*S. palustris*) and a Cordilleran form (*S. navigator*).

Sorex palustris navigator

Original Description. 1858. *Neosorex navigator* Baird, Mammals, *in Rep. Expl. Surv. Railr. to Pacific*, 8(1):11, July 14.

Type Locality. Near head of Yakima River, Cascade Mountains, Kittitas County, Washington.

Type Specimen. USNM 629/1780.

Range. Alaska, east to northwestern Northwest Territories, south to southcentral California, southern Utah, northern New Mexico, and isolated populations in the White Mountains of Arizona.

Global Distribution

Nearctic—The water shrew inhabits central Alaska and Canada south to the Sierra Nevada, Rocky, and Appalachian mountains (Hall 1981; Hutterer 2005).

Alaska Distribution

The distribution of this shrew in Alaska (map 39) is poorly understood and museum specimens are few.

Southeast Region. Specimens of this shrew are from the mainland mountains north of Haines (UAM) southward to the head of Portland Canal, and on Wrangell Island and perhaps Revillagigedo and Kupreanof islands in the Alexander Archipelago (MacDonald and Cook 2007).

Southcentral Region. Water shrew records in this region are from the West Fork of the Gulkana River, Peters Creek near Anchorage, Fort Richardson (Peirce 2003), Wasilla (USNM), and at Twin

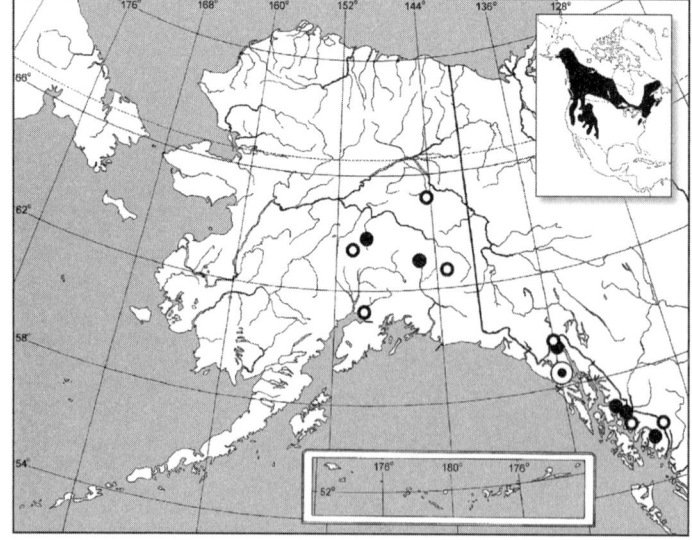

MAP 39. Distribution of American water shrew, *Sorex palustris*.

● = Specimen Record (n = 37)
○ = Marginal Records. 65.22750, –144.50028, Circle Quad., Big Windy Hot Springs (8 UAM); 62.53056, –143.25917, Nabesna Quad., Twin Lakes (1 UAM); 59.62322, –136.07497, Skagway Quad., Tohitkah Mountain (1 UAM); 56.02694, –130.07056 Bradfield Canal Quad., Salmon River, Hyder area (1 UAM); 56.26972, –132.07056, Petersburg Quad., Wrangell Island, Fools Creek (1 UAM); 61.21955, –149.69167, Anchorage Quad., Fort Richardson (1 UAM); 63.50000, –150.58333, Healy Quad., Denali NP&P, Moose Creek, 11 miles N of Denali Rd. (1 UAM).

Glacier Bay water shrew, *Sorex alaskanus*.

◉ = Type Locality
Specimen Record (n = 3): Type locality (2 USNM); 58.45000, –135.91667, Juneau Quad., Glacier Bay, Bartlett Cove (1 UAM).

Lakes on the Nabesna Road (Cook and MacDonald 2003; MacDonald and Elliot 1984).

Central Region. The northernmost specimens of this species are from Big Windy Hot Springs, approximately 160 km northeast of Fairbanks on a tributary of Birch Creek south of the Yukon River (Cook et al. 1997). Regional records also include the north side of the Alaska Range near Healy (MacDonald and Elliot 1984), and Moose Creek in Denali National Park and Preserve (UAM).

Southwest Region. Manville and Young (1965) reported the water shrew in the vicinity of Iliamna Lake, but this record remains unsubstantiated.

Habitat

Water shrews are highly aquatic and usually restricted to dense ground cover along streams, lakes, beaver ponds, and marshes from sea level to alpine areas (Banfield 1974; Beneski and Stinson 1987; Nagorsen 1996).

Status

Water shrews appear widespread, but localized to riparian areas.

Fossils

No fossils of this shrew have been reported from Alaska or northwest Canada.

Glacier Bay Water Shrew

Sorex alaskanus
Merriam, 1900

Other Common Names

None.

Systematics

Two male water shrews taken on 12 June 1899 by A. K. Fisher from Point Gustavus were described as a new subspecies of *S. navigator* (Merriam 1900). The subsequent elevation of these two specimens to full species status, *S. alaskanus*, by Jackson (1926, 1928) has been rejected by Junge and Hoffmann (1981) and Harris (1999), although Hall (1981), Beneski and Stinson (1987), George (1988), and Hutterer (2005) treated it as distinct. Preliminary molecular evidence bearing on the taxonomic status of *S. alaskanus* suggests full species status is unwarranted (K. Hildebrandt, UAM, pers. comm.).

Sorex alaskanus

Original Description. 1900. *Sorex navigator alaskanus* Merriam, *Proc. Washington Acad. Sci.*, 2:18, March 14.

Type Locality. Point Gustavus, Glacier Bay, Alaska.

Type Specimen. USNM 97713.

Distribution.

Nearctic—The Glacier Bay water shrew is known only from the two original specimens taken at Point Gustavus (map 39) in 1899 and deposited at USNM and an additional specimen collected at Bartlett Cove in 1970 (UAM 49979).

Habitat

The habitat of this shrew is probably similar to *S. palustris*.

Status

The current status of *S. alaskanus* is unknown. In 1994 it was recognized as an ESA Candidate, Category 2 species (i.e., listing was possibly appropriate but for which persuasive data on biological vulnerability and threat was not available) (Federal Register 59 FR 58982 11/15/1994).

Fossils

No fossils are known from this species.

Pribilof Island Shrew

Sorex pribilofensis
Merriam, 1895

Other Common Names

None.

Systematics

Sometime between 1840 and 1848 (in Fay and Sease 1985), I. G. Voznesenskii, of the Zoological Museum in St. Petersburg, Russia, was the first to collect shrews from the Pribilof Islands. He also collected two specimens that were labeled only as "Unalaska," possibly not referring specifically to Unalaska Island but to the Unalaska Administrative District of the Russian-American Company, which included the Pribilof Islands. In 1889, G. E. Dobson described one of Voznesenskii's "Unalaska" specimens, assuming that it was from Unalaska Island in the eastern Aleutians (where no shrews have since been taken), and assigning to it the name *Sorex hydrodromus*. Six years later and from specimens collected on St. Paul Island, C. H. Merriam (1895) formally described and named them *Sorex pribilofensis* and designated one of his specimens as the holotype. Such discrepancies have caused confusion and considerable debate in the literature on the correct name for this species (Baranova et al. 1981; Hoffmann 1964; Hoffmann and Peterson 1967; Junge and Hoffmann 1981; van Zyll de Jong 1982, 1991; Yudin 1969). *Sorex pribilofensis* has name priority given its unambiguous type locality and dental features (Hutterer 2005; Rausch and Rausch 1997), but an effort to capture and study shrews on Unalaska Island is needed.

Sorex pribilofensis

Original Description. 1889. *Sorex pribilofensis* Merriam, *N. Amer. Fauna*, 10:87, December 31.

Type Locality. St. Paul Island, Pribilof Group (in Bering Sea), Alaska.

Type Specimen. USNM 30911

Remarks. Replaces *Sorex hydrodromus* Dobson, 1889, *Ann. Mag. Nat. Hist.*, ser. 6, 4:373, November; type locality "Unalaska Islands, Aleutian Islands" but probably from St. Paul Island, Pribilof Islands according to Hoffmann and Peterson (1967). E. R. Hall (in Murie 1959) and in 1991, R. S. Hoffmann and J. A. Cook examined the two shrew specimens, housed at the Zoological Institute of the Academy of Sciences in St. Petersburg, claimed to have originated from Unalaska Island: ZISP 2389, the designated type specimen of *hydrodromus*

collected sometime between 1840 and 1848, and ZISP 2370, a pregnant female collected at Unalaska in 1848.

Distribution

Beringian—*Sorex pribilofensis* is known only from St. Paul Island, Pribilof Islands, Bering Sea, Alaska (map 40).

The status of shrews on Unalaska Island remains unclear. Although Murie (1959) reported sightings of shrews on the island, no new specimens have been secured (Peterson 1967). Two small, tricolored shrew collected in the 1920s (and housed at the USNM) from west of Unimak Pass on Tigalda Island, one of the Fox Islands of which Unalaska Island is a member, warrant study using molecular techniques.

Habitat

Preble (1923) collected this shrew in "a marshy tract, grown up to rank grasses, bordering a shallow pond," and Fay and Sease (1985) reported its capture in hillside habitats dominated by beach rye, *Elymus*. Byrd and Norvell (1993) found the highest densities in dune and grass-umbel habitats.

Status

Jackson (1928) regarded the shrews on St. Paul Island as "not uncommon," and F. H. Fay found them abundant there in August 1965 (Fay and Sease 1985). Byrd and Norvell (1993) found relatively high numbers of this shrew widely distributed on the island during the summers of 1986 and 1987. *Sorex pribilofensis* is listed as Endangered by IUCN. In 1994 it was recognized (as *S. hydrodromus*) as an ESA Candidate,

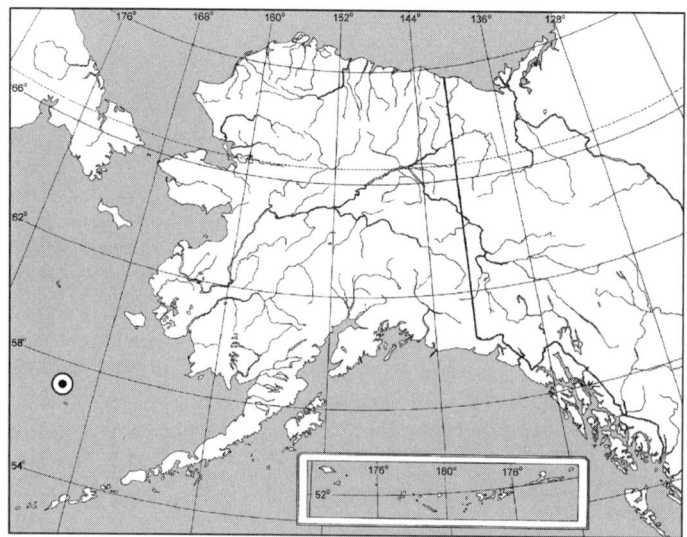

MAP 40. Distribution of Pribilof Island shrew, *Sorex pribilofensis*.

⊙ = Type Locality
Specimen Record (n = 220)

Category 2 species (i.e., listing was possibly appropriate but for which persuasive data on biological vulnerability and threat was not available) (Federal Register Doc 94-28029).

Fossils

No fossils are known.

Tundra Shrew

Sorex tundrensis
Merriam, 1900

Other Common Names

Black-backed shrew, tundra saddle-backed shrew.

Systematics

The tundra shrew was considered a subspecies of *Sorex arcticus* (Bee and Hall 1956; Hall 1981), but is now recognized as a distinct species (Hutterer 2005; Youngman 1975). Palearctic populations formerly referred to *arcticus* were included in *tundrensis* by Junge et al. (1983) and van Zyll de Jong (1983). A chromosome study by Rausch and Rausch (1993) suggested that the Nearctic and Palearctic populations are distinct species. Van Zyll de Jong (1999b) tentatively concurred, noting the similar appearance of the Siberian shrews but significant chromosomal differences. Hutterer (2005) also noted the extensive karyotype variability displayed in this species, but emphasized the similarity of a karyotype from Canada to some in Siberia as reported by Meylan and Hausser (1991) and supported by Lukáčová et al. (1996). Preliminary analyses of mtDNA sequence data from Alaska and Siberia specimens (A. Hope, pers. comm.) is supportive of a Holarctic distribution.

Various subspecies have been proposed from across this shrew's extensive range (e.g., Junge et al. 1983). According to Okhotina (1984), the nominate subspecies *tundrensis* occurs on both sides of Bering Strait. Hutterer (2005) considered this taxon monotypic. He and others (e.g., Rausch and Rausch 1993) noted that Eurasian taxa referred to as *S. tundrensis* probably make up a complex of sibling species. A thorough review of all populations using an integrated morphological, chromosomal, and molecular assessment is needed.

Sorex tundrensis tundrensis

Original Description. 1900. *Sorex tundrensis* Merriam, *Proc. Washington Acad. Sci.*, 2:16, March 14.

Type Locality. St. Michael, Norton Sound, Alaska.

Type Specimen. USNM 99286.

Range. Alaska, northwest Canada, and northeastern Russia, from Chukotka southward to the Amur River (Okhotina 1984).

Remarks. Taxonomic designations remain unsettled and in need of revision (see above).

Global Distribution

Holarctic—Shrews referred to *S. tundrensis* occur widely across Eurasia, on the Sakhalin Islands, and across the Bering Strait in Alaska and northwestern Canada (Hutterer 2005).

Alaska Distribution

This species is widely distributed throughout Alaska, but is absent from any islands (map 41).

Southeast Alaska. The discovery of this species in extreme northwestern British Columbia (Nagorsen and Jones 1981) suggests that the alpine and subalpine zones of the northern mainland of this region should be surveyed.

Southcentral Region. This species has not been documented from the Kenai Peninsula, the Wrangell–St. Elias region of Alaska, or any islands in the Gulf of Alaska. We have been unable to document shrew specimens from Kodiak-Afognak islands and agree with Clark (1958) that no shrews of any kind occur there and that Hall (1981) and others (e.g., van Zyll de Jong 1999b) are in error in reporting shrews on these islands.

Central Region. *Sorex tundrensis* occurs throughout central Alaska.

Southwestern Region. There have been only a few specimens of this species reported from this region of the state, and none is recorded from anywhere on the Kenai Peninsula, Prince William Sound, or the Copper River basin. Tundra shrews have been documented on the Alaska Peninsula as far west as Wide Bay (Orr 1939; CAS) and the lower Dog Salmon River east of Ugashik Bay (UAM).

Western Region. Records of this shrew are scattered throughout the mainland of western Alaska; none occur on any Bering Sea island.

Northern Region. Tundra shrews probably occur throughout northern Alaska but remain poorly documented.

Habitat

The tundra shrew inhabits a variety of arctic and alpine tundra habitats as well as forest,

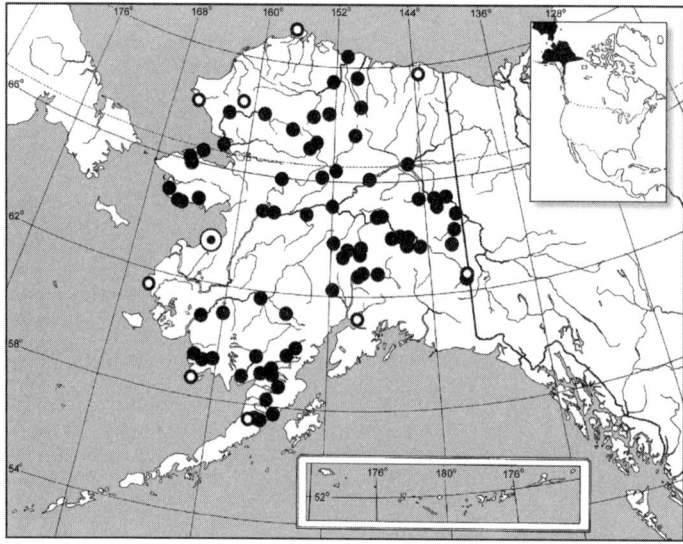

MAP 41. Distribution of tundra shrew, *Sorex tundrensis*.

⊙ = Type Locality
● = Specimen Record (n = 994)
○ = Marginal Records. 71.29056, −156.788611, Barrow Quad., Barrow (1 MVZ); 69.70722, −143.51500, Demarcation Point Quad., 6 miles NE VABM Bitty on the Jago River (2 UAM); 62.34100, −141.06483, Nabesna Quad., Scottie Creek (1 UAM); 61.17127, −149.71263, Anchorage Quad., Ft. Richardson (3 UAM); 57.42305, −157.30870, Ugashik Quad., Lower Dog Salmon River (1 UAM); 58.55000, −161.76667, Hagemeister Island Quad., Cape Peirce (4 UAM); 68.10000, −165.75000, Point Hope Quad., Ogotoruk Creek drainage (31 UAM); 68.47500, −161.47750, Misheguk Mtn Quad., 8 km W of Copter Peak (1 UAM).

shrub, bog, and marsh habitats within the taiga at lower elevations and latitudes (Banfield 1974; MacDonald 1980; Nagorsen 1996).

Status

This species is occasionally relatively common.

Fossils

Rand (1954) and others theorized that *S. tundrensis* originated in unglaciated Beringia during the Late Pleistocene (Wisconsinan). Shrew fossils clearly identified as this species are lacking (Junge et al. 1983); however, the small series of "*Sorex arcticus*" specimens recovered from Late Pleistocene–Early Holocene deposits in central Alaska (Broken Mammoth Site, central Tanana Valley) as reported by Yesner (2001) warrant further study.

Barren Ground Shrew

Sorex ugyunak
Anderson and Rand, 1945

Other Common Names

None.

Systematics

The barren ground shrew was formerly included as a subspecies of *Sorex cinereus*, but van Zyll de Jong (1976, 1991) argued that *ugyunak* was distinct from *cinereus*. He further suggested that *ugyunak* is probably conspecific

with *jacksoni* on St. Lawrence Island, Alaska, along with *leucogaster* and particularly *portenkoi* from northwestern Siberia. Under this scheme, *ugyunak*, *portenkoi*, and *leucogaster* would be subsumed under the name *Sorex jacksoni*. A preliminary analysis by Rausch and Rausch (1995) indicated that *ugyunak* may not represent a separate species from *S. cinereus*, but in a more recent study of mtDNA sequences by Demboski and Cook (2003), *cinereus* readily differentiated from *ugyunak* and the four other closely related but as yet unresolved Beringian species: *camtschatica*, *jacksoni*, *portenkoi*, and *pribilofensis*. Waltari and Cook (in prep.) compared nuclear and mitochondrial phylogenies of 101 *cinereus* and *ugyunak* in sympatry from northern Alaska and found only three individuals with signals of possible hybridization.

Geographic variation in *S. ugyunak* has not been studied in enough detail to determine differentiation at the subspecific level, although van Zyll de Jong (1983) found that some cranial measurements of animals in Alaska are larger than those east of the MacKenzie River.

Sorex ugyunak

Original Description. 1945. *Sorex cinereus ugyunak* Anderson and Rand, *Canadian Field-Nat.*, 59:62, October 16.

Type Locality. Tuktuk (Tuktuyaktok) [= Tuktoyaktuk], about 20 miles southwest of Toker Point, on Arctic coast near northeastern corner of Mackenzie River delta,

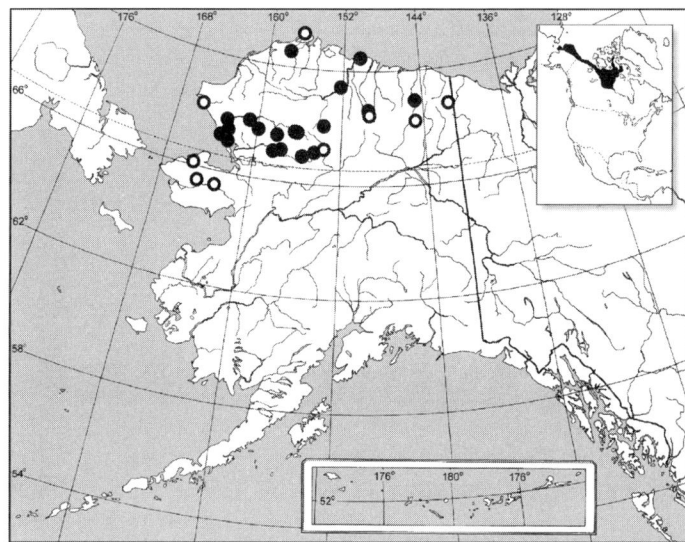

MAP 42. Distribution of barren ground shrew, *Sorex ugyunak*.

● = Specimen Record (n = 162)
○ = Marginal Records. 71.35000, –156.60000, Barrow Quad., Point Barrow (1 UAM); 68.66667, –141.25000, Table Mtn Quad., Mancha Creek (2 UAM); 68.05000, –145.00000, Mt Michelson Quad., Old John Lake (1 UAM); 68.49278, –149.49250, Philip Smith Mts Quad., Galbraith Lake airport road (1 UAM); 67.34572, –153.66025, Survey Pass Quad., S side of Takahula Lake (2 UAM); 65.38938, –163.26447, Bendeleben Quad., Kuzitrin Lake (1 UAM); 65.42500, –164.64333, Bendeleben Quad., Kougarok Airfield, mile 83.2 Taylor Highway (1 UAM); 66.12917, –165.12917, Shishmaref Quad., Grayling Creek Shelter (1 UAM); 68.10000, –165.75000, Point Hope Quad., Ogotoruk Creek drainage (21 UAM).

Mackenzie District, Northwest Territories, Canada.

Type Specimen. AMNH 31365.

Global Distribution

Nearctic (but see above)—The barren ground shrew occurs in the arctic tundra zone from northwestern Alaska to the western shore of Hudson Bay (van Zyll de Jong 1999c).

Alaska Distribution

The distribution of the barren ground shrew in Alaska is poorly understood. It appears to be restricted to the northern region, from the Brooks Range northward to the Arctic coast, and in northwestern Alaska at least as far south as the Seward Peninsula (map 42).

Habitat

This high arctic species appears to favor low sedge-grass tundra and thickets of dwarf willow and birch (van Zyll de Jong 1999c). Bee and Hall (1956) found them in a variety of tundra communities, but indicated that damp to wet communities with grasses and sedges were optimum habitat.

Status

Barren ground shrews are widespread in arctic Alaska north of tree line. Their abundance fluctuates considerably (Bee and Hall 1956; van Zyll de Jong 1999c).

Fossils

None is known.

Alaska Tiny Shrew

Sorex yukonicus
Dokuchaev, 1997

Other Common Names

None.

Systematics

The Alaska tiny shrew is a newly described species of shrew endemic to Alaska (Dokuchaev 1997). It appears to be a closely related sister taxon of *Sorex minutissimus* from the Palearctic region (Dokuchaev 1997), although unpublished mitochondrial sequence data suggest forms in far east Siberia and Alaska may be conspecific (Hope et al., in prep.).

Sorex yukonicus

Original Description. 1997. *Sorex yukonicus* Dokuchaev, *J. Mammalogy*, 78:814, August 22.

Type Locality. Crow Creek, 1¾ miles N, 2¼ miles W Beaver Creek (64°44'N, 156°50'W) near Galena, Alaska.

Type Specimen. UAM 19268.

Global Distribution

Nearctic—The Alaska tiny shrew appears widespread but scarce across Alaska and has been captured within 3 km of neighboring Yukon Territory.

Alaska Distribution

By 2005, only thirty-eight specimens of this species had been collected from a wide

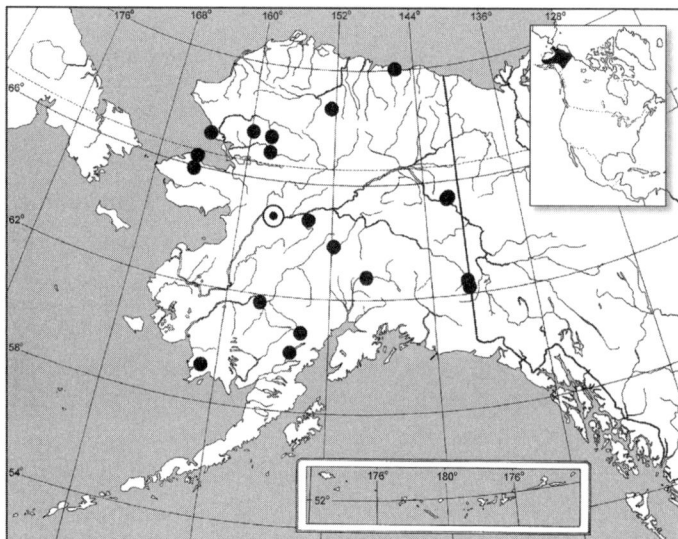

MAP 43. Distribution of Alaska tiny shrew, *Sorex yukonicus*.

⊙ = Type Locality
● = Specimen Record (n = 38; see Table 3 for a complete listing)

scattering of localities across the state (map 43 and Table 3).

Southcentral Region. A single specimen from the upper Susitna River taken in August 1982 is the earliest record of this species in Alaska and so far the only record from the region.

Central Region. Limited specimen data suggest the occurrence of this shrew throughout central Alaska and its probable occurrence in neighboring Yukon Territory.

Southwest Region. On 28 June 1999, a tiny shrew was collected near Turquoise Lake in Lake Clark National Park (N. Dokuchaev, pers. comm., 2001). In 2003, two tiny shrews were taken near the shore of Turner Bay, Lake Clark (Cook and MacDonald 2004b).

Western Region. Three tiny shrews were collected along the Goodnews River, Togiak National Wildlife Refuge (Peirce and Peirce 2000a). In 2001, four tiny shrews were pitfall trapped on the Seward Peninsula (Cook and MacDonald 2004a).

Northern Region. In 2002, a tiny shrew was collected on the north slope of the Brooks Range near Fortress Mountain (Cook and MacDonald 2004a), and in 2004 another was found dead near the Beaufort Sea coast at the mouth of the Canning River (UAM).

Habitat

Although still poorly documented, tiny shrews are found in a wide range of forested and non-forested habitats, with riparian scrub the most common habitat (Table 3).

Status

This shrew appears to be rare throughout its range, at least in comparison to the more common species of shrews captured at the same localities.

Fossils

None is known.

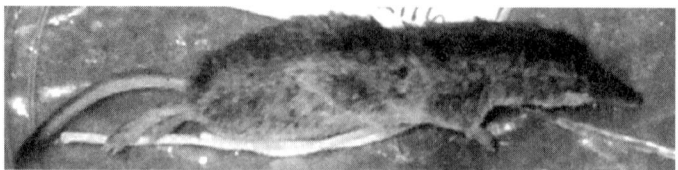

A male *Sorex yukonicus* specimen (UAM 57111; AF 52116) captured in 2001 near treeline, Kathul Mountain, upper Yukon River, Alaska. It had a total length of only 74 mm (Table 3).

Table 3. Details on all known specimens (as of August 2007) of Alaska tiny shrew, *Sorex yukonicus*. Museum catalog number (and field number), specific locality, latitude and longitude coordinates, date of capture, sex, reproductive status (T=testis size), standard measurements (total length/tail length/hind foot length = weight, in millimeters and grams), and general comments on habitat are reported.

Inst.	Cat. No.	Locality	Lat/Long	Date	Sex	Repro.	Measure	Habitat
UAM	19268	Crow Creek (Holotype)	64.73/−156.83	2 Sep 1987	M?	subadult	74/26/8.5=1.5 g	Riparian
UAM	22599	upper Susitna River (Paratype)	62.82/−149.14	23 Aug 1982	?			Riparian
UAM	29829 (AF 5267)	Little Mud River	64.75/−153.83	2 Sep 1993	F			Riparian
UAM	29830 (AF 5332)	Nulato	64.78/−157.18	26 Aug 1993	F			Riparian
UAM	45820 (AF 24378)	Goodnews River	59.30/−161.11	26 Aug 1996	F			Open low mixed shrub/sedge tussock tundra
UAM	45841 (AF 24380)	Goodnews River	59.30/−161.11	30 Aug 1996	F			Tussock tundra
UAM	45843 (AF 24379)	Goodnews River	59.30/−161.11	30 Aug 1996	F			Closed tall willow shrub
UAM	55806 (AF 46409)	Serpentine Hot Springs (Seward Pen.)	65.85/−164.70	10 Aug 2001	F	4 emb: 1R, 3L (4.6 x 4.0 CR)	72/23/8.6=2.0 g	Low scrub near hot springs
UAM	56130 (AF 46251)	Devil Mt. Lakes (Seward Pen.)	66.39/−164.48	25 Jul 2001	M	T=3.7 x 2.7	72/23/8.2=2.3	Low willow/alder
UAM	56133 (AF 46266)	Devil Mt. Lakes (Seward Pen.)	66.39/−164.48	26 Jul 2001	M		68/23/8.5=1.5 g	Low willow/alder
UAM	56157 (AF 46277)	Devil Mt. Lakes (Seward Pen.)	66.39/−164.48	27 Jul 2001	M	T=3.8 x 2.7	71/24/8.3=2.2 g	Open willow
UAM	57059 (AF 52234)	Kathul Mt. (upper Yukon)	65.37/−142.01	29 Jul 2001	F	no embryos	74/24/9=1.8 g	Open spruce forest
UAM	57104 (AF 52097)	Kathul Mt. (upper Yukon)	65.37/−142.02	25 Jul 2001	F	no embryos	73/24/9=1.8 g	Open spruce forest
UAM	57111 (AF 52116)	Kathul Mt. (upper Yukon)	65.37/−142.02	26 Jul 2001	M	T=3 x 2	74/22/9=2.3 g	Open spruce forest
UAM	57142 (AF 51995)	Kathul Mt. (upper Yukon)	65.37/−142.02	22 Jul 2001	M	T=3 x 2	[68]/23/9=2.2 wet	Closed mixed forest
UAM	57395 (AF 52000)	Kathul Mt. (upper Yukon)	65.37/−142.02	22 Jul 2001	M	T=3 x 2	79/25/9=2.2 wet	Dwarf tree woodland
UAM	59401 (AF 52249)	Kathul Mt. (upper Yukon)	65.37/−142.02	29 Jul 2001	F	young animal	71/23/9=1.6 g	Open tall scrub
UAM	57531 (AF 55119)	Carden Hills (upper Tanana)	62.31/−141.18	23 Jul 2001	M	T=2 x 1	66/23/9/4=1.9 g	Mesic grassy meadow; 1299 m elev.
UAM	57546 (AF 55142)	Braye Lakes	62.03/−141.13	26 Jul 2001	M	T=4 x 2	69/34/9/3=2.3 wet	Open spruce woodland; 1139 m elev.
UAM	58567 (AF 49777)	Glenn Creek (upper Yukon)	65.30/−142.09	10 Aug 2001	M	T<1, non-breeding	59/22/8=1.5 g	?
UAM	60167 (AF 51207)	Kandik cabin (upper Yukon)	65.38/−142.53	18 Aug 2001	M	T=5 x 4 overwinter	66/24/9=2.5 g	?

(continued on next page)

Table 3 (continued)

Inst.	Cat. No.	Locality	Lat/Long	Date	Sex	Repro.	Measure	Habitat
UAM	60281 (AF 49457)	Glenn Creek area (upper Yukon)	65.30/−142.04	8 Aug 2001	?		64/22/8=1.7 g	?
UAM	60566 (AF 49733)	Glenn Creek area (upper Yukon)	65.30/−142.04	9 Aug 2001	F		62/23/8=1.5	?
UAM	64945 (AF 61533)	Chilchukabena Lake	63.93/−151.49	14 Jul 2002	M	T=2 x 1	71/24/8=1.6 g	Streamside at edge of birch forest
UAM	72928 (AF 31547)	Red Devil (Sleetmute Quad)	61.67/−157.41	9 Aug 1999	M		73/21/9/5=2.3	?
UAM	78467 (AF 61829)	Fortress Mt. (Brooks Range)	68.53/−152.95	30 Jul 2002	M	young animal	75/24/8=1.6 g	Moist grassy swale
UAM	(IF 9024)	Lake Clark, Turner Bay	60.18/−154.565	8 Jul 2003	M	T=1 x 1	65/28/9/1=1.3 g	Lakeshore edge in willow thickets interspersed with *Iris, Elymus, Carex*
UAM	(IF 9189)	Lake Clark, Turner Bay	60.18/−154.565	9 Jul 2003	F	V=imperforate	70/25/8/xx=2.3	"
UAM	(IF 10512)	Kakagrak Hills	67.264/−163.677	8 Jul 2003	M	T=3 x 2	74/22/8/x=1.9	Open tall willow along drainage with cottongrass and *Potentilla*
UAM	(IF 10551)	Kakagrak Hills	67.264/−163.677	10 Jul 2003	M	T=4 x 2	72/22/7/x=2.2	"
UAM	(IF 10752)	Baird Mts., Salmon River	67.593/−159.815	18 Jul 2003	M	T=3 x 2	67/19/8/x=1.6	Edge of tall scrub and mesic meadow
UAM	(IF 10788)	Baird Mts., hdws. Akillik River	67.482/−158.247	23 Jul 2003	M	T=1 x 1	70/23/7/5=1.5	Open tall willow
UAM	(IF 10895)	Waring Mts.	67.018/−158.492	30 Jul 2003	M	T-1 x 1	67/22/8/x=1.5	Open low alder in tussock tundra with blueberry
UAM	(AF 68147)	Canning River Delta	70.0953/−145.7373	6 Jul 2004				?
MU	S-167362	Turquoise Lake	60.80/−153.98	28 Jun 1999				?
UWBM	39045	SE Ruby (NWR); 8 mi. S, 43.5 mi. W of Ruby; NNWR, T 10S, R **35E** [25E?]		16 Sep 1991	M	T=0.5 x 0.5	68/23/9/5=1.5 g	Riparian
UWBM	39046	SE Ruby (NWR) T 9S, R 35E		31 Aug 1991	F	subad; uterus slightly thickened	74/27/9/7=1.7 g	Riparian
UWBM	39047	4 mi. S Ruby (NWR) T 9S, R 35E		29 Aug 1992	M	T=1.5 x 1	72/26/9,5/6=1.5	Riparian

Notes

- N = 38
- Average measurements: 70/24/8.5=1.8 g (N = 30)
- Sexed individuals: 12 females, 21 males
- Reproduction: 1 pregnant female with 4 embryos (10 August 2001)
- Regional Occurrence: Central = 20 (53%); Western = 12 (32%); Southwest = 3 (8%); Northern = 2 (5%); Southcentral = 1 (2%); Southeast = 0

Order CHIROPTERA Blumenbach, 1779

Chiroptera is the second largest order of mammals after Rodentia, comprising 1,116 species and 202 genera in 18 families (Simmons 2005). Although highly speciose and diverse in form and function in the tropics, bats are found in Alaska in relatively low numbers. The division of bats into two suborders, Megachiroptera (Old World fruit bats) and Microchiroptera (all other bats), along with some other traditional taxonomic hypotheses, has been challenged by recent molecular studies.

The fossil history of bats extends back at least to the Early Eocene. The evolutionary history of Chiroptera was reviewed by Simmons and Geisler (1998).

Family VESPERTILIONIDAE Gray, 1821—evening bats

Vespertilionidae is the largest family of bats, comprising 48 genera and 407 species in 6 subfamilies (Simmons 2005). Members of the genus *Myotis* (subfamily Myotinae) have the widest distribution of any genus of bats (Nowak 1991). Six species of myotine bats in three genera have been recorded in Alaska.

Fossil remains of vespertilionid bats date back to the Middle Oligocene in Europe, Late Oligocene in North America, Middle Miocene in Africa and Asia, and Pleistocene in South America, the West Indies, and Australasia (Koopman 1984).

Adult bats in this family are relatively small to medium-sized, ranging in weight from 4 to 50 grams. All have small eyes, no noseleaf, ears with a tragus, and tails that are relatively long and extend to or beyond the edge of the tail membrane. Most are blackish, gray, or various shades of brown. Some species are migratory, while other species hibernate to survive the cold winter months. Most vespertilionids are cave dwellers, but they also seek shelter in a wide variety of situations including trees or man-made structures. Nearly all feed primarily on insects.

Little brown myotis, *Myotis lucifugus* (B. Hines)

Big Brown Bat

Eptesicus fuscus
(Palisot de Beauvois, 1796)

Other Common Names
None.

Systematics
Hall (1981) recognized five subspecies of *E. fuscus* for continental North America. Burnett (1983) later modified the boundaries of these subspecies, assigning all British Columbia populations to *E. f. bernardinus*, including a specimen taken from the northeastern region of the Province. The one Alaska specimen is reported as *E. f. pallidus* in Manville and Young (1965) and Hall (1981). Parker et al. (1997) examined this specimen to confirm that it was *E. fuscus* and not *E. nilssoni* from Siberia.

Eptesicus fuscus pallidus
Original Description. 1908. *Eptesicus pallidus* Young, *Proc. Acad. Nat. Sci. Philadelphia*, 60:408, October 14.
Type Locality. Boulder, Colorado.
Type Specimen. USNM 142526.
Range. Alberta and Manitoba to western Texas, northwestern Mexico, and southern California.

Global Distribution
Nearctic—The big brown bat ranges widely throughout North and Central America, and northern South America (Kurta and Baker 1990).

Alaska Distribution
The occurrence of this species in Alaska (not mapped) is based on one record, that of an adult female collected from a cabin at the mouth of Shaw Creek in the central region of the state on 5 September 1955 (UMMZ; Reeder 1965). The nearest specimen record of *E. fuscus* is from Pine Lake in northern Alberta, 1,600 km from Shaw Creek. Echolocation calls of this species have been reported from the Laird River watershed of British Columbia, 1,100 km from Shaw Creek (Parker et al. 1997).

Manville and Young (1965) reported the occurrence of this species in southeast Alaska, but this has not been verified by specimens or by calls (Parker et al. 1997).

Status
The one record of this bat in Alaska suggests that it may have been a vagrant well beyond its range or, as suggested by Reeder (1965), brought there by "vehicular traffic along the Alaska Highway."

Fossils
Kurtén and Anderson (1980) reported Pleistocene records from Irvingtonian and Rancholabrean deposits far south of Alaska.

Silver-Haired Bat

Lasionycteris noctivagans
(Le Conte, 1831)

Other Common Names
None.

Systematics
The genus includes only one species. *Lasionycteris noctivagans* shows little or no geographic variation and no subspecies are recognized (Hall 1981; van Zyll de Jong 1985).

Lasionycteris noctivagans
Original Description. 1891. *V[espertilio] noctivagans* Le Conte, *in* McMurtrie, *Animal Kingdom*, 1: [app.]431.
Type Locality. Eastern United States.
Type Specimen. Not known to exist.

Global Distribution
Nearctic—The silver-haired bat ranges widely in North America from southeast Alaska and southern Canada to northernmost Mexico (Kunz 1982).

Alaska Distribution
Only four specimen records of *L. noctivagans* have been reported from Alaska; all were from the southeast region (map 44). In addition, Boland (2007) reported acoustically detecting and sighting one individual on Prince of Wales Island. Reports of this species from Prince William Sound (Manville and Young 1965) are unsubstantiated (Parker et al. 1997).

This species was known for southeast Alaska from one specimen (AMNH 213141), a juvenile female found roosting on 4 November 1964 in an old gill net hanging in a shed on Canyon Island, Taku River (Barbour and Davis 1969).

Three new specimens (Parker et al. 1997), all collected in January, include a specimen (UAM 20738) found in a woodpile 15 km south of Wrangell, a specimen (UAM 30100) found clinging to the side of a house in Petersburg, and a specimen (UAM 30099) found alive in a house entryway in Ketchikan.

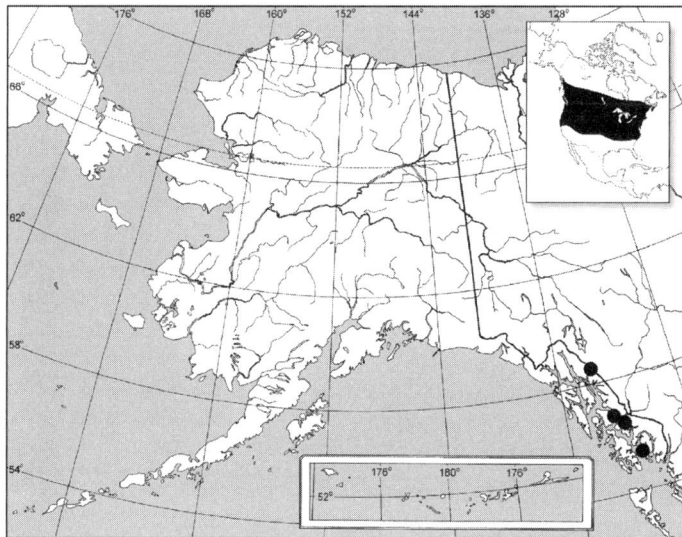

MAP 44. Distribution of silver-haired bat, *Lasionycteris noctivagans.*

● = Specimen Record (n = 4): 55.37220, –131.06670, Ketchikan Quad., Revillagigedo Island, 4 miles North Tongass Highway (1 UAM); 56.36667, –132.36667, Petersburg Quad., Wrangell Island (1 UAM); 56.75000, –132.93333, Petersburg Quad., Mitkof Island (1 UAM); Taku River, Canyon Island (1 AMNH).

Habitat

This species is closely associated with mature forests, ponds, and streams. They roost, often individually, in or behind loose bark, crevices, woodpecker holes, and bird nests in trees. Other sites, including buildings, also are used during migration. Trees are important as hibernacula; this species rarely uses caves. Confirmed records of this species in southeast Alaska are from cold-weather months (November, January), suggestive of local hibernation. All were associated with man-made structures.

Status

Lasionycteris noctivagans in southeast Alaska is probably rare and possibly vulnerable to habitat loss due to its apparent dependence upon trees and snags. The species is known to undertake regular seasonal migrations; however, this may not generally apply for populations occurring in British Columbia and southeast Alaska given the number of winter records that have been reported.

Fossils

Fossils of this species are known from the Early Blancan in Texas and the Late Pleistocene from sites in Wyoming (Kunz 1982).

California Myotis

Myotis californicus
(Audubon and Bachman, 1842)

Other Common Names

Californian myotis, California bat.

Systematics

There has not been a comprehensive study of geographic variation throughout the entire range of this species. Four subspecies are recognized (Hall 1981); one occurs in Alaska.

Myotis californicus caurinus

Original Description. 1897. *Myotis californicus caurinus* Miller, *N. Amer. Fauna,* 13:72, October 16.

Type Locality. Massett, Graham Island, Queen Charlotte Islands, British Columbia.

Type Specimen. USNM 72219.

Range. Extreme south-coastal Alaska to California.

Global Distribution

Nearctic—The California myotis occurs in western North and Central America from southeast Alaska to Guatemala (Hall 1981; Simpson 1993).

Alaska Distribution

Myotis californicus is known in southeast Alaska from only a few specimens taken in the southern Alexander Archipelago (map 45) and from a recent capture and acoustic study (Boland 2007).

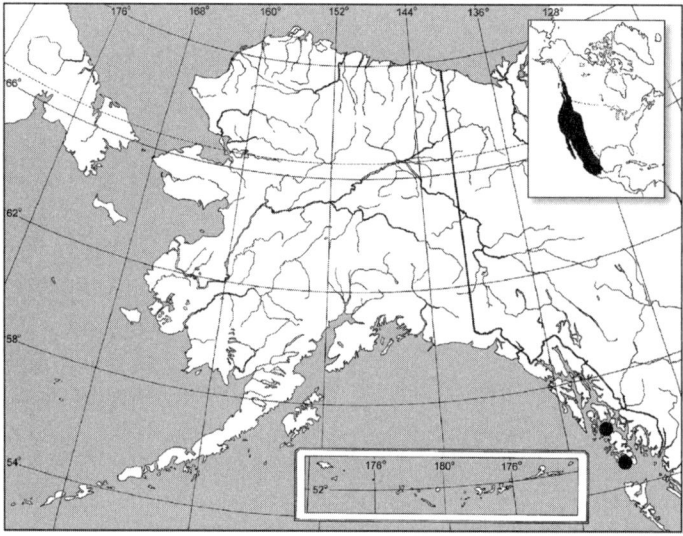

MAP 45. Distribution of California myotis, *Myotis californicus.*

● = Specimen Record (n = 13): 54.87500, –132.80139, Dixon Entrance Quad., Long Island, Howkan (2 MVZ); 56.16667, –133.31667, Petersburg Quad., Prince of Wales Island, El Capitan Cave (3 UAM); 56.18333, –133.31667, Petersburg Quad., Prince of Wales Island, El Capitan cave entrance (2 UAM); 56.16667, –133.2, Petersburg Quad., Prince of Wales Island, Twin Island Lake outlet (1 UAM); 56.17444, –133.36917, Petersburg Quad., Prince of Wales Island, Turn Creek, El Capitan area (5 UAM).

Grinnell (1918) and Miller and Allen (1928) reported two specimens of *Myotis californicus* from Howkan, Long Island (MVZ) More recently, two *M. californicus* skulls were discovered in El Capitan Cave on Prince of Wales Island (UAM 22143, 22144; Parker et al. 1997). A third live animal was collected there in February 1992 (UAM 20498). Samples of skin, hair, and feces were sequenced (B. Jacobsen, pers. comm.) from an additional eight capture-release California myotis netted at three localities in the El Capitan area, northern Prince of Wales Island in 1997–1998 (UAM 47035-37, 54529-31, 54533-34). Based on morphological characteristics and pelage, Boland (2007) recorded this species' occurrence near Juneau (4 captures) and on Mitkof Island (1 capture) and Prince of Wales Island (24 captures) in 2005 and 2006.

Habitat

This species is found in a wide range of habitats, at various elevations, from rain forest along the coast to dry semidesert in the interior. In southeast Alaska, Boland (2007) found this species primarily associated with relatively dense forested habitats, making her most frequent captures in mistnets set on narrow trails and creeks. Roosting habits are not known in Alaska. Elsewhere, day roosts in summer include rock crevices, tree cavities, tree bark, and buildings. Night roosts are flexible and can include caves and mine shafts. Females form small maternity colonies separate from the males. Small numbers are known to hibernate along the Pacific Coast, including winter records from caves in southeast Alaska (Barbour and Davis 1969; Nagorsen and Brigham 1993; Parker et al. 1997).

Status

Although there are too few records of this species to determine its status, *M. californicus* appears to occur in low densities in southeast Alaska (Boland 2007).

Fossils

Holocene material from this species is known from Klein Cave, Texas (Kurtén and Anderson 1980).

Keen's Myotis

Myotis keenii
(Merriam, 1895)

Other Common Names

Keen bat, Keen's long-eared bat, Keen's long-eared myotis.

Systematics

At one time *Myotis keenii* included two subspecies: *M. k. keenii* from the west coast of North America, and *M. k. septentrionalis* of east-central North America (Fitch and Shump 1979; Hall 1981). Morphological studies by van Zyll de Jong (1979, 1985), however, concluded that the two were separate species. Keen's myotis is genetically very similar to long-eared myotis, *M. evotis,* and may be conspecific (Burles et al. 2004). No subspecies are currently recognized (van Zyll de Jong 1985).

Myotis keenii

Original Description. 1895. *Vespertilio subulatus keenii* Merriam, *Am. Nat.,* 29:860, September.

Type Locality. Masset, Graham Island, Queen Charlotte Islands, British Columbia.

Type Specimen. USNM 72922.

Global Distribution

Nearctic—Keen's myotis is restricted to the west coast of North America, ranging from southeast Alaska to the Olympic Peninsula of Washington (Burles et al. 2004).

Alaska Distribution

Myotis keenii is rare in Alaska, being known only from a few localities in the southeast region (map 46).

The occurrence of this species in the state was based until recently on a single specimen collected at Wrangell (Miller 1897; USNM 187394). A second specimen of Keen's myotis was collected in 1993 at Turn Creek, Prince of Wales Island (UAM 23338).

Based on DNA sequences of the mitochondrial cytochrome *b* gene (Cook, unpubl.), the specimen of *M. keenii* reported by MacDonald and Cook (1996) and Parker and Cook (1996) from Hoonah, Chichagof Island, in 1994 (UAM 29831) may have been misidentified. Corroborative evidence is also needed for one collected in Ketchikan on Revillagigedo

Island in August 1999 (UAM 55944), and another found rabid on northern Prince of Wales Island in mid-July 2006 (Castrodale 2006; D. Parker McNeill, pers. comm.).

In a study of bats in southeast Alaska, Boland (2007) documented the occurrence of Keen's myotis from as far north as Juneau (2 captures), and in the Alexander Archipelago on Mitkof Island (2 captures), Wrangell Island (1 capture), and Prince of Wales Island (34 captures). Species' identity for all captures was confirmed by tissue biopsy (wing punch) and mtDNA analysis.

Habitat

Nagorsen and Brigham (1993) suggested that so little information is currently available on this species that little can be said about its habitat affinities other than it is assumed to be closely associated with coastal forest habitats. A summary of known habitat requirements was provided by Burles et al. (2004).

In southeast Alaska, Keen's myotis is primarily associated with coniferous forests, and females appear to be primarily associated with old-growth forests and preferentially with cedar trees (*Thuja, Chamaecyparis*) in riparian areas for day roosts (Boland 2007).

It is not known where this species hibernates during winter months, but Parker and Cook (1996) presumed that it is nonmigratory. Limited information from coastal British Columbia suggest high-elevation caves are the most likely location for their hibernation (Burles et al. 2004).

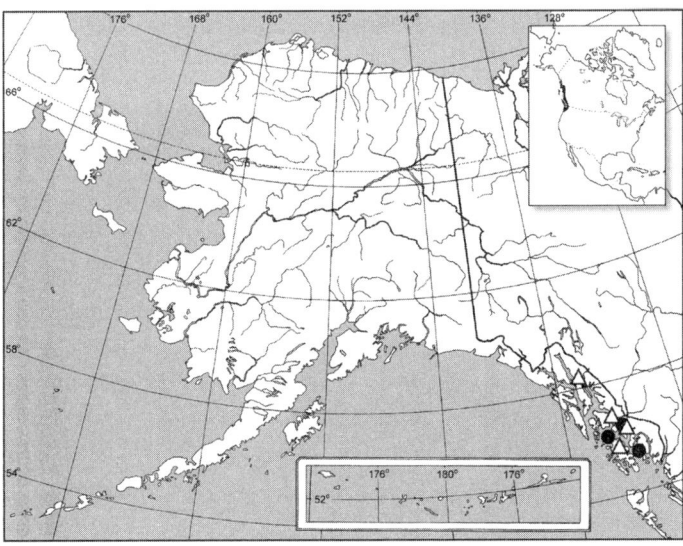

MAP 46. Distribution of Keen's myotis, *Myotis keenii.*

● = Specimen Record (n = 3): 55.75000, –131.50000, Ketchikan Quad., Revillagigedo Island, Ketchikan (1 UAM); 56.46670, –132.43330, Petersburg Quad., Wrangell Island, Ft. Wrangell (1 USNM); 56.16667, –133.28333, Petersburg Quad., Prince of Wales Island, Turn Creek, upstream from Rd 15 bridge (1 UAM).

△ = Tissue Biopsy Record (from Boland 2007: wing punches without specific localities): Juneau (n = 2); Mitkof Island (n=2); Wrangell Island (n=1); Prince of Wales Island (n=34).

Status

Myotis keenii is represented in museum collections by very few specimens (van Zyll de Jong and Nagorsen 1994). Whether this is an indication that this species is actually rare, and thus a species of concern for conservation, is unknown.

As pointed out by Boland (2007), given this species' limited range and apparent low densities, all existing populations may be critical to its persistence. The close association of the females with old-growth for day roosts may make them especially vulnerable to loss of habitat.

Like all of Alaska's bats, *M. keenii* is managed by ADFG as a nongame species with no closed season or bag limits. In Canada, *M. keenii* is listed by COSEWIC as Data Deficient.

Fossils

No fossils are known from the west coast of North America (Kurtén and Anderson 1980).

Little Brown Myotis

Myotis lucifugus
(Le Conte, 1831)

Other Common Names

Little brown bat.

Systematics

Six subspecies were recognized by Hall (1981), with two occurring in Alaska. There is some confusion over the designation and distribution of the subspecies of *M. lucifugus* from the central regions of the state. Youngman (1975) referred populations from western Alberta, south-central Northwest Territories, northern British Columbia, and interior Alaska to *M. l. pernox* based on their larger size and darker coloration when compared to *M. l. lucifugus* and *M. l. alascensis*. Hall (1981) appeared to follow Youngman's assessment. In Hall's (1981) text account of the species, but not on the accompanying map, he shows *pernox* restricted to a relatively small area in western Alberta and *lucifugus* inclusive of all northwestern North America, including Alaska. Generally, subsequent authors (e.g., Fenton and Barclay 1980; van Zyll de Jong 1985) have accepted Youngman's conclusion, but not all agree (e.g., Fenton 1999; Nagorsen 1990).

Preliminary genetic studies reported by Burles et al. (2004) indicate significant differ-ences, possibly at the species level, between coastal and continental populations.

Myotis lucifugus alascensis
Original Description. 1897. *Myotis lucifugus alascensis* Miller, *N. Amer. Fauna*, 13:63, October 16.
Type Locality. Sitka, Baranof Island, Alexander Archipelago, Alaska.
Type Specimen. USNM 77416.
Range. West coast of North America from southeast Alaska to California, most of British Columbia, and extreme western Alberta.

Myotis lucifugus pernox
Original Description. 1911. *Myotis pernox* Hollister, *Smithson. Misc. Collect.*, 56(26):4, December 5.
Type Locality. Henry House, Alberta.
Type Specimen. USNM 174134.
Range. Central Alaska and northwestern Canada.

Global Distribution

Nearctic—*Myotis lucifugus* has the broadest distribution of any bat in North America, ranging from central Alaska to central Mexico (Hall 1981; Nagorsen 1990).

Alaska Distribution

The little brown bat is the most common and widely distributed bat in Alaska, and the only species known to occur further north and west of the southeast region of the state (map 47). The relatively small number of widely scattered specimen records probably does not reflect the regional abundance or range limits of this species within the state (Parker et al. 1997).

Southeast Region. This species is probably found all along the mainland and throughout the Alexander Archipelago but is not adequately documented (MacDonald and Cook 1996; Parker et al. 1997). Specimens of this species extend from Hyder to Yakutat along the mainland, and in the Alexander Archipelago from Admiralty, Baranof, Chichagof, Dall, Grant, Mitkof, Prince of Wales, Revillagigedo, Ring, and Wrangell islands (MacDonald and Cook 2007). Maternity colonies have been found at Barlett Cove, Hoonah, Loring, Ketchikan, and Hyder (Parker et al. 1997; UAM).

Southcentral Region. Most specimens of *M. lucifugus* have been taken near the coast. A single specimen collected inland near McCarthy, along with several unidentified bats seen far up the Copper River (Parker

RECENT MAMMALS OF ALASKA

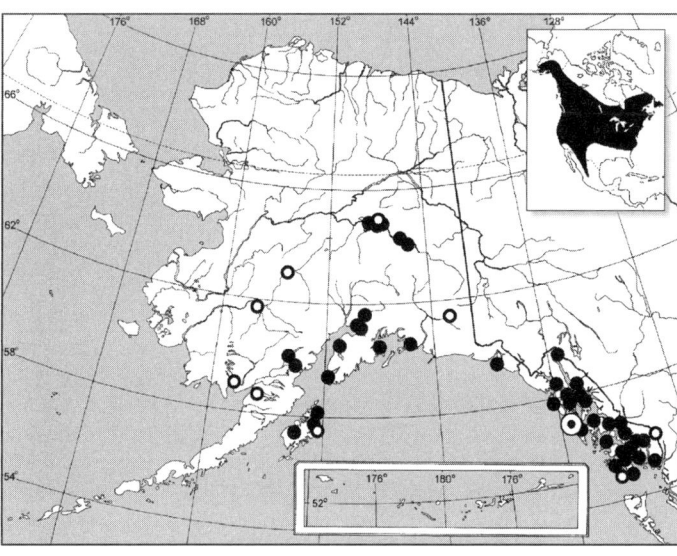

MAP 47. Distribution of little brown myotis, *Myotis lucifugus.*

⊙ = Type Locality
● = Specimen Record (n = 397)
○ = Marginal Records. 64.91667, −147.25000, Fairbanks Quad., Smallwood Creek (1 UAM); 61.43333, −142.51667, McCarthy Quad., Whispering Cave, E of Lime Springs (2 UAM); 55.91667, −130.01667, Ketchikan Quad., Hyder, house attic (30 UAM); 54.80000, −132.85000, Dixon Entrance Quad., Dall Island, Essowah Lakes (1 UAM); 57.61667, −152.11667, Kodiak Quad., Kodiak Island, Chiniak (18 UAM); 58.68835, −156.66129, Naknek Quad., King Salmon (2 UAM); 59.45600, −158.31600, Dillingham Quad., Dillingham (1 UAM); 61.77333, −157.32639, Sleetmute Quad., Red Devil (18 UAM); 62.95830, −155.59170, McGrath Quad., McGrath (1 FMNH).

et al. 1997), suggest a region-wide occurrence. A maternity colony was located in a building between Wasilla and Anchorage (Parker et al. 1997).

Central Region. The northernmost specimens of this species are from near Fairbanks, and the westernmost from near Sleetmute (Parker et al. 1997). In addition to specimens, there are reports of *M. lucifugus* sighted near Fort Yukon and Nulato (Hall 1981). A maternity colony of at least seventy bats was located in a building near Salcha River (Whitaker and Lawhead 1992).

Southwest Region. Little brown bats may occur throughout the year on Kodiak and Afognak islands, and at the base of the Alaska Peninsula at least as far south as King Salmon and Iliamna Lake (Parker et al. 1997).

Western Region. Parker et al. (1997) reported the summer observation of bats, presumably *M. lucifugus*, along the lower Yukon River near Holy Cross.

Habitat

This species occurs in numerous habitats across its broad range, but is especially found through the forested regions of Alaska in a wide variety of sites and situations, including roosts and maternity colonies in man-made structures (Parker et al. 1997). In the temperate rain forests of southeast Alaska, old-growth forest and riparian habitats appear to be used extensively (Parker et al. 1996). The vast karst system in this region may be an important resource for *M. lucifugus* and probably other species as well (MacDonald and Cook 1996).

Status

Myotis lucifugus is the most common and widespread bat species in the state. They appear common in the narrow belt of temperate forest along the state's southern coasts as far west as Kodiak Island. This species occurs throughout the year in southern Alaska, but it is not known whether little brown bats overwinter at more northern latitudes or if they migrate elsewhere to seek suitable hibernacula (Parker et al. 1997).

Fossils

Fossils of Pleistocene age have been reported from a number of sites in North America, all from well south of Alaska (Kurtén and Anderson 1980).

Long-Legged Myotis

Myotis volans
(H. Allen, 1866)

Other Common Names

Hairy-winged myotis, long-legged bat.

Systematics

Hall (1981) recognized four subspecies in North America; only one has been found in Alaska.

Myotis volans longicrus

Original Description. 1886. *Vespertilio longicrus* True, *Science* 8:588, December 24.

Type Locality. Vicinity of Puget Sound, Washington.

Type Specimen. USNM 15263/22480.

Range. Pacific coast from southeast Alaska to California and east to Alberta, Canada.

Global Distribution

Nearctic—The long-legged myotis occurs throughout western North America, from southeast Alaska and northern British Columbia to Mexico (Hall 1981; Warner and Czaplewski 1984).

Alaska Distribution

This species is documented in Alaska by only a few specimens collected in the Alexander Archipelago of southeast Alaska (MacDonald and Cook 2007; Parker et al. 1997) (map 48). Boland (2007) reported long-legged myotis on Wrangell Island (1 capture) and Prince of Wales Island (9 captures) during the summers of 2005 and 2006. External characters were used to determine their identities.

A single specimen of *M. volans* (MVZ 186; Miller and Allen 1928) collected at Mole Harbor, Admiralty Island, was the first record of this species in Alaska. Three specimens from Wrangell Island (UAM 19756, 19757; ADFG uncatalogued) and one specimen from Prince of Wales Island (UAM 24822) are now known (MacDonald and Cook 2007).

Habitat

Myotis volans in southeast Alaska is primarily associated with coniferous forests (Boland 2007). Elsewhere in its range, it is found in a wide variety of habitats at various elevations. Habitat preferences are poorly understood in southeast Alaska but, as with other myotid bats in the region, this species may prefer old-growth forest and riparian habitats. All but one of the ten long-legged myotis reported by Boland (2007) were netted over creeks. No day roosts or nurseries have been discovered in southeast Alaska. Elsewhere, summer day roosts and nursery colonies have been found in rock crevices in cliffs, cracks in the ground and in trees, under tree bark, and in attics of buildings; trees are probably the most commonly used sites for nursery colonies. Caves and mine tunnels may be used during winter (Barbour and Davis 1969; Warner and Czaplewski 1984), but this has not been documented in Alaska.

Status

Boland (2007) considered this species very rare in southeast Alaska.

Fossils

Fossils of this species have been reported from Late Pleistocene deposits in Wyoming (Anderson 1968). An Early Holocene occurrence from Texas was reported by Roth (1972).

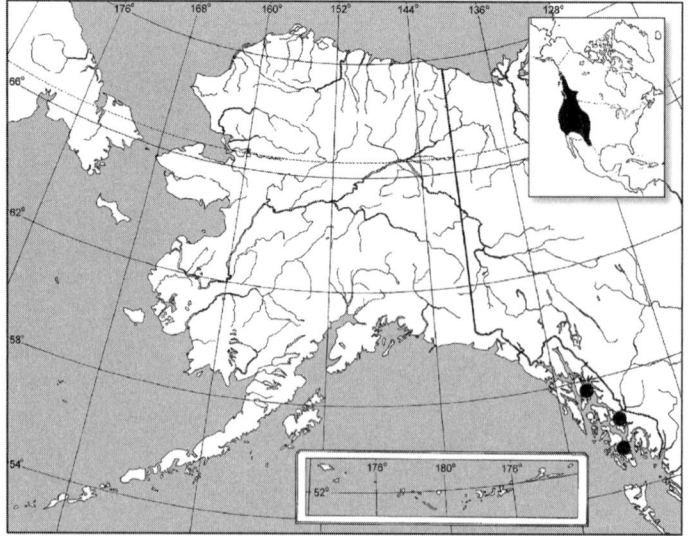

MAP 48. Distribution of long-legged myotis, *Myotis volans*.

● = Specimen Record (n = 4): 55.33333, −132.50000, Craig Quad., Prince of Wales Island, Polk Inlet Forest Service camp (1 UAM); 56.47806, −132.38722, Petersburg Quad., Wrangell Island, Mt Dewey trailhead (2 UAM); 57.66670, −134.05000, Sitka Quad., Admiralty Island, Mole Harbor (1 MVZ).

Order CARNIVORA Bowdich, 1821

Worldwide, the carnivores, or meat eaters, comprise 286 species and 126 genera in two suborders, Feliformia and Caniformia (Wozencraft 2005). Carnivora was once divided into two different suborders, Pinnipeda (primarily marine species of seals and walrus) and Fissipeda (all other Carnivora), but this split is no longer recognized, except informally (Rice 1998). Carnivores comprise the most speciose order of mammals in Alaska.

Family FELIDAE Fischer von Waldheim, 1817—cats

Disagreements on the classification of cats below the family level have been contentious, especially with regard to questions focused on the number of genera. Wozencraft (2005) recognized two subfamilies, fourteen genera, and forty species of cats, worldwide. Two species, Canadian lynx and the newly documented cougar, occur in Alaska; both are members of the subfamily Felinae.

The geologic range of felids is Late Eocene to Recent in North America and Eurasia, and Late Pliocene to Recent in South America (Stains 1984).

Cats are skilled, usually solitary hunters. All have a lithe, muscular body, a rounded, shortened head, furry feet, large eyes with pupils that contract vertically, and a tongue covered with sharp-pointed, recurved, horny papillae. A baculum is absent or vestigial. The dentition of felids is well adapted to seize and cut their prey. The incisors are small and unspecialized; the canines long, sharp, and slightly recurved; and the carnassials, which cut the food, are large and well developed.

Canadian lynx, *Lynx canadensis* (B. Hines)

Canadian Lynx

Lynx canadensis
Kerr, 1792

Other Common Names

Lynx.

Systematics

Matyushkin (1979), Werdelin (1981), and García-Perea (1992) recognized the validity of the genus *Lynx*, whereas Groves (1982), Hemmer (1978), and McKenna and Bell (1997) considered *Lynx* a subgenus of *Felis*. Matyushkin (1979) and Werdelin (1981) suggested that the three Holarctic lynxes (*canadensis*, *lynx*, and *pardinus*) are specifically distinct, whereas Kurtén and Rausch (1959), Ewer (1973), Corbet and Hill (1986), and Tumlison (1987) regarded them as conspecific under *Lynx*. Youngman (1993) suggested that the differences between modern Palearctic and Nearctic forms were only in size and, therefore, a poor indicator of specific distinctness. Yom-Tov et al. (2007) found that skull (and presumably body) size of lynx in Alaska was significantly and negatively related to population size.

Genetic studies (Johnson et al. 2004; Mattern and McLennan 2000) lend support to the monophyletic ancestry of *Lynx* and the specific distinctiveness of *canadensis*, *lynx*, and *pardinus*, in addition to *rufus*. In a molecular genetic study comparing seventeen lynx populations in North America, the most unique and isolated population found was from the Kenai Peninsula (Schwartz et al. 2002).

Hall (1981) recognized two subspecies of lynx in North America; one occurs in Alaska. Van Zyll de Jong (1975) and Werdelin (1981) considered the designation of subspecies unwarranted.

Lynx canadensis canadensis

Original Description. 1792. *Lynx canadensis* Kerr, *The Animal Kingdom . . .* , 1:157.

Type Locality. Eastern Canada [= Quebec].

Type Specimen. Not known to exist.

Range. Northern North America, excluding Newfoundland.

Remarks. Includes *Lynx canadensis mollipilosus* Stone, 1900, *Proc. Acad. Nat. Sci. Philadelphia*, 52:48, March 24, type from Wainwright Inlet, Point Barrow, Alaska.

Global Distribution

Nearctic (but see above)—The Canadian lynx ranges across the boreal forest zone of North America (Hall 1981; Tumlison 1987).

Alaska Distribution

The Canadian lynx occurs primarily in interior forests of Alaska (map 49).

Southeast Region. The Canadian lynx is a probable uncommon resident on the northern mainland of southeast Alaska and an occasional visitor to major river corridors elsewhere along the mainland.

Specimens from the region are limited to Yakutat (MVZ) and Taku Inlet (USNM).

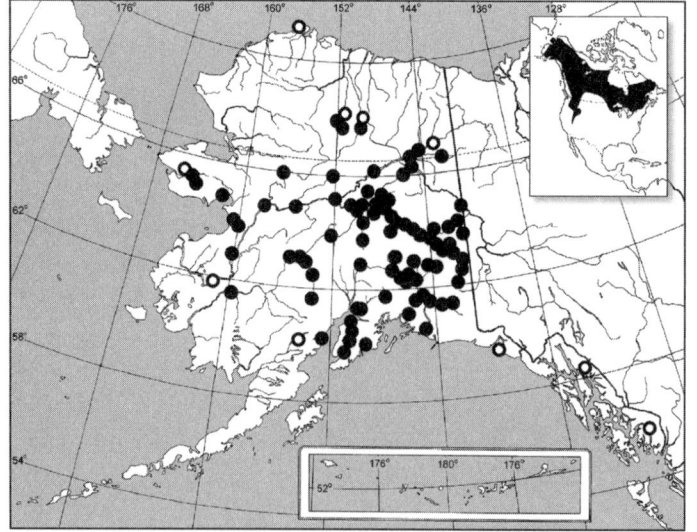

MAP 49. Distribution of Canadian lynx, *Lynx canadensis*.

● = Specimen Record (n = 5603)
○ = Marginal Records. 71.12500, −156.78750, Barrow Quad, Barrow (22 USNM, 1 MVZ); 68.13333, −151.75000, Chadler Lake Quad., Anaktuvuk Pass (11 PSM); 67.66667, −149.83333, Chandalar Quad., Dietrich River (1 UAM); 67.03333, −142.71667, Coleen River Quad., Porcupine River, 5 mi. below Coleen River (2 USNM); Juneau Quad., Taku Inlet (1 USNM); 59.78333, −130.96667, Ketchikan Quad., mouth of Chickamin River (1 trapped, 1 seen by SOM in 1973–74); 59.55000, −139.73333, Yakutat Quad., Yakutat area (2 MVZ); 60.21925, −154.24974, Lake Clark Quad., Lake Clark (1 UAM); 61.56667, −159.57500, Russian Mission Quad., Russian Mission (1 USNM); 65.46667, −165.78333, Teller, Igloo, Seward Peninsula (2 UAM).

Furbearer harvest reports (ADFG 2004) indicated two lynx trapped near Gustavus, four near Yakutat, and eighteen in the Chilkat Valley between 2000 and 2003. A number of lynx have been seen (and a skeleton found) at Glacier Bay (MacDonald and Cook 1996). Individual sightings have also been reported from Hyder, Chickamin River, Grant Creek, Unuk River, and Taku River (MacDonald and Cook 2007).

Alfred M. Bailey, in January 1920, reported seeing the carcasses of two lynx killed on Douglas Island close to Juneau, apparently the only island record (MacDonald and Cook 1996).

Southcentral Region. The lynx is relatively uncommon along the northern Gulf Coast. Lynx immigrated to the region from interior populations along the larger river systems during years of snowshoe hare scarcity inland (Crowley 2004a). They are occasionally found in the Prince William Sound area along the coast east of Cordova and in the drainages of the Lowe River (ADFG 1978). Lynx are cyclically abundant on the Kenai Peninsula (Selinger 2004).

Central Region. This species occurs throughout the region.

Southwest Region. The lynx is present in low numbers along the Alaska Peninsula as far west as Port Heiden (ADFG 1978).

Western Region. Lynx are absent from the Yukon-Kuskokwim Delta and generally are confined to the forested areas along the Kuskokwim River. They also are found along the river systems of western Alaska, apparently expanding their range onto the Seward Peninsula and into the lower Noatak and Kobuk valleys during peak hare abundance (ADFG 1978). Prime habitat includes the lower Koyuk, Niukluk, Unalakleet, lower Noatak, Selawik, and upper Kobuk river systems (ADFG 1978).

Northern Region. Few lynx are found regularly in this region, but they have been known to periodically disperse in numbers all the way to the Arctic coast from the Interior (Bee and Hall 1956). Lynx expanded their range into this region during the late 1990s following a snowshoe hare irruption in the Colville River drainage (Carroll 2004a).

Habitat

This species is a denizen of northern forests with dense understory, roaming into tundra and coastal rain forest areas during periods of prey scarcity (Banfield 1974).

Status

Canadian lynx are common, periodically abundant in Alaska. Their populations fluctuate in close association with snowshoe hare populations, their preferred prey. This species is listed in CITES-Appendix II.

Fossils

Fossil records suggest that *Lynx* species originated in North America (MacFadden and Galiano 1981; Martin 1989). The fossil history of Nearctic species is poorly known. *Lynx canadensis* has been recorded from Sangamonian deposits in Alberta and Utah, and from Wisconsinan in Alaska, Idaho, Wyoming, and Yukon Territory (Kurtén and Anderson 1980; Youngman 1993). Size is the primary difference between modern Old and New World species; the Palearctic form is larger. Youngman (1993) reported a Pleistocene transitional form in eastern Beringia.

Cougar

Puma concolor
(Linnaeus, 1771)

Other Common Names

Mountain lion, puma.

Systematics

This species was placed in the genus *Puma* by Pocock (1917), Weigel (1961), Hemmer (1978), and Kratochvíl (1982). Some thirty subspecies are generally recognized throughout the species' broad range (Young and Goldman 1946). Hall (1981) recognized fifteen subspecies in North America. There has not been a recent revision of subspecies. The taxonomic affinity of the Alaska animals has not been evaluated, but they would likely be related to either *P. c. missoulensis* from the interior of British Columbia or *P. c. oregonensis* from farther south along the British Columbia coast.

Puma concolor missoulensis?

Original Description. 1943. *Felis concolor missoulensis* Goldman, *J. Mammalogy*, 24:229, June 7.

Type Locality. Sleeman Creek, about 10 mi. SW Missoula, Missoula Co., Montana.

Type Specimen. USNM 262116.

Range. Western Canada, North Dakota, Montana, and Idaho.

Puma concolor oregonensis?

Original Description. 1832. *Felix* [sic] *oregonensis* Rafinesque, *Atlantic Jour.*, 1:62, June 20.

Type Locality. Ohanapecosh River, Mount Rainier National Park, Pierce County, Washington.

Type Specimen. None designated.

Range. Western Oregon, Washington, and British Columbia, northward along coast at least to the Skeena River.

Global Distribution

Nearctic and *Neotropical*—The cougar has the widest distribution of any wild mammal in the Western Hemisphere, extending from Yukon Territory (Jung and Merchant 2005; Youngman 1975) and southeast Alaska (MacDonald and Cook 2007) to southern Argentina and Chile (Beier 1999).

Alaska Distribution

The growing number of sightings and voucher specimens of cougar from various localities in southeast Alaska, including islands in the Alexander Archipelago (map 50), along with a number of unsubstantiated sightings in east-central Alaska, suggest that this species may be recolonizing the state (see "Fossils" below).

Southeast Region. The occurrence of cougar in southeast Alaska was first substantiated by the collection of a male on the east side of Wrangell Island along Blake Channel opposite Aaron Creek on 25 November 1989 (UAM 18551).

On 20 April 1998, a lone cougar was seen and photographed at Myers Chuck, Cleveland Peninsula (Fig. 19), and in December of 1998, a trapper captured a male cougar in a wolf snare at Totem Bay, Kupreanof Island (UAM 50544).

In recent years there have been a number of unconfirmed reports of cougar in the region (as well as several unsubstantiated sightings farther north in southcentral Alaska). These include Gustavus (1958); Twelve-mile Arm, Prince of Wales Island (summer 1992); Cleveland Peninsula (1998, 2003); Mitkof Island (1999); Snettisham (2000); Skagway (in 2002); Revillagigedo Island (2004); Haines (dates unknown) (Home 1973; MacDonald and Cook 1996; Woodford 2004; L. Carson, pers. comm., 1992; M. Brown, pers. comm., 1992; J. Hunley, pers. comm., 2003; *Ketchikan Daily News*, 17 August 2004).

Central Region. Youngman (1975) and Jung and Merchant (2005) documented seven cougar sightings and one specimen in Yukon Territory, Canada, between 1955 and 2000, adding credence to unsubstantiated sightings of this cat in adjacent central Alaska (Cook and MacDonald 2003; MacDonald and Cook 2001; Nowak 1976; Russell 1978). Perhaps cougars are expanding their range in relation to the abundance of Sitka black-tailed deer in southeast Alaska, and to the recent westward expansion of mule and white-tailed deer (Hoefs 2001) in Yukon Territory.

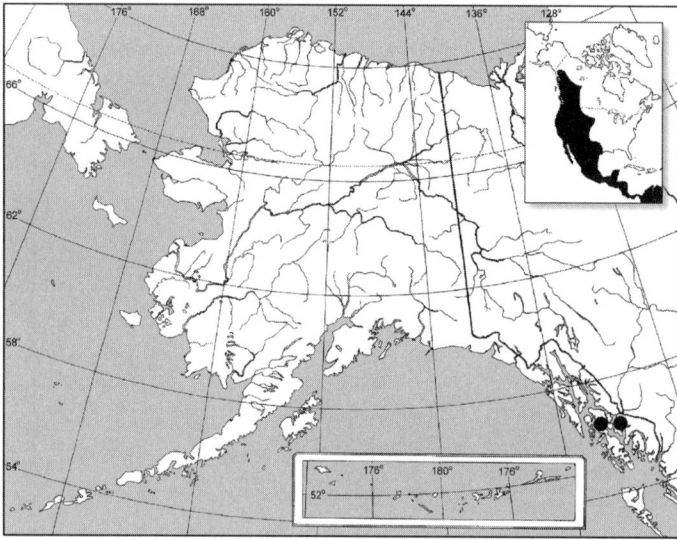

MAP 50. Distribution of cougar, *Puma concolor*.

● = Specimen Record (n = 2): 6.41667, −132.25000, Petersburg Quad, Wrangell Island, Blake Channel opposite Aaron Creek (1 UAM); 56.46889, −133.43083, Petersburg Quad., Kupreanof Island, Little Totem Bay (1 UAM).

Habitat

Although generally associated with mountainous terrain, canyons, and rimrock, the broadranging cougar can be found from sea level to high in the mountains. It uses many habitats ranging from desert scrub, grasslands, and swamps to temperate and tropical rain forests (Banfield 1974; Beier 1999; Currier 1983).

Status

The current status of this species in Alaska is unknown.

Fossils

Kurtén and Anderson (1980) and Currier (1983) reviewed the fossil history of this species. Apparently, the cougar was a member of the eastern Beringian fauna during the Late Pleistocene. Youngman (1993) identified Pleistocene fossil remains of *P. concolor* in eastern Beringia from a deposit at Cripple Creek, Alaska, and another at Bluefish Cave 3 in Yukon Territory. The Yukon specimen was

Figure 19. Nightime photo of a cougar at Myers Chuck, Cleveland Peninsula, southeast Alaska, taken on 20 April 1998 by resident Jacque Hunley. Tracks and scat were also noted.

radiocarbon (AMS) dated at 18,970±1490 years ago. In addition, Harington (1977) reported a fourth premolar as belonging to a cougar or some other large cat from a Late Pleistocene deposit in Yukon Territory.

Family CANIDAE Fischer von Waldheim, 1817—wolf, coyote, and foxes

The family Canidae includes thirteen genera and thirty-five species worldwide (Wozencraft 2005); four species occur in Alaska.

Fossil records of wild canids range in age from Late Eocene to Recent in North America and Europe, Early Eocene to Recent in Asia, Pleistocene to Recent in South America, and Early Miocene to Recent in Africa (Stains 1984).

Members of this family are generalized hunters that range in size from the highly social wolves to the smaller, more solitary foxes. Wild canids possess long legs, digitigrade four-toed hind feet, nonretractile claws, large and sharp canines, and other morphological adaptations well suited to the swift pursuit and agile capture of prey.

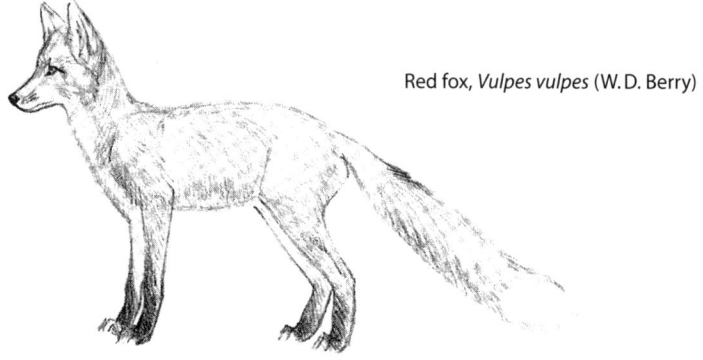

Red fox, *Vulpes vulpes* (W. D. Berry)

Coyote

Canis latrans
Say, 1823

Other Common Names
None.

Systematics
Hall (1981), following Jackson (1951), recognized nineteen subspecies of *C. latrans*; one occurs in Alaska.

Canis latrans incolatus
Original Description. 1934. *Canis latrans incolatus* Hall, *Univ. California Publ. Zool.*, 40:369, November 5.

Type Locality. Isaacs Lake, 3000 ft., Bowron Lake region, British Columbia.

Type Specimen. MVZ 43898.

Range. Alaska and northwestern Canada.

Global Distribution
Nearctic—Coyotes range widely across North and Central America (Hall 1981). In the past two hundred years, they have expanded into eastern North America in response to habitat changes and extirpation of the wolf, *C. lupus* (Bekoff 1977; Nagorsen 1990).

Alaska Distribution
According to Dufresne (1946), ADFG (1978), and others, the coyote first arrived in Alaska shortly after the turn of the last century, first along the mainland of southeast Alaska, followed by their rapid expansion first in east-central Alaska around 1930 and later in southcentral Alaska, where they persist in highest numbers today. Coyotes are occasionally reported (but rarely preserved as specimens) as far west as the Alaska Peninsula and as far north as the North Slope (ADFG 1978) (map 51).

Southeast Region. The coyote is an infrequent visitor to the river valleys and adjacent coastlines along the southern mainland of this region and a probable resident of the northern mainland north of the Taku River. During the mid-1960s, coyotes were reported as very common to abundant in the Yakutat area (MacDonald and Cook 1996). Barten (2004a) reported coyotes becoming common near Gustavus and in the foothills of the Chilkat Mountains. The only island record is the report of an adult female coyote trapped on Mitkof Island near Dry Straits in the winter of 1983–1984 (MacDonald and Cook 1996).

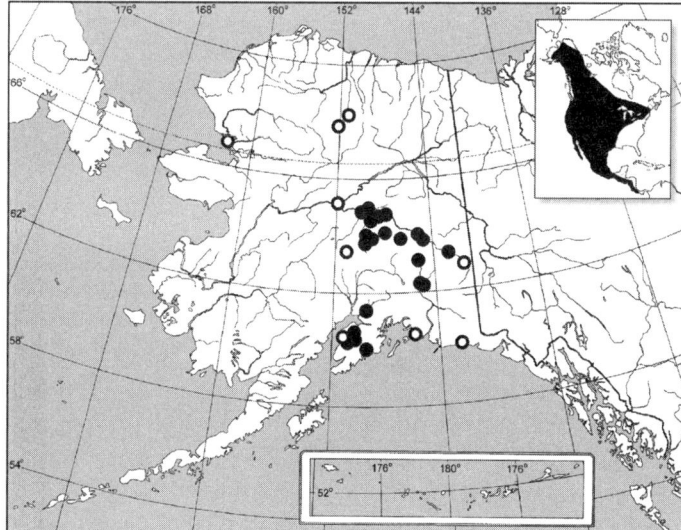

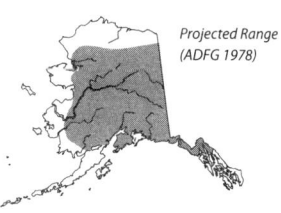

MAP 51. Distribution of coyote, *Canis latrans*.

Projected Range (ADFG 1978)

● = Specimen Record (n = 233
○ = Marginal Records. 68.08333, −151.46667, Chandler Lake Quad., Ben Creek, 11.2 km E, 4 km S of Anaktuvuk Pass (1 UAM); 62.96667, −141.93333, Nabesna Quad., Northway (1 UAM); 60.06667, −142.41667, Bering Glacier Quad., Cape Yakataga (1 UAM, 1 LACM); 60.55000, −145.75000, Cordova Quad., Cordova (1 USNM); 60.45000, −151.10000, Kenai Quad., 100 meters E of Arc Lake trail (1 UAM); 63.47500, −150.87500, Mt McKinley Quad., Wonder Lake (1 MVZ); 65.16670, −152.06670, Tanana Quad., Tanana (1 USNM); 66.90000, −162.58333, Kotzebue Quad., Kotzebue (1 USNM); 67.73333, −152.41667, Wiseman Quad., confluence of John River and Hunt Fork (1 UAM).

Southcentral Region. Coyotes are common on the Kenai Peninsula and the Matanuska and Copper River valleys, and fairly common throughout the rest of the region (ADFG 1978; Rhode and Barker 1953). Along the coast, they are found primarily east of Prince William Sound.

Central Region. East-central Alaska is considered good habitat for this species. Coyotes have been trapped occasionally as far west as McGrath on the Kuskokwim River (ADFG 1978).

Southwest Region. Coyotes are relatively scarce on the Alaska Peninsula, with most reports coming from the northern portions of the region (ADFG 1978), where they have become more common in recent years (Woolington 2004a).

Western Region. This species is generally absent from this region. According to ADFG (1978), they can be found in the upper portions of the Kobuk and Noatak river drainages. The capture of two coyotes in the Unalakleet River drainage in December 1999 came as an unprecedented surprise to local residents (Gorn 2004). USNM houses a coyote specimen (USNM 264185) taken near Kotzebue in 1938.

Northern Region. Coyotes are rare in northern Alaska, with recent reports of their occasional occurrence coming only from the extreme south-central portion of the region (ADFG 1978; Carroll 2004a). Individuals have been reported as far north as Point Barrow (Bee and Hall 1956; Dufresne 1942, 1946; Jackson 1951).

Habitat

The versatile coyote can thrive in diverse habitats ranging from warm deserts and wet grasslands to colder climates at high elevations and latitudes (Bekoff 1999). In Alaska, they prefer broken and open country (ADFG 1978).

Status

Although not especially abundant on a statewide basis, coyotes have been reported to be common in some areas, particularly in the drainages of the Tanana, Copper, Matanuska, and Susitna rivers and on the Kenai Peninsula (ADFG 1978).

Fossils

Whether the coyote was a member of the east Beringian fauna during the Pleistocene or a postglacial immigrant is unclear. Guthrie (1968b) listed the occurrence of this species in Pleistocene deposits from Cripple Creek near Fairbanks, and Harington (1989) reported Pleistocene fossils from Old Crow River, Yukon Territory. In a thorough review of these and other materials, Youngman (1993) failed to locate a specimen that he could identify positively as a coyote, although one specimen he examined, an undated humerus from the Dawson area, was questionable. Youngman (1993) identified a canid dentary from Cripple Creek (and perhaps the same one reported as a coyote by Guthrie 1968b) as *Cuon* sp., the dhole or Asiatic wild dog.

Wolf

Canis lupus
Linnaeus, 1758

Other Common Names

Gray wolf, timber wolf, tundra wolf.

Systematics

Eight Palearctic subspecies have been described (Mech 1970). Wozencraft (2005) recognized thirty-seven subspecies worldwide, which included *C. l. familiaris*, the domestic dog. Hall (1981) followed Goldman (1944) and recognized twenty-four subspecies of wolves in North America; four occur in Alaska (five including feral dog). Some of the subspecies recognized by Goldman (1944) were described on the basis of few specimens. Jolicoeur (1959) and Mech (1974) concluded that there are far too many subspecies recognized, leading Mech (1999) to distinguish only five North American forms (two in Alaska). Nowak (1983, 1995) also proposed major taxonomic changes, concluding that the North American populations of *C. lupus* consisted of five groups, with two of these (the large northern morph and small southern morph) ranging across much of the continent.

Relatively little genetic variation was uncovered in initial surveys across North American wolf populations (Brewster and Fritts 1995; Wayne et al. 1995); however, Pacific Northwest coastal wolves were not analyzed until recently.

Weckworth et al. (2005) assessed variation at eleven microsatellite loci and across sequences of the mitochondrial control region with an emphasis on coastal Alaska wolf populations. Coastal wolf populations were found to be genetically distinctive from continental wolves. In addition, high levels of genetic diversity were found within the southeast region of the state. Differentiation of these southeast Alaska wolves from wolf populations elsewhere is apparently due to distinctive biogeographic histories, lineage sorting, and contemporary barriers to gene flow. Morphological research was equivocal with regard to whether coastal wolves differ from continental populations (Nowak 1995).

Canis lupus alces

Original Description. 1941. *Canis lupus alces* Goldman, *Proc. Biol. Soc. Washington*, 54:109, September 30.

Type Locality. Kachemak Bay, Kenai Peninsula, Alaska.

Type Specimen. USNM 147471.

Range. Known only from the Kenai Peninsula.

Canis lupus familiaris

Original Description. 1758. *Canis familiaris* Linnaeus, *Syst. nat.*, 10th ed. , 1:38.

Type Locality. Sweden.

Type Specimen. None designated. *C. familiaris* was designated as the Type Species for the genus *Canis* by Linnaeus.

Range. Nearly worldwide.

Remarks. *Canis familiaris* has page priority over *C. lupus* in Linnaeus's 1758 edition of *Systema Naturae*, but both were published simultaneously, and *C. lupus* has been universally used for this species.

Canis lupus ligoni

Original Description. 1937. *Canis lupus ligoni* Goldman, *J. Mammalogy*, 18:39, February 11.

Type Locality. Head of Duncan Canal, Kupreanof Island, Alaska.

Type Specimen. USNM 243323.

Range. Southeast Alaska.

Canis lupus pambasileus

Original Description. 1905. *Canis pambasileus* Elliot, *Proc. Biol. Soc. Washington*, 18:79, February 21.

Type Locality. Susitna River, region of Mount McKinley, Alaska.

Type Specimen. CMNH 13481.

Range. Western and central Alaska, and adjacent Yukon Territory.

Canis lupus tundrarum

Original Description. 1912. *Canis tundrarum* Miller, *Smithsonian Miscl. Coll.*, 59(15):1, June 8.

Type Locality. Point Barrow, Alaska.

Type Specimen. USNM 16748.

Range. Northern Alaska and Yukon Territory.

Global Distribution

Holarctic—Wolves ranged throughout the Northern Hemisphere north of about 20° N latitude, one of the broadest distributions of any mammal (Mech 1974). Deliberate extermination and loss of habitat have considerably restricted their current range, especially in the mid-latitudes of North America and Europe. The species has been reintroduced to limited areas at lower latitudes in North America.

Alaska Distribution

Wolves are widely distributed in Alaska, occurring throughout the mainland, on Unimak Island in the Aleutians, and on all of the major southern islands in Southeast (map 52).

Southeast Region. Wolves occur throughout the mainland of southeast Alaska and on islands in the Alexander Archipelago south of Frederick Sound (excluding Coronation, Forrester, and undoubtedly some of the smaller, more isolated islands without an adequate prey base). We are unaware of any specimen-documented records of this species from any of the islands north of Frederick Sound, although there have been several convincing sightings of this animal on Admiralty Island in recent years (MacDonald and Cook 2007).

Wolves were introduced experimentally to Coronation Island in 1960 and 1963; none remained there by the early 1970s (Burris and McKnight 1973; Klein 1996).

Southcentral Region. Wolves have never been common along the coastal mainland and occur only occasionally on Hawkins and Hinchinbrook islands of Prince William Sound (ADFG 1973; Crowley 2003). Wolves disappeared from the Kenai Peninsula shortly after 1900 (about the time caribou were eliminated). They reappeared in the mid-1960s and by the early 1970s were distributed over most of the Kenai Peninsula (ADFG 1973; Peterson et al. 1984).

Central Region. *Canis lupus* is well distributed throughout central Alaska.

Southwest Region. Wolves occur in low-to-moderate numbers on the Alaska Peninsula and Unimak Island (ADFG 1973; Sellers 2003). Nelson (1887) reported the killing of two wolves on Akun Island in 1830, farther west in the Aleutian Chain. This species is absent from the Kodiak Archipelago (but see feral dog, Appendix 7) and islands south of the Alaska Peninsula.

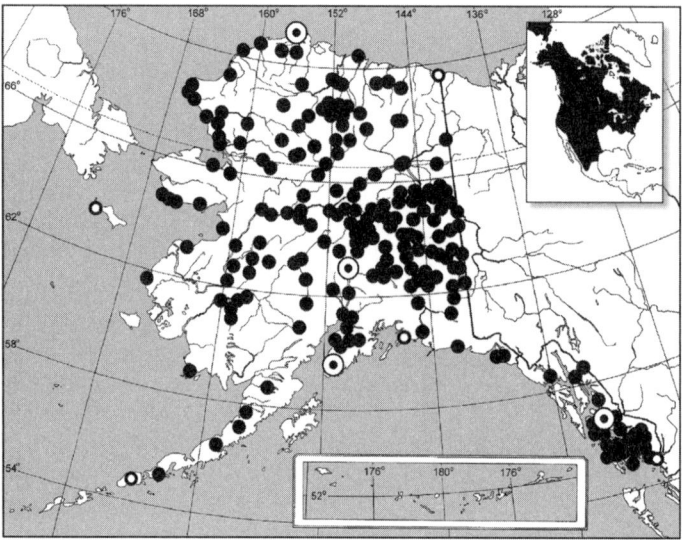

MAP 52. Distribution of wolf, *Canis lupus*.

◉ = Type Locality
● = Specimen Record (n = 4171)
○ = Marginal Records. Demarcation Point Quad., Okerokovik River (1 UAM); 54.76667, –130.78333, Prince Rupert Quad., Nakat (3 UAM); Cordova Quad., Hawkins Island (1 AMNH); Unimak Quad., Unimak Island, along the Bering Sea (1 USNM); 63.77972, –171.74111, St Lawrence Quad., Saint Lawrence Island, 41–50 miles from Gambell (1 UAM).

Western Region. Wolves are rarely found on the Yukon-Kuskokwim Delta. Occasionally, they are seen on Nunivak Island (ADFG 1973). Murie (1936) reported the occurrence of a wolf on St. Lawrence Island during 1927, and in April 1994 a wolf was collected on the island south of Gambell (UAM 30678).

Northern Region. Wolves range throughout northern Alaska, where they are closely tied to the seasonal movements and abundance of caribou (ADFG 1973; Ballard et al. 1997).

Habitat

Wolves thrive in a wide variety of climates and terrains. Prior to their extermination throughout most of their southern range in North America, they occupied many terrestrial habitats from subtropical Mexico to the tundra bordering the Arctic Ocean (Mech 1974). In Alaska, they are found where suitable prey populations exist, from the dense coastal rain forests to the open arctic tundra.

Status

Wolves are currently widespread in Alaska and their populations are generally stable and in healthy condition. During winter 1989–1990, the wolf population of Alaska was estimated at 5,900 to 7,200 animals within 700 to 900 packs (Stephenson et al. 1995). A later ADFG estimate (http://www.wc.adfg. state. ak.us/index.cfm?adfg=wolf.main, accessed 8

Aug. 2007) was 7,700 to 11,200 wolves in the state.

In the past, Alaska wolves were subjected to bounties, unrestricted hunting and trapping, poison, and aerial gunning. Recent efforts to actively reduce wolf numbers in some areas of the state have been contentious. *Canis lupus* is listed under the ESA as Endangered in Mexico and the Lower 48 states except Minnesota, where it is considered Threatened (NatureServe 2007). Wolves in Alaska are not listed in the ESA and are of Least Concern by IUCN.

The Alexander Archipelago wolf, *C. l. ligon*, was considered a Subspecies of Concern by West (1991) and Suring et al. (1992) and identified by the USFS (1997) as a subspecies with special management concerns. Conservation assessments by Person et al. (1996) and Schoen (2007) indicate significant changes in management are needed to ensure the long-term viability of this coastal endemic. For island populations, such changes need to include a combination of conservative harvest regulations and large roadless reserves (Person and Russell 2008). The wolf is a Management Indicator Species for the Tongass National Forest (USFS 1997).

Fossils

The wolf first appeared in North America during the Late Irvingtonian. In Alaska, the species was present in the Illinoian, and perhaps earlier (Kurtén and Anderson 1980). It was one of the most widely distributed land mammals in the Late Pleistocene (Kurtén 1968; Mech 1970; Nowak 1973). Wolf remains are well represented in the large carnivore fauna of eastern Beringia during the Late Pleistocene (Guthrie 1968b). A Pleistocene wolf skull from central Alaska was dated at 18,610±165 years before present (Guthrie 1990b). Savinetsky et al. (2004) reported the recovery of wolf bones (along with brown bear and caribou remains) from Late Holocene archaeological sites on the easternmost Aleutian Island of Akun. No remains of *C. lupus* have been recovered from cave deposits in southeast Alaska (Heaton and Grady 2003).

Arctic Fox

Vulpes lagopus
(Linnaeus, 1758)

Other Common Names
Polar fox, white and blue fox color phases.

Systematics
Bobrinskii et al. (1965), Youngman (1975), McKenna and Bell (1997), Baker et al. (2003), and Wozencraft (2005) considered *Alopex* congeneric with *Vulpes*. Van Gelder (1978) considered it a subgenus of *Canis*. Genetic studies (Geffen et al. 1992) found *Alopex* closely related to *Vulpes vulpes* and *V. velox*, suggesting separate generic status based solely on morphological adaptations to living in the Arctic was inappropriate.

Wozencraft (2005) accepted three subspecies as valid across this species' entire range: *pribilofensis* (on the Pribilof Islands), *beringensis* (Bering Island, Commander Islands), and *fuliginosus* (described by Bechstein in 1799 from Iceland); no mention is made of the nominate subspecies, *lagopus*. Both Rausch (1953b) and Youngman (1975) recognized only the nominate form occurring across continental North America and Eurasia. Hall (1981) also recognized *lagopus* as a Holarctic continental form but, in addition, included *ungava* from eastern continental Canada as a valid form. According to Hall (1981), three subspecies occur in Alaska.

Vulpes lagopus hallensis
Original Description. 1900. *Vulpes hallensis* Merriam, *Proc. Washington Acad. Sci.*, 2:15, March 14.
Type Locality. Hall Island, Bering Sea, Alaska.

Type Specimen. USNM 98067.
Range. Hall, St. Lawrence, and St. Matthew islands.

Vulpes lagopus lagopus
Original Description. 1758. *Canis lagopus* Linnaeus, *Syst. Nat.*, 10th ed., 1:40.
Type Locality. Lapland.
Type Specimen. Not known to exist.
Range. Circumpolar distribution.
Remarks. See "Systematics," above.

Vulpes lagopus pribilofensis
Original Description. 1902. *Vulpes pribilofensis* Merriam, *Proc. Biol. Soc. Washington*, 15:171, August 6.
Type Locality. St. George Island, Pribilof Islands, Alaska.
Type Specimen. USNM 30651/42624.
Range. Restricted to the Pribilof Islands.

Global Distribution
Holarctic—The arctic fox has a circumpolar distribution in Europe, Asia, and North America (including Greenland) as well as Iceland, Spitsbergen, and other islands in the Bering Sea, North Atlantic, and Arctic oceans (Youngman 1993).

Alaska Distribution
Arctic foxes occur naturally along the Arctic coast of Alaska as far south as the northwestern shore of Bristol Bay, and on Diomede, Hall, King, St. Lawrence, St. Matthew, Pribilof, and Nunivak islands in the Bering Sea (ADFG 1978; Bailey 1993; Chesemore 1968a; Preble 1923; Rausch and Rausch 1968) (map 53). Individuals are occasionally found considerable distances inland from the coast in the Brooks Range (Bee and Hall 1956; Manville and Young 1965; Rausch 1951).

During the period of extensive fur farming in Alaska, arctic foxes were released on many of the Aleutian Islands (including Attu Island; see Black 1984; Buskirk and Gipson 1980; Nelson 1887), islands off the Alaska Peninsula, and along the Gulf Coast to southeast Alaska (Bailey 1993). The industry collapsed in the 1930s and many introduced populations (especially southeast Alaska) are no longer extant.

Habitat
Arctic foxes are well adapted to life in the cold, harsh arctic environment. They are found in arctic tundra, along rocky beaches, and far out and widely dispersed on the frozen pack ice where they scavenge carcasses left from polar bear kills (ADFG 1978; Banfield 1974). They

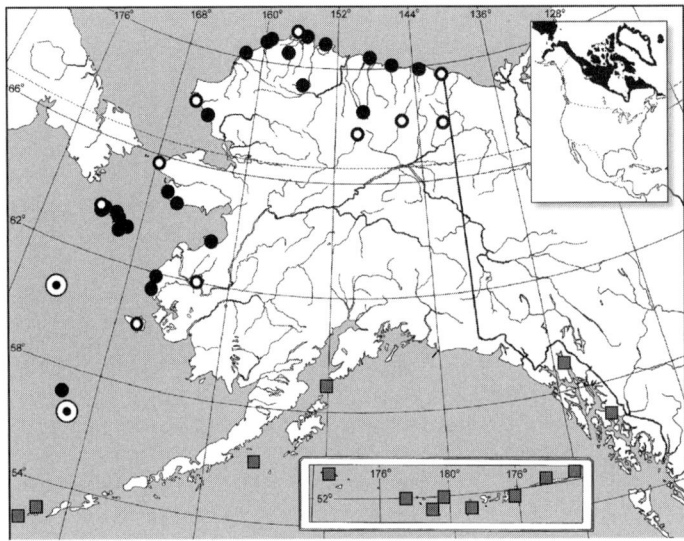

MAP 53. Distribution of arctic fox, *Vulpes lagopus*.

⊙ = Type Locality
● = Native
■ = Non-native Specimen Record (n = 3123)
○ = Marginal Records. 71.25000, –156.85000, Barrow Quad., Point Barrow (1 UAM); 69.06667, –141.31667, Demarcation Pt Quad., Demarcation Point (1 USNM); Coleen Quad., 60 mi. N Rampart House, Porcupine River (1 USNM); 68.13333, –145.53333, Arctic Quad., Arctic Village (1 USNM); 67.46667, –150.08333, Chandalar Quad., Hammond River (1 UAM); 62.05000, –163.16667, Kwiguk Quad., Andreafsky (15 USNM); 60.00000, –166.00000, Nunivak Island Quad., Nunivak Island (17 MSB); 63.77972, –171.74111, St Lawrence Quad., Saint Lawrence Island, Gambell (75 UAM); 65.00000, –168.00000, Teller Quad., 6 mi N of Wales (1 UAM); 68.13333, –165.96667, Point Hope Quad., Cape Thompson (1 UAM).

prefer to den in light, sandy soil along river-banks, on small hillocks, and occasionally in talus (Anderson 1999).

Status

Arctic foxes are common, sometimes abundant, with populations fluctuating considerably between years in relation to forage availability such as densities of lemmings (ADFG 1978). The release of mainly arctic foxes to over 450 islands in Alaska had catastrophic effects on island avifauna populations (Bailey 1993; Murie 1959). Eradication efforts by USFWS have now eliminated nonnative arctic foxes from all but six islands in AMNWR (Shemya, Tanaga, Kanaga, Adak, Atka, Chuginadak), and bird populations are rebounding (Ebbert and Byrd 2002; Gibson and Byrd 2007; Williams et al. 2003).

Fossils

Fossils of *V. lagopus* have been found in Late Riss deposits in Eurasia (Rausch 1994) and in Wisconsinan deposits in North America (Kurtén and Anderson 1980). Youngman (1993) examined 207 undated fossil remains of arctic foxes from eastern Beringia deposits that included sites from central (Fairbanks area, Porcupine River, Tofty) and northern (Teshekpuk Lake) Alaska. These fossils were within the range of measurements of modern western arctic specimens. Fossils of arctic foxes have been reported from caves on Prince of Wales Island in southeast Alaska that date from Middle Wisconsinan through the last glacial maximum (Heaton and Grady 2003).

Red Fox

Vulpes vulpes
(Linnaeus, 1758)

Other Common Names

Color phase names include cross fox, silver fox, and black fox.

Systematics

The North American red fox was long recognized as a separate species, *Vulpes fulva*, until Rausch (1953a) and others provided evidence that it is conspecific with the Eurasian *V. vulpes*. Thirty-three subspecies of Eurasian red foxes have been described (Ellerman and Morrison-Scott 1966; Wozencraft 2005); *V. v. beringiana* is the race that occurs across the Bering Strait from Alaska in far eastern Siberia. Hall (1981) recognized ten subspecies of North American red foxes; four occur in Alaska.

Vulpes vulpes abietorum

Original Description. 1900. *Vulpes alascensis abietorum* Merriam, *Proc. Washington Acad. Sci.*, 2:669, December 28.

Type Locality. Stuart Lake, British Columbia.

Type Specimen. USNM 71197.

Range. Southern Yukon and Northwest Territories, interior British Columbia (and adjacent coastal southeast Alaska), and northern Alberta.

Remarks. Youngman (1975) found this race indistinguishable from *V. v. alascensis*.

Vulpes vulpes alascensis

Original Description. 1900. *Vulpes alascensis* Merriam, *Proc.Washington Acad. Sci.*, 2:668, December 28.

Type Locality. Andreafski, about 110 km above the delta of Yukon River, Alaska.

Type Specimen. USNM 21420.

Range. Most of Alaska, Yukon Territory, and Northwest Territories.

Vulpes vulpes harrimani

Original Description. 1900. *Vulpes harrimani* Merriam, *Proc. Washington Acad. Sci.*, 2:14, March 14.

Type Locality. Kodiak Island, Alaska.

Type Specimen. USNM 99626.

Range. Kodiak Archipelago.

Vulpes vulpes kenaiensis

Original Description. 1900. *Vulpes kenaiensis* Merriam, *Proc.Washington Acad. Sci.*, 2:670, December 28.

Type Locality. Kenai Peninsula, Alaska.

Type Specimen. USNM 96145.

Range. Known only from the type locality.

Global Distribution

Holarctic—Red foxes range broadly across North Africa, Europe, Asia, and most of North America (Larivière and Pasitschniak-Arts 1996).

Alaska Distribution

Natural populations of red fox are found throughout most of mainland Alaska, the Kodiak Archipelago, and on several far eastern Aleutian Islands (map 54). Commercial fox farming resulted in the introduction of red foxes to many North Pacific islands (Bailey 1993).

Southeast Region. Red foxes are uncommon residents along the mainland of southeast Alaska north of the Taku River. They are rare to uncommon visitors along the southern mainland (usually in or near major river valleys).

Red foxes were introduced for commercial harvest on Cleft, Dry, Kupreanof, Passage, and Sokoi islands between 1894 and 1929; none is known to have survived (Bailey 1993).

Recent reports of red foxes on Chichagof Island and Baranof Island (MacDonald and Cook 2007) lack verification.

Southcentral Region. Red fox populations are sparse on the Kenai Peninsula and along the Gulf Coast of this region, occurring most frequently in the major drainages that connect to interior mainland areas where they are common (ADFG 1978). Apparently, red foxes were abundant on the Kenai Peninsula and along the coast prior to 1930, but quickly disappeared as coyotes established and rapidly increased in subsequent years (Selinger 2004). According to Bailey (1993), red foxes were introduced to at least nine islands in this region, namely Aurora, Bettles, Dangerous, Dutch Group, Hesketh, Kalgin, Kanak, Starinkof, and Wingham islands. Red foxes apparently occurred naturally on Kayak and Wingham islands. Augustine Island in Cook Inlet,

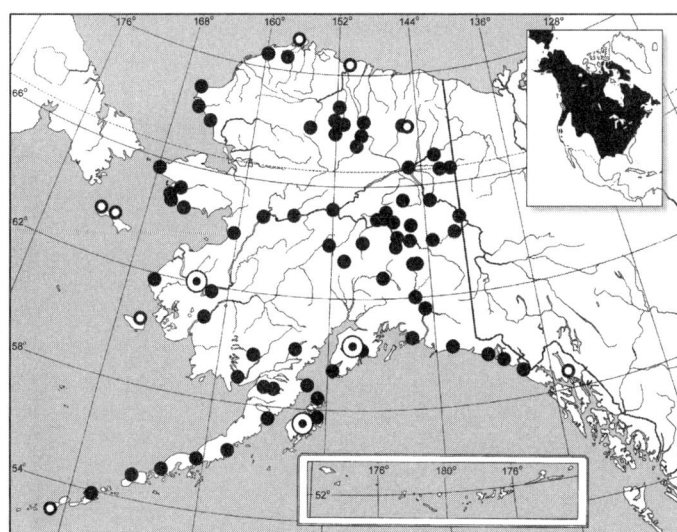

MAP 54. Distribution of red fox, *Vulpes vulpes*.

⊙ = Type Locality
● = Specimen Record (n = 1183)
○ = Marginal Records. 71.29100, –156.78900, Barrow Quad., Barrow (3 UAM); 70.25000, –151.00000, Harrison Bay Quad., Colville River Delta (1 PSM); 68.05000, –145.00000, Arctic Quad., Old John Lake (1 UAM); 58.64717, –134.90833, Juneau Quad., Cowee Creek near N end of Juneau road system (1 UAM); 52.93972, –168.85972, Samalga Island Quad., Umnak Island, Nikolski (1 USNM); 60.38880, –166.18330, Nunivak Island Quad., Nunivak Island, Mekoryuk (1 PSM); 63.70000, –170.48330, St Lawrence Island Quad., St. Lawrence Island, Savoonga (1 UAM); St Lawrence Island Quad., St. Lawrence Island, Gambell (1 USNM).

though never reportedly stocked, supports red foxes.

Central Region. The red fox is found extensively throughout central Alaska.

Southwest Region. Red foxes are found on the Alaska Peninsula and are indigenous to the eastern Aleutians, at least as far west as Umnak Island (Murie 1959). Bailey (1993) speculated that foxes may have naturally ranged as far west as the Islands of Four Mountains. Red foxes are probably indigenous to some of the larger islands south of the Alaska Peninsula, including Nagai, Unga, Korovin, and Popof islands in the inner Shumagins (Bailey 1993). This species may also be a native to three of the Pavlof Islands and Sanak Island, although stocking by Russians or Aleuts could not be discounted (Bailey 1993). This species is native to Kodiak, Afognak, Shuyak, Sitkalidak, and Sitkinak islands (Bailey 1993). Red fox bones were abundant in Aleut middens on Kodiak Island (Murie 1959). Red foxes reach the Walrus Islands in Bristol Bay by ice in winter (Bailey 1992). In the Aleutians, red foxes are known to have been introduced to the following islands: Adak, Adokt-Koschekt, Anocknock, Chuginadak, Escelsior (Baby Islands), Great Sitkin, Kanaga, Kositka, Rat Islands, Signals (Ugalga), Tangaman, and west Ulak (Bailey 1993). South of the Alaska Peninsula, red foxes were released on Andronica, Big Koniuji, Chowiet (South Semidi), Clifford, Deer, Dolgoi, Iliasik, Jacob, Korovin, Kiukpalik, Mary, Nagai (prior indigenous population), Popof (prior indigenous population), Sarana, and Ukolnoi islands (Bailey 1993). Red foxes were introduced to at least eighteen islands in the Kodiak Archipelago, beginning with Long Island in 1880. They have persisted on only Geese, Ugak, and Marmot (possible prior indigenous population) islands (Bailey 1993).

Western Region. Red foxes are present throughout western mainland Alaska, but predominate inland in low-lying areas (ADFG 1978). In northwestern Alaska in winter, they are reported to congregate along the coast and in river valleys to feed on carrion. This species occurs on Nunivak Island (ADFG 1978). Murie (1936) reported that red foxes are occasionally seen on St. Lawrence Island. Rausch (1953b) considered the red fox a very rare visitor to the island (probably from neighboring Siberia) and secured several skins from local residents (MSB, PSM).

Northern Region. Red foxes are found throughout northern Alaska, but populations are highest along riparian drainages in the mountains and foothills of the Brooks Range and lowest on the coastal plains and along the coast (ADFG 1978).

Habitat

The red fox lives in a wide variety of habitats ranging from tundra to boreal forests, steppe, and temperate deserts (Larivière and Pasitschniak-Arts 1996).

Status

Red foxes are common south of the arctic tundra, and most abundant in central Alaska and on the coastal areas south of Norton Sound, including the Alaska Peninsula (ADFG 1978). Bailey (1993) documented the introduction of red foxes to sixty Alaska islands. Due to natural mortality and active eradication efforts, only one island remains inhabited by red foxes (Ebbert and Byrd 2002).

Fossils

The earliest records of red foxes for North America are thought to be of Kansan age (Graham 1972). Undated fossil remains in eastern Beringia have been reported from Yukon Territory (Harington 1977; Kurtén and Anderson 1980; Youngman 1993) and central Alaska (Dixon 1984; Péwé 1975; Youngman 1993). Fossils of Early Holocene age have been discovered in cave deposits on Prince of Wales Island in southeast Alaska (Heaton and Grady 2003).

Family URSIDAE Fischer von Waldheim, 1817—bears

Ursids comprise eight living species in five genera with no subfamilies currently recognized (Wozencraft 2005). Three bear species, all in the nominate genus *Ursus*, occur in Alaska.

The fossil record of bears extends to the Late Eocene in Europe, Late Miocene in North America, Early Miocene in Asia, Late Pliocene in South America, and Pleistocene in northwestern Asia (Stains 1984).

Bears are the largest land-dwelling members of Carnivora. They are found nearly worldwide from the Arctic to the tropics. Except for the polar bear, all are omnivorous and some strongly herbivorous. Bears are large, powerfully built mammals with large heads, short tails, and small eyes and ears. All limbs have five digits that end in well-developed claws. They walk plantigrade, with the heel of the foot touching the ground. Many are capable climbers. Most are brown, black, or white in color; some have white marks on the chest or face. Bears of some species accumulate fat reserves in the fall and go into dormancy through most of the winter.

Brown bear, *Ursus arctos* (W. D. Berry)

American Black Bear

Ursus americanus
Pallas, 1780

Other Common Names

Black bear, glacier bear.

Systematics

Hall (1981) recognized sixteen subspecies of black bear, of which four occur in Alaska. The validity of some of these, particularly Pacific coastal forms, including the so-called "glacier bear" (*U. a. emmonsii*), is questionable and in need of revision (Nagorsen 1990).

Two distinct mtDNA lineages (coastal and continental) of black bear have been identified. A recent study using nuclear DNA sequence data and larger sample sizes generally agreed with the mitochondrial data from southeast Alaska (Peacock 2004). The coastal lineage extends along the Pacific Coast from Kupreanof Island in southeast Alaska to Northern California, while the continental lineage is more widespread, occurring from central Alaska to the East Coast (Byun et al. 1997; Stone and Cook 2000; Wooding and Ward 1997). Contact between these lineages is along the central mainland of southeast Alaska in the vicinity of Frederick Sound (Cook et al. 2001; MacDonald and Cook 2007; Peacock 2004; Stone and Cook 2000). Coastal bears have a deep, complex tenure in the region that extends onto a few islands, with pre- and postglacial fossil remains of black bears from cave deposits on Prince of Wales Island in southeast Alaska (Heaton and Grady 2003), and in early postglacial deposits on Haida Gwaii (Wigen 2005).

Ursus americanus americanus

Original Description. 1780. *Ursus americanus* Pallas, *Spicilegia zoologica*, . . . , fasc. 14:5.
Type Locality. Eastern North America.
Type Specimen. Not known to exist.
Range. Northwestern Alaska and across Canada, to the eastern United States.

Ursus americanus emmonsii

Original Description. 1895. [*Ursus americanus*] var. *emmonsii* Dall, *Science*, n.s., 2:87, July 26.
Type Locality. St. Elias Alps, near Yakutat Bay, Alaska.
Type Specimen. None designated.
Range. Glacier Bay region northward to Prince William Sound.

Ursus americanus perniger

Original Description. 1910. *Ursus americanus perniger* J. A. Allen, *Bull. Amer. Mus. Nat. Hist.*, 28:115, April 30.
Type Locality. [Mountains south of Chugachik Bay, opposite] Homer, Kenai Peninsula, Alaska.
Type Specimen. AMNH 17790.
Range. Restricted to the Kenai Peninsula by Hall (1981).
Remarks. A renaming of *U. a. kenaiensis* J. A. Allen.

Ursus americanus pugnax

Original Description. 1911. *Ursus americanus pugnax* Swarth, *Univ. California Publ. Zool.*, 7:141, January 12.
Type Locality. Rocky Bay, now Bobs Bay, Dall Island, Alaska.
Type Specimen. MVZ 8332.
Range. Southeast Alaska.

Global Distribution

Nearctic—Black bears range across most of North America, from central Alaska into the mountains of northern Mexico (Larivière 2001; Rogers 1999).

Alaska Distribution

Black bears occur over most of the forested areas of the state (map 55). They are generally absent from Alaska's North Slope, the Seward Peninsula, the Yukon-Kuskokwim Delta, the Alaska Peninsula beyond Lake Iliamna, the Kodiak Island group, and the islands in southeast Alaska north of Frederick Sound (ADFG 1973).

Southeast Region. Black bears are found along the mainland coast of southeast Alaska and on most of the islands in the Alexander Archipelago south of Frederick Sound. We are unaware of any reports of black bears on the southern islands of Annette, Duke, Mary, Warren, Coronation, or Forrester islands, or on any island north of Frederick Sound except Pleasant Island close to the mainland in Icy Strait (MacDonald and Cook 2007). Kuiu Island supports the highest density of black bears documented in North America (Peacock 2004).

Southcentral Region. This species is common throughout most of the mainland and on a few islands in Prince William Sound (ADFG 1973), including Hawkins Island (Heller 1910), but not Hinchinbrook or Montague islands (Johnson 1994). It is also absent from Kayak and Middleton islands

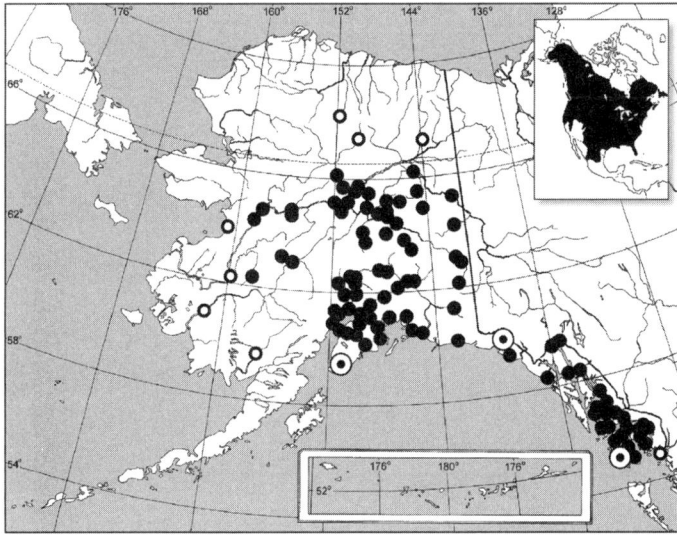

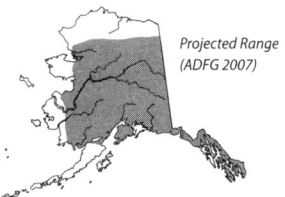

MAP 55. Distribution of American black bear, *Ursus americanus*.

Projected Range (ADFG 2007)

⊙ = Type Locality
● = Specimen Record (n = 1218)
○ = Marginal Records. 68.12500, –151.75000, Chandler Lake Quad., Inukpasugruk Creek, near head of John River (1 USNM); 67.25833, –150.18333, Wiseman Quad., Coldfoot (1 UAM); 67.36667, –143.80000, Coleen Quad., Small Lake (1 UAM); 54.80000, –130.73333, Prince Rupert Quad., Nakat Inlet (1 UAM); 59.31667, –157.55000, Dillingham Quad., Kokwok (1 USNM); 60.79167, –161.75000, Bethel Quad., Bethel (2 USNM); 62.20000, –159.76667, Holy Cross Quad., Holy Cross (6 USNM); 63.87361, –160.78333, Unalakleet Quad., Unalakleet (1 USNM).

along the North Gulf Coast (Crowley 2005a). Black bears are abundant on the Kenai Peninsula and in the upper Cook Inlet area (ADFG 1973).

Central Region. Black bears are found throughout central Alaska.

Southwest Region. Suitable black bear habitat is limited in this region, and they rarely occur below the sparse spruce forests near Iliamna Lake (ADFG 1973). There is an unsubstantiated report of a black bear at Kutik Lake near the northern border of Katmai National Park and Preserve (Cook and MacDonald 2006). Black bears are not found on any islands in this region, including the Kodiak Archipelago.

Western Region. Black bears are of marginal occurrence in western Alaska, with no consistent records from the Seward Peninsula or the Yukon-Kuskokwim Delta (ADFG 1973). They are found in the Kobuk River drainage (Dean and Chesemore 1974) to at least as far west as the Great Kobuk Sand Dunes (Cook and MacDonald 2004a).

Northern Region. Black bears are rare north of the Brooks Range (ADFG 1973). Rausch (1951) reported a specimen killed near the mouth of Inukpasukruk Creek at the head of the John River, approximately 5 km south of the Continental Divide.

Habitat

Black bears are generally restricted to forested habitats, but depending on the season, may be found from sea level to alpine areas. They prefer semi-open areas composed primarily of fruit-bearing shrubs and herbs, lush grasses, and succulent forbs. Extensive open areas are generally avoided (ADFG 1973; Larivière 2001).

Status

This species is listed under CITES-Appendix II. The black bear is a Management Indicator Species for the Tongass National Forest. Limited movement of black bears among islands of the Alexander Archipelago suggests a need for their judicious management (MacDonald and Cook 2007).

Fossils

Kurtén and Anderson (1980) listed Alaska as a Rancholabrean site for *U. americanus*. Guthrie (1968b), however, stated that this species has not been found as a Pleistocene fossil in central Alaska deposits and may be a recent southern immigrant, a conclusion shared by Youngman (1975). Dixon (1984) identified possible black bear remains from Porcupine River cave deposits that dated near the Late Wisconsin–Early Holocene transition (older than 8,000 ybp). Mid-Wisconsin and Late

Wisconsin–Early Holocene fossil remains of black bears have been found in cave deposits on Prince of Wales Island in southeast Alaska (Heaton 1995; Heaton and Grady 1992, 1993, 2003). Wigen (2005) reported early postglacial black (as well as brown) bear remains on Haida Gwaii, and Nagorsen et al. (1995) described Early Holocene black bears of exceptionally large size from Vancouver Island. Graham (1991) found Late Pleistocene black bears generally larger than modern bears.

Brown Bear

Ursus arctos
Linnaeus, 1758

Other Common Names

Grizzly bear. Merriam (1918) gave common names to his many described taxa.

Systematics

Some 232 Recent and 39 fossil "species" and "subspecies" have been proposed for this taxon (Erdbrink 1953; Merriam 1918), a formidable and confusing situation that Kurtén and Anderson (1980) considered "a waste of systematic effort which, as far as we know, is unparalleled." Most authorities now recognize *U. arctos* as one Holarctic species. Wozencraft (2005) recognized fourteen Old and New World subspecies (seven in Alaska), whereas the Integrated Taxonomic Information System (http://www.itis.gov) currently recognizes only two (*U. a. arctos* Linnaeus, 1758, and *U. a. pruinosus* Blyth, 1854).

Recent genetic studies of brown bears indicate that the traditional morphology-based taxonomy of brown bears is highly discordant with bear phylogeography as indicated by geographic patterns of mtDNA variation. In a detailed population genetic survey using a suite of nuclear genetic markers (Paetkau et al. 1998), coastal Alaska "brown bears" were found to be part of the continuous continental distribution of brown bears, and not genetically isolated from the physically smaller "grizzly bears" of the interior. The bears of Admiralty, Baranof, and Chichagof islands in the Alexander Archipelago, which in previous mtDNA studies were shown to have undergone little or no female-mediated gene flow with mainland populations (Cronin et al. 1991; Shields and Kocher 1991; Talbot and Shields 1996), were found to share nuclear alleles with mainland bears (Paetkau et al.

1998). By contrast, brown bears from Kodiak Island appeared to have experienced little or no genetic exchange with continental populations in recent generations (Paetkau et al. 1998). Further work is needed.

Leonard et al. (2000) measured mitochondrial DNA sequence variation in seven permafrost-preserved brown bear specimens from central Alaska and northwestern Canada. They concluded that approximately thirty-six thousand years ago, the eastern Beringian brown bear populations had higher genetic diversity than any extant North American population, but by fifteen thousand years ago genetic diversity was comparable to modern populations. The older, genetically diverse Beringian population contained sequences from clades now restricted to localized regions within North America.

A recent study using ancient-DNA techniques on bone samples (n=36) of brown bears from eastern Beringia by Barnes et al. (2002) found a marked degree of genetic structure in populations over the past sixty thousand years, with major phylogeographic changes occurring 35 to 21 ka BP, before the last glacial maximum, and little change after this time. Their data also suggest that a distinct clade of brown bears currently restricted to the Admiralty, Baranof, and Chichagof islands in southeast Alaska was formerly more diverse and widespread across eastern Beringia and farther south on Prince of Wales Island near the end of the Pleistocene.

Mitochondrial DNA data from Talbot and Shields (1996), Waits et al. (1998), and others indicated that the brown bear are paraphyletic with respect to the polar bear, *Ursus maritimus*.

Ursus arctos alascensis

Original Description. 1896. *Ursus alascensis* Merriam, *Proc. Biol. Soc. Washington*, 10:74, April 13.
Type Locality. Unalaklik [= Unalakleet] River, Alaska.
Type Specimen. USNM 76466.
Range. Southcentral, central, western, and northern Alaska.
Remarks. Includes *kidderi, kenaiensis, phaeonyx, sheldoni, alexandrae, innuitus, internationalis, toklat, tundrensis, cressonus, eximius, nuchek,* and *holzworthi*.

Ursus arctos beringianus

Original Description. 1853. *Ur*[*sus*]. *arctos* var. *beringiana* Middendorff, *Sibir. Reise*, 2, 2:4, Pl. 1.

Type Locality. Great Shantar Island, Sea of Okhotsk.

Type Specimen. Not known.

Range. Eastern Palearctic.

Remarks. A Siberian form, in which *U. arctos kolymensis* Ognev, 1924 is a synonym, has occurred on St. Lawrence Island (Hall 1981; Howell 1940; Rausch 1953b).

Ursus arctos dalli

Original Description. 1896. *Ursus dalli* Merriam, *Proc. Biol. Soc. Washington*, 10:71, April 13.

Type Locality. Yakutat Bay (NW side), Alaska.

Type Specimen. USNM 75048.

Range. Northern mainland of southeast Alaska, from Yakutat area south to about Glacier Bay.

Remarks. Includes *nortoni, townsendi*, and *orgiloides*. Paetkau et al. (1998) found the brown bears of coastal Alaska genetically similar to interior populations, and suggested the designation *U. a. dalli* be dropped in favor of *U. a. horribilis*.

Ursus arctos gyas

Original Description. 1902. *Ursus dalli gyas* Merriam, *Proc. Biol. Soc. Washington*, 15:78, March 22.

Type Locality. Pavlof Bay, Alaska Peninsula.

Type Specimen. USNM 91669.

Range. Western Alaska Peninsula.

Remarks. Includes *merriami*.

Ursus arctos middendorffi

Original Description. 1896. *Ursus middendorffi* Merriam, *Proc. Biol. Soc. Washington*, 10:69, April 13.

Type Locality. Kadiak [= Kodiak] Island, Alaska.

Type Specimen. USNM 54793.

Range. Kodiak Archipelago.

Remarks. Includes *kodiaki*.

Ursus arctos sitkensis

Original Description. 1896. *Ursus sitkensis* Merriam, *Proc. Biol. Soc. Washington*, 10:73, April 13.

Type Locality. Near Sitka, Alaska.

Type Specimen. USNM 187891.

Range. Alexander Archipelago and northern mainland of southeast Alaska.

Remarks. Includes *eulophus, eltonclarki, orgilos, caurinus, shirasi, insularis, neglectus*, and *mirabilis*.

Ursus arctos stikeenensis

Original Description. 1914. *Ursus stikeenensis* Merriam, *Proc. Biol. Soc. Washington*, 27:178, August 13.

Type Locality. Tatletuey Lake, near head Skeena River, northern British Columbia.

Type Specimen. USNM 202794.

Range. Northern and coastal British Columbia and adjacent southern mainland of southeast Alaska.

Remarks. Includes *tahltanicus, pervagor, chelan, hoots, kwakiutl, warburtoni, chelidonias, atnarko*, and *crassodon*.

Global Distribution

Holarctic—Brown bears are found in North America and Eurasia. Historically, this species ranged over most of western and central North America from the Arctic Ocean to central Mexico (Guilday 1968). Their distribution in Alaska shows little change from historic times (Pasitschniak-Arts 1993).

Alaska Distribution

Ursus arctos is distributed over most of Alaska (map 56). It is absent only from the Aleutian Islands west of Unimak Island and from the islands south of the Alaska Peninsula and islands south of Frederick Sound in the Alexander Archipelago.

Southeast Region. Brown bears occur along the entire coastal mainland of southeast Alaska (especially along the major river systems), and on most of the islands of the Alexander Archipelago north of Frederick Sound. They are occasionally seen but have not become established on islands close to the mainland south of Frederick Sound, specifically Etolin, Mitkof, Revillagigedo, and Wrangell islands (MacDonald and Cook 2007).

Southcentral Region. Brown bears occur on some larger islands and much of the mainland in this region. Major island populations are on Hinchinbrook, Montague, and Kayak islands; a few bears also inhabit Hawkins Island. Brown bears are generally absent from western Prince William Sound (ADFG 1973).

Central Region. This species occurs primarily in alpine and subalpine areas, and sporadically in forested lowland areas, throughout central Alaska.

Southwest Region. Brown bears are numerous along the Alaska Peninsula and on Unimak Island, but are absent from the Shumagins and other islands off the southern mainland coast (ADFG 1973). They inhabit all of the major islands in the Kodiak Archipelago, and are absent only from the lower southwestern Trinity Islands and other smaller

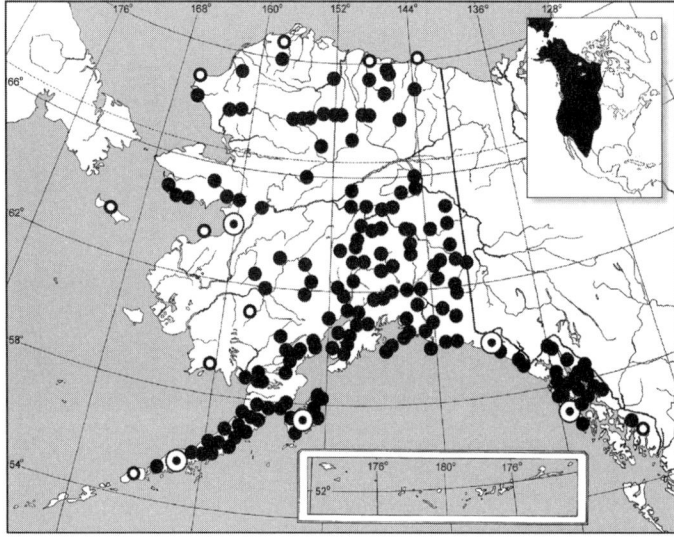

MAP 56. Distribution of brown bear, *Ursus arctos*.

⊙ = Type Locality
● = Specimen Record (n = 2700)
○ = Marginal Records. 70.48056, −157.41667, Meade River Quad., Meade River (5 UAM); 70.32100, −148.03300, Beechey Point Quad., Deadhorse, Sagavaninktok drainage (1 UAM); 70.11667, −143.66667, Barter Island Quad., Barter Island (1 USNM); 55.81667, −130.91667, Ketchikan Quad., Chickamin River (1 UAM); 54.75000, −165.00000, False Pass Quad., Unimak Island (2 USNM); 59.00000, −160.30000, Goodnews Bay Quad., Togiak Bay (1 USNM); 61.00000, −158.00000, Taylor Mts Quad., Holitna–Kogrukluk Rivers (1 UAM); 63.53333, −162.66667, St. Michael Quad., Stuart Island (1 UAM); 63.50000, −170.50000, St Lawrence Quad., St. Lawrence Island (1 USNM); 68.88333, −166.21667, Point Hope Quad., Cape Lisburne (1 USNM).

islands some distance offshore (Clark 1958)

Western Region. Brown bears are distributed throughout the hills and mountains of this region, but seldom roam onto the Yukon-Kuskokwim Delta (ADFG 1973). Brown bears from far eastern Siberia have occurred infrequently on St. Lawrence Island (Geist 1934; Howell 1940; Rausch 1953b; Schwarz 1940).

Northern Region. *Ursus arctos* ranges throughout arctic Alaska, but is most numerous in the foothills of the Brooks Range and along major river valleys (ADFG 1973; Rausch 1953a).

Habitat

In Alaska, brown bears are most common in areas of open tundra and grassland. Even where they occur in forested areas, as in southeast Alaska, substantial mountain meadows, muskegs, sedge flats, and other grassland areas are present (ADFG 1973; Schoen and Gende 2007). Den sites are often on hillsides.

Status

Alaska contains more than 70 percent of North America's population of brown bears (Eide and Miller 1994). Brown bears are probably as abundant now in Alaska as during earlier times, except where they have been displaced by man, such as near human population centers and where livestock interests conflict (ADFG 1973). Poisoning efforts by federal agents in the 1950s caused brown bear populations to decline and remain depressed into

the 1970s in some areas of the state (Reynolds 2005). The number of brown bears in Alaska has been estimated at thirty-two thousand to forty-three thousand bears (Servheen 1990). This species is listed under CITES-Appendix II. It is a Species of Concern in southeast Alaska (Suring et al. 1992) and a Management Indicator Species for the Tongass National Forest (MacDonald and Cook 2007). Waits et al. (1998) recommended evolutionarily significant unit (ESU) status for the conservation of brown bears in two areas of the state: (1) the Admiralty, Baranof, and Chichagof islands of southeast Alaska, and (2) mainland Alaska and Kodiak Island.

Fossils

The earliest records of *U. arctos* are from China, about five hundred thousand years ago (Pasitschniak-Arts 1993). The species is believed to have a limited history in North America, appearing in eastern Beringia perhaps only fifty to seventy thousand years ago (Leonard et al. 2000). Until recently, there was no reliable record of the species south of the Wisconsinan ice prior to the last glaciation (Kurtén and Anderson 1980). That changed with the discovery of a twenty-six-thousand-year-old brown bear fossil from central Alberta (Matheus et al. 2004). This fossil demonstrated that brown bears migrated south of Beringia much earlier than previously thought. Sequences of mtDNA recovered from the specimen showed it belonged to the same phylogroup as brown bears living in southern Canada and the northern United States today.

Ursus arctos remains have been discovered in limestone cave deposits on Prince of Wales Island in the Alexander Archipelago of southeast Alaska and these ranged in age from Middle Wisconsin (35,365±800 ybp) to Early Holocene (Heaton 1995; Heaton and Grady 1993, 2003). Brown bear (and black bear) bones have also been found in cave deposits on Dall and Coronation islands (Heaton and Grady 2003). These islands are south of Frederick Sound, where today only black bears occur. Brown bear remains recently discovered just south of the Alexander Archipelago on Haida Gwaii ranged in age from last glacial to early postglacial (Wigen 2005). Only black bears occur there now.

Polar Bear

Ursus maritimus
Phipps, 1774

Other Common Names

White bear, ice bear, nanuk or nanook.

Systematics

This species was revised by Wilson (1976). Gromov and Baranova (1981) placed it in the subgenus *Thalarctos*. No subspecies are recognized for living populations, although six apparently distinct populations have been identified, including one that encompasses western Alaska and eastern Siberia, and another in northern Alaska and western Canada (DeMaster and Stirling 1981). Manning (1971) found considerable geographic variation in size in Recent populations, with individuals smallest in eastern Greenland and largest in "Alaska south" (St. Paul and St. Lawrence islands). The polar bear is a sister species to the brown bear, *U. arctos* (Goldman et al. 1989; Shields and Kocher 1991), branching off the brown bear lineage during the Late Pleistocene (Amstrup 2003). In captivity, crosses between *U. arctos* and *U. maritimus* have produced fertile offspring (Davis 1950).

Cronin et al. (1991) and Scribner et al. (1997) found low divergence in mtDNA haplotypes among polar bear populations from Alaska and Canada. Using sixteen highly variable microsatellite loci, Paetkau et al. (1999) determined that polar bears throughout the Arctic (sixteen populations) were very similar.

Ursus maritimus

Original Description. 1774. *Ursus maritimus* Phipps, *Voyage toward the North Pole*, p. 185.
Type Locality. Spitzbergen, Norway.
Type Specimen. Not known to exist.

Global Distribution

Holarctic—Polar bears are circumpolar in distribution with the southern limits of their range determined by the distribution of arctic pack ice and annual landfast ice during winter (DeMaster and Stirling 1981). They have been reported within 300 km of the North Pole (Reeves et al. 1992). Individual bears have been seen some distance inland from the coast of Alaska and Yukon Territory (Bee and Hall 1956; Manville and Young 1965; Youngman 1975).

Alaska Distribution

Polar bears are frequently found throughout the drifting ice zone in the Beaufort and Chukchi seas off the coast of northern and northwestern Alaska (map 57). Some bears drift south of Bering Strait in winter, occasionally reaching St. Lawrence Island and even the Kuskokwim Delta (ADFG 1973; Lentfer 1994). Polar bears occasionally wander far inland, as happened in the early spring of 2008 when one appeared (and was subsequently shot) near the Interior village of Fort Yukon (Smetzer 2008).

Historically, considerable numbers of polar bears lived year-round on St. Matthew Island in the Bering Sea. In 1874 federal agent H. W. Elliott (1881) estimated 250 to 300 bears present on the island. Twenty-five years later, members of the Harriman Expedition found no bears inhabiting the island, only their deep-worn trails and bones. These bear trails were still visible in 1954 (Rausch and Rausch 1968). Ray (1971) reported polar bear records from the Pribilof Islands, including a skull of an individual shot on St. Paul Island (Crossen et al. 2003).

In addition, there are questionable records (in Manville and Young 1965) of polar bear in the Gulf of Alaska from Kodiak Island, Prince William Sound (in the *Voyages of Captain James Cook* of white bear skins seen there in 1778; Murie 1959), and Yakutat (USNM 76577 is an undated skull with mandibles with "Alaska?" written on one side of the specimen tag and "Yakutat" written on the other side). The collector recorded is F. E. Frobese, who may have been a turn-of-the-last-century miner and Alaska pioneer from the Nome area. The discovery of ringed seal (*Phoca hispida*) remains from caves on Prince of Wales Island in south-east Alaska that date from the height of the last glacial (Heaton and Grady 2003) leads one to speculate that the range of this seal's main predator may have also extended southward along the northeastern Pacific Coast during this period of arctic-like conditions.

Habitat

Polar bears are associated with unstable sea ice in areas where there is an abundance of ringed seals (*Phoca hispida*) or bearded seals (*Erignathus barbatus*), their primary prey (ADFG 1973; DeMaster and Stirling 1981). They are powerful swimmers and great travelers (Durner and Amstrup 1995). According to ADFG (1973), polar bears were most abundant within about 300 km of the Alaska coast. Alaska polar bears den most commonly on offshore islands and associated heavy, stable ice from the mouth of the Colville River east to about Brownlow Point. Denning in the Beaufort Sea was reported by Amstrup and Gardner (1994). They occasionally den on shorefast ice and river bottoms from the Kuparuk River west and south along the Alaska coast to the Point Hope area. Most dens are within 40 km of the coast. Bears of western Alaska may have come from dens on Wrangel Island, Russia (ADFG 1973).

Status

Research on polar bears in northern Alaska was summarized by Amstrup and Gardner (1991). Amstrup and DeMaster (1988) estimated three thousand to five thousand polar bears statewide. The southern Beaufort Sea population experienced growth during the late 1970s and 1980s and then stabilized during the 1990s (Amstrup et al. 2001). The size of the Chukchi-Bering seas population has not been accurately determined.

The most important denning area on land in Alaska is the Arctic National Wildlife Refuge (Reeves et al. 1992), an area where human activities associated with oil and gas exploration and extraction could pose a serious threat (Lentfer 1994). Polar bears are currently protected under international agreement (see Lentfer 1974; Prestrud and Stirling 1994). In 1972, the Marine Mammal Protection Act ended responsibility of the State of Alaska for polar bear management, which eliminated the non-Native harvest of this species. Drastic changes under way in the Arctic due to warming conditions are changing the seasonal pack ice and endangering polar bears and other arctic wildlife (ACIA 2005). The polar bear was assessed as Vulnerable by IUCN in 2007 (http://www.iucnredlist.org). On 9 January 2007, the Fish and Wildlife Service proposed to list this species as Threatened. Failure by the Service to meet its one year listing deadline prompted a suit by several conservation organizations. On 14 May 2008, the Department of the Interior listed the polar bear as a threatened species under the ESA.

Fossils

Polar bears are thought to have originated from a segment of the Siberian population of brown bears, *Ursus arctos*, which was isolated

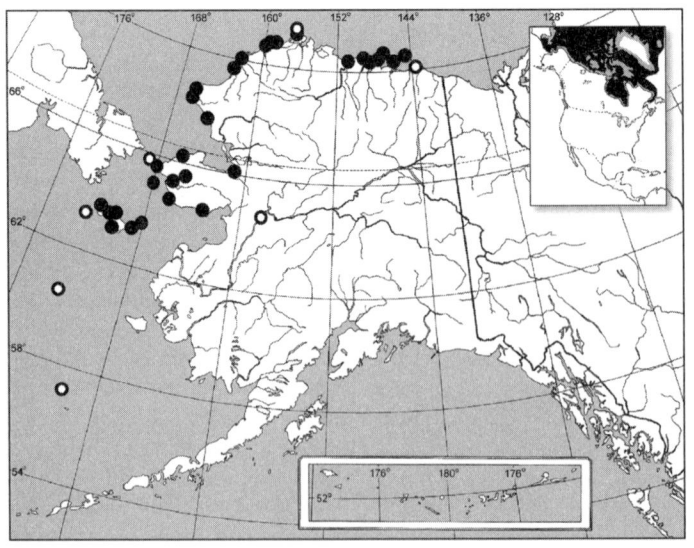

MAP 57. Distribution of polar bear, *Ursus maritimus*.

⊙ = Type Locality
● = Specimen Record (n = 1039)
○ = Marginal Records. 71.50500, −156.46583, Barrow Quad., 8 miles N of Point Barrow (1 UAM); 70.13028, −143.61750, Barter Island Quad., Barter Island, Kaktovik (2 UAM); Yukon River, near Nulato (1 USNM); 57.16667, −170.25000, Pribilof Islands Quad., St. Paul Island, NE portion (1 USNM); 60.40000, −172.70000, St Matthew Quad., St. Matthew Island (1 CAS); 63.38667, −173.17000, Bering Sea, St. Lawrence Quad., Saint Lawrence Island, 30 miles SW of Gambell, (1 UAM); 65.71667, −168.91667, Bering Strait, Teller Quad., Little Diomede Island (5 UAM).

during the glacial advances of the mid-Pleistocene (Kurtén 1964). One Late Pleistocene subspecies, *U. m. tyrannus*, has been recognized (Kurtén 1964).

Fragmentary skeletal remains of polar bears have occasionally been reported from both St. George and St. Paul islands (see Crossen et al. 2003). Recent investigations on St. Paul Island (Veltre et al. 2008) have recovered a mid-Holocene assemblage that includes polar bear, woolly mammoth, caribou, and arctic fox. Polar bear remains of similar age have also been recovered from an archaeological site on Unalaska Island in the eastern Aleutians (Crockford and Frederick 2007).

Polar bear, *Ursus maritimus* (S. O. MacDonald)

Family OTARIIDAE Gray, 1825—eared seals

Members of Otariidae, the eared seals, are found nearly worldwide except in the Arctic Ocean and Antarctica. The family is comprised of sixteen species in seven genera (Wozencraft 2005). Three species in three genera occur in Alaska.

The geologic range of otariids is Early Miocene to Recent in Pacific North America, Miocene to Recent in Europe and Asia, Pliocene to Recent in South America, and Pleistocene to Recent in Australia, New Zealand, Japan, and Africa (Stains 1984).

All otariids have small but noticeable external ear flaps. They have long, slender bodies, long necks, and long, supple fore limbs and rotating hind limbs that make them more agile on land than the earless seals. All have coarse coats of guard hair, and the fur seals also have dense underfur. Males are significantly larger than females. Pelagic most of the year, eared seals are highly gregarious and form large herds on their breeding grounds.

Northern fur seal, *Callorhinus ursinus* (B. Hines)

Guadalupe Fur Seal

Arctocephalus townsendi
Merriam, 1897

Other Common Names

Guadalupe fur-seal.

Systematics

The Guadalupe fur seal is a monotypic species (Belcher and Lee 2002) composed of a single genetically distinct stock (Forney et al. 2000). Scheffer (1958) and Hall (1981) considered *A. townsendi* conspecific with the Juan Fernández fur seal, *A. philippii*.

Arctocephalus townsendi

Original Description. 1897. *Arctocephalus townsendi* Merriam, *Proc. Biol. Soc.Washington* 11:178.
Type Locality. Guadalupe Island, Baja California, Mexico.
Type Specimen. USNM 83617.

Global Distribution

Prior to their near extermination during the 1800s, Guadalupe fur seals ranged from Monterey Bay, California, to the Revillagigedo Islands, Mexico (Hanni et al. 1997). Currently, their breeding range is limited almost exclusively to Guadalupe Island off Baja California. Little is know about this fur seal's whereabouts during the nonbreeding season (Reeves et al. 2002).

Alaska Distribution

In July 2007, an emaciated male fur seal was discovered adrift in Kachemak Bay off Point Pogibshi (map 58). It was rescued and taken to the Alaska SeaLife Center in Seward, where it was determined with the help of DNA testing to be a Guadalupe fur seal. Following rehabilitation at the center, it was flown south and released off San Simeon, California. The unprecedented occurrence of this species in Alaska waters may have been linked to warm water temperatures and the unusual number of strandings of Guadalupe fur seals reported from Oregon and Washington that same summer (Armstrong 2007).

Habitat

During the breeding season, this species favors rocky shores with large cliffs and caves (Belcher and Lee 2002).

Status

The Guadalupe fur seal is an extralimital vagrant to Alaska waters. The species underwent one or more severe bottlenecks due to commercial sealing in the late nineteenth century (Bernardi et al. 1998), and was considered extinct up until its rediscovery in 1954. Subsequently protected, the population has grown steadily at an average annual rate of nearly 14 percent (Forney et al. 2000). In 1993, the population was estimated at about 7,400 animals (Gallo-Reynoso 1994).

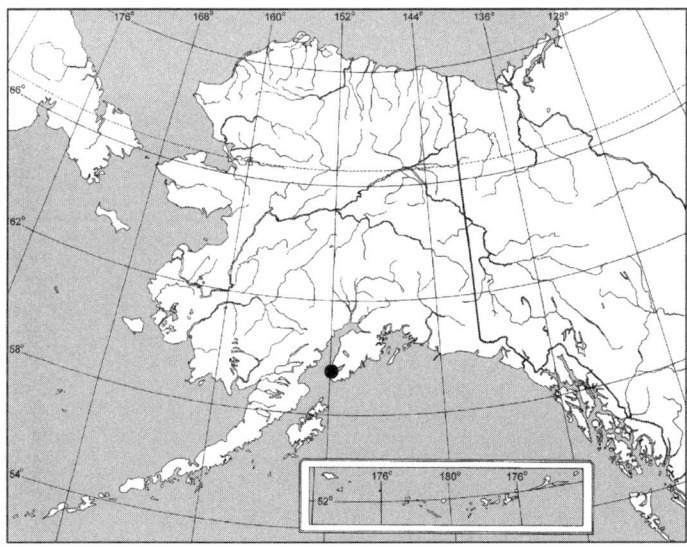

MAP 58. Distribution of Guadalupe fur seal, *Arctocephalus townsendi*.

● = Specimen Record (n = 1; captured and later translocated to California)

Arctocephalus townsendi is fully protected by the State of California, and is listed as Threatened under the ESA, Vulnerable by IUCN, and as an Appendix 1 species under CITES. The species and the Island of Guadalupe are protected under Mexican law (Bonner 1994).

Fossils

Late Holocene subfossils are known from archaeological shell midden deposits on the east shore of San Francisco Bay (Broughton 1999).

Northern Fur Seal

Callorhinus ursinus
(Linnaeus, 1758)

Other Common Names

Alaska fur seal, Pribilof fur seal, North Pacific fur seal.

Systematics

Gardner and Robbins (1998) pointed out that *Otoes* and *Halarctus* are the earliest available generic names for northern and southern fur seals, respectively. The northern fur seal is a monotypic species (Hall 1981; Scheffer 1958).

Callorhinus ursinus

Original Description. 1758. *Phoca ursina* Linnaeus, *Syst. Nat.*, 10th ed., 1:37.

Type Locality. Bering Island, Commander Islands.

Type Specimen. None designated.

Global Distribution

The northern fur seal ranges across the subarctic waters of the North Pacific Ocean and Bering and Okhotsk seas, and into the Sea of Japan (Fiscus 1978). During the summer, northern fur seals are concentrated in rookeries located on the Pribilof Islands, the Commander Islands, Robben Island in the Sea of Okhotsk, the Kuril Islands, San Miguel Island of southern California, and Bogoslof Island in the eastern Aleutian Islands (Reijnders et al. 1993).

Alaska Distribution

Approximately 66 percent of the world's northern fur seal population breed on the Pribilof Islands of St. Paul and St. George in the Bering Sea (Zimmerman 1994a) (map 59). Vagrant fur seals have been reported from the western Beaufort Sea (Bee and Hall 1956). A small breeding colony has recently become established on Bogoslof Island, north of Umnak Island in the eastern Aleutians (Reijnders et al. 1993) by seals from both the Commander and Pribilof islands (Loughlin and Miller 1989). Most female and young fur seals leave the Bering Sea by late November and usually remain offshore along the continental shelf as far south as Southern California and Japan until March. Adult males leave the rookeries from late August through early October and are believed to remain near the Aleutians during the winter (Reijnders et al. 1993).

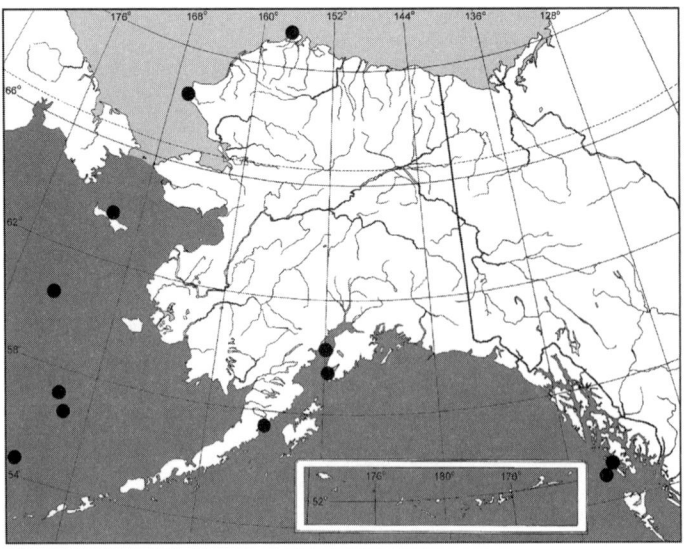

MAP 59. Distribution of northern fur seal, *Callorhinus ursinus*.

■ = Projected Range (after Angliss and Outlaw 2007)
● = Specimen Record (n = 1408)

Habitat

Essentially an animal of the open sea, the northern fur seal hauls out on islands only to breed. Northern fur seals spend from seven to ten months at open sea feeding mostly at night on a variety of schooling fish and squid (Zimmerman 1994a).

Status

This species has been hunted commercially since 1786. Pelagic hunting, which had greatly decimated the species, was abolished in 1911 by international treaty. The total number of northern fur seals has since recovered to an estimated 1.2 million animals and is considered stable (NMFS 1993; Reijnders et al. 1993; Wynne 1993). In 1985, commercial harvesting of fur seals on the Pribilof Islands was terminated (Zimmerman 1994a) and the annual subsistence harvest by Alaska Natives is less than two thousand animals (Wynne 1993). This species is currently listed as Vulnerable by IUCN, and Threatened by COSEWIC.

Fossils

This genus was represented in the Late Pliocene of Southern California (Berta and Deméré 1986). Fossils from a bluff north of Kougechuck Creek on the Seward Peninsula are the only known Pleistocene occurrence of *Callorhinus* (Kurtén and Anderson 1980).

Steller's Sea Lion

Eumetopias jubatus
(Schreber, 1776)

Other Common Names

Steller sea lion, northern sea lion.

Systematics

The genus *Eumetopias* includes only *E. jubatus*. No subspecies have been described (Hall 1981; Scheffer 1958).

Brunner (2002) found little variation between a small sampling of Steller's sea lion skulls from eastern and western Alaska, but identified variation between Alaska, Japan, Russia, and California populations.

Bickham et al. (1996) and Baker et al. (2005) assessed geographic variation in the mitochondrial control region and noted that populations of Steller's sea lion from southeast Alaska and Oregon were distinctive from

populations found farther west in the Gulf of Alaska and Aleutians.

Trujillo et al. (2004) compared variation at six nuclear microsatellite loci to mtDNA sequence variation at three geographic scales (rookeries, regions, and stocks). Population structure was not well defined, and there was no obvious phylogeographic pattern to the distribution of microsatellite alleles. That finding contrasts with the clear phylogeographic pattern revealed by control-region sequences of mtDNA in which two well-differentiated stocks were identified, eastern and western.

Eumetopias jubatus

Original Description. 1776. *Phoca jubata* Schreber, *Die Saugthiere . . . theil* 3, heft 17, Pl. 83b and p. 300.

Type Locality. North Pacific Ocean.

Type Specimen. Not known to exist.

Remarks. Monotypic.

Global Distribution

Steller's sea lions range along the North Pacific Rim from northern Japan to California, with centers of abundance and distribution in the Gulf of Alaska and Aleutian Islands, respectively (Loughlin et al. 1987). Individuals are occasionally found on Herschel Island in the Beaufort Sea (Rice 1998).

Alaska Distribution

Sea lions are not known to migrate, but individuals disperse widely in the nonbreeding season, occurring as far north as St. Matthew, St. Lawrence, Round, and Diomede islands in the Bering Sea and along the coast in Bristol Bay at Cape Newenham (Reeves et al. 1992) (map 60). Throughout the Aleutian Islands, adult and subadult males move north in the late summer and return when the ice forms (Gentry and Withrow 1978). Most reproduction occurs at scattered rookeries along the central coast of the Gulf of Alaska and in the central Aleutian Islands (Angliss and Outlaw 2007; Reeves et al. 1992). The largest rookery worldwide is now on Forrester Island in southeast Alaska (MacDonald and Cook 1996). The northernmost rookery of Steller's sea lions is Seal Rocks in Prince William Sound (Loughlin et al. 1987). Historically, large numbers of sea lions breed on the islands of St. Paul and St. George in the Pribilof Islands; only about three hundred births occur each year at nearby Walrus Island, down from three thousand in 1960 (Reeves et al. 1992).

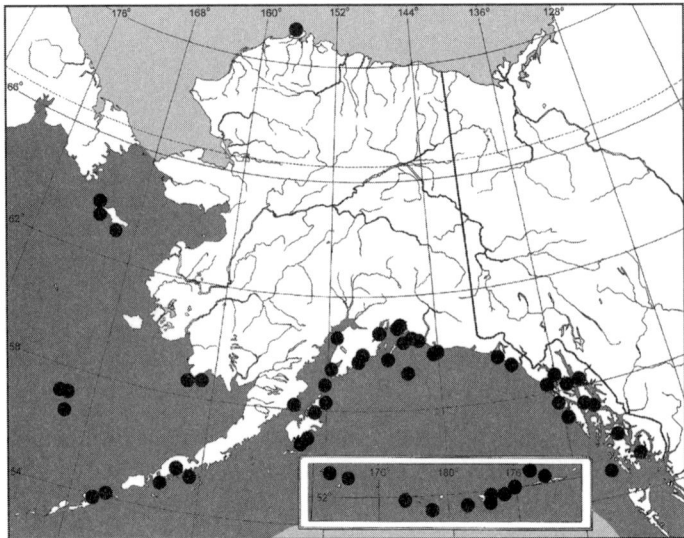

MAP 60. Distribution of Steller's sea lion, *Eumetopias jubatus*.

■ = Projected Range (after Angliss and Outlaw 2007)
● = Specimen Record (n = 523)

Habitat

Steller's sea lions forage mostly near shore and over the continental shelf (Reeves et al. 1992). Walleye pollock (*Theraga chalcogramma*) is an important part of their diet in Alaska, but this fish is commercially exploited (Lowry et al. 1989). This is the only otariid that habitually hauls out on sea ice (Rice 1998).

Status

There has been a dramatic decline of Steller's sea lions in the Gulf of Alaska and Aleutian Islands since the mid-1970s (Braham, Everitt, et al. 1980). There has not been a concomitant decline in the eastern stock (Angliss and Outlaw 2007). The decline of western stocks appeared to have spread eastward to the Kodiak Island area during the late 1970s and early 1980s, and then westward to the central and western Aleutian Islands during the early and mid-1980s. The greatest declines since the 1970s occurred in the eastern Aleutian Islands and western Gulf of Alaska, but declines also occurred in the central Gulf of Alaska and central Aleutian Islands. Counts of Steller's sea lions at trend sites for this western United States stock decreased 40 percent from 1991 to 2000 and then increased 5.5 percent from 2000 to 2002, and at a similar rate between 2002 and 2004. These were the first region-wide increases for the western stock since standardized surveys began in the 1970s.

However, the 2004 count was still 7.4 percent below the 1996 count and 32.6 percent below the 1990 count. The long-term, average decline for 1991–2004 is 3.1 percent per year (Angliss and Outlaw 2007). Angliss and Outlaw (2007) gave a minimum abundance estimate of 38,988 western Steller's sea lions in 2004–2005, and 44,885 eastern sea lions in 2005.

The primary causes of Steller's sea lion decline remain clouded. Lowery et al. (1989) suggested deliberate killing by fishermen, disease, incidental take by fisheries, and reduced food supply as factors that may have contributed most to the decline. More recently, NRC (2003) favored "top-down" hypotheses that encompassed factors that kill sea lions independently of the environment's capacity to support sea lion populations, including predator switching by killer whales (or sharks) to target sea lions, increasing incidental take (or disturbance) through capture or entanglement in fishing gear, subsistence harvesting more than estimated, underestimation of sea lion shooting, and increasing mortality from pollution and disease, independent of nutrition.

The eastern stocks (east of 144°W) of *E. jubatus* are listed as Threatened and the western stocks are listed as Endangered under the ESA (1990). This species is listed as endangered by IUCN, of special concern by COSEWIC, and a species of concern by ADFG.

Fossils

The fossil record of otariids was reviewed by Loughlin et al. (1987). Fossil remains of *Eumetopias* are known from St. Paul Island (possibly Sangamonian in age) and Shishmaref Inlet on the Seward Peninsula (Kurtén and Anderson 1980).

California Sea Lion

Zalophus californianus
(Lesson, 1828)

Other Common Names

California sealion.

Systematics

Three subspecies of *Z. californianus* are generally recognized for three disjunct populations (Scheffer 1958; but see Rice 1998). The nominate subspecies is occasionally sighted in Alaska waters.

Genetic differences have been found between the United States stock and the Gulf of California stock (Maldonado et al. 1995).

Zalophus californianus californianus

Original Description. 1828. *Otaria californiana* Lesson, *Dictionaire classique d'histoire naturalle*, 13:420

Type Locality. In the vicinity of San Francisco Bay, California.

Type Specimen. None designated.

Range. West coast of North America from the Mexico-Guatemala border to southern Alaska.

Global Distribution

The California sea lion is restricted to the Pacific Ocean in three allopatric populations: the Sea of Japan (now in the IUCN Red List as probably extinct), the west coast of North America, and the Galapagos Islands (Nagorsen 1990).

Alaska Distribution

During the past three decades over fifty California sea lions have been documented in Alaska, with an increased number of observations recorded in recent years (Maniscalco et al. 2004) (map 61). Sightings have been of both sexes, during all seasons of the year (but mostly during the spring), and have ranged as far south as Forrester Island in southeast Alaska, as far north as Prince William Sound, to as far west as St. Paul Island (UAM 83388) in the Bering Sea. Maniscalco et al. (2004) speculated that their apparent increase in Alaska waters may be due to increasing populations within their southern breeding range, increasing competition for food, and changes in environmental conditions.

Status

The *Z. c. californianus* population was estimated at about 160,000 animals in 1989 (Majluf 1999; Reijnders et al. 1993). Recent estimates

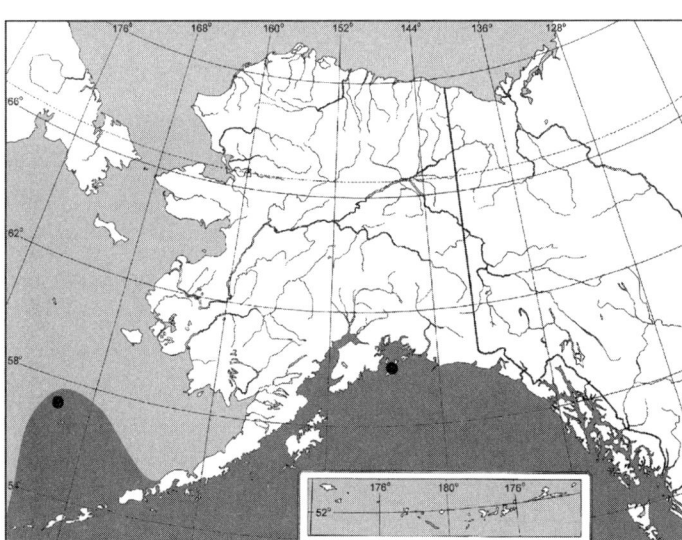

MAP 61. Distribution of California sea lion, *Zalophus californianus*.

■ = Projected Range (after Angliss and Outlaw 2007)
● = Specimen Record (n = 3): 57.1083, −170.2917, Pribilof Islands Quad., St. Paul Island, Reef Point (1 UAM); Montague Island, Prince William Sound (1 USNM); Alaska, no specific locality (1 MVZ).

CARNIVORA

for the entire population ranged from 237,000 to 244,000, and are increasing (Carretta et al. 2007).

Fossils

Kurtén and Anderson (1980) reported Late Pleistocene fossils from California.

Steller's sea lion, *Eumetopias jubatus* (B. Hines)

Family ODOBENIDAE Allen, 1880—walrus

The Odobenidae (Wozencraft 2005), considered a subfamily of Otariidae by some authors (Mitchell and Tedford 1973; Wozencraft 1989) and a subfamily of Phocidae by others (McKenna and Bell 1997), contains only one species.

The relatively well-known fossil record of walruses dates back to at least the Miocene, some twenty-two million years ago (Deméré 1994).

An excellent swimmer and diver, the ponderous-looking walrus is well adapted to shallow arctic waters and ice floes. The stout head and massive neck supports a large, ever-growing pair of tusks, a diagnostic trait present in both sexes. Males are larger than females and can reach up to 1,700 kg and measure 3.7 m in body length. An adult is grayish brown in color and nearly hairless. A dense, stiff covering of whiskers grows on its upper lip. Like eared seals, a walrus has hind flippers that can rotate forward for locomotion on land and ice. It lacks external ears and has nails on all five toes. Walruses are social animals, gathering in large mixed herds when feeding and migrating. They breed mainly during the dark month of February, deep within the pack ice (Fay 1999).

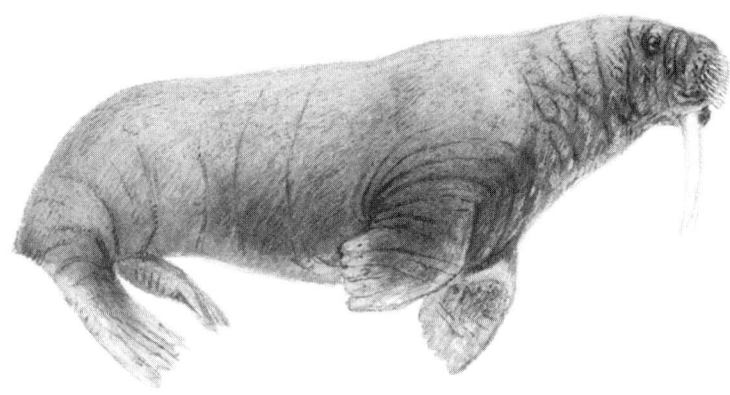

Walrus, *Odobenus rosmarus* (W. D. Berry)

Walrus

Odobenus rosmarus
(Linnaeus, 1758)

Other Common Names

Pacific walrus (subspecies name).

Systematics

Fay (1985) followed Repenning and Tedford (1977) in assigning the genus *Odobenus* to the superfamily Otarioidea and family Odobenidae, rather than to the superfamily Canoidea and family Otariidae, as proposed by Mitchell (1968a). Two subspecies, an Atlantic Arctic form, *O. r. rosmarus*, and a Pacific Arctic form, *O. r. divergens*, are commonly recognized (Hall 1981; Fay 1982, 1985; Smirnov 1929). A third subspecies has been recognized, the Laptev Sea walrus, *O. r. laptevi* (Chapskii 1940; Heptner et al. 1976; IUCN 2007), although this is not generally accepted (Fay 1985; Reijnders et al. 1993; but see Rice 1998).

The two geographically isolated subspecies exhibit distinct differences in mtDNA haplotypes and have slight differences in cranial morphology and tusk characteristics (Cronin et al. 1994).

Odobenus rosmarus divergens

Original Description. 1815. *Trichechus obesus* Illiger, *Abh. Preuss. Akad.Wiss.*, Berlin, 1804–1811, p. 64, *a nomen nudum*.

Type Locality. Chukchi Sea, about 35 mi. SW Icy Cape, Alaska.

Type Specimen. Not known to exist.

Range. Bering, Chukchi, Beaufort, and East Siberian seas.

Global Distribution

Odobenus rosmarus has a circumpolar distribution, occurring in two distinct areas in the North Atlantic, and discontinuously in the North Pacific and Arctic oceans from the Laptev Sea eastward to the Beaufort Sea and southward to the Bering Sea (Fay 1985).

Alaska Distribution

Walruses occur seasonally in the Bering, Chukchi, and western Beaufort seas of Alaska (map 62). Most of the animals migrate with the movement of pack ice. At least twelve thousand males, however, remain in northern Bristol Bay on or near Round Island from March through October and then migrate northward in the fall to the St. Lawrence Island area, where they join the rest of the herd to spend the winter and spring in the ice pack (Burns 1994c). During fall migration, large groups haul out to rest on Big Diomede Island and the Punuk Islands. The population's distribution has been changing in recent years, as thousands now summer around the Diomede Islands, King Island, and Arakamchechen Island, areas where they were historically scarce or absent (Reeves et al. 1992). There have been rare sightings of walruses south of

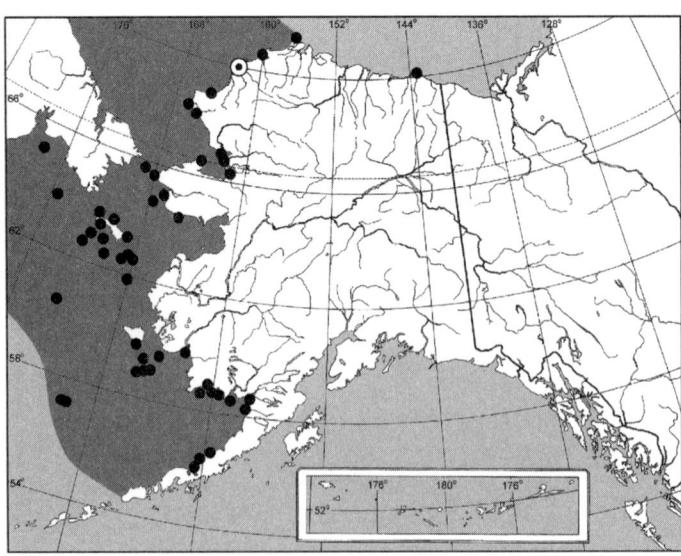

MAP 62. Distribution of walrus, *Odobenus rosmarus*.

■ = Projected Range (after Angliss and Outlaw 2007)
● = Specimen Record (n = 2328)

the Alaska Peninsula near Kodiak Island, and in Cook Inlet and Yakutat Bay (Fay 1982).

Habitat

The gregarious walrus uses moving pack ice for resting, pupping, and molting and also hauls out on secluded rocky shores and islands. Walruses feed mainly on a variety of clams and other bottom-dwelling invertebrates found on the relatively shallow Bering-Chukchi Platform (Burns 1994c).

Status

Ninety percent of the world's walrus population, about 230,000 animals, is in the Bering and Chukchi seas (Reeves 1978a; Reijnders et al. 1993). The Pacific walrus population is believed to have doubled between 1960 and 1980. The only historically documented haul-out sites in the Pacific that have not yet been reoccupied by walruses are in the Pribilof Islands on Walrus Island (since 1891) and

Northeast Point on St. Paul Island (Reeves et al. 1992).

As with other ice breeding pinnipeds, walruses are threatened by shrinking sea ice due to global warming (Kelly 2001). This species is listed under Appendix III of CITES.

Fossils

The fossil history of walruses was summarized by Fay (1985) and Deméré (1994). Fossils closely resembling the modern walrus are known from numerous Pleistocene sites found from southern Quebec and Scandinavia to North Carolina and France. Walruses reinvaded the North Pacific region via the Arctic Ocean, possibly during the last (Sangamon) interglacial period. Pelukian-age fossils, some fifty to one hundred thousand years old, have been found on the shores of the Bering and Chukchi seas. Fossils, probably of mid- to late Wisconsin age, have been found as far south as Tokyo, Vancouver Island, and San Francisco.

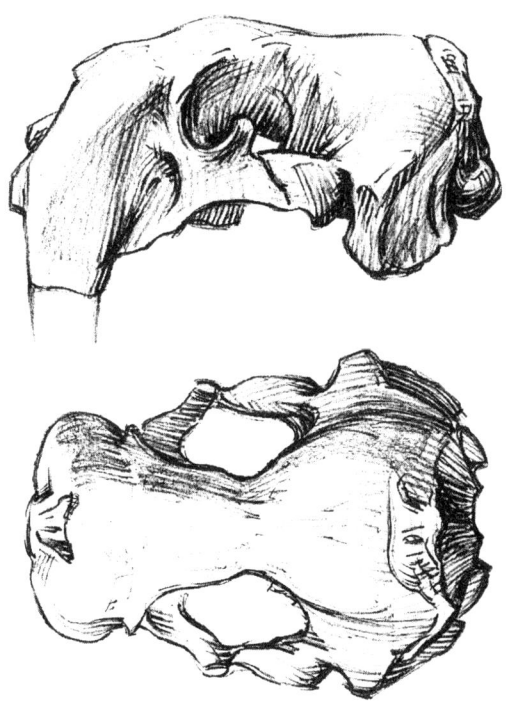

Skull of walrus, *Odobenus rosmarus* (W. D. Berry)

Family PHOCIDAE Gray, 1821—hair seals

The traditional separation of the phocids into four subfamilies (Kellogg 1922; Simpson 1945) has received little support from recent studies. King (1966), de Muizon (1982), Berta and Wyss (1994), and others presented evidence that recognize only two, Phocinae and Monachinae, a conclusion adopted by Rice (1998) and supported by Higdon et al. (2007). Wozencraft (2005) considered the taxonomy of the phocids too "unsettled" to permit the recognition of subfamilies. Nineteen species in thirteen genera comprise the entire family; eight species and seven genera occur, or have recently occurred, in Alaska. *Phoca, Pusa, Histriophoca,* and *Pagophilus* have been considered a monophyletic clade by some authors (e.g., Burns and Fay 1970; Wyss 1988), but recent morphological and molecular (mtDNA) analyses found *Pagophilus* and *Histriophoca* constituting one clade, while *Phoca, Pusa,* and *Halichoerus* (gray seal) constituted a second (Wozencraft 2005; but see Davis et al. 2004)

The geographic range of phocids is Middle Miocene to Recent in North America, Late Oligocene to Late Miocene in Europe, Pliocene in South America, Oligocene to Recent in Asia, and Early Pleistocene to Recent in Africa (Stains 1984).

The earless, hair, or true seals are the most specialized, most numerous, and most widespread family of marine carnivores. All have small fore flippers and posteriorly projecting hind flippers that make them less proficient on land but highly adept locomoting through water. They have streamlined bodies, broad heads, large luminous eyes, and ear openings that lack external flaps or pinnae. The fur is stiff and, in some genera, distinctly marked.

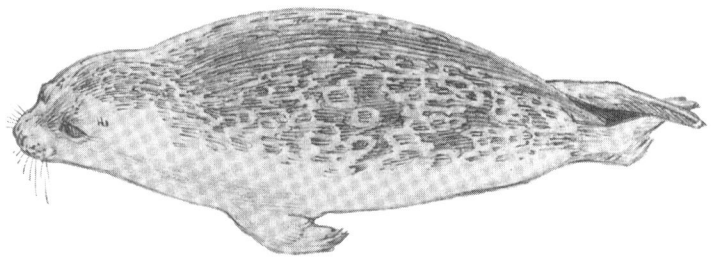

Ringed seal, *Pusa hispida* (W. D. Berry)

Hooded Seal

Cystophora cristata
(Erxleben, 1777)

Other Common Names
Crested seal, bladdernose seal.

Systematics
Cystophora is a monotypic genus in the tribe Cystophorini (Burns and Fay 1970; Kovacs and Lavigne 1986). No subspecies are recognized.

Cystophora cristata
Original Description. 1777. [*Phoca*] *cristata* Erxleben, *Systema regni animalis . . . ,* 1:590.
Type Locality. Southern Greenland or Newfoundland.
Type Specimen. None designated.

Global Distribution
Hooded seals inhabit the deep waters of the North Atlantic, migrating seasonally from feeding areas around Greenland to the pack ice of Labrador and Davis Strait where they breed (Boness 1999b). They are solitary throughout most of the year (J. E. King 1983).

Alaska Distribution
Hooded seals tend to wander and sometimes appear in unexpected places (Reeves et al. 1992), including Alaska waters (map 63). Several young males were reported in the western Beaufort Sea near Prudhoe Bay in the 1970s (Burns and Gavin 1980). Between July and October 1979, a single hooded seal was seen (and photographed) on the west side of Prince of Wales Island near Craig and Klawock in the southeast region of the state (Fay 1995).

Habitat
Hooded seals are associated with drifting pack ice during much of the year (Kovacs and Lavigne 1986; Reeves et al. 1992).

Status
The hooded seal is a vagrant to Alaska waters.

Fossils
There is no demonstrable fossil record for the hooded seal in North America (Ray 1983).

Bearded Seal

Erignathus barbatus
(Erxleben, 1777)

Other Common Names
None.

Systematics
Erignathus is a monotypic genus. Analysis of mtDNA sequences indicated that *E. barbatus* is more closely related to the tropical and warm-temperate monk seals (*Monachus*) and the antarctic Weddell seal (*Leptonychotes weddelli*) than it is to other arctic phocids (Cleator and Stirling 1999). Two subspecies are commonly recognized (Hall 1981); one occurs in

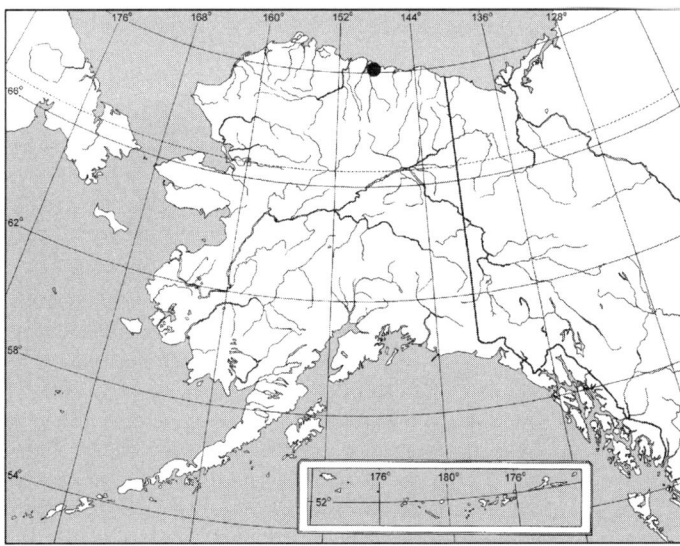

MAP 63. Distribution of hooded seal, *Cystophora cristata*.

● = Specimen Record (n = 1): 70.4333, –148.61667, Beechey Point Quad., Sagwon, Sagavanirktok River (1 UAM).

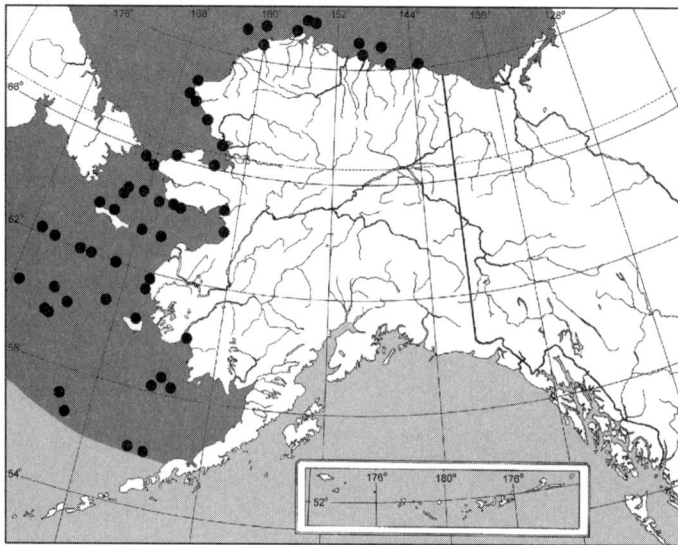

MAP 64. Distribution of bearded seal, *Erignathus barbatus*.

■ = Projected Range (after Angliss and Outlaw 2007)
● = Specimen Record (n = 1472)

Alaska. Because the geographic limits of these races are vague and morphological distinctions slight, this species needs to be revised (Reeves et al. 1992; Youngman 1975).

Erignathus barbatus nauticus

Original Description. 1811. *Phoca nautica* Pallas, *Zoographia Rosso-Asiatica* . . . , 1:108.

Type Locality. Okhotsk Sea.

Type Specimen. Not known to exist.

Range. Laptev Sea east to central Canadian Arctic.

Global Distribution

Bearded seals are found in the ice-covered seas of the subarctic and, to a lesser extent, the arctic (Burns 1978). In the Pacific Ocean they range as far south as Hokkaido, Japan, and in the Atlantic as far south as northeastern Newfoundland (Cleator and Stirling 1999).

Alaska Distribution

The bearded seal has a wide but patchy distribution throughout the Bering, Chukchi, and Beaufort seas (map 64). In late winter and early spring, they are found north of the ice edge and south of Bering Strait, with the largest concentrations occurring near St. Lawrence Island, in the ice 60 to 100 km north of the ice front, and west of St. Matthew Island. In summer, they are found scattered across the broken margin of the multiyear ice in the Chukchi and western Beaufort seas (ADFG 1973; Reeves et al. 1992). In the fall, young seals are sometimes found in open water and up the mouths of some rivers (Burns 1978; Reeves et al. 1992).

Habitat

Bearded seals are associated with sea ice year-round, moving north in summer and south in winter (Burns 1978, 1981b). Typically found alone or in small groups, they prefer areas of moving ice and open water that is less than 150 to 200 m deep (Cleator and Stirling 1999). This species tends to avoid shorefast ice and thick, unbroken drift ice (Wynne 1993). Their primary food is bottom-dwelling invertebrates (Burns 1978, 1981b; Lowry et al. 1980, 1982).

Status

The northern Pacific population has been estimated to consist of 300,000 to 450,000 animals, with the Bering Sea portion totaling about 250,000 seals (Reijnders et al. 1993). Angliss and Outlaw (2007) concluded that there is no reliable population abundance estimate or trends data for the Alaska stock of bearded seals.

In Alaska, the hunting of bearded seals by Natives is not monitored or regulated. At the current level of harvest, however, Reijnders et al. (1993) concluded that there is no risk of overexploitation.

Because the arctic climate is changing drastically (Johannessen et al. 2004), bearded seals, along with other sea ice–dependant mammals, will be vulnerable to reductions in sea ice; however, there are insufficient data to make reliable predictions of the effects of warming on these animals (Angliss and Outlaw 2007).

Fossils

Fossils of this species have been reported from Middle Pleistocene and Wisconsinan deposits on the Seward Peninsula of Alaska (Kurtén and Anderson 1980).

Ribbon Seal

Histriophoca fasciata
(Zimmermann, 1783)

Other Common Names

Banded seal.

Systematics

Various authors (Carr and Perry 1997; de Muizon 1982; Rice 1998; Scheffer 1958; Wozencraft 2005) place the ribbon seal in the genus *Histriophoca*. Burns and Fay (1970), among others (see Rice 1998), included it in the genus *Phoca*. Studies based on morphology (de Muizon 1982) and mtDNA (Árnason et al. 1995; Mouchaty et al. 1995; Perry et al. 1995) supported a close relationship of the ribbon seal with harp seal. Fedoseev (1984) found only weak morphological differences between populations in the western and eastern parts of the Bering Sea. No subspecies have been designated.

Histriophoca fasciata

Original Description. 1783. *Phoca fasciata* Zimmerman, *Geographische Geschichte . . .* , 3:277.
Type Locality. Kurile Islands [Russia].
Type Specimen. None designated.

Global Distribution

The ribbon seal inhabits the seasonally ice-covered seas of the North Pacific region occurring mainly in the Bering Sea and Sea of Okhotsk (Burns and Fay 1970; Reijnders et al. 1993).

Alaska Distribution

In Alaska, ribbon seals are found principally in the Bering Sea (map 65). Vagrants have occurred south of the Aleutians into the North Pacific Ocean, including near Cordova in Prince William Sound. North of the Bering Strait they range into the Chukchi Sea, and a few reach the western Beaufort Sea (Kelly 1988a; Reeves et al. 1992; Rice 1998). The identification by Loring (1902) of a skin believed to be that of a ribbon seal from Iliamna Lake is most likely in error.

Habitat

Ribbon seals concentrate mainly in the ice front of the Bering Sea pack during late winter and spring to pup, nurse, and molt. When the sea ice recedes northward and melts in May and June, they become pelagic, with their center of abundance probably remaining in the Bering Sea during the open water period

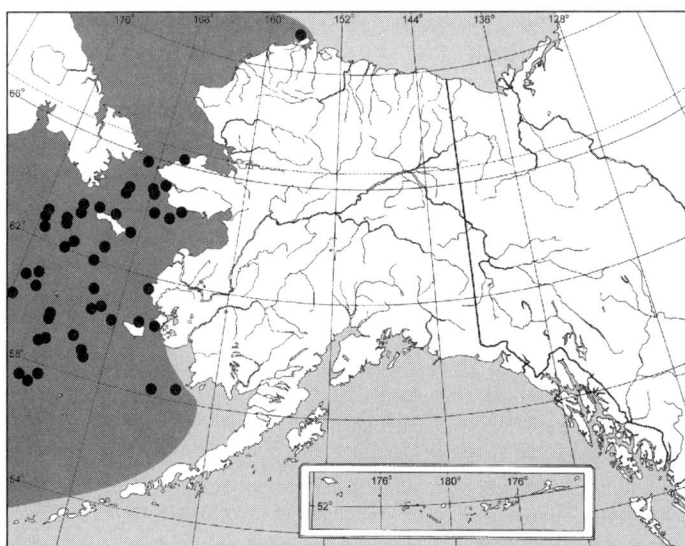

MAP 65. Distribution of ribbon seal, *Histriophoca fasciata*.

■ = Projected Range (after Angliss and Outlaw 2007)
● = Specimen Record (n = 292)

CARNIVORA

187

(Burns 1970). Their main prey is fish, especially pollock (*Theragra chalcogramma*) (Boness 1999a; Frost and Lowry 1980).

Status

The pelagic nature of this species makes them difficult to study and survey. Following the cessation of heavy harvesting by Soviet sealers from 1961 through 1967, the number of ribbon seals has increased dramatically and is now thought to be approaching the pre-exploitation level (Burns 1994b). Burns (1981a) estimated the worldwide population of ribbon seals at 240,000 in the mid-1970s, with an estimate for the Bering Sea at 90,000 to 100,000.

Fossils

Repenning (1983) reported a Late Pleistocene fossil of ribbon seal from the Gubik Formation near Teshekpuk Lake in northern Alaska. Kurtén and Anderson (1980) mention a possible ribbon seal fossil from near Barrow dated at 28,000 ybp.

Northern Elephant Seal

Mirounga angustirostris
(Gill, 1866)

Other Common Names
None.

Systematics

The genus *Mirounga* has two species: *M. leonina*, in the Southern Hemisphere, and *M. angustirostris*, in the Northern Hemisphere. *M. angustirostris* is monotypic (Hall 1981; Stewart and Huber 1993).

Mirounga angustirostris
Original Description. 1866. *Macrorhinus angustirostris* Gill, *Proc. Chicago Acad. Sci.*, 1:33, April.
Type Locality. St. Bartholomews Bay, Lower California, Mexico.
Type Specimen. USNM 4704 (lectotype).

Global Distribution

Northern elephant seals range widely in the North Pacific when they are not ashore during brief periods to breed and molt (Stewart and Huber 1993). They breed in rookeries from Baja California to California. During the non-breeding season, the females of this species remain off the Washington and Oregon coast,

whereas males and juveniles range farther north. Generally the adults remain far offshore while migrating and foraging whereas juveniles and subadult males are found often in nearshore waters (DeLong 1978; Stewart and Huber 1993).

Alaska Distribution

Sightings and strandings of elephant seals are becoming more frequent in the Gulf of Alaska and along the Alaska Peninsula and eastern Aleutian Islands (Reeves et al. 1992; Stewart and Huber 1993; Wynne 1993). Specimens have been preserved from Prince of Wales Island, the Kenai Peninsula, and the Alaska Peninsula (map 66).

Habitat

Elephant seals use sandy beaches to pup, breed, and molt. During much of the year, they remain pelagic, foraging away from the coast and diving continually (Reeves et al. 2002).

Status

There has been an almost exponential growth in populations of northern elephant seals since their reduction to near extinction in the late 1800s. The species is listed as Least Concern by IUCN and not at risk by COSEWIC.

Fossils

The earliest records of this genus in the North Pacific are from Late Pleistocene deposits of Southern California (Stewart and Huber 1993).

Harp Seal

Pagophilus groenlandicus
(Erxleben, 1977)

Other Common Names
Greenland seal, saddle-backed seal.

Systematics

Although some authorities (e.g., Burns and Fay 1970; Wozencraft 1993) have suggested that the genus *Pagophilus* should be recognized only at the subgeneric level, others (Carr and Perry 1997; Chapskii 1955; Corbet 1978; Rice 1998; Wozencraft 2005) continue to recognize *Pagophilus* at the generic level. No subspecies are usually recognized; however, the White Sea population has been treated as a separate subspecies by Russian taxonomists (Rice 1998).

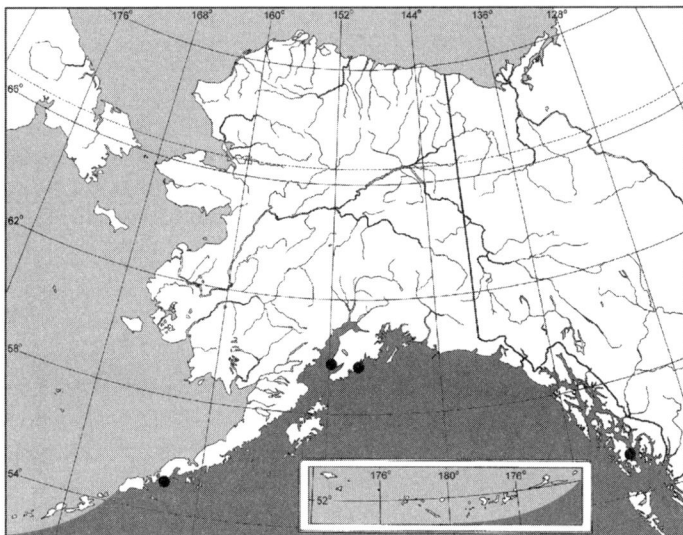

MAP 66. Distribution of northern elephant seal, *Mirounga angustirostris*.

■ = Projected Range (after Angliss and Outlaw 2007)
● = Specimen Record (n = 4): 55 N 162 W, Cold Bay Quad., King Cove (1 UAM); 59.77733, –151.86450, Seldovia Quad., 0.5 miles from Anchor Point boat launch (1 UAM); Craig Quad., Prince of Wales Island, Kasaan (1 USNM); Kenai Peninsula, between Blying Sound and Gore Point (1 UWBM).

Pagophilus groenlandicus

Original Description. 1777. [*Phoca*] *groenlandica* Erxleben, *System regni animalis* . . . , 1:588.
Type Locality. Greenland or Newfoundland.
Type Specimen. Not known to exist.

Global Distribution

Harp seals inhabit the North Atlantic in three populations or stocks that are defined by their distinct whelping areas (Sergeant 1976).

Alaska Distribution

The occurrence of the harp seal in Alaska is based on the following quote from Murie (1936):

> "Natives describe a seal which closely resembles the saddle-backed seal from the north Atlantic, and recognized pictures of it. None has been taken since Mr. Geist's acquaintance with the island, but it is reported to have been taken occasionally in rather recent years."

We are not aware of any other reports or specimens of this species from Alaska.

Habitat

Harp seals are associated with sea ice throughout much of their range (Reeves et al. 1992).

Fossils

Repenning (1983) reported the fossil of a *P. groenlandicus* of Late Pliocene age (possibly unknown, see Youngman 1993:147) from the Gubik Formation at Ocean Point, northern Alaska. Remains of this species have also been recovered from Late Pleistocene deposits in Ontario and New Brunswick (Kurtén and Anderson 1980; Miller 1990).

Specimens

None.

Spotted Seal

Phoca largha
Pallas, 1811

Other Common Names

Larga seal.

Systematics

Scheffer (1958) considered *largha* conspecific with *vitulina*; however, McLaren (1966) regarded them as specifically different. Shaughnessy and Fay (1977) and Burns et al. (1984) demonstrated that the two were morphologically, ecologically, socially, and reproductively distinct. Recent genetic studies comparing homologous sequences (Árnason et al. 1995; Carr and Perry 1997) indicated that the degree of differentiation between *largha* and *vitulina* was less than that between conspecific individuals of any other species of phocid seals examined. Therefore, *P. largha*

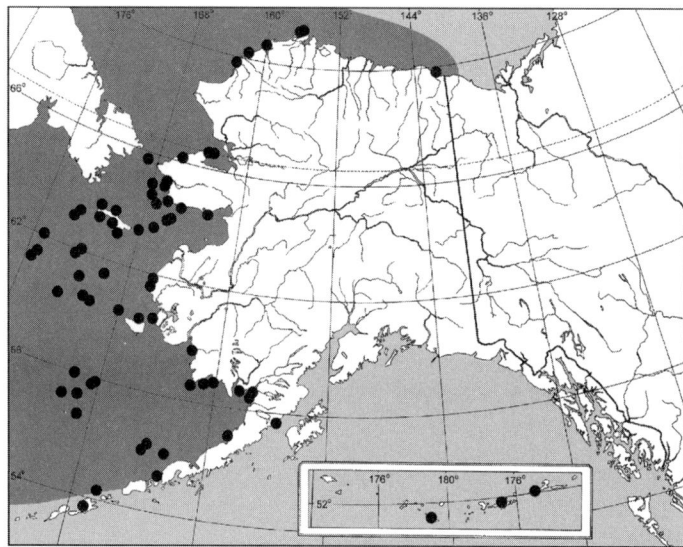

MAP 67. Distribution of spotted seal, *Phoca largha*.

■ = Projected Range (after Angliss and Outlaw 2007)
● = Specimen Record (n = 629)

is perhaps best considered a subspecies of *P. vitulina*, a view more recently supported by Higdon et al. (2007). *Phoca largha* was included in a phylogenetic analysis of northern hair seals based on 458-base pair sequences from the mitochondrial cytochrome *b* gene (Mouchaty et al. 1995). Atlantic and Pacific populations of harbor seals, *P. vitulina*, were found to be more closely related to each other than either was to *P. largha*. No subspecies are recognized.

Phoca largha

Original Description. 1811. *Phoca largha* Pallas, *Zoogr. Rosso-Asiat.*, 1:113.

Type Locality. East coast of Kamchatka.

Type Specimen. None designated (topotypes at BMNH; see Shaughnessy and Fay 1977:396).

Global Distribution

The spotted seal occurs in the Bering Sea, Chukchi Sea, and along the Arctic coast of Alaska, and from the Sea of Okhotsk south to Hokkaido, the Po Hai Sea, and the northwest Yellow Sea (Reijnders et al. 1993).

Alaska Distribution

During summer and early autumn, spotted seals can be found in Alaska waters from the southern Bering Sea near the Aleutian Islands and Bristol Bay to the Beaufort Sea near the Alaska-Yukon border (map 67). Satellite tagging studies (Lowry et al. 1998, 2000)

indicate that spotted seals migrate south from the Chukchi Sea in October and pass through the Bering Strait in November. They overwinter in the Bering Sea along the ice edge and make east-west movements along the edge.

Habitat

The spotted seal is closely associated with sea ice from late autumn to early summer and bears its young on the drifting pack ice. It lives in coastal waters during the ice-free months (Burns 1978; Quakenbush 1988).

Status

Reliable estimates of spotted seal population abundance and trends are currently not available (Angliss and Outlaw 2007). In the late 1970s, the Bering Sea population of spotted seals was estimated at 200,000 to 250,000 (Burns 1994a). Popov (1982) estimated the Bering Sea population at 135,000, and Fedoseev et al. (1988) estimated it at about 100,000. Burkanov et al. (1988), however, considered these figures to be overestimated.

Fossils

A number of phocid fossils have been reported from sites on the Seward Peninsula (Kurtén and Anderson 1980). One of Late Pleistocene age from near Deering was identified as *P. vitulina* but might be *P. largha*. A twenty-eight-thousand-year-old specimen (identified as either *P. fasciata* or *P. vitulina*) was found in the Barrow area (Kurtén and Anderson 1980).

Harbor Seal

Phoca vitulina
Linnaeus, 1758

Other Common Names

Common seal (Europe), hair seal (southern Alaska).

Systematics

The taxonomy of harbor seals in the North Pacific and Bering Sea has been controversial. The ice-associated spotted seal, *P. largha*, of the Bering and Okhotsk seas, has only recently been widely recognized as a full species (but see spotted seal account). Four, sometimes five, subspecies of harbor seals are recognized across its broad range. The eastern Pacific form of harbor seal, *richardii* (spelled *richardsi* by some authors), and the western Pacific form, *stejnegeri*, generally are recognized as valid subspecies (Reeves et al. 1992; Reijnders et al. 1993; Rice 1998). The eastern Pacific subspecies occurs in Alaska but may intergrade with *P. v. stejnegeri* in the Aleutian Islands (Shaughnessy and Fay 1977). Additionally, mtDNA and nuclear DNA data (Burg et al. 1999; Lamont et al. 1996) indicate the existence of at least three populations of harbor seals in the Pacific: (1) Japan, Russia, Alaska, and northern British Columbia; (2) southern British Columbia and Puget Sound; and (3) outer coasts of Washington, Oregon, and California. The existence of the two nominal subspecies in the Pacific Ocean was not supported.

Other phylogenetic analyses using mtDNA data revealed extensive macrogeographic subdivision along a distributional transect from Japan to Alaska (O'Corry-Crowe and Westlake 1997; Westlake and O'Corry-Crowe 2002), but the two currently recognized subspecies, *richardii* of North America and *stejnegeri* of Asia, did not represent discrete mtDNA clades. The greatest differentiation detected was along the Commander-Aleutian Island chain. Differentiation between the Kodiak Archipelago and Prince William Sound, and between Bristol Bay and the Pribilof Islands, indicated that current management stocks are inappropriate.

Taxonomic separation of Atlantic and Pacific harbor seals is based on disjunct distributions rather than on morphological evidence (Reeves et al. 1992).

Phoca vitulina richardii

Original Description. 1864. *Halicyon richardii* Gray, *Proc. Zool. Soc. London*, p. 28.

Type Locality. Vancouver Island, British Columbia.

Type Specimen. BMNH 1864.2.19.1 (see Shaughnessy and Fay 1977).

Range. Eastern Pacific from the Aleutian Islands to California.

Remarks. The name *richardsi* used by some authors (e.g., Shaughnessy and Fay 1977) is considered an invalid spelling according to Rice (1998) and Wozencraft (2005).

Global Distribution

The vast range of this seal encompasses much of the continental and island coasts of the temperate Northern Hemisphere (Reeves et al. 1992).

Alaska Distribution

Harbor seals occur along the Alaska coast from British Columbia north to Kuskokwim Bay and the Pribilof Islands westward throughout the Aleutian Islands (map 68). Most are associated closely with coastal waters, although they are occasionally observed farther offshore (Kinkhart and Pitcher 1994). During salmon migration, harbor seals can be found considerable distances upstream along major rivers and in lakes. At Iliamna Lake in southwestern Alaska, seals are present year-round and are probably resident (Kinkhart and Pitcher 1994).

Habitat

Most harbor seals are associated closely with coastal waters. Aggregations of harbor seals haul out on intertidal ledges, rocky islets, reefs, mudflats, isolated beaches, and in some areas, on glacial and sea ice. Some live mainly in freshwater, either far up rivers or in lakes (Reeves et al. 1992).

Status

Since the mid-1970s, the number of harbor seals has declined in the central and western Gulf of Alaska, but has so far remained relatively stable in the eastern Gulf of Alaska (R. J. Small et al. 1998). At Tugidak Island near Kodiak, numbers have declined 90 percent from approximately eleven thousand seals to one thousand (Pitcher 1990). Harbor seals have declined dramatically in Prince William Sound over the past few decades (Westlake and O'Corry-Crowe 2002). Trend data from aerial counts in southeast Alaska conducted during the period 1983–2001 indicated significant increase in the Ketchikan area, relative

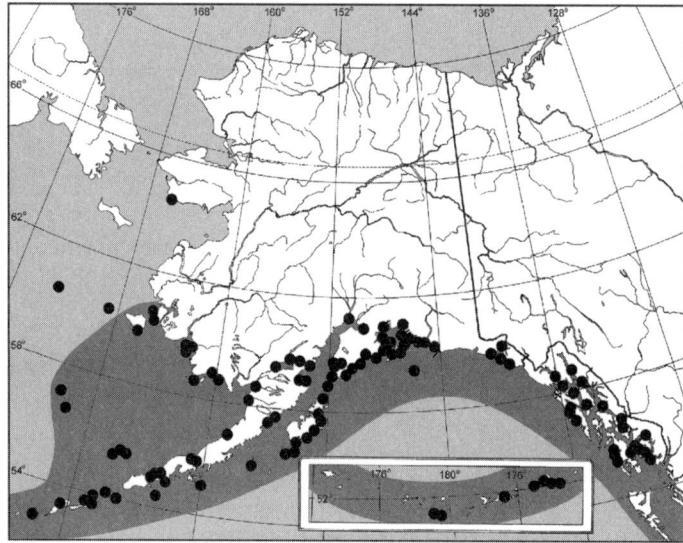

MAP 68. Distribution of harbor seal, *Phoca vitulina.*

■ = Projected Range (after Angliss and Outlaw 2007)
● = Specimen Record (n = 1391)

stability near Sitka, and a recent decline in Glacier Bay (R. J. Small et al. 2003). The total Alaska population probably ranges between two hundred and three hundred thousand animals, and the annual harvest of harbor seals by Alaska Natives is about twenty-five hundred to four thousand (Kinkhart and Pitcher 1994). The Copper River Delta, the mouths of the Stikine and Taku rivers, and portions of Bristol Bay are areas with notable harbor seal–salmon fishery conflicts.

Fossils

The earliest known occurrence of *P. vitulina* is from an Early Pleistocene deposit less than two million years old in Oregon (Repenning et al. 1979). A Late Pleistocene specimen reported as *P. vitulina* from Deering on the Seward Peninsula (Kurtén and Anderson 1980) should be reexamined because it may be the more northern species, *P. largha.*

Ringed Seal

Pusa hispida
(Schreber, 1775)

Other Common Names
None.

Systematics

Some authors (Baker et al. 2003; Carr and Perry 1997; Hall 1981; Rice 1998; Scheffer 1958; Wozencraft 2005) place the ringed seal in the genus *Pusa*; others (Frost and Lowry 1981; Honacki et al. 1982; J. E. King 1983; O'Corry-Crow and Westlake 1997; Wozencraft 1993) as a member of the genus *Phoca.* Higdon et al. (2007) suggest that *Pusa* be included within a redefined *Phoca*, possibly as a subgenus. Several subspecies have been recognized, mainly on the basis of geographical isolation (Reeves et al. 1992; Reijnders et al. 1993; Rice 1998); the nominate subspecies occurs in Alaska.

A phylogenetic study based on sequences of the mitochondrial cytochrome *b* gene found a sister-taxon relationship between the ringed seal and the gray seal, *Halichoerus grypus*, of the North Atlantic (Mouchaty et al. 1995). In captivity, these two species have successfully mated (Mohr 1952).

Pusa hispida hispida

Original Description. 1775. *Phoca hispida* Schreber, 1775. *Die Säugthiere . . .* , theil 3, heft 13, Pl. 86.

Type Locality. Coasts of Greenland and Labrador.

Type Specimen. Not known to exist.

Range. Arctic sea and coasts of Alaska, Canada, Greenland, and Europe (including Russia).

Global Distribution

Ringed seals are very widespread throughout the arctic and subarctic regions of the world, occurring as far south as Newfoundland and as far east as the Baltic Sea in the North Atlantic Ocean, and throughout the Bering Sea and as far south as the Sea of Japan in the Pacific

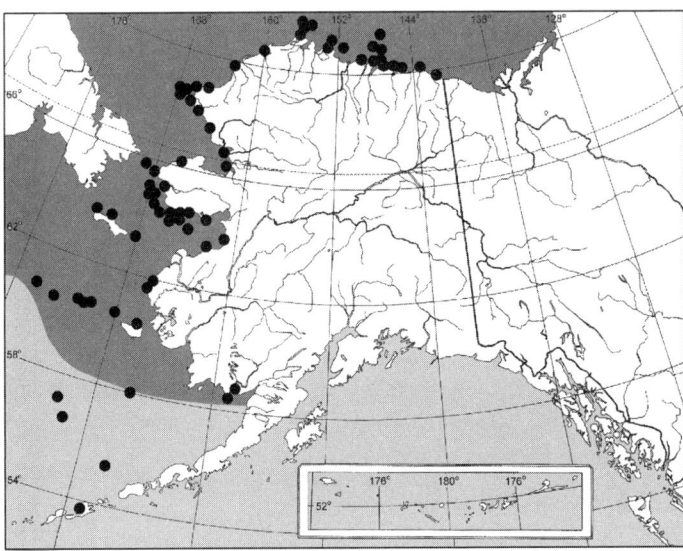

MAP 69. Distribution of ringed seal, *Pusa hispida*.

■ = Projected Range (after Angliss and Outlaw 2007)
● = Specimen Record (n = 3172)

(Reijnders et al. 1993). Their range also includes several freshwater lakes (Perry 1999).

Alaska Distribution

This arctic species is found throughout the ice-covered regions of the Bering, Chukchi, and Beaufort seas of Alaska (map 69). During winter and spring, ringed seals occur in the Bering Sea as far south as the limit of sea ice, usually at least to Nunivak Island or Bristol Bay, rarely to the Aleutians (Eley 1994; Kelly 1988b; Reeves et al. 1992).

Habitat

The ringed seal is associated with sea ice throughout the year. In winter, ringed seals occupy the landfast or shore ice, and in other seasons they migrate with the annual advance and retreat of the pack ice (Burns 1978). Ringed seals occupying the Bering and southern Chukchi seas in winter apparently are migratory, but details of their movements are unknown (Angliss and Outlaw 2007). Females pup in snow dens on either landfast or drifting pack ice during March and April (Eley 1994).

Status

Reliable abundance and trends estimates for the entire Alaska stock of ringed seals are currently not available (Angliss and Outlaw 2007). According to Kelly (1988b), the Alaska population was roughly 1 to 1.5 million animals.

They are most abundant in the landfast ice north of the Bering Strait (Burns 1978). The estimated annual harvest of ringed seals in Alaska declined from between seven thousand and fifteen thousand during the 1960s and early 1970s to two thousand to three thousand by the late 1970s (Reeves et al. 1992). The annual take in Alaska is currently about ninety-five hundred seals (Angliss and Outlaw 2007). Kelly (2004) associated advancing spring temperatures and snow melt due to global warming with an observed increase in lair abandonment by ringed seals and predicted increased juvenile mortality rates via exposure to freeze-thaw conditions and predation.

Fossils

There are Pleistocene records of this seal from sites on the Seward Peninsula of Alaska (Kurtén and Anderson 1980), and a Late Pleistocene record from Teshekpuk Lake Gubik Formation in northern Alaska (Repenning 1983). An older specimen (possibly Pliocene) tentatively referred to the subgenus *Pusa* was found in the "Malaspina District" of Alaska (Kurtén and Anderson 1980). The fossil bones and teeth of ringed seals from the last glacial maximum (from about 24,150 to 13,690 ybp) have been found in a cave on Prince of Wales Island in southeast Alaska (Heaton and Grady 2003).

CARNIVORA

Family MUSTELIDAE Fischer von Waldheim, 1817—weasels

Weasels and their kin form the largest and most diverse family of Carnivora, comprising twenty-two genera and fifty-nine species in two subfamilies (Wozencraft 2005). North America's skunks are now considered members of a separate family, Mephitidae. Mustelids are found on all continents except Antarctica and Australia (Nowak 1991). Alaska is home to nine species, as recent molecular studies (Carr and Hicks 1997; Stone and Cook 2001; Stone et al. 2002) support splitting the American marten (*Martes*) once again (Merriam 1890) into two separate species, *M. americana* and *M. caurina*.

The geological range of this diverse family is Early Oligocene to Recent in North America, Eocene to Recent in Europe, Middle Oligocene to Recent in Asia, Early Pliocene to Recent in Africa, and Late Pliocene to Recent in South America (Stains 1984).

Mustelids are mainly flesh eaters. Most have a long and slender body, short legs with five digit feet, and curved, nonretractable claws (the partially retractable claws of the fisher are an exception). Most have long tails and short, often rounded ears. In nearly all genera, well-developed anal scent glands are present. Many species are agile climbers, and otter and mink are skillful swimmers. Male mustelids are usually larger than females and possess a baculum. The dense, luxurious fur of many mustelid species has long been prized in the commercial fur trade, as well as for domestic garment manufacture.

Ermine, *Mustela erminea*, in winter pelage (W. D. Berry)

Sea Otter

Enhydra lutris
(Linnaeus, 1758)

Other Common Names
None.

Systematics
Molecular sequence (Masuda and Yoshida 1994) and morphologic data (Berta and Morgan 1985) placed *Lutra* as the closest living relative of *Enhydra*. In a combined analysis of molecular and morphologic data, however, *Enhydra* was placed as a sister taxon to *Lutra* and the short-clawed otter *Aonyx* (Bryant et al. 1993; Dragoo and Honeycutt 1997). Three subspecies of the sea otter, *E. lutris*, have traditionally been recognized: *lutris*, *nereis*, and *gracilis* (Davis and Lidicker 1975; Hall 1981; but see Roest 1973). Wilson et al. (1991) conducted a complete reappraisal of sea otter taxonomy that was based on material from throughout the species' range and concluded that three subspecies be recognized: the nominate form, *E. l. lutris*, from the Kuril Islands north to the Commander Islands; a new subspecies, *E. l. kenyoni*, from the Aleutian Islands eastward to Washington; and *E. l. nereis*, from along the California coast and off St. Nicolas Island. Following their classification scheme, one subspecies occurs in Alaska.

Mitochondrial DNA variation among subspecies and populations of sea otters throughout their range was studied by Cronin et al. (1996). The California subspecies, *E. l. nereis*, appeared monophyletic, while *E. l. lutris* and *E. l. kenyoni* did not.

Gorbics and Bodkin (2001) used several types of data to identify three sea otter stocks in Alaska: southeast, southcentral, and southwest. Recent analyses of mitochondrial and nuclear DNA by Cronin et al. (2002) were consistent with recognition of three stocks.

Enhydra lutris kenyoni
Original Description. 1991. *Enhydra lutris kenyoni* Wilson, *J. Mammalogy*, 72:33, February 13.
Type Locality. Amchitka Island, Alaska.
Type Specimen. USNM 527045.
Range. Throughout the Aleutian Islands and southward in the eastern Pacific to Washington.

Global Distribution
North Pacific—Sea otters once ranged along most of the rim of the North Pacific from the northern Japanese archipelago to central Baja California, including the Kuril, Commander, Aleutian, and Pribilof islands (Reeves et al. 1992). Between 1751 and 1911, the distribution was reduced to thirteen remnant populations. Several of these subsequently declined to extinction (see Kenyon 1969 for a thorough review). Today much of the original range is occupied from the Kuril and Aleutian islands to Prince William Sound. Translocations have reestablished sea otters in southeast Alaska, British Columbia, and Washington.

Alaska Distribution
Sea otters occur along the southern coasts of Alaska (map 70).

Southeast Region. Sea otters were completely extirpated from southeast Alaska (Rotterman and Simon-Jackson 1988). Fred H. Gray (1915) noted five otters shot at Forrester Island in the spring of 1899, three in 1903, and one near Cape Pole, Kosciusco Island, in 1904. He reported three seen in Blake Channel, between Wrangell Island and mainland, in about 1910, and one seen at Forrester Island in 1912.

Sea otters were reintroduced to southeast Alaska in the late 1960s from Amchitka Island and Prince William Sound (Pitcher 1989; Riedman and Estes 1991). From the original 412 animals released, the regional total had grown to about twelve thousand animals by 1994 (Agler et al. 1995).

Southcentral Region. The remnant sea otter population that survived in southwestern Prince William Sound and near Kayak Island rebounded following the initiation of protection measures (Rotterman and Simon-Jackson 1988). Severe losses (an estimated twenty-eight hundred animals according to Raloff 1993) were experienced in the Prince William Sound area following the disastrous oil spill of the *M/V Exxon Valdez* (Bayha and Kormendy 1990; Reeves et al. 1992). Sea otters in Prince William Sound have not increased appreciably since 1994 (USFWS 2002a).

Southwestern Region. Small groups of sea otters survived overhunting along exposed coasts of several of the Aleutian Islands, southern Bristol Bay, and the Bering Sea north of Unimak Island, the south coast of the Alaska Peninsula, the Sanak and Shumagin islands, and Kodiak Island area (Kenyon 1969). Five otters from Amchitka Island were moved to Attu Island in 1956 (Kenyon and Spencer 1960).

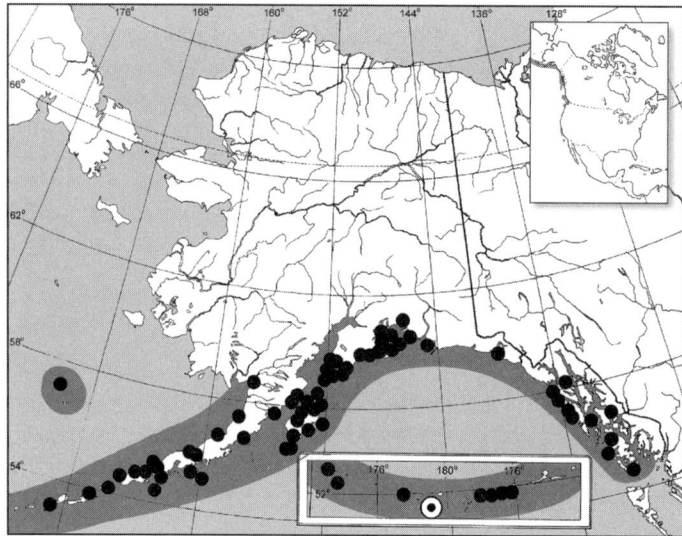

MAP 70. Distribution of sea otter, *Enhydra lutris*.

■ = Projected Range (after Angliss and Outlaw 2007)
● = Specimen Record (n = 2708)

Western Region. Sea otters were present throughout the Pribilof Islands at the time of discovery but they soon were extirpated by Russian hunters (Hanna 1923; Preble 1923). Transplant efforts were conducted there in 1955, 1959, and again in 1968 to reestablish a permanent population (Burris and McKnight 1973; Estes 1980; Rotterman and Simon-Jackson 1988). USFWS (1994) estimated thirty otters present in the Pribilof Islands in 1991. Whether these were descendants of the reintroduced animals or animals that dispersed from the Alaska Peninsula remains unknown.

Northern Region. Sea ice limits the northern permanent range of this marine species (Estes 1980). Individuals have been reported, however, in the northern Bering Sea (Lensink 1960, 1962) and Arctic Ocean (Bee and Hall 1956).

Habitat

Sea otters inhabit shallow coastal waters that provide an adequate supply of food and preferably the shelter of kelp beds for rest and sleep (Rotterman and Simon-Jackson 1988). They can spend their entire lives at sea but may come out to rest on land in areas where population densities are high and they are not disturbed (Estes 1980). The sea otter is considered a "keystone species" because of their effectiveness in limiting the abundance and distribution of their invertebrate prey, which in turn have integral ecological roles in these marine ecosystems (Estes 1999; Riedman and Estes 1991).

Status

Sea otters played a key role in Russian colonization of Alaska and the west coast of North America. Except for the population in central California, they were nearly hunted to extinction between Prince William Sound and Baja California (Kenyon 1969). In Alaska, they were given legal protection in 1899, and transplant efforts were initiated in the mid-1950s. Most of their historic range from Prince William Sound westward is now recolonized (Rotterman and Simon-Jackson 1988).

A precipitous decline of sea otters in southwest Alaska appears to have begun in the mid- to late 1980s. Widespread declines were found throughout the Aleutian Islands, with the greatest decreases occurring in the central Aleutians (USFWS 2002b). Elsewhere in the range of the southwest stock, declines of 93–94 percent are documented from the south coast of the Alaska Peninsula, 27–49 percent for the north Alaska Peninsula, and by 40 percent since 1994 in the Kodiak Archipelago (USFWS 2002b). Overall, sea otters from the Aleutians to Cook Inlet have declined an estimated 56–68 percent over the past ten to fifteen years, and recent surveys indicate the decline is continuing. The USFWS listed this Distinct Population Segment as Threatened in

August 2005 under the ESA. In late December 2006, the USFWS was sued to designate critical habitat.

Cause of the sea otter decline remains uncertain, although vulnerability to oil spills, conflicts, and incidental take in commercial fisheries, and potential increases in killer whale predation have been implicated.

Sea otters in southcentral Alaska are generally stable to slightly increasing (USFWS 2002a). The population trend in southeast otters is uncertain, although the Glacier Bay population increased 185 percent (to 1,590 otters) between 2000 and 2001 (Bodkin et al. 2002; USFWS 2002c).

Fossils

The modern sea otter arose in the North Pacific at the beginning of the Pleistocene (Berta and Sumich 1999). Estes (1980) reviewed the fossil history of *Enhydra*. Remains of *E. lutris* have been taken from Wisconsinan deposits from Southern California. The only fossil of *Enhydra* from Beringia was recovered from the Gubik Formation at Ocean Point near Barrow, Alaska (Repenning 1983). The age and identity of this specimen has been contentious. Kurtén and Anderson (1980) tentatively suggested that it was of Yarmouthian (Middle Pleistocene) age, whereas Repenning (1983) suggested its age as Late Pliocene. Youngman (1993), after conversation with Repenning, stated the age of this specimen (as well as its specific identity) remained unknown. Late Holocene semifossils of *E. lutris* have been recovered from a number of archaeological sites in the Aleutians and coastal Siberia (Savinetsky et al. 2004).

Wolverine

Gulo gulo
(Linnaeus, 1758)

Other Common Names
None.

Systematics

Degerbøl (1935) and Kurtén and Rausch (1959) demonstrated that Eurasian and North American forms of wolverines are conspecific, and recent molecular work corroborates their close relationship (Frances 2008; Tomasik and Cook 2005). Kyle and Strobeck (2001) investigated population structure of wolverines from northwestern Alaska to eastern Manitoba by examining 461 individuals typed at twelve microsatellite loci. Their results suggested little genetic structuring of populations in the northern regions of this species' range. A slightly higher level of genetic structure was detected, however, in samples from the Nome area of the Seward Peninsula in comparison to other Alaska regions. Tomasik and Cook (2005) analyzed mtDNA sequence data and hinted that the wolverines of southeast Alaska, the Kenai Peninsula, and Nunavut (Canada) were distinctive, although Frances (2008) did not identify the Kenai Peninsula population as different.

Heptner and Naumov (1974) distinguished three Eurasian subspecies of *G. gulo*. Kurtén and Rausch (1959) recognized only two subspecies of wolverines with all North American populations included in a single race, *G. g. luscus*. Hall (1981), however, recognized four North American races, of which two occur in Alaska.

Gulo gulo katschemakensis

Original Description. 1918. *Gulo katschemakensis* Matschie, *Gesell. Naturforsch. Freunde*, Berlin, p. 151, July 30.

Type Locality. Katschemak [= Katchemak] Bay, Kenai Peninsula, Alaska.

Type Specimen. Not known to exist.

Range. Known only from the Kenai Peninsula.

Remarks. Tomasik and Cook (2005) found that the occurrence of common and widespread mtDNA haplotypes on the Kenai Peninsula is not consistent with subspecies status.

Gulo gulo luscus

Original Description. 1776. [*Ursus*] *luscus* Linnaeus, *Syst. Nat.*, 10th ed., 1:47.

Type Locality. Hudson Bay.

Type Specimen. Not known to exist.

Range. Broadly distributed across Alaska, Canada, and the northwestern contiguous United States.

Remarks. Includes *Gulo hylaeus* Elliot, 1905, *Proc. Biol. Soc. Washington*, 18:81, February 21, type from Susitna River, in the region of Mt. McKinley, Alaska (see Rausch 1953a).

Global Distribution

Holarctic—*Gulo gulo* occur broadly across northern Eurasia and northern North America, including many of the islands in the arctic archipelago of northern Canada (Pasitschniak-Arts and Larivière 1995). In North America they range as far south as Northern California.

Alaska Distribution

Wolverines occur throughout mainland Alaska and on Unimak Island and several islands in the Alexander Archipelago (map 71).

Southeast Region. Wolverines are found on the mainland and in the Alexander Archipelago on Etolin, Fair, Kuiu, Kupreanof, Mitkof, Revillagigedo, and Wrangell islands (MacDonald and Cook 2007).

Southcentral Region. Wolverines occur on the mainland and may possibly visit some of the nearshore islands in Prince William Sound such as Bainbridge Island, where there are a few recent records (D. W. Crowley, ADFG, pers. comm.). On the Kenai Peninsula, they are found most commonly in the Kenai Mountains (Selinger 2004).

Central Region. The wolverine occurs throughout the region.

Southwest Region. *Gulo gulo* occurs throughout the length of the Alaska Peninsula and on Unimak Island.

Western Region. The wolverine inhabits the entire mainland of this region.

Northern Region. Wolverines are found throughout the region, but are more abundant in the mountains and foothills of the Brooks Range than along the coast or on the coastal plains (Bee and Hall 1956).

Habitat

Wolverines range widely over great distances and habitat types and can be found from sea level to the tops of mountains (Banci and Harestad 1990; Gardner 1985; Gardner et al.

1986; Magoun 1985; Whitman et al. 1986). Reproductive dens are usually long, complex snow tunnels with no associated trees or boulders (Magoun and Copeland 1998).

Status

Populations of this species from the contiguous United States were listed as Vulnerable by the IUCN. In Alaska, between 548 and 1,037 wolverines were harvested annually from 1971 to 1977 (Schreiber et al. 1989). Wolverines, while never abundant anywhere in the state, are most numerous in the southcentral region of Alaska (ADFG 1973). The current status of wolverines on Unalaska Island is not known (ADFG 1973); none has been trapped on the island since 1980 (Butler 2004a). The population of wolverines on the Kenai Peninsula is considered small and perhaps in decline (Schreiber et al. 1989).

Fossils

The origin of the modern wolverine can be traced to the Early Miocene. The Holarctic *Gulo schlosseri* of the Middle Pleistocene is considered the immediate ancestor of *G. gulo* (Kurtén 1963; Kurtén and Anderson 1980; but see Bryant 1987). Fossil remains of *G. gulo* occurred in deposits of Irvingtonian age and younger in North America (Kurtén and Anderson 1980; Rausch 1994). Youngman (1993) examined seventy-five Pleistocene fossils of wolverines from eastern Beringia. A specimen from Sixtymile, Yukon Territory, dated at 41,420±1100 years ago. Wolverine specimens from the Late Pleistocene of

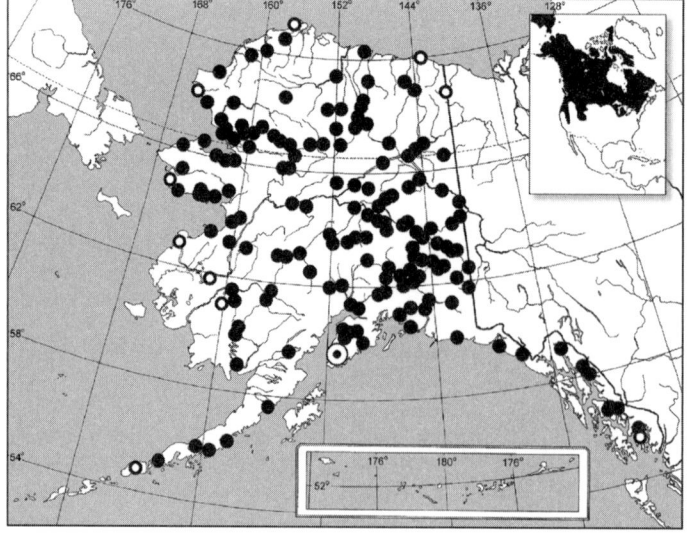

MAP 71. Distribution of wolverine, *Gulo gulo*.

⊙ = Type Locality
● = Specimen Record (n = 1681)
○ = Marginal Records. 71.29100, −156.78900, Barrow Quad., Barrow (1 UAM); Barter Island Quad., Between Barter I. and Jago River (2 UAM); Table Mt Quad, Firth River and 141 degrees longitude (1 USNM); 55.75000, −131.50000, Ketchikan Quad., Revillagigedo Island, Ketchikan area (1 UAM); Unimak Quad., Unimak Island, N Creek (2 USNM); Russian Mission Quad., Nyac, Upper Tuluksak River (1 UAM); Russian Mission Quad., Yukon River, Russian Mission (1 UAM); Kwiguk Quad., mouth of Yukon River (1 USNM); 64.58333, −166.25000, Nome Quad., Sinuk River (4 UAM); 68.13333, −165.96667, Point Hope Quad., Cape Thompson, stream valley 2 mi SE Ogoturuk Creek (1 UAM).

the Fairbanks region are especially large (Anderson 1977). Heaton and Grady (2003) reported finding a single molar fragment that appeared to match only *G. gulo* from a cave on Prince of Wales Island in southeast Alaska. That specimen was dated Late Pleistocene or Early Holocene.

North American River Otter

Lontra canadensis
(Schreber, 1777)

Other Common Names

River otter, land otter, northern river otter, Canadian otter.

Systematics

The genus was revised from *Lutra* to *Lontra* by van Zyll de Jong (1972, 1987). Some authors have opposed the use of *Lontra* following the work of Sokolov (1973), a misinterpretation that is not justified (Kellnhauser 1983).

Van Zyll de Jong (1972) and Hall (1981) recognized seven subspecies; three occur in Alaska.

Patterns of genetic variation among otter populations of eastern North America did not corroborate current subspecific designations, and numerous translocations have crossed subspecies' range boundaries (Serfass et al. 1998). A genetic study that includes western populations has yet to be published.

Lontra canadensis kodiacensis

Original Description. 1935. *Lutra canadensis kodiacensis* Goldman, *Proc. Biol. Soc. Washington*, 48:180, November 15.

Type Locality. Uyak Bay, Kodiak Island, Alaska.

Type Specimen. USNM 98142.

Range. Restricted to the Kodiak Archipelago.

Lontra canadensis mira

Original Description. 1935. *Lutra canadensis mira* Goldman, *Proc. Biol. Soc. Washington*, 48:185, November 15.

Type Locality. Kasaan Bay, Prince of Wales Island, Alaska.

Type Specimen. USNM 127888.

Range. Southeast Alaska and coastal British Columbia.

Remarks. *Lontra c. periclyzomae* is currently considered an endemic subspecies to Haida Gwaii, British Columbia, on the basis of morphological differences (van Zyll de Jong 1972).

Lontra canadensis pacifica

Original Description. 1898. *Lutra hudsonica pacifica* Rhoads, *Trans. Amer. Philos. Soc.*, n.s., 19:429, September.

Type Locality. Lake Keechelus, 3000 ft., Kittitas Co., Washington.

Type Specimen. ANSP 7616.

Range. Northwestern North America from Alaska to Minnesota, Colorado, and California.

Remarks. Includes *Lutra canadensis optiva* Goldman 1935, *Proc. Biol. Soc. Washington*, 48:179, November 15, type from Zaikof Bay, Montague Island, Alaska (USNM 137320); *Lutra canadensis yukonensis* Goldman 1935, *Proc. Biol. Soc. Washington*, 48:180, November 15, type from Unalakleet, Norton Sound, Alaska (USNM 21480); and *Lutra canadensis extera* Goldman 1935, *Proc. Biol. Soc. Washington*, 48:181, November 15, type from Nagai Island, Shumagin Islands, Alaska (USNM 12485).

Global Distribution

Nearctic—The range of the northern river otter once extended throughout Canada and the United States. It is now extirpated or rare throughout most of the central and eastern United States, although reintroductions have expanded its range in recent years (Larivière and Walton 1998).

Alaska Distribution

River otters are found throughout most of Alaska south of the Brooks Range (map 72).

Southeast Region. River otters are common along the coastal and inland waters of southeast Alaska (MacDonald and Cook 2007).

Southcentral Region. Northern river otters occur throughout the region and are abundant in Prince William Sound (ADFG 1978).

Central Region. *Lontra canadensis* is found across the region and into the major drainages of the Brooks Range.

Southwest Region. River otters occur on the mainland, some adjacent islands east of the Alaska Peninsula, and Unimak Island (Butler 2004a).

Western Region. River otters occur throughout the mainland of this region. They are abundant in the Yukon-Kuskokwim Delta and are occasionally found on Nunivak Island (ADFG 1978). On the Seward Peninsula, they are present primarily in the upper drainages of the Kuzitrin, Buckland, and Koyuk rivers. Farther north, they are

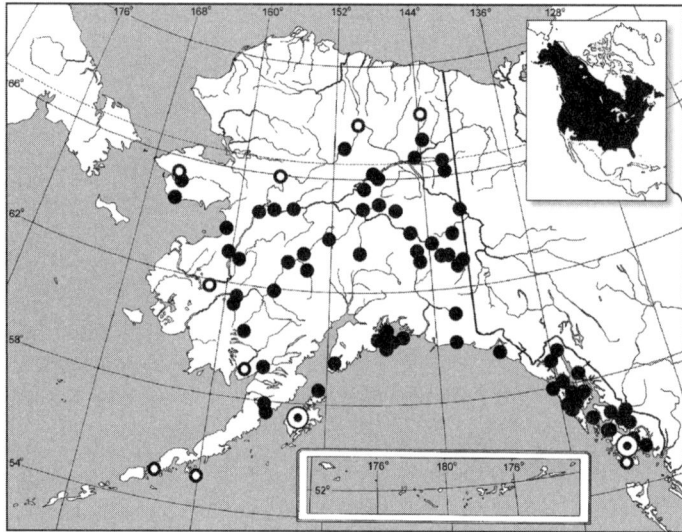

MAP 72. Distribution of North American river otter, *Lontra canadensis*.

⊙ = Type Locality
● = Specimen Record (n = 413)
○ = Marginal Records. 68.05000, –143.80000, Table Mt Quad., Sheenjek Lake (1 UAM); 54.87083, –132.80133, Dixon Entrance Quad., Dall Island, Howkan (7 MVZ); Port Moller Quad., Nagai Island, Popof Strait (1 USNM); Cold Bay Quad, east base Frosty Peak (1 USNM); Russian Mission Quad., Russian Mission (1 USNM); 65.15000, –165.06667, Teller Quad., Igloo (8 UAM); 66.00000, –156.93333, Shungnak Quad., North Fork Huslia River (1 UAM).

frequently found in the lower and middle Noatak River, the Salmon River, the upper Selawik River, and the upper Kobuk River (ADFG 1978).

Northern Region. River otters are present in this region only in scattered populations. They are found occasionally in the western part and, according to Bee and Hall (1956) and ADFG (1978), less often in the rest of the region. Magoun and Valkenburg (1977) provided twenty-four locality records (none documented by specimen) from the eastern Brooks Range and North Slope, and suggested that river otters had been extending their range northward since the 1950s. In summer 2001, river otter tracks were noted along a stream in the Prudhoe Bay oilfields (C. T. Seaton, ADFG, pers. comm.), and in July of that year, a river otter was found drowned in a fish sampling net set in Lions Bay on the Beaufort Sea coast approximately 11 km west of the Canning River (Haskell 2006; specimen not preserved).

Habitat

The semiaquatic river otter inhabits a wide variety of coastal marine and freshwater habitats (Banfield 1974; Ceballos-G 1999; Larivière and Walton 1998).

Status

Alaska populations of this species are relatively stable, especially in coastal areas (ADFG 1978). In southeast Alaska, they are a Management Indicator Species of the Tongass

National Forest (MacDonald and Cook 1996). *Lontra canadensis* is listed in Appendix II of CITES.

Fossils

Larivière and Walton (1998) reviewed the fossil record of this species. Ancestral river otters probably entered North America from Eurasia across the Bering Land Bridge during the Early Irvingtonian. The oldest fossils of *L. canadensis* are from Middle Pleistocene (Irvingtonian) deposits in the northeastern United States (Kurtén and Anderson 1980). Repenning (1977; in Youngman 1993) reported the first river otter fossils of Pleistocene age from Beringia. Yesner (2001) listed twelve otter specimens of Late Pleistocene and Early Holocene age from central Alaska.

American Marten

Martes americana
(Turton, 1806)

Other Common Names

American pine marten, pine marten.

Systematics

The holarctic martens, *M. melampus*, *M. martes*, *M. zibellina*, and *M. americana*, are closely related (Anderson 1970; Hagmeier 1961; Stone and Cook 2001; but see Clark et al. 1987; Youngman 1975). Before 1953, two species of marten, *M. americana* and *M. caurina*, were recognized in North America (Youngman and Schueler 1991). They were subsequently synonymized under *M. americana* (Hagmeier 1961; Wright 1953). Giannico and Nagorsen (1989) suggested that *caurina* and *americana* may intergrade in southeast Alaska; however, their study did not distinguish introduced from native populations.

Recent mitochondrial and nuclear DNA studies by Carr and Hicks (1997), Demboski et al. (1999), McGowan et al. (1999), Stone (2000), Cook et al. (2001), Stone et al. (2002), and M. Small et al. (2003) suggest that the "*caurina*" and "*americana*" groups represent two distinct species.

Demboski et al. (1999), Stone et al. (2002), and M. Small et al. (2003) analyzed geographic variation of mitochondrial and nuclear genes across mainland and island populations of marten to shed light on hybridization patterns, population structure, and evolutionary histories of the distinctive "*caurina*" and "*americana*" groups of *M. americana* in Alaska. Hybridization between *caurina* and *americana* was documented in two regions of sympatry (Kuiu Island in southeast Alaska and southern Montana). Northern insular populations of *caurina* exhibited higher among-population differentiation and lower within-population variability relative to northern populations of *americana*. Greater divergence among insular *caurina* populations may reflect longer isolation in coastal forests that were fragmented in the Early Holocene. Lower differentiation among northern *americana* populations and close relationship to other continental *americana* populations likely reflects recent expansion into the Pacific Northwest coast and/or continued gene flow among populations. Population differentiation in *caurina* was attributed to habitat fragmentation (i.e., rising sea level); oceanic straits pose significant barriers to gene flow among *caurina* populations. Extant populations of *caurina* are found in Alaska only on Admiralty and Kuiu islands.

Clark et al. (1987) tentatively recognized eight subspecies that were divided into two groups: "*americana*" (a continental group comprising five subspecies) and "*caurina*" (a west coast group comprising three subspecies). Anderson (1970) and Hagmeier (1958) emphasized that the subspecies designations were arbitrary because morphological variation was clinal or discordant (but see Dillon 1961) and based on too few samples. Hall (1981) recognized seven subspecies within the range of *M. americana*; two occur in Alaska.

Martes americana actuosa

Original Description. 1900. *Mustela americana actuosa* Osgood, *N. Amer. Fauna*, 19:43, October 6.

Type Locality. Fort Yukon, Alaska.

Type Specimen. USNM 6043.

Range. Interior Alaska and northwestern Canada.

Martes americana kenaiensis

Original Description. 1903. *Mustela americana kenaiensis* Elliot, *Field Columb. Mus.*, Publ. 72, Zool. Ser., 3:151, March 20.

Type Locality. Kenai Peninsula, Alaska.

Type Specimen. FMNH 9847.

Range. Southern coastal Alaska.

Global Distribution

Nearctic (but see "Systematics")—*M. americana* occurs throughout the boreal forest regions of North America (Clark et al. 1987).

Alaska Distribution

Marten inhabit most of the forested regions of Alaska (map 73).

Southeast Region. American marten are found on the coastal mainland of southeast Alaska and on islands in the Alexander Archipelago in close proximity to the mainland. Natural populations occur on Admiralty, Etolin, Gravina, Kuiu, Kupreanof, Mitkof, Revillagigedo, Woewodski, and Wrangell islands (MacDonald and Cook 2007). Marten of unknown status and affinity have also been reported from Gravina Island (L. Carson, Ketchikan, pers. comm.), Annette Island (J. Moran, Metlakatla, pers. comm.), and Zarembo Island (trapper reports in 2003–2005; R. Lowell, ADFG, Petersburg, pers. comm.; N. Dawson, pers. comm.).

In 1934, marten from Behm Canal and Thomas Bay on the mainland were introduced on Prince of Wales Island and Baranof Island. Between 1949 and 1952, marten were introduced successfully on Chichagof Island with stock taken from Baranof Island, Revillagigedo Island, the Stikine River drainage, Wrangell Island, Mitkof Island, and near Anchorage (Burris and McKnight 1973; Elkins and Nelson 1954). In addition, UAM has specimens of marten from the vicinity of Baranof Island from Kruzof, Otstoia, Catherine, Partofshikof, and Yakobi islands. These specimens likely originated from undocumented transplants.

Southcentral Region. Marten inhabit the mainland portion of this region; they are not found on the islands of Prince William Sound (ADFG 1978), except perhaps Hawkins Island (D. Crowley, ADFG, pers. comm.). Marten are moderately abundant in eastern portions of the Kenai Peninsula but rare elsewhere (Selinger 2004). Marten were translocated to Kayak and Patterson islands in the early 1940s (Elkins and Nelson 1954), but Burris and McKnight (1973) could not verify these efforts or determine their current status.

ADFG (1978) indicated marten occurring on Kayak Island, and former local resident M. E. Isleib (pers. comm., 1978) considered them "common" there.

Central Region. Marten occur in suitable habitat throughout the central region.

Southwest Region. Marten are restricted to the northern portion of this region in mature spruce forests (ADFG 1978). In 1952, twenty marten from the Lake Minchumina area were released on Afognak Island (Burris and McKnight 1973). ADFG (1978) indicated that marten became established on the island but because of limited habitat their numbers remained low.

Western Region. Marten occur in forested areas of the Kobuk and lower Noatak drainages, with a westward expansion in range noted during the 1990s (Dau 2004a). They inhabit the southwestern portion of the Seward Peninsula but their status there or elsewhere on the peninsula is poorly known (Gorn 2004). Marten are a peripheral

species to the Yukon-Kuskokwim Delta region (Seavoy 2004a).

Northern Region. Marten usually are absent north of the Brooks Range, although individuals occasionally wander over the divide (ADFG 1978; Bee and Hall 1956).

Habitat

American marten are adapted to a variety of forested habitats (Strickland et al. 1982). Optimum habitat elements appear to be mature old-growth spruce communities with well-established understory and ground cover to support microtine rodents and other prey (Buskirk 1983; Buskirk and MacDonald 1984; Clark et al. 1987; Flynn and Blundell 1992; Lensink et al. 1955; Schoen et al. 2007).

Status

Overtrapping and loss of habitat from clearcut logging and forest fires have been the greatest causes of marten decline across its range (Hagmeier 1956, 1961; Strickland et al. 1982). Before game laws early this century, marten numbers in southeast Alaska were greatly reduced by overtrapping (MacDonald and Cook 1996). High-quality habitat is still present in interior Alaska and marten populations are healthy throughout the area.

Fossils

The American marten is known from several Late Pleistocene deposits in the eastern United States and from western sites, including Yukon Territory in eastern Beringia (Kurtén and

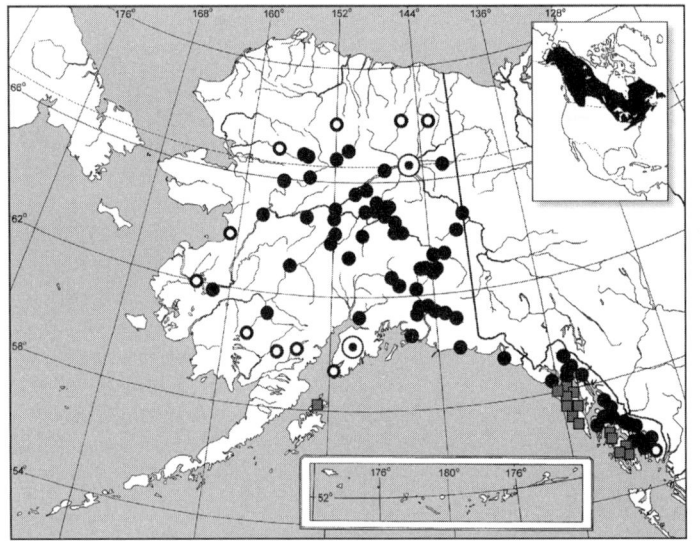

MAP 73. Distribution of American marten, *Martes americana*.

⊙ = Type Locality
● = Native, ▣ = Non-native Specimen Record (n = 4470)
○ = Marginal Records. Wiseman Quad., John River, Hunt Fork (1 USNM); 68.133, −145.533, Arctic Quad., Arctic Village (2 PSM); 68.000, −143.000, Table Mtn Quad., Upper Coleen River (45 UAM); 55.033, −130.967, Ketchikan, Kah Shakes (1 UAM); Seldovia Quad., Seldovia (2 AMNH); 60.231, −154.399, Lake Clark Quad., Lake Clark, Dice Bay (3 UAM); Dillingham Quad., Kakhtul River, 30 mi. up (5 USNM); 60.330, −157.783, Taylor Mts Quad., Game Management Unit 17B (1 UAM); Kwiguk Quad., Andraeffsky (1 USNM); Unalakleet Quad., near Unalakleet, 60 mi. from St. Michael (10 FMNH); Ambler River Quad., Ambler (1 MSB).

Anderson 1980; Youngman 1993). Georgina (2001) reported marten remains from Early Holocene (perhaps Late Pleistocene) deposits in a Lime Hills cave, central Alaska.

This species probably reached North America via the Bering Land Bridge in the Late Rancholabrean (Wisconsinan) and spread eastward. This population, constituting the "*americana*" subspecies group, became isolated in eastern North America and only after the ice sheets retreated did it reinvade Alaska and western Canada. A second invasion of marten directly from Siberia was proposed to have populated the West Coast, Sierra Nevada, and Rocky Mountains (Clark et al. 1987; Kurtén and Anderson 1980). These populations, the "*caurina*" group, purportedly showed more similarities in cranial and dental characteristics to *M. zibellina* than to the eastern subspecies (Anderson 1970; Hagmeier 1961). Phylogenetic analyses of nuclear and mitochondrial DNA sequences do not support this scenario, but instead suggest that *M. americana* and *M. caurina* are sister species (Stone et al. 2002).

Pacific Marten

Martes caurina
(Merriam, 1890)

Other Common Names
Named subspecies include Queen Charlotte marten and Vancouver Island marten.

Systematics
After 1890, when Merriam described *Martes caurina* from coastal Washington, the extant marten of North America were considered to belong to two species: *Martes americana* and *M. caurina*. A latter revision by Rhoads (1902) supported Merriam's contention. Subsequently, these two polytypic forms were synonymized under *Martes americana* by Wright (1953). Recent mtDNA studies by Carr and Hicks (1997), Demboski et al. (1999), McGowan et al. (1999), Stone (2000), Cook et al. (2001), and Stone et al. (2002), and a nuclear DNA study by M. Small et al. (2003) indicate that the "*caurina*" and "*americana*" groups represent two distinct species, a view we follow in these accounts.

The taxonomic status of named subspecies of *caurina* remains unresolved. Clark et al. (1987) included three subspecies of western martens within the "*caurina*" group: *caurina*

(Merriam, 1890), *humboldtensis* Grinnell and Dixon, 1926, and *nesophila* (Osgood, 1901). The subspecies *origens* (Rhoads, 1902), *sierrae* Grinnell and Storer, 1916, *vancouverensis* Grinnell and Dixon, 1926, and *vulpina* (Rafinesque, 1819) were considered synonyms of *caurina* by Clark et al. (1987). Their submergence of *vulpina*, the oldest named taxon, was made without comment in regard to date priority.

The subspecies *nesophila* was restricted by Swarth (1911a) and Hall (1981) to insular populations on the Queen Charlotte Islands and Kuiu Island in southeast Alaska. Giannico and Nagorsen (1989) demonstrated that this race is strongly differentiated from other coastal populations, but suggested that *nesophila* should be applied only to Haida Gwaii (Queen Charlotte Islands) populations. Their analysis of skull morphology of southeast Alaska marten unfortunately included only material from the mainland or from island populations that were the result of introductions of American marten from multiple mainland stocks. They did not include material from Kuiu Island or Admiralty Island (see MacDonald and Cook 2007:Fig. 11), the only two island localities of *M. caurina* in southeast Alaska that were later identified from mitochondrial and nuclear molecular data by Stone and Cook (2001), Stone et al. (2002), and M. Small et al. (2003).

Preliminary findings from an expanded genetic study comparing mitochondrial and nuclear markers of marten from throughout North America suggest that each North Pacific island population is distinctive and that Kuiu, Admiralty, Queen Charlotte, and Vancouver islands harbor a significant portion of the genetic diversity reported for populations of *M. caurina* throughout the species' range (Dawson 2008). Admiralty and the Queen Charlottes have unique genetic signatures found nowhere else. Phylogenetic analysis of mtDNA and nuclear sequences of a nematode that parasitizes marten (Koehler 2006) corroborates the distinctive signatures of each of the insular *caurina* populations.

Hall (1981) recognized seven subspecies now within the range of *M. caurina* and two of these occur in southeast Alaska. The *caurina* marten on Admiralty Island have not been included in previous assessments, but based on preliminary genetic data, they may constitute an undescribed subspecies.

Martes caurina

Original Description. 1890. *Mustela caurina* Merriam, *N. Amer. Fauna*, 4:27, October 8.

Type Locality. Near Grays Harbor, Grays Harbor County, Washington.

Type Specimen. USNM 186450.

Range. Southeast Alaska southward along the coast (including Haida Gwaii and Vancouver islands) to California and eastward to Wyoming, Montana, and Idaho (Hall 1981; Stone et al. 2002).

Global Distribution

Nearctic—Southeast Alaska southward along the coast (including Haida Gwaii and Vancouver islands) to California and eastward to Wyoming, Montana, and Idaho and likely south to New Mexico (Hall 1981; Stone et al. 2002).

Alaska Distribution

Natural populations of Pacific marten in southeast Alaska have been documented on Kuiu and Admiralty islands (map 74). Credible, but as yet undocumented, reports of marten on a number of outer islands (Dall, Heceta, Marble, Orr, Tuxekan) may have been derived from introduced *M. americana* stocks on Prince of Wales or may be surviving remnants of the older, more insular *M. caurina* lineage.

Habitat

Marten are closely tied to forested habitats, particularly mature and old-growth conifer forests with complex physical ground structure (Buskirk and Powell 1994). Habitat use by insular populations of Pacific marten is limited, however (Burles et al. 2004; Nagorsen et al. 1991). On Haida Gwaii, marten are assumed to use all low-elevation late successional and old-growth forests, with use of higher-elevation forest unknown (Burles et al. 2004). Foster (1965) and Cowan (1989) suggested these island marten make extensive use of marine shorelines for foraging.

Status

Overtrapping and loss of habitat from clear-cut logging and forest fires have been the greatest causes of marten decline across its range (Hagmeier 1956, 1961; Strickland et al. 1982). Marten in southeast Alaska were greatly reduced in number by overtrapping before game laws were instituted (MacDonald and Cook 1996).

Fossils

Youngman and Schueler (1991) considered the fossil marten, *M. nobilis*, of eastern Beringia (Youngman 1993) indistinguishable from the modern "*caurina*" subspecies group except for its larger size. They suggested that specimens of *M. nobilis* are large Pleistocene chronomorphs of *M. americana*, similar to the large specimens of Pleistocene *M. foina* and *M. martes* from Europe (Anderson 1970).

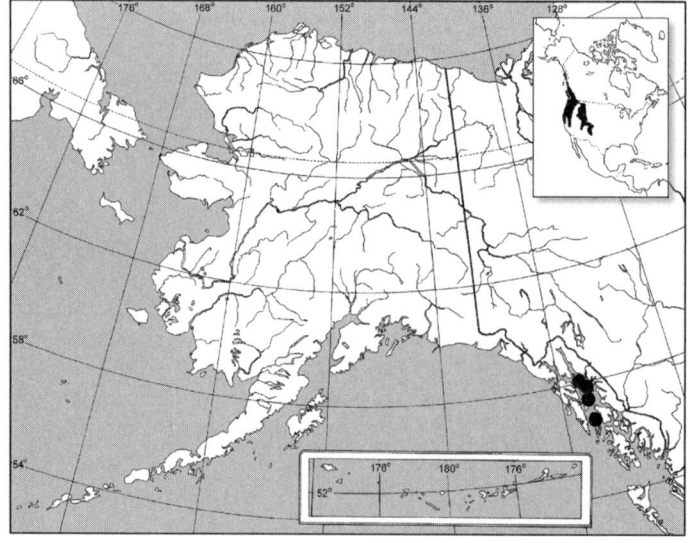

MAP 74. Distribution of Pacific marten, *Martes caurina*.

● = Specimen Record (n = 127)

Fisher

Martes pennanti
(Erxleben, 1777)

Other Common Names

Pekan, Pennant's marten, fisher cat.

Systematics

Stone and Cook (2001) placed the fisher in the subgenus *Pekania* and suggested that *M. pennanti* and *Gulo gulo* may form a monophyletic group, which would make *Martes* paraphyletic. Goldman (1935) recognized three subspecies, but Hagmeier (1958, 1959) recommended that no subspecies be recognized. Nagorsen (1990) concluded that geographic variation throughout the entire range of this species should be assessed to evaluate the validity of the subspecies recognized by Hall (1981), of which one occurs in Alaska. In a range-wide assessment of fisher based on mtDNA sequences, Drew et al. (2003) found minimal divergence among twelve haplotypes that roughly corresponded to the geographic limits of the nominal subspecies.

Genetic and paleontological evidence suggests that fishers evolved in eastern North America and have expanded westward to the Pacific Coast since the last glaciation (Graham and Graham 1994; Wisely et al. 2004).

Martes pennanti pacifica

Original Description. 1898. *Mustela canadensis pacifica* Rhoads, *Trans. Amer. Philos. Soc.*, n.s., 19:435, September.

Type Locality. Lake Keechelus, Kittitas County, Washington.

Type Specimen. ANSP 8074.

Range. Mainland southeast Alaska southward along the Pacific Coast to California.

Global Distribution

Nearctic—Formerly the fisher had an extensive range across northern Canada and the western and northeastern United States. Their present range is reduced in the United States from what it was before the European settlement of North America (Powell 1999).

Alaska Distribution

The fisher occurs only in the southeast region of Alaska (map 75).

The occurrence of fishers in southeast Alaska was confirmed by voucher specimens collected from the Taku River in 1994, a young male in 1996 near Juneau, and a mature female in 1997 from the Besi Creek area about 40 km north of Juneau. From these and other reports in the region (MacDonald and Cook 2007; Woodford 2006), it appears that this species is a rare to uncommon visitor or perhaps resident of the mainland of southeast Alaska.

This animal has been found as far north as southeastern Yukon Territory and may be

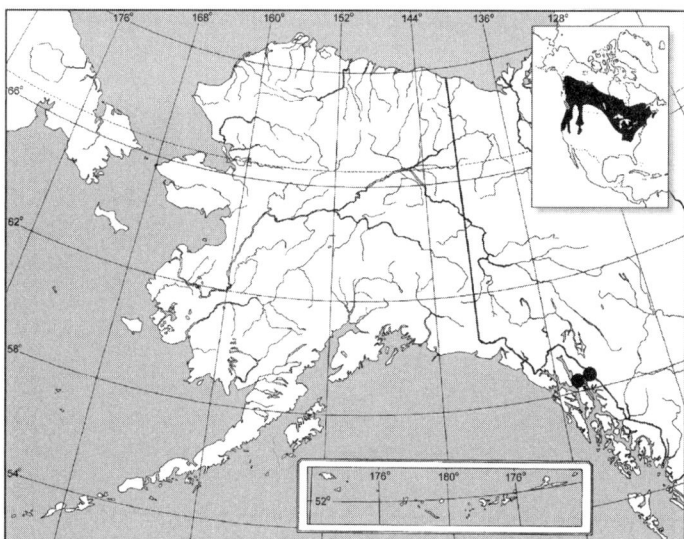

MAP 75. Distribution of fisher, *Martes pennanti*.

● = Specimen Record (n = 3): 58.5500, –133.6833, Taku River Quad., Taku River, Canyon Island (1 UAM); 58.5833, –134.8333, Juneau Quad. (1 UAM); 58.6167, –134.9167, Juneau Quad., E of Sunshine Cove (1 UAM).

expanding its range (Youngman 1975). A trapper's sighting of a fisher near Northway in east-central Alaska (D. Frederick, pers. comm., 2002) could not be verified.

Habitat

Fishers prefer the extensive, continuous canopy of dense, lowland forests and spruce-fir forests. They avoid open areas and forest with little overhead cover (Powell 1981).

Status

The status of the fisher in Alaska is unclear. It is probably a rare to uncommon resident of mainland southeast Alaska, at least from the Juneau area southward.

Fossils

Powell (1981) reviewed the fossil history. The probable ancestors to modern fishers are known from the Pliocene of China. *Martes divuliana*, the first true fisher, is known only from the Middle Pleistocene of North America and probably colonized North America via the Bering Land Bridge (Anderson 1970; Kurtén 1971). Pleistocene specimens of *M. pennanti* have been reported from the Yukon Territory and from Fairbanks Creek in central Alaska (Harington 1977; Jopling et al. 1981; Youngman 1993). Those discoveries dispel Youngman's (1975) suggestion that this species was a postglacial immigrant to eastern Beringia. Fossils have also been recorded in Late Pleistocene deposits located farther south than the range they occupied prior to their broad extirpation following the arrival of Europeans (Kurtén and Anderson 1980). There are no morphological differences between Late Pleistocene fishers and modern fishers (Anderson 1970).

Ermine

Mustela erminea
Linnaeus, 1758

Other Common Names

Short-tailed weasel, stoat.

Systematics

Nearctic populations of *M. erminea* were last revised by Eger (1990) and reviewed by C. M. King (1983). Hall (1981) recognized twenty subspecies of Nearctic ermine, of which seven occur in Alaska. A smaller number of subspecies are recognized in the Palearctic. Eger (1990) also recognized seven subspecies in the

state and seven in the southeast region. In her assessment of geographic variation in thirteen craniometric characters, she identified the Queen Charlotte endemic, *M. e. haidarum*, as also occurring in the Alexander Archipelago of southeast Alaska.

Fleming and Cook (2002) uncovered substantial geographic variation in the mitochondrial cytochrome *b* gene in Holarctic ermines, with three distinctive lineages detected (Fleming and Cook 2002). Southeast Alaska is the only region in the world where all three lineages converge. One lineage apparently is restricted to the Prince of Wales Archipelago and the Queen Charlotte Islands and this endemic may provide a signal for the hypothesized North Pacific Coastal refugium during the Pleistocene. Another lineage has a Holarctic distribution. It occurs only on Admiralty Island and near Yakutat in southeast Alaska, then into Yukon Territory and westward across Alaska, Siberia, and into Europe. The third lineage is widespread and endemic to North America. This lineage is found on Baranof and Chichagof islands and along the mainland and nearshore islands of southeast Alaska, then into east-central Alaska near Eagle, eastward across Canada to the East Coast and southward to California and New Mexico. An expanded analysis needs to be completed.

Mustela erminea alascensis

Original Description. 1896. *Putorius richardsoni alascensis* Merriam, *N. Amer. Fauna*, 11:12, June 30.

Type Locality. Juneau, Alaska.

Type Specimen. USNM 74423.

Range. Mainland southeast Alaska from Lynn Canal south to Portland Canal.

Mustela erminea arctica

Original Description. 1896. *Putorius arcticus* Merriam, *N. Amer. Fauna*, 11:15, June 30.

Type Locality. Point Barrow, Alaska.

Type Specimen. USNM 14062/23010.

Range. Northern Canada and nearly all of Alaska south to upper Lynn Canal (Hall 1951).

Remarks. Contact of two morphologically (Eger 1990) and genetically (Fleming and Cook 2002) distinct groups of ermine near Eagle in east-central Alaska suggest separate evolutionary histories that call to question current subspecific designations and relationships, particularly between this taxon and the broad-ranging Canadian subspecies, *richardsonii*.

Mustela erminea celenda

Original Description. 1944. *Mustela erminea celenda* Hall, *Proc. Biol. Soc. Washington*, 57:38, June 28.

Type Locality. Kasaan Bay, Prince of Wales Island, Alaska.

Type Specimen. USNM 130987.

Range. Prince of Wales, Long, and Dall islands, Alexander Archipelago.

Mustela erminea initis

Original Description. 1944. *Mustela erminea initis* Hall, *Proc. Biol. Soc.Washington*, 57:37, June 28.

Type Locality. Saook Bay, Baranof Island.

Type Specimen. MVZ 289.

Range. Baranof and Chichagof islands, Alexander Archipelago.

Mustela erminea kadiacensis

Original Description. 1896. [*Putorius arcticus*] subspecies *kadiacensis* Merriam, *N. Amer. Fauna*, 11:16, June 30.

Type Locality. Kadiak [= Kodiak] Island, Alaska.

Type Specimen. USNM 65290.

Range. Kodiak Island.

Mustela erminea salva

Original Description. 1944. *Mustela erminea salva* Hall, *Proc. Biol. Soc.Washington*, 57:39, June 28.

Type Locality. Mole Harbor, Admiralty Island, Alaska.

Type Specimen. MVZ 74641.

Range. Admiralty Island.

Mustela erminea seclusa

Original Description. 1944. *Mustela erminea seclusa* Hall, *Proc. Biol. Soc.Washington*, 57:39, June 28.

Type Locality. Port Santa Cruz, Suemez Island, Alaska.

Type Specimen. MVZ 31232.

Range. Known only from type locality.

Global Distribution

Holarctic—Ermine range broadly across the boreal and arctic regions of Eurasia and North America (Nagorsen 1990).

Alaska Distribution

The ermine is found throughout Alaska and on many North Pacific Coast islands (map 76).

Southeast Region. This species is widely distributed throughout the region, occurring along the entire mainland coast and probably on most of the islands in the Alexander Archipelago. Ermine have been documented on the following islands: Admiralty, Annette, Baranof, Chichagof, Dall, Douglas, El Capitan, Etolin, Heceta, Kuiu, Kupreanof, Long, Mitkof, Prince of Wales, Revillagigedo, Suemez, Wrangell, and Zarembo (MacDonald and Cook 2007).

Southcentral Region. Ermine are found throughout the mainland and on Hinchinbrook Island (Heller 1910). Recent reports of ermine on Evans (E. Lance, pers. comm.) and other islands in Prince William Sound lack verification.

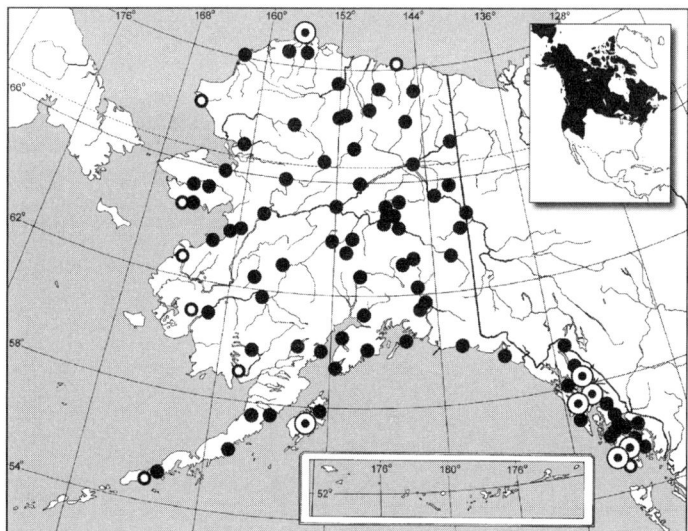

MAP 76. Distribution of ermine, *Mustela erminea*.

⊙ = Type Locality
● = Specimen Record (n = 1269)
○ = Marginal Records. 0.1833, −146.0500, Flaxman Island Quad., Flaxman Island (3 USNM); 54.8708, −132.8014, Dixon Entrance Quad., Long Island, Howkan (1 FMNH); 54.8542, −163.4083, False Pass Quad., Unimak Island, False Pass (3 USNM); 59.1167, −158.6000, Dillingham Quad., Aleknagik (1 USNM); 60.9167, −162.5000, Baird Inlet Quad., Nunapitchuk (4 UAM); Kwiguk Quad., mouth of Yukon River (1 USNM); 64.5000, −165.4167, Nome Quad., Nome (13 UAM); 68.1333, −165.9667, Point Hope, Cape Thompson (1 UAM).

Central Region. This species is widespread throughout central Alaska.

Southwest Region. Ermine occur throughout the mainland, as well as on Kodiak and Afognak islands. Clark (1958:576) stated they are found "on all the large islands" in this group. According to Murie (1959) and ADFG (1978), they are found in the Aleutians only on Unimak Island. The skull of an ermine from Umnak Island (USNM 268425) collected by anthropologist A. Hrdlička in 1938 appears old and discolored and suggestive of one unearthed from an archaeological site.

Western Region. Ermine are found only on the mainland of western Alaska.

Northern Region. This species occurs throughout northern Alaska (including a specimen from Flaxman Island, a coastal barrier island in the Beaufort Sea).

Habitat

Ermine are found in a wide variety of habitats ranging from forest to tundra and from sea level to above tree line. They favor ecotonal boundaries near meadows, shrubby stream banks, lakeshores, and beaver ponds, where they stay close to the cover of rock talus, shrub thickets, stumps, and logs and prey on small mammals.

Status

Ermine are common and widespread and their populations fluctuate dramatically, depending upon the population of lemmings, voles, and other small mammal prey species. Limited range of the lineage endemic to the Prince of Wales Archipelago and Haida Gwaii (where it is considered Threatened as *M. e. haidarum*; see Appendix 6) suggests the need to carefully monitor the conservation status of insular populations.

Fossils

The fossil history of this species was reviewed by Kurtén and Anderson (1980) and King (1983). Fossils of *M. erminea* can be traced back to the Late Pliocene in Europe. The species probably reached North America in the Late Blancan or Early Irvingtonian (Hall 1936, 1951). Specimens of Late Pleistocene and Early Holocene age have been recovered from deposits in Yukon Territory and central Alaska (Georgina 2001; Kurtén and Anderson 1980; Youngman 1993). Macpherson (1965), Youngman (1975), and Eger (1990) concluded that the subspecies *M. e. arctica* owes its dis-

tinctiveness to past isolation in the Beringian refugium.

Least Weasel

Mustela nivalis
Linnaeus, 1766

Other Common Names

Weasel, mouse-weasel.

Systematics

There has been considerable confusion regarding the taxonomic status of this species and its subspecies, particularly in Europe (Sheffield and King 1994). Allen (1933) provided the first evidence that *M. nivalis* is a Holarctic species. Reichstein (1957) was the first to advocate the name *nivalis* in place of *rixosa* for all least weasels. For North America, Youngman (1975), however, suggested that the short-tailed form in Alaska, Yukon Territory, and part of the Northwest Territories belong to *M. nivalis*, while populations of the long-tailed form occupying the remainder of North America belong to *M. rixosa*. He also placed *nivalis* in the subgenus *Mustela*.

Reig (1997) examined skull variation in samples from North America, Central Europe, and Siberia and concluded that the Old World subspecies *subpalmata* warrants consideration as a separate species. Reig also suggested that subspecies *M. nivalis rixosa* of the eastern United States and adjacent southern Canada may be specifically distinct from *M. n. eskimo* of Alaska and adjacent Canada, a view not supported in a preliminary analysis of mtDNA sequences within a small sample of North American least weasels (M. Fleming, pers. comm.).

Up to seventeen subspecies have been described in the Palearctic (Ellerman and Morrison-Scott 1951; Frank 1985). Sheffield and King (1994) accepted four of five Palearctic subspecies listed by Corbet (1978). Hall (1981) recognized four Nearctic subspecies, of which one is found in Alaska.

Mustela nivalis eskimo

Original Description. 1900. *Putorius rixosus eskimo* Stone, *Proc. Acad. Nat. Sci. Philadelphia*, 52:44, March 24.

Type Locality. Point Barrow, Alaska.

Type Specimen. ANSP 848.

Range. Alaska, extreme northwestern Canada.

Global Distribution

Holarctic—Least weasels range across Eurasia and northern North America (Sheffield and King 1994).

Alaska Distribution

Least weasels are found throughout mainland Alaska and on Unimak Island (map 77). The status and distribution of the species in the southeastern corner of the state lacks documentation and needs clarification.

Southeast Region. Reports of least weasel in the Glacier Bay area (MacDonald and Cook 2007) and elsewhere along the mainland (ADFG 1978) have not been substantiated.

Southcentral Region. There are specimens of this species only from the head of Cook Inlet and near Homer on the Kenai Peninsula (T. McDonough, ADFG, pers. comm., 2009). Least weasels remain undocumented from the Chugach and Wrangell mountains and vicinities. ADFG (1978) reported them throughout the interior areas of this region and the late M. E. Isleib (pers. comm., 1978) provided trapper reports of *M. nivalis* from the Copper River Delta.

Central Region. Least weasels are found throughout central Alaska.

Southwest Region. *Mustela nivalis* is found along the mainland of the Alaska Peninsula as well as on Unimak Island (Murie 1959).

Western Region. *Mustela nivalis* occurs throughout the mainland of this region.

Northern Region. Least weasels are found throughout northern Alaska.

Habitat

Least weasels occur in a wide variety of forest and tundra habitats, but favor meadows, marshes, and riparian situations where small rodent prey are abundant.

Status

The status of *M. nivalis* is unknown, but it is probably uncommon and sparsely distributed throughout much of its range except on the Arctic Slope of northern Alaska, where it is abundant during periods of high rodent populations (ADFG 1978).

Fossils

This species is known from Late Pleistocene faunas in Europe, Africa, and Asia (Sheffield and King 1994). In North America, fossils have been reported from ten sites of Rancholabrean age including several outside their current range (Kurtén and Anderson 1980; Savage and Russell 1983; Sheffield and King 1994). Harington (1989) and Youngman (1993) reported Wisconsinan fossils of *M. nivalis* from cave deposits in Yukon Territory.

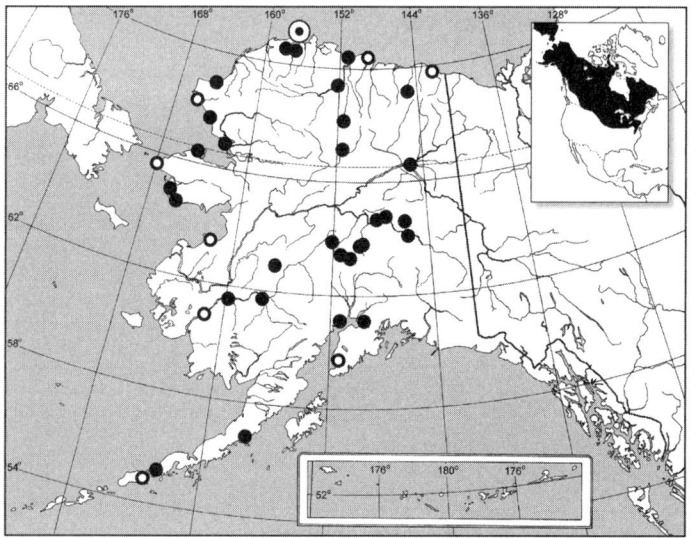

MAP 77. Distribution of least weasel, *Mustela nivalis*.

⊙ = Type Locality
● = Specimen Record (n = 156)
○ = Marginal Records. 70.4889, −149.1583, Beechey Point Quad., 3 mi. W of Beechey Point (1 AMNH); 69.8167, −142.0833, Demarcation Pt Quad., Egaksrak River, Egasak benchmark, 1 mi. SW (1 UWBM); 59.68217, −151.42083, Seldovia Quad., east of Homer (1 UAM); 54.8542, −163.4083, False Pass Quad., Unimak Island, False Pass (1 USNM); 60.7917, −161.7500, Bethel Quad., Bethel (3 USNM); 65.6167, −168.0833, Teller Quad., Wales (1 DMNH); 68.0833, −165.7500, Point Hope Quad., Ogotoruk Creek region (2 UAM).

American Mink

Neovison vison
(Schreber, 1777)

Other Common Names
Mink.

Systematics
Removal of *vison* from the genus *Mustela* follows Wozencraft (2005) according to Abramov (1999) and others. Of the fifteen subspecies listed by Hall (1981), five occur in Alaska.

Neovison vison aniakensis
Original Description. 1964. *Mustela vison aniakensis* Burns, *Canadian Jour. Zool.*, 42:1073, November.
Type Locality. Vicinity of Aniak, along Salmon River, Alaska.
Type Specimen. USNM 337939.
Range. Yukon-Kuskokwim Delta region of Alaska.

Neovison vison energumenos
Original Description. 1896. *Putorius vison energumenos* Bangs, *Proc. Boston Soc. Nat. Hist.*, 27:5, March.
Type Locality. Sumas, British Columbia.
Type Specimen. MCZ B3555.
Range. Western Canada, mainland southeast Alaska to California and New Mexico.
Remarks. Youngman (1975) synonymized *aniakensis* and *melampeplus* with *energumenos*.

Neovison vison ingens
Original Description. 1900. *Lutreola vison ingens* Osgood, *N. Amer. Fauna*, 19:42, October 6.
Type Locality. Fort Yukon, Alaska.
Type Specimen. USNM 6530.
Range. Alaska north of the Alaska Range and Yukon-Kuskokwin Delta, eastward into extreme northern Yukon and Northwest Territories.

Neovison vison melampeplus
Original Description. 1903. *Putorius vison melampeplus* Elliot, *Field Columb. Mus. Publ.* 74, Zool. Ser., 3:170, May 2.
Type Locality. Kenai Peninsula, Alaska.
Type Specimen. FMNH 9844.
Range. Southcentral and southwestern Alaska.

Neovison vison nesolestes
Original Description. 1909. *Lutreola vison nesolestes* Heller, *Univ. California Publ. Zool.*, 5:259, February 18.
Type Locality. Windfall Harbor, Admiralty Island, Alaska.
Type Specimen. MVZ 201.
Range. Alexander Archipelago of southeast Alaska.

Global Distribution
Nearctic—The American mink ranges across most of North America north of Mexico with the exception of arid southwestern United States (Hall 1981; Larivière 1999).

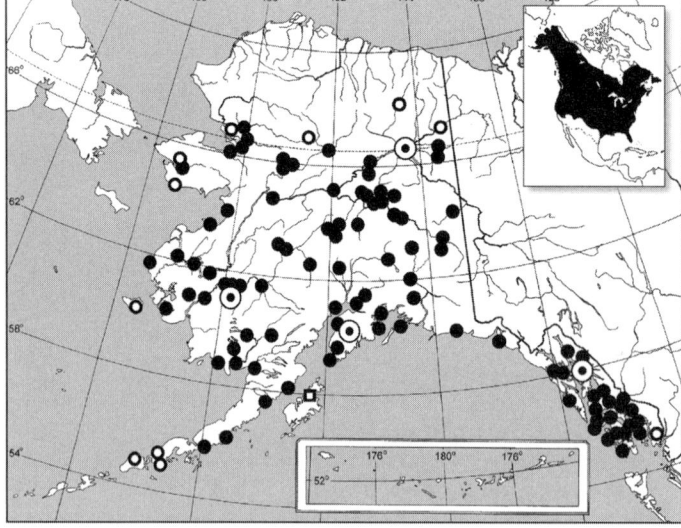

MAP 78. Distribution of American mink, *Neovison vison*.

⊙ = Type Locality
● = Native, ☐ = Non-native Specimen (n = 2335)
○ = Marginal Records. 68.1333, –145.5333, Arctic Quad., Arctic Village (1 UAM); 67.1667, –142.1667, Coleen Quad., Porcupine River (3 UAM); Ketchikan Quad., Boca de Quadra (2 MVZ); 57.9833, –152.9167, Kodiak Quad., slough, (2 UAM; no date; presumably from introduced stocks); 55.0000, –162.3333, False Pass Quad., Deer Island (2 UAM); Unimak Quad., Unimak Island (no preserved specimens but see Murie 1959); 55.3333, –162.6667, Cold Bay Quad., Russell Creek, Cold Bay (11 UAM); 60.0000, –166.0000, Nunivak Island Quad., Nunivak Island (27 UAM); 64.5000, –165.4167, Nome Quad., Snake River, Nome (1 UAM); 65.1500, –165.0667, Teller Quad., Igloo (1 UAM); Selawik Quad., Kobuk River Delta (1 UMMZ); 67.0000, –160.4333, Selawik Quad., Kobuk River near Kiana (2 UAM); 66.9667, –154.7500, Hughes Quad., Nutuvukti Lake (1 USNM).

Alaska Distribution

The American mink inhabits much of Alaska south of the Brooks Range (map 78).

Southeast Region. Mink occur throughout southeast Alaska, usually in close association with marine and freshwater ecosystems on the mainland and probably on most of the islands in the Alexander Archipelago. Mink records in the archipelago are incomplete, however. Mink raised at the once-active Petersburg Fur Experimental Farm were translocated to Strait Island in 1956 (Burris and McKnight 1973).

Southcentral Region. Mink are found throughout the region but are most abundant along the Gulf Coast and Prince William Sound (ADFG 1978). Among the islands of Prince William Sound, Heller (1910) noted mink on Disc, Eleanor, Esther, Hinchinbrook, and Knight islands. Mink from the Petersburg Fur Experimental Farm were introduced, apparently successfully, on Montague Island in 1951 (Burris and McKnight 1973).

Central Region. This species occurs across the region and in some of the drainages along the south side of the Brooks Range (Bee and Hall 1956; Rausch 1951).

Southwest Region. Mink inhabit the Alaska Peninsula and Unimak Island (Murie 1959). They are apparently absent from the Shumagins and other Gulf Coast islands, including the Kodiak Archipelago. There was an unsuccessful attempt in 1952 by the Alaska Game Commission to estab-lish from the Petersburg Fur Experiment Farm (Burris and McKnight 1973).

Western Region. Mink occur throughout the mainland of western Alaska and on Nunivak Island (UAM; USNM; ADFG 1978).

Northern Region. Mink have been occasionally reported from this region, but have not been substantiated with specimens (Bee and Hall 1956; Rausch 1950, 1951, 1953a).

Habitat

Mink are found in close association with coastal marine and freshwater ecosystems, preferring saltwater beaches and riparian habitats of lakeshores, marshes, and stream banks (Banfield 1974; Larivière 1999).

Status

Mink populations in Alaska reach their highest densities along the rich coastal areas of southeastern and southcentral Alaska. Lower densities and greater fluctuations characterize interior populations (ADFG 1978).

Fossils

Although records of *N. vison* extend back to the Irvingtonian, they are uncommon in Pleistocene faunas (Kurtén and Anderson 1980). The only known fossil mink from Beringia is from Fairbanks Creek, Alaska (Youngman 1993). Youngman (1975) proposed that morphological divergence between *M. v. ingens* and other North American mink was due to isolation in Beringia.

American mink, *Neovison vison* (B. Hines)

Family PROCYONIDAE Gray, 1825—raccoons

The procyonid family comprises six genera and fourteen species (Wozencraft 2005). Most species of this exclusively New World family are found in tropical areas, with only three occurring in North America north of Mexico (Wilson and Ruff 1999). One species, *Procyon lotor*, occurs as an island exotic in southeast Alaska.

Procyonids are moderate-sized, primarily tree-living carnivores that probably share a common ancestry with the canids (Kurtén and Anderson 1980). The earliest fossil records of this family date from at least the Early Oligocene in North America and Europe (Stains 1984).

Raccoon

Procyon lotor
(Linnaeus, 1758)

Other Common Names
None.

Systematics
The source for extant raccoon populations in southeast Alaska is Indiana (Burris and McKnight 1973), where the nominate subspecies *P. l. lotor* occurs (Hall 1981).

Global Distribution
Nearctic—Raccoons are found from southern Canada to Panama, and have been introduced to numerous localities across Eurasia (Lotze and Anderson 1979)

Alaska Distribution
Raccoons are not native to Alaska. The closest indigenous populations are found at Kingcome Inlet in British Columbia (Nagorsen 1990). Raccoons also were introduced to Haida Gwaii (Queen Charlotte Islands) in the early 1940s, and colonized the two major islands and at least thirty-five smaller islands of the archipelago (Hartman and Eastman 1999).

Southeast Region. Eight melanistic raccoons from Indiana were released by private individuals on Singa Island, Sea Otter Sound, in October 1941, spreading to nearby El Capitan and several other islands in this area (Burris and McKnight 1973; Scheffer 1947). Melanistic raccoons have been sighted and likely still occur on El Capitan Island. In 1999, they were sighted near Staney Creek, Prince of Wales Island, and in the Shakan Strait area, Kosciusko Island (MacDonald and Cook 2007).

In 1950, raccoons of unknown origin were released or escaped on Japonski Island near Sitka, with a few eventually spreading to nearby Baranof Island. Individuals were occasionally seen around the garbage dump at the Sitka airport on Japonski Island up until the early 1970s, when the airport was extended and the dump covered. None has been reported in the Sitka area since (MacDonald and Cook 1996).

Southwest Region. Sometime prior to 1936, fur farmers released raccoons from several Midwestern states to Long Island, near Kodiak. There were reports of raccoons on the island in 1936 (Murie 1959) and in 1948 (Manville and Young 1965). No raccoons or their sign have been reported from anywhere in the Kodiak Archipelago since about 1990 and are now presumed extirpated (L. van Daele, ADFG, pers. comm., 2006).

Habitat
Raccoons are found in woody areas along streams, lakes, and marine coastlines.

Status
Raccoons are believed extirpated from near Kodiak and Sitka, but are still present in unknown numbers on islands in Sea Otter Sound. Since their introduction to Haida Gwaii, British Columbia, these nonnative predators have spread to many islands in the archipelago where they are heavily impacting colonies of seabirds and other ground-dwelling species (Harfenist et al. 2000; Hartman and Eastman 1999). This species should be actively managed in Alaska.

Fossils
Remains of this species have been found in Upper Pleistocene deposits in the southeastern United States (Kurtén and Anderson 1980).

Specimen Records
1 ("Alaska" 1950, LSUMZ).

Foraging raccoon, *Procyon lotor* (http://itech.pjc.edu).

CARNIVORA

Order ARTIODACTYLA Owen, 1848

The Artiodactyla, or even-toed ungulates, comprise 10 families, 89 genera, and 240 species world-wide (Grubb 2005). The largest and most diverse family is Bovidae (143 species in 50 genera), followed by Cervidae (51 species, 19 genera). In Alaska, the ungulate fauna is eight species in eight genera and two families.

This order includes eighteen extinct families, with the first appearing in the fossil record in the Early Eocene, about fifty-four million years ago (Nowak 1991). By the start of the Pliocene, some seven million years before present, all modern families are represented (Shackleton 1999).

Family CERVIDAE Goldfuss, 1820—antlered ungulates

Members of the deer family occur almost worldwide and in a variety of habitats from forests to deserts and tundra. Deer were introduced by humans to many other places where the family does not naturally occur. The cervids of Alaska consist of four genera and species in three subfamilies (Grubb 2005). Wapiti or elk, *Cervus canadensis,* were reintroduced to Alaska from nonnative stocks while Sitka black-tailed deer, *Odocoileus hemionus sitkensis*, were transplanted from the southeast region of the state to sites further north along the coast.

Fossil remains of North American cervids date to at least the Early Miocene (Simpson 1984).

In North America, male cervids, and in the case of caribou the females also, are distinguished by antlers that are replaced each year. All possess long, thin legs, small feet, and limb joints specialized for speed to escape predators. Cervids have long skulls that lack upper incisors. They have facial glands and lack a gallbladder. The stomach is four-chambered and ruminating.

Moose, *Alces americanus* (W. D. Berry)

Moose

Alces americanus
(Clinton, 1822)

Other Common Names

American moose, elk (Eurasia).

Systematics

Differences in karyotype and anatomical features (Boyeskorov 1999) suggested that moose found from the Yenisei River in central Siberia eastward into North America are specifically distinct from *Alces alces* of western Eurasia, a view further developed by Grubb (2005).

Four subspecies have been recognized in North America (Hall 1981); two are found in Alaska. Little variation was detected in North American moose by Cronin (1992). In contrast, Hundertmark et al. (2003) examined the phylogeography of North American moose based on mtDNA variation and found evidence of restricted gene flow among regional populations in the past, consistent with the distribution of four subspecies of moose in North America.

Alces americanus andersoni

Original Description. 1950. *Alces americana andersoni* Peterson, *Royal Ontario Mus. Zool.*, Life Sci. Occas. Pap., 9:1, May 25.

Type Locality. Sec. 27, T. 10. R. 16, Sprucewood Forest Reserve, 15 mi. E. Brandon, Manitoba.

Type Specimen. ROM 20068.

Range. Eastern Yukon Territory, central British Columbia, and southeast Alaska, eastward to Michigan.

Remarks. This subspecies makes contact with *gigas* on the mainland in the vicinity of the Taku River (Hundertmark et al. 2006; Klein 1965).

Alces americanus gigas

Original Description. 1899. *Alces gigas* Miller, *Proc. Biol. Soc. Washington*, 13:57, May 29.

Type Locality. North side of Tustumena Lake, Kenai Peninsula, Alaska.

Type Specimen. USNM 86166.

Range. Alaska, Yukon Territory, and northwestern British Columbia.

Global Distribution

Holarctic—*Alces americanus* is broadly distributed across the boreal regions of North America and central Siberia east of the Yenisei River (Boyeskorov 1999; Grubb 2005).

Alaska Distribution

Moose occur across most of Alaska and in recent years have been expanding their range (map 79).

Southeast Region. Moose began moving into southeast Alaska from interior British Columbia via transcoastal river corridors around the turn of the twentieth century (Klein 1965), with many of the populations becoming established in the early to mid-1900s and, in the case of the Gustavus Forelands, not until the 1960s (Barten

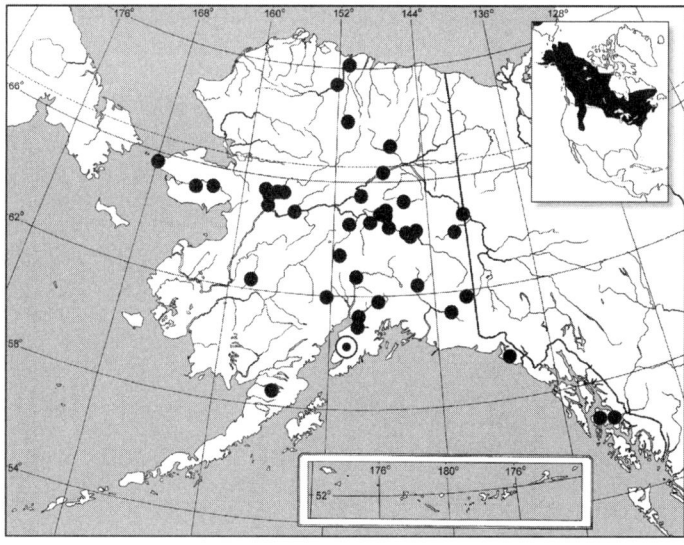

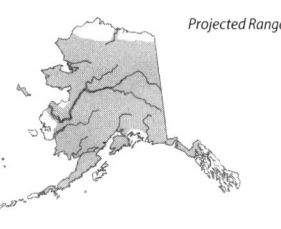

MAP 79. Distribution of moose, *Alces americanus*.

Projected Range

⊙ = Type Locality
● = Specimen Record (n = 249)

2004b). Moose continue to expand their range in southeast Alaska and coastal British Columbia (Darimont et al. 2005).

The moose population in southern southeast Alaska is localized in the Unuk River drainage. Moose were introduced to the Chickamin River drainage in 1963–1964 but none has been seen there in recent years (Porter 2004a). Moose are occasionally reported from Revillagigedo Island, Cleveland Peninsula, along the south end of the mainland near Portland Canal, and Hyder (MacDonald and Cook 2007; Porter 2004a). A moose population of unknown size and composition now inhabits the central portion of Prince of Wales Island (Porter 2004a).

Moose populations found along the central mainland are concentrated near Thomas Bay and along the Stikine River. Small numbers also occur around Virginia Lake, Mill Creek, and Aaron Creek (Lowell 2004a).

Farther up the coast, moose occur along the Taku, Whiting, and Speel rivers. In recent years, moose and their sign have been seen regularly in the Port Houghton area (Barten 2004b). Moose in the Berners Bay area are the result of transplants there in 1958 and 1960 (Burris and McKnight 1973). Moose were first documented west of Lynn Canal in 1962 on the Bartlett River, in the Chilkat Mountain range in 1963, the Endicott River and St. James Bay in 1965, and in the Gustavus Forelands in 1968. These animals probably originated from the Chilkat Valley population near Haines, where moose were first seen around 1930 (Barten 2004b). Moose were first documented along the lower Alsek River in the late 1920s or early 1930s, and slowly expanded their range westward into the Yakutat and Malaspina forelands to the glaciers of Icy Bay (Barten 2004c).

Moose have also expanded their range into the central Alexander Archipelago and now appear to be well distributed on Wrangell, Mitkof, and Kupreanof islands. Their numbers also appear to be increasing on Etolin, Zarembo, and Kuiu islands (Lowell 2004b).

Moose have been seen swimming Icy Strait, including one that was preyed on by a killer whale. They have been reported from Chichagof Island (MacDonald and Cook 1996).

Southcentral Region. Moose populations along most of the North Gulf Coast are from translocations of calves from the Kenai Peninsula, Anchorage, and Matanuska-Susitna area to the Copper River Delta during 1949–1958 (Crowley 2004b). Eastward expansion resulted in the arrival of moose to the Bering River area by the late 1960s, and to Cape Yakataga by the mid-1970s. Small populations are native only in the Lowe River drainage and Kings Bay (Crowley 2004b). Moose populations on the Kenai Peninsula, and particularly the Kenai National Moose Range, have benefited from numerous fires, beginning in the 1920s (most human caused), and direct habitat manipulations (ADFG 1973). Moose are present in suitable habitat throughout the interior portions of this region. Moose calves translocated to Kalgin Island from mainland stocks during 1957–1959 (Burris and McKnight 1973) increased to high enough numbers by the early 1980s to cause severe degradation of habitat (Del Frate 2004).

Central Region. Moose are found throughout central Alaska.

Southwest Region. Moose were scarce on the Alaska Peninsula before the mid-1900s, but they increased dramatically and spread southwest during the 1950s and 1960s. The scarcity of suitable habitat south of Port Moller limits further expansion (Butler 2004b). Moose are also relatively new inhabitants in the Bristol Bay area, possibly migrating into the area from middle Kuskokwim River drainages during the last century. Moose are now common along the Nushagak/Mulchatna rivers and all of their major tributaries. They are also found throughout the Wood/Tikchik Lakes area. Moose have successfully extended their range westward into the Togiak and Kulukak river drainages, where a viable population has become established (Woolington 2004b).

Western Region. According to ADFG (1973), moose occur in low numbers on the Yukon-Kuskokwim Delta and other tundra areas of western Alaska. Moose are thought to have begun migrating to the Yukon-Kuskokwim Delta during the mid- to late 1940s. The Yukon population occupies most of the available riparian habitat and the population is growing. The Kuskokwim population is small and is still in the process of colonizing the available riparian habitat.

Most of the Yukon-Kuskokwim Delta is not suitable as winter habitat for moose (Seavoy 2004b). Since the 1940s and 1950s, this species has expanded its range and numbers on the Seward Peninsula, and by the late 1960s much of the suitable habitat contained moose. Moose numbers grew rapidly in the 1960s through the early 1980s and peaked in the mid-1980s; these populations are currently declining (Persons 2004). Moose began to colonize northwestern Alaska during the 1920s and expanded their range to the Chukchi Sea coast by the mid- to late 1940s (Dau 2004b).

Northern Region. Since about 1940, moose populations have increased in size and have become well established in northern Alaska (Chesemore 1968b). Nearly all moose are confined to riparian habitat along river corridors during winter. During summer, many moose move into small tributaries and hills surrounding riparian habitat, and some disperse as far as the foothills of the Brooks Range and across the coastal plain. The largest winter concentrations of moose are found in the inland portions of the Colville River drainage (Carroll 2004b).

Habitat

Moose are associated with a wide variety of forest, shrub (particularly willow), and wetland habitats at various elevations. In Alaska, moose traditionally move between mountains and adjoining lowlands on a seasonal basis (ADFG 1973; Franzmann 1981; Peterson 1955). Wildfire plays an important role in providing this animal with forage (Shackleton 1999).

Status

Moose are one of Alaska's abundant large mammals. Their populations, which naturally fluctuate over time, are currently in healthy condition and, in some areas, expanding. Habitat is considered the primary limiting factor of moose because they need large quantities of forage (Franzmann 2000). Alaska's moose may have reached peak levels in the 1960s (Rearden 1981). The number of Alaska's moose in the early 1990s was estimated between 144,000 and 166,000 animals (Franzmann 2000).

Widespread spruce bark beetle (*Dendroctonus rufipennis*) infestations commencing in the 1990s have impacted more than 500,000 ha of spruce forests on the Kenai Peninsula (http://www.borough.

kenai.ak.us/sprucebeetle). Since 2001, infestation rates are decreasing as the number of unaffected trees becomes scarce (USDA et al. 2002). Spruce mortality and salvage logging efforts will affect the quality of moose habitat on a large scale, but the nature of the effect remains uncertain (McDonough 2004).

Fossils

The extinct species *A. latifrons* was the largest known cervid from Middle to Late Pleistocene in northern Eurasia, Alaska, and Yukon Territory (Franzmann 1981). The extant species was first recorded from Riss glaciation deposits in Europe and from Late Pleistocene in Siberia (Kurtén 1968), reaching eastern Beringia during the Late Wisconsinan in Alaska and spreading southward into North America in the very Late Wisconsinan–Early Holocene (Kurtén and Anderson 1980). Recent genetic data suggest that moose appeared in North America about fifteen thousand years ago (Hundertmark et al. 2002). Moose were one of the least common ungulates in the fossil assemblage of interior Alaska during the Late Pleistocene (Guthrie 1968b).

Wapiti

Cervus canadensis
(Erxleben, 1777)

Other Common Names

Elk, eastern red deer.

Systematics

Many authors consider all New World and Old World elk or red deer to be a single species, *C. elaphus* (Bryant and Maser 1982; Grubb 2005). Geist (1998) recommended that the elk from North America and eastern Asia be recognized as specifically distinct from the red deer, a conclusion supported by the genetic studies of Randi et al. (2001) and Baker et al. (2003). An analyses of mtDNA (cytochrome *b*) sequences (Ludt et al. 2004) of red deer from across their range lent strong support to the existence of two species with three subspecies in Asia and North America (wapiti or eastern red deer), four subspecies in Eurasia (red deer or western red deer), and an additional one or two subspecies in Central Asia (Tarim group). These data also point to a Central Asia origin of the clade.

Up to six subspecies have been described in North America (Bryant and Maser 1982).

There has been no comprehensive study of geographic variation throughout the entire North American range of this species and the division of *C. elaphus* into subspecies has been questioned (Groves and Grubb 1987). Cronin (1992) found little variation in mtDNA among populations in Mongolia, Alberta, California, Montana, Colorado, Oregon, British Columbia, and Washington. Cronin (1992) speculated that low variation in mtDNA among North American populations and subspecies may reflect founder effects following colonization south from Beringia in the Late Pleistocene. Taxonomy of existing North American wapiti has been complicated by historical extinction and translocations (Polziehn et al. 1998). Wapiti translocated to Alaska were from *C. c. nelson* and *C. c. roosevelti* stocks (Burris and McKnight 1973).

Cervus canadensis nelsoni

Original Description. 1935. *Cervus canadensis nelsoni* V. Bailey, *Proc. Biol. Soc. Washington*, 38:188, November 15.

Type Locality. Yellowstone National Park, Wyoming.

Type Specimen. USNM 49722/124656.

Range. Originally extended along the Rocky Mountains from northern New Mexico to northern British Columbia and Alberta.

Cervus canadensis roosevelti

Original Description. 1897. *Cervus roosevelti* Merriam, *Proc. Biol. Soc. Washington*, 11:272, December 17.

Type Locality. Mt. Elain, on ridge between heads of Hoh, Elwha, and Soleduc rivers, near Mt. Olympus, Mason Co., Washington.

Type Specimen. USNM 91579.

Range. Originally extended from the San Francisco area of California north to southwestern British Columbia.

Global Distribution

Holarctic—Wapiti or eastern red deer occur across parts of eastern Eurasia and North America. The original North American range has been greatly reduced (Bryant and Maser 1982).

Alaska Distribution

Wapiti inhabited central Alaska well into the Holocene (Guthrie 1966). The two populations that occur in the state today (map 80) are from introductions within the last eighty years.

Southeast Region. There have been a number of attempts to introduce wapiti to southeast Alaska (Burris and McKnight 1973), beginning in 1926 and 1927 with the release of seven animals (from the state of Washington) on Kruzof Island. These did not persist.

Three attempts were made to introduce wapiti to Revillagigedo Island, the first in 1937 (Washington stock), then again in 1963 and 1964 (from the Afognak Island herd, which was originally from Washington). Animals were also released on Gravina Island in 1962, and on Annette Island in 1963, both from Afognak or

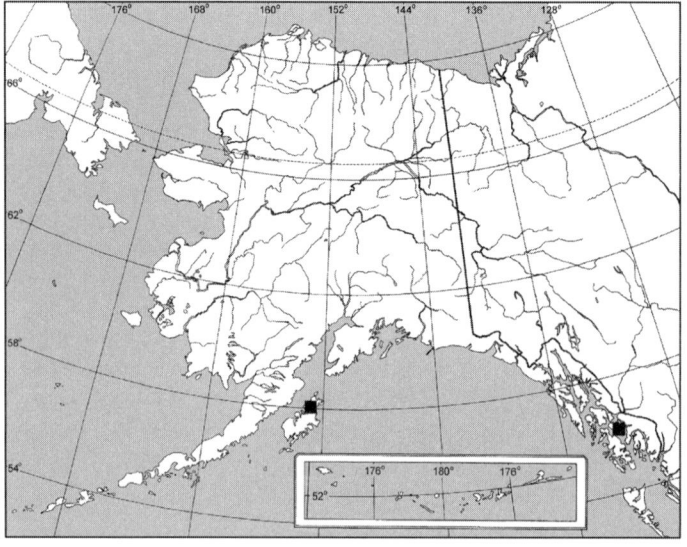

MAP 80. Distribution of wapiti, *Cervus canadensis*.

■ = Non-native Specimen Record (n = 2): 58.06667, –152.83333, Afognak Quad., Afognak Island, mouth of Afognak River (1 UAM); 56.05833, –132.58333, Petersburg Quad., Etolin Island, Rocky Bay (1 UAM)

Raspberry island stocks; like all of the previous attempts, it failed.

In 1987, fifty wapiti from Oregon were released on Etolin Island (Burris and McKnight 1973). Since then the Etolin population has continued to increase and extend its range by establishing a breeding population on nearby Zarembo Island. By June 2003 the number of wapiti on both of these islands was estimated at 350–450 animals (Lowell 2004c). Wapiti sightings have been reported from Bushy, Deer, Kupreanof, Mitkof, Prince of Wales, and Wrangell islands and the Cleveland Peninsula, raising concerns about potential negative effects that the increasing wapiti population may have on the native Sitka black-tailed deer (Kirchhoff and Larsen 1998; Lowell 2004c).

Southwest Region. Eight wapiti calves from Washington were released on Afognak Island in 1929. Animals subsequently spread to Raspberry Island. Overall population numbers reached a peak of approximately 1,200 to 1,500 animals by 1965 and subsequently underwent a sharp decline to 450 animals by 1972 (Burris and McKnight 1973). Throughout most of the 1990s the population had recovered to a minimum of 1,400 animals, but winter mortality during the late 1990s reduced overall numbers to 740 to 860 animals by 2003 (van Daele and Crye 2004a).

Wapiti have been reported on Shuyak and Whale islands. In 2002, there were unverified reports of wapiti on the northern coast of Kodiak Island, suggesting that a small number of wapiti may now reside on the island (van Daele and Crye 2004a).

Habitat

Across its broad range, *C. canadensis* utilizes a wide variety of habitats in both lowlands and mountains (Nowak 1991). The biology of wapiti introduced to Alaska is essentially the same as that of naturally occurring populations. On Afognak Island, wapiti forage on a broad variety of grasses, sedges, forbs, and shrubs. In winter months, they feed almost exclusively on browse, with elderberry (*Sambucus*) and willow (*Salix*) the most important plants. Winter range on Afognak Island has been characterized by dense stands of Sitka spruce (*Picea sitchensis*) interspersed with extensive shrub thickets of alder (*Alnus*), elderberry, and willow (ADFG 1973).

Status

See regional accounts.

Fossils

Wapiti were members of the Pleistocene–Early Holocene fauna of Beringia (Guthrie 1966, 2006). Fossil records of wapiti south of the last continental ice sheets lack substantiation (Kurtén and Anderson 1980).

Mule Deer

Odocoileus hemionus
(Rafinesque, 1817)

Other Common Names

Sitka black-tailed deer, Sitka deer.

Systematics

Most authors place mule and black-tailed deer in a single species (Anderson and Wallmo 1984); however, Cronin (1991, 1992) surveyed mitochondrial variation in this species and reported a distinctive coastal form. From eight to eleven subspecies have been proposed (Cowan 1936, 1956; Hall 1981; Wallmo 1981); one occurs in Alaska and another has occurred sporadically.

The generic name *Dama* as employed by Hall (1981) was preoccupied (Grubb 1993; ICZN 1960).

Odocoileus hemionus sitkensis

Original Description. 1898. *Odocoileus columbianus sitkensis* Merriam, *Proc. Biol. Soc. Washington*, 12:100, April 30.
Type Locality. Sitka, Alaska.
Type Specimen. USNM 74383.
Range. Southeast Alaska and northern coastal British Columbia.

Odocoileus hemionus hemionus

Original Description. 1817. *Cervus hemionus* Rafinesque, *Amer. Monthly Mag.*, 1:436, October.
Type Locality. Sioux River, South Dakota.
Type Specimen. None designated.
Range. Yukon Territory and sporadically in east-central and southeast Alaska (see below), southward to New Mexico.

Global Distribution

Nearctic—The mule deer ranges across western North America from Yukon Territory and southeast Alaska to Mexico (Anderson and Wallmo 1984).

Alaska Distribution

Indigenous populations of mule deer in Alaska are confined to the Alexander Archipelago in the southeast region of the state; they have been introduced to the Yakutat area, Prince William Sound, and the Kodiak Archipelago (map 81).

Southeast Region. Sitka black-tailed deer are found throughout most of southeast Alaska. They are most numerous on islands in the Alexander Archipelago and seem to have little trouble crossing from island to island even over wide expanses of coastal waters. As a result, they are found on nearly every island in the archipelago except remote Forrester Island (MacDonald and Cook 1996).

Deer are unknown along the coastal mainland northward from Cape Spencer except in the Yakutat area, where they were successfully transplanted to islands in Yakutat Bay in 1934 (from Rocky Pass stocks; Burris and McKnight 1973; MacDonald and Cook 2007). Other transplants included the Taiya Valley near Skagway in 1951, 1952, and 1956 (all unsuccessful), and on Sullivan Island, Lynn Canal, in 1951–1954 (animals still present). Seven female and six male deer were translocated to north Kupreanof Island in 1979 to increase (unsuccessfully) deer numbers there (Franzmann 1988).

The larger subspecies of mule deer, *O. h. hemionus*, from interior British Columbia has been reported from the Stikine River and near Hyder in 1991 (MacDonald and Cook 2007).

Southcentral Region. Deer from the Sitka area were moved to Hinchinbrook and Hawkins islands in Prince William Sound from 1917 through 1923. This effort proved successful and resulted in the spread of deer throughout the islands and, to a lesser extent, along the mainland of the sound (Burris and McKnight 1973). In subsequent years, there have been a number of reports of deer and tracks on the outer coast of the Kenai Peninsula west of Cape Fairchild (ADFG 1973), and at the head of Turnagain Arm as far west as Anchorage (Crowley 2005b). As early as 1972, but especially between 1980 and 1984, at least nineteen sightings of deer, presumably all *O. h. sitkensis*, were reported in the Copper River Basin north of Prince William Sound. These entered the basin either over Thompson Pass or through the Copper River canyon (Roberson 1986). The subsequent lack of sightings in the basin suggests these may be temporary dispersals and not permanent range extensions.

Central Region. There have been a number of reliable but unsubstantiated reports of mule deer, presumably *O. h. hemionus*, in east-central Alaska. During the late 1980s, deer sightings were reported in the Tanana River Valley near Tok, Salcha, and Fairbanks (*Fairbanks Daily News-Miner*, 15 March 1989, 23 October 1989). Hoefs (2001) assumed that deer crossed into Interior

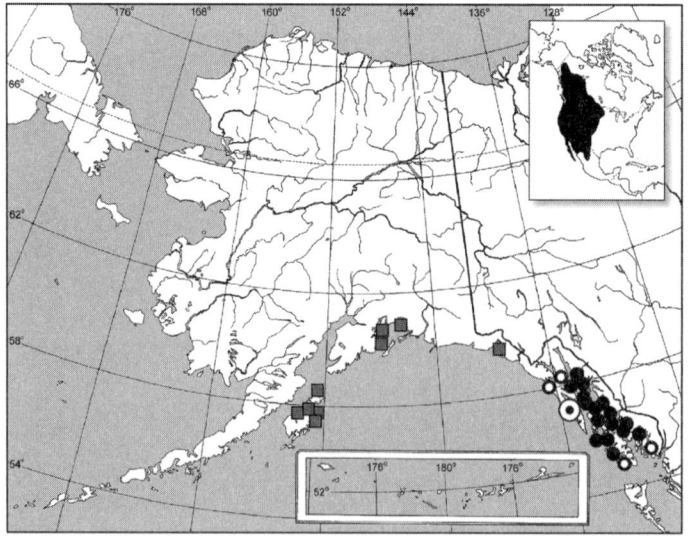

MAP 81. Distribution of mule deer, *Odocoileus hemionus*.

⊙ = Type Locality
● = Native, ■ = Non-native Specimen Record (n = 292)
○ = Marginal Record. Juneau Quad., Pleasant Island (4 CMNH); 54.9833, –130.9333, Prince Rupert Quad., Very Inlet (1 UAM); Dixon Entrance Quad., Long Island, Howkan (1 CAS).58.0000, 136.5000, Sitka Quad., Yakobi Island (3 AMNH).

Alaska from Yukon Territory in the late 1970s to early 1980s following the valley of the Ladue River.

Southwest Region. Van Daele (2005) suggests deer may have been introduced to the Kodiak Archipelago as early as 1900, but according to Burris and McKnight (1973), the deer population in the Kodiak Archipelago originated from three transplants. The first was in 1924, when fourteen deer from the Sitka area were released on Long Island. These were supplemented with two more from Prince of Wales Island in 1930. Nine deer captured in the Rocky Pass area west of Petersburg were released on Kodiak Island in 1934. By the late 1960s, deer had spread to adjacent Afognak Island. Deer have since been observed on Sitkinak Island and there is an unconfirmed report of deer tracks on Tugidak Island in the mid-1980s (ADFG 1995).

Habitat

Sitka black-tailed deer inhabit the wet rain forests of coastal Alaska. In summer, they can be found high in alpine meadows and subalpine shrub thickets. In fall and winter, they move down to lower elevations and beaches to forage. Critical winter range for these far northern populations of deer includes adequate stands of uneven-aged old-growth forests that are more than two hundred years old (Schoen and Kirchhoff 2007; Suring et al. 1992; Wallmo 1981).

Status

Deer populations along coastal Alaska fluctuate widely depending upon the severity of winters (ADFG 1973). The most serious problem for deer is the permanent loss of quality winter range due to clearcut logging (Merriam et al. 1994; Suring et al. 1992). In addition, Farmer et al. (2006) demonstrated that important factors increasing mortality of deer in the Alexander Archipelago included use of second-growth forest, level terrains, muskegs, young clearcuts, and landscapes that were accessible by roads.

Fossils

The earliest fossil remains of *Odocoileus* in North America are mid- to Late Pliocene. Most remains of *O. hemionus* are Rancholabrean in age, and are from scattered sites located south of the Wisconsin glacial ice (Kurtén and Anderson 1980). Skeletal material of deer recovered from cave deposits in southeast Alaska have all been from Holocene deposits (Heaton and Grady 1993, 2003).

Caribou

Rangifer tarandus
(Linnaeus, 1758)

Other Common Names

Reindeer (Eurasia and domesticated strains).

Systematics

Taxonomy of caribou and reindeer has been problematic. Banfield (1961) recognized one holarctic species for both caribou and reindeer, *Rangifer tarandus*. Moreover, he reduced the number of named forms in North America to six subspecies; one, *R. t. granti*, was recognized for all Alaska and Yukon populations (as *R. t. groenlandicus* by Youngman 1975). Hall (1981) recognized an additional subspecies, *R. t. caribou*, from east-central, southcentral, and, hypothetically, southeast Alaska. Feral reindeer of Eurasian origin are considered *R. t. sibiricus* by Grubb (2005) and others.

Cronin (1992) assessed intraspecific variation in mtDNA with restriction enzymes to determine relationships among populations and subspecies. He found considerable geographic variation, although subspecies were not discernible as distinct mtDNA assemblages. He speculated that the high level of variation in mtDNA of Alaska caribou (seven genotypes among fourteen individuals) may reflect large historical populations in Beringia during the Pleistocene. Alternatively, feral reindeer may have increased genetic diversity, although allele frequencies of serum transferrin found by Røed and Whitten (1986) indicated minimal genetic introgression between reindeer and caribou in Alaska.

We tentatively follow Banfield (1961) and modifications by Geist (1998) in recognizing a single species, but an assessment of geographic variation is needed using modern techniques and larger sample sizes. Grubb (2005) provided an overview of currently recognized subspecies.

Rangifer tarandus caribou

Original Description. 1788. [*Cervus tarandus*] *caribou* Gremlin, *Syst. nat.*, 13th ed., 1:177, February 7.

Type Locality. Quebec City, Quebec.

Type Specimen. None designated.

Range. Extreme east-central Alaska and southern Yukon Territory to Newfoundland.

Remarks. According to Hall (1981) includes *Rangifer stonei* J. A. Allen, *Bull. Amer. Mus.*

Nat. Hist., 14:143, 28 May 1901, type from Kenai Peninsula, Alaska (AMNH 17609).

Rangifer tarandus granti
Original Description. 1902. *Rangifer granti* J. A. Allen, *Bull. Amer. Mus. Nat. Hist.*, 16:122, March 31.

Type Locality. West end of Alaska Peninsula, opposite Popof Island, Alaska.

Type Specimen. AMNH 17593.

Range. Alaska and Yukon Territory to extreme northwestern Northwest Territories.

Rangifer tarandus sibiricus
Original Description. 1866. *Rangifer tarandus sibiricus* Murray, *Geographical distribution of mammals*, p. 334 (description and figure on pp. 153, 155).

Type Locality. Siberia. According to Hollister (Smithsonian Misc. coll56(35):7, 1912), West Siberia, near Beresov.

Type Specimen. Not known to exist.

Range. Siberian and east European tundra zone; islands of Asiatic arctic seas

Remarks. Flerov (1933) included *asiaticus*; Grubb (2005) also included *chukchenis*, *lenensis*, *taimyrensis*, and *yakutskensis*. Semidomesticated and sometimes feral reindeer were first introduced into northwestern Alaska in 1891 and at various times until 1902 from the Chukchi Peninsula and Gulf of Anadir. Reindeer from the Buckland (Seward Peninsula)

herds considered to be largely descended from the Anadir stock were driven across northern Alaska in 1933 to northwestern Canada and the Mackenzie River Delta, Richards Island, spreading later to lower Anderson River (Anderson 1946).

Global Distribution
Holarctic—Caribou (reindeer) are circumpolar and once ranged across the arctic tundra and boreal forest zones of Eurasia and North America. Most of the world's wild caribou herds are in Alaska and Canada (ADFG 1973).

Alaska Distribution
Caribou range over many areas of Alaska (map 82). Thirty-one relatively discrete herds, or populations, have been identified (Fig. 20). Herds are recognized if the females in a general area have a distinct calving ground (Bergerud 2000).

Southeast Region. Caribou rarely occur on the northern mainland of southeast Alaska, with reports of single animals observed near Haines (in about 1990) and Glacier Bay (in the late 1950s and in 1967) (MacDonald and Cook 1996). Caribou may have regularly inhabited the Haines area in the past (Murie 1935).

Southcentral Region. Caribou formally described as a separate species, *R. stonei*, by J. A. Allen in 1901 were extirpated from the

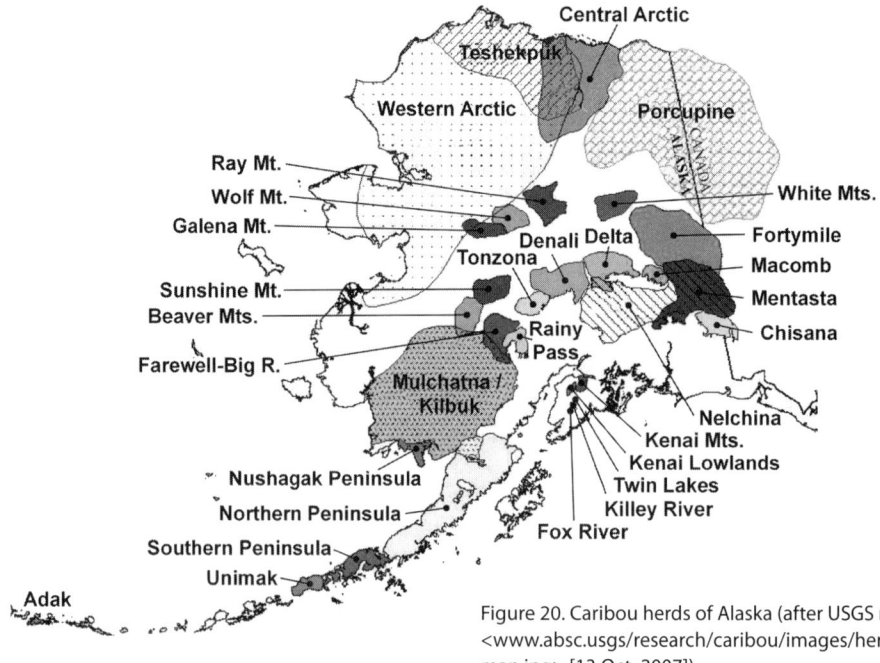

Figure 20. Caribou herds of Alaska (after USGS map <www.absc.usgs/research/caribou/images/herd.map.jpg> [13 Oct. 2007]).

Kenai Peninsula by about 1913, apparently due to overhunting and habitat loss from extensive fires (Spencer and Hakala 1964). Between 1965 and 1986, caribou from the Nelchina herd were successfully introduced back onto the peninsula (Burris and McKnight 1973; Franzmann 1988). The Mentasta, Chisana, and Nelchina portions of the Fortymile herd and the McKinley herd occur in southcentral Alaska (ADFG 1973).

Central Region. The central region of Alaska is permanent or periodic home to the Beaver, Mulchatna, Delta, Fortymile, McKinley, Porcupine, and Arctic caribou herds (ADFG 1973).

Southwest Region. According to ADFG (1973), the Alaska Peninsula herd extends from Naknek Lake southwest to Unimak Island. Three subpopulations have been defined: one north of Port Moller, one south of Port Moller, and one ephemeral population on Unimak Island, which Butler (2005) considered a separate herd based on fidelity to calving grounds on the island and patterns of genetic variation. In the mid- to late 1800s, caribou were considered abundant on coastal areas near Bristol Bay. Domestic reindeer were introduced to the Alaska Peninsula in 1932, but the industry lasted only a few years. Because of overgrazing and other factors, caribou and reindeer numbers reached a low level during the 1940s and the few remaining reindeer herds were abandoned at this time.

Twenty-three caribou calves were released on Adak Island in 1958 and 1959. Adak caribou now range over the entire island (ADFG 1973; Burris and McKnight 1973), and between 1998 and 2005 experienced a three-fold increase to over twenty-seven hundred animals (Williams and Tutiakoff 2005).

Western Region. Caribou were present in large numbers throughout the Yukon-Kuskokwim area of western Alaska in the mid-1800s and apparently began to decline during the 1870s. At its peak, the herd probably ranged from the Alaska Peninsula to the Seward Peninsula along the Bering Sea. Caribou were also numerous on Nunivak Island. Caribou essentially disappeared from the entire region by 1895. Primary factors associated with their decline and disappearance included uncontrolled commercial hunting and degradation of wild caribou range by domestic reindeer (ADFG 1973; Murie 1935). Twenty-nine reindeer from Nunivak Island were released on St. Matthew Island in 1944 (Klein 1959). Their numbers had increased to at least four hundred to five hundred animals by 1954, when it became evident that winter range was being severely degraded. The population peaked at about six thousand animals in the summer of 1963 and then crashed the following winter. Only about forty animals remained on the island by the summer of 1966 (Klein 1968). The majority of the herd died

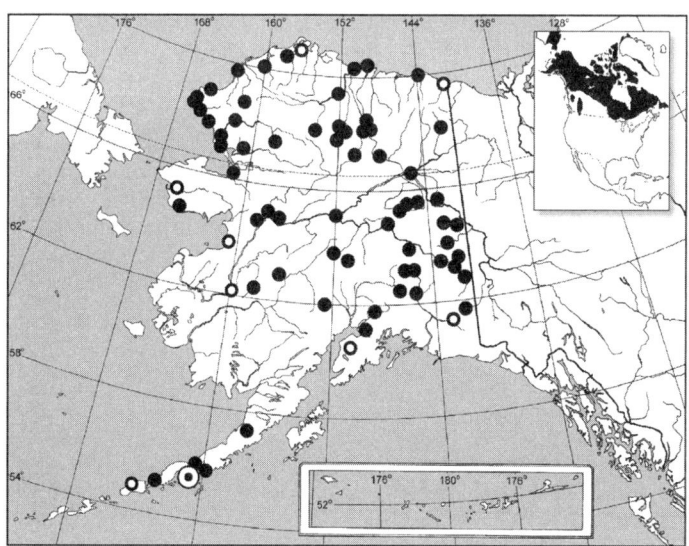

MAP 82. Distribution of caribou, *Rangifer tarandus*.

⊙ = Type Locality
● = Specimen Record (n = 1319)
○ = Marginal Records. Marginal Record. Mead River Quad., 25 mi. S of Pt. Barrow (2 USNM); Demarcation Pt Quad., Demarcation Point (2 AMNH); McCarthy Quad., McCarthy (1 USNM); Kenai Quad., Kenai Peninsula, Game Mgt. Unit 15 (1 FMNH); Unimak Quad., Unimak Island (38 USNM); Holy Cross Quad., Holy Cross (1 USNM); Unalakleet Quad., Unalakleet (2 USNM); 65.33330000, −166.41670000, Teller Quad., Teller Mission (1 UAM).

during the severe winter of 1963–1964. By the 1980s, the St. Matthews Island herd had completely died out. Reindeer introduced on the Pribilof Islands are owned by the Aleut Corporation.

Northern Region. The Arctic herd occupies much of northwestern Alaska, and historically may have ranged into the western region well south of the Seward Peninsula (ADFG 1973). In the early 1800s, caribou were apparently abundant in coastal areas, including north and east to the Colville River. By the early 1880s, few caribou were found in northwestern Alaska except on the north coast. Not until the late 1930s was it apparent that a distinct northwestern calving area had formed along the upper Utukok and Colville rivers. The Arctic herd now usually winters south of the Brooks Range, often as far south as the base of the Seward Peninsula, the middle Koyukuk, and the Ray Mountains, and as far east as the Wiseman area, where overlap sometimes occurs with the Porcupine herd. Some also winter on the North Slope in coastal areas. The Porcupine herd ranges over the eastern half of northern Alaska, parts of north-central Alaska, and nearly to the Mackenzie River in Canada (ADFG 1973).

Habitat

Caribou are social and nomadic animals. The spring calving area is the traditional center of distribution of a caribou herd (ADFG 1973). Calving grounds are invariably above timberline or along the coast in windswept rolling hills or foothills dominated by heaths, sedges, and forbs. Summer ranges may take caribou high into the mountains to seek relief from biting insects and to forage in alpine tundra habitats. With first snows, they begin to seek lower elevations and may travel long distances to find adequate supplies of lichens, sedges, and browse plants in a variety of boreal forest and tundra habitats. Different herds occasionally mix on winter ranges (Valkenburg 1999).

Status

Through protection and limited hunting, most caribou populations in Alaska are in healthy condition. Alaska caribou numbers fluctuate and have grown substantially from a low of about seventy-five thousand total animals in the state in the 1970s to almost one million by 1995 (Bergerud 2000). On the Seward Peninsula, the rapid growth of the wild caribou population has resulted in increased conflicts with the domestic reindeer industry (Revkin 2001).

The importation of domestic reindeer to Alaska first began as a "trial run" of sixteen animals to Amaknak Island in the Fox Islands, eastern Aleutians, in 1891, followed the next year with 171 animals (along with five Siberian herders) to Port Clarence on the Seward Peninsula (see Rausch and Baldwin 2002). Reindeer abundance in Alaska was highest in the early 1930s. About 640,000 reindeer are estimated to have existed then, with a precipitous decrease in the number beginning probably in 1932. In 1940, only 250,000 reindeer were estimated to exist. By 1950, only 25,000 reindeer were recorded. Reasons advanced for the decline were a combination of overstocked ranges, lack of care in herding, predation by wolves, and large losses to migrating caribou.

Since 1960, there has been increased interest in developing a native reindeer industry (Stern et al. 1980). Reindeer pastures include the entire Baldwin and Seward peninsulas, as well as permitted areas near Shaktoolik, Stebbins, and on St. Lawrence Island. There are also small herds in Palmer, Delta Junction, and the Kenai Peninsula. Reindeer herds currently occur on the islands of Nunivak, Umnak, St. Paul, and St. George. Reindeer were removed from Hagemeister Island in the early 1990s (Ebbert and Byrd 2002). There are an estimated 17,650 reindeer in fourteen herds in Alaska with about 10,000 on the Seward Peninsula (Jernsletten and Klokov 2002).

Fossils

According to Kurtén and Anderson (1980), the earliest records of *Rangifer* in North America are from Irvingtonian and Illinoian deposits in western and central Alaska, respectively. During the Wisconsinan, caribou also occurred south of the ice sheets. Banfield (1961) referred all Pleistocene specimens to the extant species, *R. tarandus*. Kurtén and Anderson (1980) suggested that *Rangifer* probably originated in Beringia or in the mountains of northeastern Asia. Caribou bones have been reported from middens on Kodiak Island (Kellogg 1936), from north of the Alaska Peninsula on Amak Island (Murie 1959), and from pre- and postglacial cave deposits in the Alexander Archipelago of southeast Alaska (Heaton and Grady 2003).

Family BOVIDAE Gray, 1821—horned ungulates

Classification of the Bovidae remains inconsistent and controversial, with much of the debate revolving around relationships at the higher taxonomic levels (e.g., Groves 1981; Janis and Scott 1987; Matthee and Davis 2001; Simpson1984).

The great majority of bovids are native to Africa and southern and central Asia. Four of the five species of North American bovids occur in Alaska, with two species, muskox and bison, reintroduced after their disappearance from the state in the nineteenth century.

The geological range of this family is Early Miocene to Recent in Europe and Africa, Middle Miocene to Recent in Asia, and Pleistocene to Recent in North America (Simpson 1984).

Bovids, unlike cervids, have true horns, which are a permanent bony outgrowth of the frontal bone covered by a horny sheath. Another difference is the presence of a gallbladder in most bovid species. Bovids are ruminants and possess a four-chambered stomach.

Dall's sheep, *Ovis dalli* (W. D. Berry)

American bison

Bison bison
(Linnaeus, 1758)

Other Common Names

American buffalo, plains bison, wood bison.

Systematics

Mitochondrial and ribosomal DNA analyses, together with reproductive, cranial, and other molecular data, strongly indicate that the genus *Bison* should be treated as a synonym of *Bos* rather than as a distinct genus in the tribe Bovini (Miyamoto et al. 1989; Wall et al. 1992). *Bison* includes two extant species, *B. bison*, the Late Holocene bison of North America, and *B. bonasus*, the European bison or wisent (Meagher 1986). Van Zyll de Jong (1986) presented evidence suggesting the two are conspecific.

Subspecies of North American bison are poorly defined. Two subspecies of *B. bison* have been recognized by most authors (Hall 1981; McDonald 1981; Skinner and Kaisen 1947) based on cranial measurements, particularly those of highly variable horn cores. Other skull and external characteristics, as well as genetic perspectives, do not, however, support a distinction between these two morphs (Bork et al. 1991; Polziehn et al. 1996; Shackleton 1999; Shackleton et al. 1975; Strobeck et al. 1993). Geist (1991) suggested that because differences between woodland and plains bison are environmentally induced, subspecific status is not warranted. Wilson and Strobeck (1999) investigated variability in eleven microsatellite loci of bison and concluded that wood bison were historically independent from plains bison. Grubb (2005) did not recognize any subspecies. Systematics of existing populations of *B. bison* have been complicated by admixture of herds (Reynolds et al. 1982; van Zyll de Jong 1986).

Bison bison athabascae

Original Description. 1898. *Bison bison athabascae* Rhoads, *Proc. Acad. Nat. Sci. Philadelphia*, 49:498, January 18.

Type Locality. Within 50 mi. SW Fort Resolution, Mackenzie.

Type Specimen. NMC 299.

Range. Recently occurred from central Alaska southeastward to Alberta.

Bison bison bison

Original Description. 1758. [*Bos*] *bison* Linnaeus, *Syst. Nat.*, 10th ed., 1:72.

Type Locality. Ancient "Quivira," central Kansas.

Type Specimen. None designated.

Range. Once found from Alberta to northern Mexico. Introduced to several localities in Alaska.

Global Distribution

Nearctic—The original range of *B. bison* covered much of the North American continent, including central Alaska (Hall 1981; Meagher 1986; Stephenson et al. 2001; van Zyll de Jong 1986).

Alaska Distribution

Wood bison (*B. b. athabascae*) persisted along the Tanana and upper Yukon river drainages of east-central Alaska into the 1900s (Stephenson et al. 2001) (map 83).

As a result of transplants, small herds of plains bison (*B. b. bison*) are today found in several areas within the state. Twenty bison from Montana were released in the central Tanana Valley near Delta Junction in 1928 and 1930. The herd peaked at about five hundred animals in the early 1940s, but leveled at less than three hundred animals by the 1970s. The Delta herd is currently managed to maintain a population size of approximately 360 animals (DuBois 2004).

The small herd of bison along the South Fork of the Kuskokwim near Farewell are the result of transplants from the Delta herd in 1965 and 1968. This herd reached a population size of 350 animals by 1999, and remained stable through 2003 (Boudreau 2004).

The Copper River herd, currently numbering 110 animals, originated from a translocation to the Nabesna Road in 1950 of 17 bison from Delta Junction. They moved and eventually settled along the Dadina and Chetaslina rivers (Tobey 2004a). The small Chitina River herd (fifty bison in 2003; Tobey 2004b) originated either as a result of dispersal from the Copper River herd or from a transplant effort of questionable success in 1962 of twenty-five Delta animals to May Creek (ADFG 1973).

A herd of free-ranging bison maintained by hunting at approximately one hundred animals occurs on Popof Island in the Shumagins. This feral herd was established in 1954 from a failed ranching venture by a private individual who shipped in five or six animals from Montana (Fall et al. 1993).

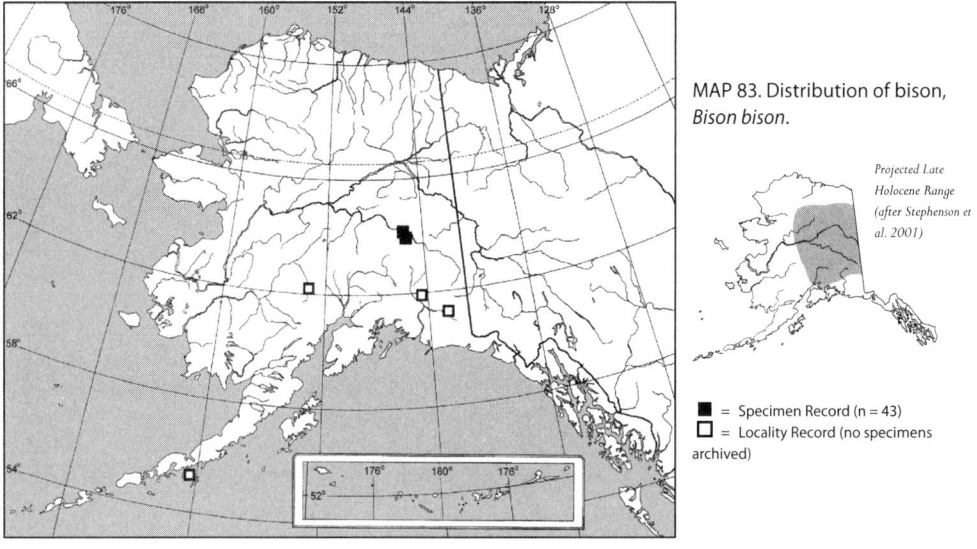

MAP 83. Distribution of bison, *Bison bison*.

Projected Late
Holocene Range
(after Stephenson et al. 2001)

■ = Specimen Record (n = 43)
☐ = Locality Record (no specimens archived)

Small domestic herds are located at Healy and near Kodiak (Griffin and Johnson 1994).

Habitat

Bison use a variety of open and forested habitats, grazing on grasses and sedges found in meadows and early successional habitats (Larter and Gates 1991). Historically, the wood bison was associated with northern forests and parklands, while the plains form inhabited the grasslands farther south. Introduced populations occur in floodplain sedge flats, near extensive river bar systems and deltas where rivers emerge from windy mountain passes, and to some extent on grassy habitats that are successional after forest fires (Guthrie 1990a, 2001a).

Fossils

Bison probably originated in Eurasia in the Late Pliocene and dispersed to North America by way of the Bering Land Bridge possibly during a transition from Irvingtonian to Rancholabrean fauna (Harington 1984; McDonald 1981). They first appeared in the fossil record of Beringia in the Pleistocene (Guthrie 1980; McDonald 1981). Large-horned forms similar to steppe bison (*B. priscus*) prevailed until near the end of the Pleistocene. Large-horned bison in Siberia, Alaska, and northwestern Canada were replaced or evolved into modern, small-horned bison, including the wood bison (*B. b. athabascae*), during the Late Pleistocene and first half of the Holocene. Bison were abundant on the Alaska landscape during most of the

last one hundred thousand years (Guthrie 1968b, 1990a), and wood bison were the last form to occupy Alaska and adjacent regions (Harington 1977; Skinner and Kaisen 1947; Stephenson et al. 2001; van Zyll de Jong 1986). Shapiro et al. (2004) used ancient DNA to reconstruct the biogeographic history of bison throughout the Late Pleistocene and Holocene. Their analyses portrayed a large diverse population living throughout Beringia until around 37,000 years before present, when changes associated with the onset of the last glacial cycle correlated with the timing of their decline.

Mountain Goat

Oreamnos americanus
(de Blainville, 1816)

Other Common Names

Rocky Mountain goat.

Systematics

Mountain goats are the only North American representatives of the Old World tribe Rupicaprini (goat-antelopes) (Rideout and Hoffmann 1975). Hall (1981) recognized four subspecies in North America; two occur in Alaska. Cowan and McCrory (1970), however, considered this a monotypic species and concluded that the small amount of geographic variation in cranial characters did not warrant subspecific designation. Few specimens have been available to adequately evaluate the taxonomy of this species (Shackleton 1999).

Oreamnos americanus columbiae

Original Description. 1912. *Oreamnos americanus columbiae* Hollister, *Proc. Biol. Soc. Washington*, 25:186, December 24.

Type Locality. Shesley Mountains, British Columbia.

Type Specimen. AMNH 19838.

Range. Northwestern Canada and southeast Alaska.

Remarks. A renaming of *Oreamnos montanus columbianus* J. A. Allen, 1904.

Oreamnos americanus kennedyi

Original Description. 1900. *Oreamnus* [sic] *kennedyi* Elliot, *Field Columb. Mus.*, Publ. 46, Zool. Ser., 3:3, June 20.

Type Locality. Mountains at mouth Copper River, opposite Kayak Island, Alaska.

Type Specimen. FMNH 19067.

Range. Southcentral Alaska.

Global Distribution

Nearctic—The mountain goat ranges in western North America from southcentral Alaska, Yukon Territory, and Northwest Territories southward through British Columbia to Montana and Idaho in the northern Rocky Mountains and Washington and Oregon in the Cascade Mountains (Nagorsen 1990).

Alaska Distribution

Mountain goats inhabit the southern mountain ranges of Alaska (map 84).

Southeast Region. Mountain goats are found in suitable habitat along the entire mainland coast of Southeast Alaska. The only island record of natural occurrence is an individual observed on Wrangell Island for several years (Klein 1965). The highest counts of mountain goats have been made in the vicinities of Tracy Arm and the Peabody Mountains, southeast of Ketchikan (ADFG 1973).

This species was translocated onto Baranof Island from Tracy Arm stocks in 1923 (Burris and McKnight 1973), where it continues to expand its range across the island (Mooney 2004). Transplant attempts on Chichagof Island in 1954 and 1955 were failures (Franzmann 1988; L. Johnson, pers. comm., 1994). A successful transplant of mountain goats to Revillagigedo Island occurred in 1983 at Swan Lake (Smith and Nichols 1984) and in 1991 at upper Mahoney Lake (Porter 2004b). The Swan Lake population now numbers about 120 to 160 animals, and the upper Mahoney Lake population is estimated at 100 to 140 animals and expanding (Porter 2004b).

Southcentral Region. Mountain goats range along the coastal mountains as far west as Cook Inlet and on the Kenai Peninsula. Natural populations also occur on Bainbridge, Culross, and Knight islands in western Prince William Sound (Crowley 2004c). The northern limit of their range extends into the Talkeetna Mountains and into

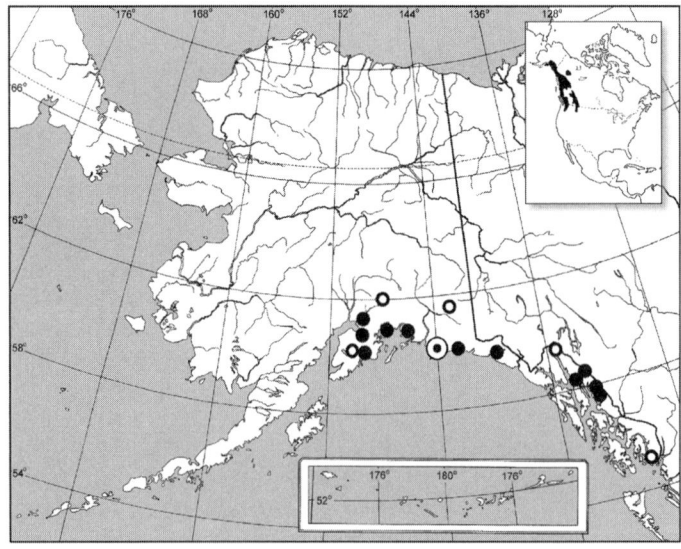

MAP 84. Distribution of mountain goat, *Oreamnos americanus*.

⊙ = Type Locality
● = Specimen Record (n = 130)
○ = Marginal Record. Talkeetna Mountains, near Tazlina Lake (1 UAM); McCarthy Quad., Wrangell Mountains, Nizina Glacier (2 USNM); Skagway Quad., Skagway, 20 mi. from (1 USNM); 55.1667, –130.6667, Ketchikan Quad., mainland, southeast of Ketchikan, Boca de Quadra (1 UAM); Kenai Quad., Kenai Peninsula, Twin Lake (1 PSM).

the southern drainages of the Wrangell Mountains (Coltrane 2004).

In 1983, two male and ten female goats were translocated from surrounding areas on the Kenai Peninsula to Cecil Rhode Mountain to restock a depleted population (Smith and Nichols 1984).

Southwest Region. This species was successfully introduced to Kodiak Island in the vicinity of Ugak Bay from Kenai Peninsula and Anchorage area stocks during 1952–1953. During the 1980s the population continued to increase from an estimated 150 to more than 400 animals, and new pockets of goats were observed on the southern end of the island. Goats now occupy most of the suitable habitat on Kodiak Island, and there are confirmed reports of goats as far south as Kaguyuk Bay and Akalura Lake. In 2002, the entire population was an estimated fourteen hundred goats and increasing (van Daele and Crye 2004b).

Habitat

Hjeljord (1973), Schoen and Kirchhoff (1981), and Fox et al. (1989) described habitat use by mountain goats in Alaska. Mountain goats, in contrast to mountain sheep (*Ovis*), seem to thrive in wet coastal climates. Their habitat is generally restricted to mountains, often in steep and rugged terrain. In summer, they most frequently are found in alpine meadows, moving up into the higher snowfields on warm days to rest on patches of snow. In winter, mountain goats can extend from sea level to timberline (ADFG 1973; Shackleton 1999).

Status

Populations of this species in Alaska appear relatively stable (ADFG 1973; Ballard 1977). The Talkeetna Mountains provide only marginal habitat for mountain goats, and the population there remains chronically low (Coltrane 2004). The mountain goat is a Management Indicator Species of the Tongass National Forest (MacDonald and Cook 2007).

Fossils

Fossils of *O. americanus* are rare. Remains from Sangamonian interglacial deposits (ninety thousand years ago) near Quesnel Forks, British Columbia, are the earliest record (Kurtén and Anderson 1980; Shackleton 1999). Fossil remains dating from about twelve thousand years ago have recently been recovered from cave deposits on northern Vancouver Island (Nagorsen and Keddie 2000). Fossils of the much smaller *O. harringtoni* from the

Quaternary have been found in deposits from Nevada and Mexico (Rideout and Hoffmann 1975). Harington (1971) believed that a species of *Oreamnos* entered North America from Eurasia before the Wisconsinan glaciation. None existed in eastern Beringia during the height of the Wisconsin glaciation (Cowan and McCrory 1970; Hoffmann and Taber 1967).

Muskox

Ovibos moschatus
(Zimmermann, 1780)

Other Common Names
None.

Systematics

The muskox is included in the subfamily Caprinae (Grubb 2005). A phylogenetic study based on cytochrome *b* sequences (Groves and Shields 1996, 1997) suggested a close relationship of *Ovibos* to the mountain goat (*Oreamnos*) of North America and to the goral and serow (*Naemorhedus*) of Asia, but did not support close kinship with the morphologically similar takin (*Budorcas*) of central Asia.

The genus *Ovibos* is monotypic. Three subspecies have been recognized (Allen 1913; Hall 1981); one occurs in Alaska. Tener (1965) concluded that the Hudson Bay muskox, *O. m. niphoecus*, was not a valid subspecies and that differences between *O. m. wardi* of the high Canadian Arctic and *O. m. moschatus* were "suggestive of incipient subspeciation." A recent study of intraspecific variation of mtDNA based on control region sequences from all three subspecies found little genetic diversity within the species and did not support subspecific designations (Groves 1997).

Ovibos moschatus moschatus
Original Description. 1780. *Bos moschatus* Zimmermann, *Geographische Geschichte . . .* , 2:86.

Type Locality. Between Seal and Churchill rivers, Manitoba.

Type Specimen. None designated.

Range. Northern Alaska and northern continental Canada.

Global Distribution

Nearctic (but see below)—Muskoxen historically ranged from northern Alaska to Hudson Bay, on the northern and western islands of the Canadian Arctic, and in Greenland (Allen

1913; Hall 1981; Lent 1988). The same species, or a close relative, was present in northern Eurasia during the Late Pleistocene and may have survived in Siberia until about two thousand years ago (Corbet 1978; Tener 1965). Remains aged at several thousand years are commonly found on the Taimyr Peninsula in northwestern Siberia, and *O. moschatus* may have persisted there until one hundred to two hundred years ago (Uspenski 1984).

Alaska Distribution

Muskoxen from the Beringian refugium (Harington 1970, 1977) persisted along the arctic coastal and foothills zones in Alaska until their extirpation in the late 1800s or early 1900s (Allen 1912; Lent 1998). Today there are several introduced populations in various parts of the state, including a reintroduction into their historic range in northeastern Alaska (map 85).

Reestablishment of muskoxen in Alaska started in the early 1930s (Burris and McKnight 1973; Klein 1988; Spencer and Lensink 1970). Thirty-one muskoxen originating from Greenland were transplanted to Nunivak Island in western Alaska during 1935–1936. From these stocks, animals were released on Nelson Island in 1967 and 1968. Some of these subsequently emigrated to inhabit the Yukon-Kuskokwim Delta. During the period 1969–1981, Nunivak animals were moved onto historic range in the Camden Bay area near Barter Island, to Kavik River on the western edge of the Arctic National Wildlife Refuge to the Seward Peninsula, and to Cape Thompson on the northwest Arctic coast.

In 1975, muskoxen from Nunivak Island and Banks Island, Canada, were taken to Russia and successfully introduced on the eastern Taimyr Peninsula and Wrangel Island (Uspenski 1984).

Since introduction, wandering bull muskoxen have shown up at localities as distant as the upper Yukon River above Circle, Nulato, Ruby, and Galena. In addition, a few bull muskoxen and some small groups have recently been sighted at the Gisasa, Kateel, and Hogatza rivers (Lenart 2005).

Efforts to domesticate muskoxen were initiated at the University of Alaska Fairbanks in the early 1950s. Some of this captive herd was moved to Unalakleet by 1977 (Smith 1984). Captive animals persist in Fairbanks and at the Muskox Farm in Palmer.

Habitat

Ovibos moschatus is a social and gregarious inhabitant on the arctic tundra. In the summer it prefers moist habitats and riparian vegetation where sedges and in some situations, shrubs, play a major role in its summer diet (Lent 1978, 1988). In the winter it may shift to hilltops, slopes, and plateaus, where winds keep snow levels to a minimum (Nowak 1991).

Status

Muskox abundance and distribution have fluctuated in response to long-term shifts in climate (Vibe 1967). Throughout much of their range, populations are increasing (Lent 1978). The world population of muskoxen has been estimated at about 162,000 animals and over 3,000 reside in Alaska (Klein 2000). Current surveys (ADFG 2005) estimate about 600 muskoxen on Nunivak Island, 300 on Nelson Island, fewer than 100 on the Yukon-Kuskokwim Delta, 2,000 on and near the Seward Peninsula, 370 in northwestern Alaska, and around 500–600 in northeastern Alaska. Successful transplants to the Arctic National Wildlife Refuge (ANWR) resulted in a population that was well established and showing significant growth within the historic range of the species (Jingfors and Klein 1982; Reynolds and Ross 1984). However, beginning in the late 1990s the eastern North Slope population began to decline, and by 2006, muskoxen were again absent from ANWR and greatly reduced in number (to about 200–250 animals) elsewhere in the region (Valkenburg 2007). Causes for this recent decline remain unclear. Lent (1978) speculated that increased development and extraction of energy resources from ANWR and other areas in the region may impact muskoxen over the long term.

Fossils

Muskoxen originated in Asia in the Late Pliocene or Early Pleistocene (Harington 1961). The primitive muskox *Soergelia* was widespread in Eurasia and also is reported from a few Pleistocene (Kansan) sites in North America (Harington 1977; Kurtén and Anderson 1980). This lowland-adapted species was probably replaced by more specialized forms (*Symbos*, *Praeovibos*) (Harington 1977). In North America, the woodland muskox, *Symbos*, occurred in plains and savannahs from central Alaska to southern New Jersey and persisted until about eleven thousand years ago. *Ovibos* was first present in Alaska in Illinoian time, and the genus extended widely

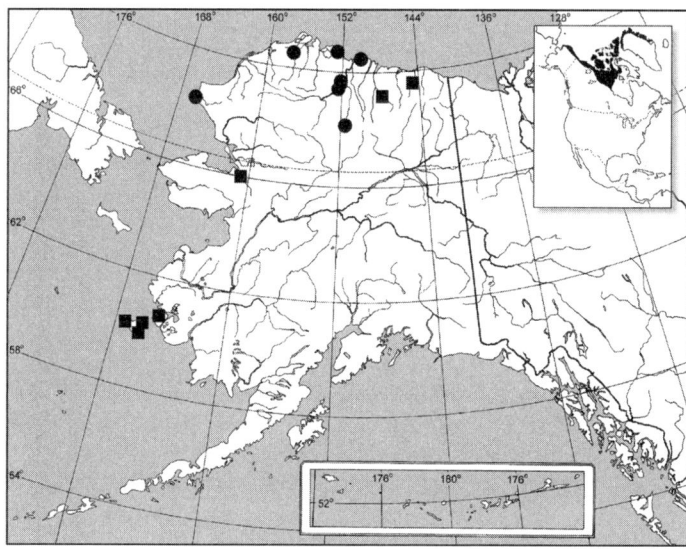

MAP 85. Distribution of muskox, *Ovibos moschatus*.

■ = Specimen Record (post-reintroduction; n = 38)

● = Specimen Record (pre-reintroduction; n = 5): Harrison Bay Quad., inland from Cape Halkett –152 degrees (1 MVZ); 70.5103, –149.8600, Beechey Point Quad., Oliktok Point, east of Colville River delta (1 UAM); 69.7158, –152.2492, Umiat Quad., Kikiakrorak River, National Petroleum Reserve (1 UAM); Umiat Quad., near Umiat (1 LACM); Chandler Lake Quad., Anaktuvuk Pass (1 USNM); 68.3333, –166.7500, Point Hope Quad., house ruin at Point Hope (1 UAM).

to the south during the Wisconsinan glaciation (Harington 1970, 1977).

Dall's Sheep

Ovis dalli
Nelson, 1884

Other Common Names

Thinhorn sheep, Stone's sheep, Fannin's sheep.

Systematics

There have been varying opinions on the taxonomic status of North American and Siberian sheep (Chernyavskyi 1962; Youngman 1975). Rausch (1963) and Youngman (1975) considered *O. dalli* conspecific with the snow sheep, *Ovis nivicola*, of eastern Siberia. Korobitsyna et al. (1974), however, found chromosomal differences between the two (fifty-four chromosomes in *nivicola* compared to fifty-two in *dalli* and *canadensis*). Other authors (e.g., Heptner et al. 1988; Lydekker 1898) considered the two Nearctic sheep, *dalli* and *canadensis*, to be conspecific with *nivicola*, with the name *canadensis* having priority. Most authors (Anderson 1946; Bowyer and Leslie 1992; Chernyavskyi 1962; Grubb 2005) concur with Cowan (1940) in recognizing three separate species.

The two North American species are recognized based on morphological evidence (Cowan 1940; Ramey 1993). Groves and Shields (1997) reported a close phylogenetic relationship between *dalli* and *canadensis* based on mtDNA sequences. Ramey (1993) found that an *O. dalli* haplotype from northern British Columbia was more similar to *O. canadensis* than to *O. dalli* haplotypes. However, this result was based on a single sample from British Columbia and difficult to interpret. Loehr et al. (2006) used mtDNA sequences from 223 *O. dalli* and *O. canadensis* from across their range in North America and found extensive overlap between the two species.

Three subspecies of *O. dalli* are generally recognized (Cowan 1940; Hall 1981), with one, *O. d. dalli*, occurring throughout most of Alaska, Yukon, and extreme northwestern British Columbia. The British Columbia form, *O. d. stonei*, is probably the subspecies that occurs marginally in the Haines area (Shackleton 1999). Worley et al. (2004) provided some support in nuclear DNA for the classification of Dall's sheep into the subspecies *O. d. stonei* and *O. d. dalli*. However, according to Loehr et al. (2006), mtDNA and the clinal nature of color morphology did not support this grouping. The validity of the subspecies *kenaiensis* is questionable (Grubb 2005) as it may represent the extreme end of a cline in size in *dalli* (see Bowyer and Leslie 1992).

Ovis dalli dalli

Original Description. 1884. *Ovis montana dalli* Nelson, *Proc. U.S. Nat. Mus.*, 7:13, June 3.

Type Locality. Mountains S of Fort Yukon on west bank of Yukon River, Alaska; probably Tanana Hills.

Type Specimen. Cotypes USNM 13266/20786 and USNM 13265/20787.

Range. Alaska (excluding the Kenai Peninsula), Yukon Territory, extreme western Northwest Territories, and extreme northwestern British Columbia.

Ovis dalli kenaiensis

Original Description. 1902. *Ovis dalli kenaiensis* J. A. Allen, *Bull. Amer. Mus. Nat. Hist.*, 16:145, April 23.

Type Locality. Head of Sheep Creek, Kenai Peninsula, Alaska.

Type Specimen. AMNH 17609.

Range. Confined to the Kenai Peninsula.

Remarks. Osgood (1909) and Nichols (1978) regarded this smaller form synonymous with *O. d. dalli*, but Cowan (1940) considered it a valid subspecies.

Global Distribution

Nearctic (but see "Systematics" above)—The range of the Dall's sheep includes Alaska, Yukon Territory, Northwest Territories, and northern British Columbia (Bowyer and Leslie 1992).

Alaska Distribution

Dall's sheep inhabit the major mountain ranges in Alaska, including the Kenai, Chugach, Wrangell, Talkeetna, Alaska, and Brooks ranges as well as the White Mountains, Tanana Hills, and Ogilvie Mountains that extend into Alaska along the international border (map 86).

Southeast Region. Dall's sheep are found adjacent to southeast Alaska in British Columbia on the drier western slopes of the St. Elias Mountains and the Coast Mountains north of Haines and Skagway (Klein 1965; Nichols 1978).

The occurrence of Dall's sheep in southeast Alaska is based on a single female collected from the Kelsall River Valley, northwest of Haines, and from occasional sightings near there in the vicinity of Mount Raymond (MacDonald and Cook 2007).

Southcentral Region. Dall's sheep are absent from the coastal strip that borders the Gulf of Alaska. They are found, however, in the Kenai Mountains west and north of the Harding Ice Field from the head of Kachemak Bay to Turnagain Arm. Greatest concentrations of Dall's sheep occur in the arc from Kenai Lake through Skilak Lake to Tustumena Lake. They inhabit both the north and south slopes of the entire Wrangell Mountain range, and are common in the Nutzotin Mountains. The Alaska Range is the most extensive expanse of Dall's sheep habitat within the state (ADFG 1973).

Central Region. Sheep are generally limited to only the highest, most rugged peaks in the Tanana Hills–White Mountain complex.

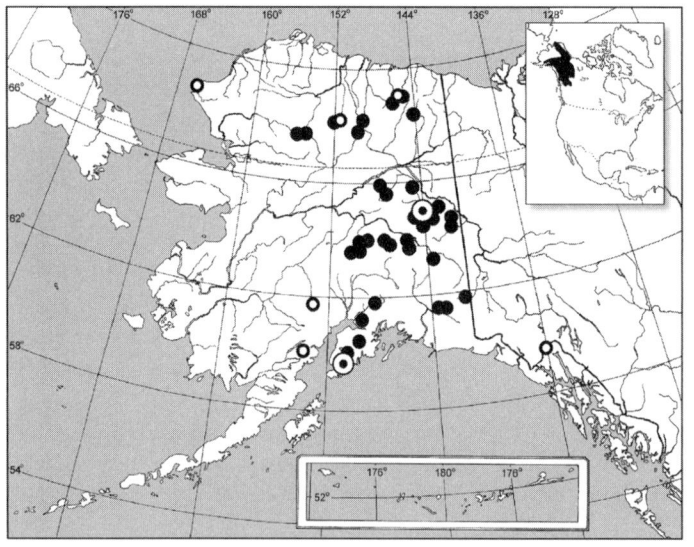

MAP 86. Distribution of Dall's sheep, *Ovis dalli*.

⊙ = Type Locality
● = Specimen Record (n = 525)
○ = Marginal Records. 69.3167, –145.0500, Mt Michelson Quad., Lake Peters (1 MVZ); Skagway Quad., Kelsall River Valley (1 collected by ADFG in early 1990s but specimen not located; see MacDonald and Cook 1996); Aleutian Range, Chigmit Mountains (2 USNM); South Fork Kuskokwim River (2 USNM); 68.7167, 166.1833, Point Hope Quad., Cape Lewis, 10 mi. S of Cape Lisburne (1 USNM); 68.1333, –151.7500, Chandler Lake, Anaktuvuk Pass (1 USNM).

They also occur along the Charley River on cliffs just above the river (ADFG 1973). Sheep also occur westward among the higher peaks of the Ogilvie Mountains north of the Yukon River along the Alaska-Yukon border.

Southwestern Region. Sheep are found in the mountains along the east side of Lake Clark near Port Alsworth, northward (ADFG 1973; Cook and MacDonald 2006).

Between 1964 and 1967, three attempts were made to translocate Dall's sheep from the Kenai Peninsula to Kodiak Island. Scattered sightings in the mid-1970s suggested a few persisted there into the 1980s (Franzmann 1988), but these are no longer extant (L. van Daele, pers. comm., 2007).

Western Region. Sheep occur along the southern slopes of the Brooks Range north of the Noatak River to as far west as the Lisburne Hills (Dau 2005), and have an erratic distributional pattern in the Baird and Schwatka mountain ranges to the south (Dean and Chesemore 1974).

Northern Region. This species is widely distributed in this region along the entire length of the Brooks Range (ADFG 1973; Bee and Hall 1956).

Habitat

Dall's sheep generally inhabit subalpine grass–low shrub habitats in dry, mountainous terrain (Bowyer and Leslie 1992). Sheep expand their range to include suitable habitat during the summer months, and become restricted in late winter to snow-free parts of their distribution (ADFG 1973). The presence of mineral licks is essential to Dall's sheep during the spring.

Status

Populations of Dall's sheep in Alaska are generally healthy (ADFG 2005; Whitten 1994). Hoefs (1989) estimated the total sheep population for Alaska at about 72,650 animals, including 1,500 on the Kenai Peninsula, while Valdez and Krausman (1999) estimated a total of about 100,000 sheep in the state. In Canada, *O. d. dalli* is included on the British Columbia Blue List of species at risk (Shackleton 1999).

Fossils

Ovis first appeared in Villafranchian deposits in Eurasia (Kurtén and Anderson 1980). The earliest records of *Ovis* (probably *dalli* but specific designation not clearly established) in North America are from Illinoian deposits near Fairbanks (Péwé and Hopkins 1967). Sheep fossils from Late Pleistocene deposits in eastern Alaska and Yukon Territory are common (Dixon 1984; Guthrie 1968b; Harington 1978, 1980a; Porter 1988; Weber et al. 1981). A twenty-three-thousand-year-old horn core from near Dawson indicated mountain sheep were present in eastern Beringia during the peak of the Wisconsinan glaciation (Kurtén and Anderson 1980), and that they were larger than present-day sheep (Guthrie 1984). Savinetsky et al. (2004) reported the recovery of *Ovis* bones (along with *Lepus* and *Rangifer* remains) from a Late Holocene archaeological site on St. Lawrence Island.

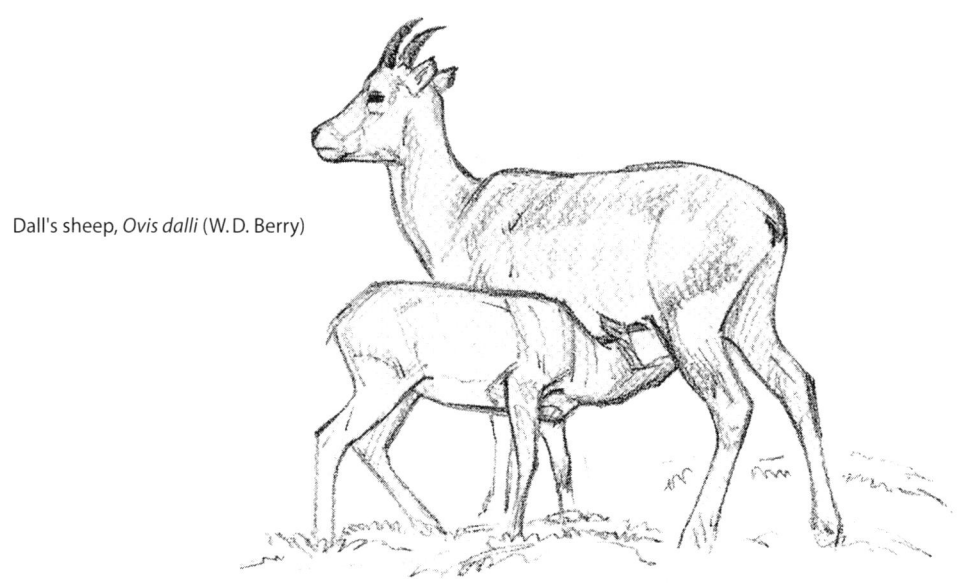

Dall's sheep, *Ovis dalli* (W. D. Berry)

Order CETACEA Brisson, 1762

Cetaceans date back some fifty-five to sixty million years. Their closest living relatives are the artiodactyls (even-toed ungulates) and recent genetic and paleontological studies suggest close evolutionary relationships with the hippopotamus (Reeves et al. 2002).

An overview of the evolutionary history of the Cetacea was presented by Rice (1998). Two suborders are generally recognized: Mysticeti, the baleen whales in four extant families, and Odontoceti, the toothed whales in seven families (Mead and Brownell 2005). Subfamilies and geologic ranges of all cetacean taxa were discussed by Fordyce and Barnes (1994) and Fordyce (2002).

Family BALAENIDAE Gray, 1821—right whales

According to Rice (1998), the family of right whales, Balaenidae, comprises two living species in a single genus, *Balaena*. Mead and Brownell (2005), however, recognized two genera, *Balaena* (bowhead) and *Eubalaena* (right whales), with the later comprising three separate species based primarily on recent genetic studies. Two species occur in Alaska waters.

Fossils of Balaenids range from Middle Miocene to Pleistocene in western North America (Rice 1984).

Balaenids are massive, robust animals with a large head, huge tail, no dorsal fin, no ventral pleats, and very long, narrow baleen (Reeves et al. 2002). The largest individuals approach 20 m in length and 100 tons in weight; females are larger than males. Balaenids are specialized "skim" feeders, swimming slowly with their mouth open through concentrations of planktonic crustaceans (Rice 1984). Bowheads remain near sea ice in high latitudes year-round, while right whales make much more extensive seasonal movements into warm water (Clapham 1999a). Both species occur singly or in small groups that are not persistent (Reeves et al. 2002).

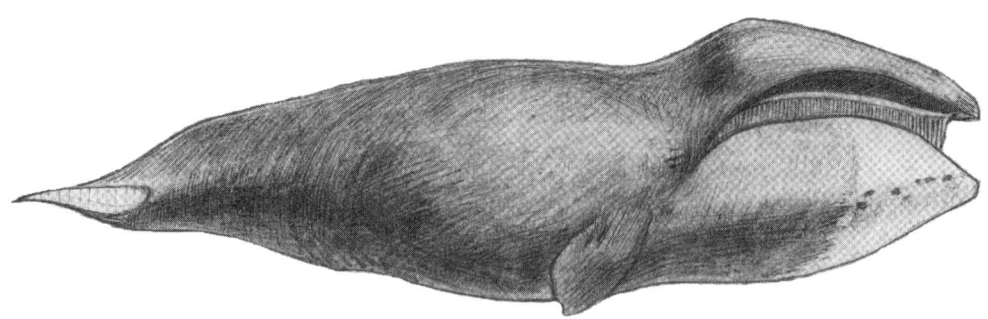

Bowhead, *Balaena mysticetus* (W. D. Berry)

Bowhead

Balaena mysticetus
Linnaeus, 1758

Other Common Names

Arctic right whale, Greenland right whale, great polar whale.

Systematics

Rice (1998) and others suggested that all the world's right whales (including the bowhead, *Balaena mysticetus*) are a single species, *Balaena glacialis*. Northern and Southern Hemisphere populations would be separated into two subspecies, *B. g. glacialis* and *B. g. australis*, respectively. A molecular study of the mitochondrial cytochrome *b* gene by Árnason and Gullberg (1994) suggested a close relationship between the bowhead and right whales; however, Rosenbaum et al. (2000) found mtDNA data consistent with recognition of *Balaena* and *Eubalaena* as distinct genera.

No adequate study of geographic variation of *B. mysticetus* has been completed. Although several separate stocks have been recognized for management purposes (Angliss and Outlaw 2007; Rugh et al. 2003), the divergence of bowhead stocks is unproven with most evidence for stock separation circumstantial (Moore and Reeves 1993). The largest population, and the only stock that is found within U.S. waters, is the western arctic stock, also known as the Bering-Chukchi-Beaufort stock (Rugh et al. 2003) or Bering Sea stock (Burns et al. 1993).

No subspecies are recognized (Rice 1998). There has been a long-standing controversy over the identity of certain whales off northern Alaska called *ingotok* or *ingutuq* (also *inito*) by Eskimo whalers (Bailey and Hendee 1926), and *poggy* by nineteenth-century American whalers in the Sea of Okhotsk (Reeves and Leatherwood 1985). Braham, Durham et al. (1980) suggested these whales were small, fat bowheads, usually female, while Jarrell (1981) argued that they are yearling bowheads, and Fetter and Everitt (1981) suggested that the differences may be due to congenital defect.

Balaena mysticetus

Original Description. 1758. [*Balaena*] *mysticetus* Linnaeus, *Syst. Nat.*, 10th ed., 1:75.

Type Locality. Greenland Sea.

Type Specimen. None; based on the Greenland right whale of whalers and authors (see Hershkovitz 1966:194).

Global Distribution

Bowhead whales formerly occurred throughout the colder, ice-bound waters of the Northern Hemisphere (Mead and Brownell 2005). Bowheads were largely removed from the North Atlantic by commercial whaling. The majority of bowheads are now restricted to the Bering, Chukchi, and Beaufort seas, with small numbers occurring in the Sea of Okhotsk, Davis Strait, Hudson Bay, and the offshore waters of Spitsbergen (Angliss and Outlaw 2007). Stray bowheads have been reported from Japan, the Gulf of St. Lawrence, and Massachusetts (Mead and Brownell 2005).

Alaska Distribution

The western arctic population of the bowhead is found within and near the pack ice of the Bering Sea during winter (map 87). In summer, these whales follow leads in ice northward to feeding grounds in the Chukchi and Beaufort seas (Angliss and Outlaw 2007; Leatherwood and Reeves 1983; Moore and Reeves 1993). They summer mainly east of Barrow to Amundsen Gulf (IUCN 1991). In autumn, some whales are thought to migrate from the Beaufort Sea to as far west as Herald and Wrangel islands and into the eastern part of the East Siberian Sea before they move south through the Bering Strait (Braham, Durham, et al. 1980). There is some question of whether a separate "substock" of bowheads regularly remains in the Bering and Chukchi seas during summer (Reeves and Leatherwood 1985). One population of bowheads appears to consist of year-round residents of the Sea of Okhotsk (Berzin and Doroshenko 1981).

Habitat

The bowhead is closely associated with pack ice in arctic and subarctic waters (Leatherwood and Reeves 1983). In the Beaufort Sea of Alaska, bowheads select outer continental shelf and slope habitats (200–2,000 m) independent of ice cover during summer. During fall, they shift to shallower shelf habitat (less than 200 m) with light ice conditions (Moore et al. 2000). The lead system near Point Barrow appears to be an important feeding habitat during spring (Angliss and Outlaw 2007; Shelden and Rugh 1995). When bowheads cannot find open water, they push up hummocks in thin ice or break holes in ice up to 0.6 m thick. They filter feed at all depths on copepods, euphausiids, and other invertebrates (Carroll 1994). Lowry and Frost (1984) suggested that the bowheads migrate to the Beaufort Sea

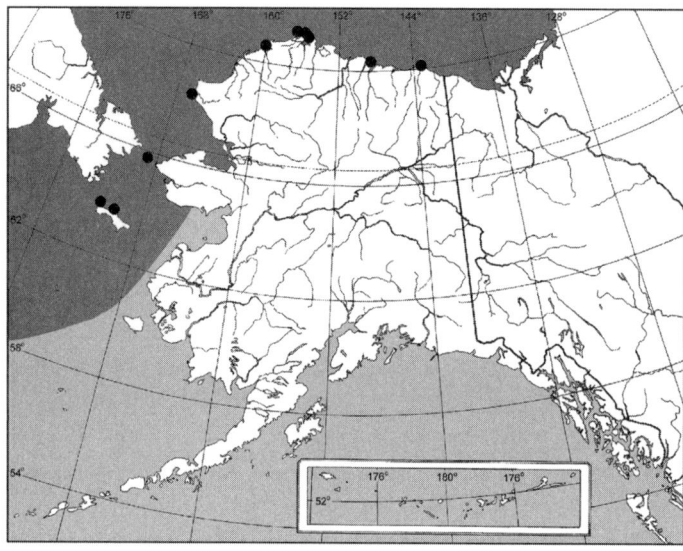

MAP 87. Distribution of bowhead, *Balaena mysticetus*.

■ = Projected Range (after Angliss and Outlaw 2007)

● = Specimen Record (n = 334)

because there is less competition for abundant food there.

Status

Reduced more than 80 percent by commercial whaling in the eighteenth and nineteenth centuries, bowheads now number about ten to twelve thousand individuals worldwide. The western arctic stock is the only substantial population remaining, numbering an estimated 10,545 and increasing in recent years toward carrying capacity (Angliss and Outlaw 2007). Bowheads are fully protected everywhere except in Alaska. The number of bowheads landed by Alaska Natives was reported to be forty-two in 1999, thirty-five in 2000, forty-nine in 2001, thirty-seven in 2002, and thirty-five in 2003 (Suydam and George 2004). *Balaena mysticetus* is listed in Appendix 1 of CITES, and under the ESA as Endangered. The western arctic population is listed as Conservation Dependent by IUCN, and of Special Concern by COSEWIC and ADFG.

In 2000, NMFS was petitioned to designate critical habitat for the western arctic bowhead stock. Petitioners asserted that the nearshore areas from the U.S.-Canada border to Barrow, Alaska, should be considered critical habitat. In 2001, NMFS found the petition to have merit, but in 2002 decided not to designate critical habitat for this population. NMFS found that designation of critical habitat was not necessary because the population is known to be approaching its pre–commercial whaling population size, the population is increasing, there are no known habitat issues that are slowing the growth of the population, and activities that occur in the petitioned area are already managed to minimize impacts to the population.

Meanwhile, increasing oil and gas development in the Arctic has led to higher risk of various forms of pollution to bowhead whale habitat, including oil spills, toxic and nontoxic waste, and noise due to higher levels of traffic as well as exploration and drilling operations (Angliss and Outlaw 2007).

Another element of concern is the potential for arctic climate change, which will heavily affect high latitudes. Ice-associated animals, such as the bowhead whale, may be sensitive to changes in arctic weather, sea-surface temperatures, or ice extent, and the concomitant effect on prey availability. There are insufficient data to make reliable predictions of the effects of arctic climate change on bowhead whales (Angliss and Outlaw 2007), but there is evidence that there has been a shift in regional weather patterns in the Arctic (Tynan and DeMaster 1997).

Fossils

Fossils of *Balaena* have been described from deposits of Late Miocene, Pliocene, or Pleistocene age from Europe and North America. One of these, *B. prisca*, is similar enough to the modern bowhead that they may be conspecific (Kenney 2002).

North Pacific Right Whale

Eubalaena japonica
(Lacépède, 1818)

Other Common Names

Right whale, black whale, black right whale, northern right whale, Pacific right whale.

Systematics

Systematics of the right whales has been contentious. Eschricht and Reinhardt (1866) established the distinction between the right whales of temperate waters and the bowhead, *B. mysticetus*, of the arctic pack ice. Right whales (*Eubalaena* spp.) and the bowhead (*Balaena*) whale often have been included in the same genus (e.g., Rice 1998), but many recent classifications recognize them as distinct genera (e.g., Baker et al. 2003; Mead and Brownell 2005). MtDNA data are consistent with recognition of *Balaena* and *Eubalaena* as distinct genera (Rosenbaum et al. 2000).

There has been a long-standing debate regarding the taxonomic status of the three geographically disjunct populations of right whales. Rice (1998) and others regarded all the world's right whale populations as conspecific, with the Northern and Southern Hemisphere populations tenuously separated into two subspecies, *B. g. glacialis* and *B. g. australis*, respectively. Rosenbaum et al. (2000) used mtDNA data to argue that the North Atlantic, the North Pacific, and the Southern Hemisphere populations represent three distinct species (*glacialis*, *japonica*, and *australis*, respectively), distinctions subsequently followed by Mead and Brownell (2005) and Angliss and Outlaw (2007). The three species concept was further supported by analyses of mtDNA sequence data of their ectoparasitic lice (Kaliszewska et al. 2005).

Two stocks of North Pacific right whales are currently recognized: a Sea of Okhotsk stock and an Eastern North Pacific stock (Angliss and Outlaw 2007).

Eubalaena japonica

Original Description. 1818. *Balaena japonica* Lacépède, *Mém. Mus. Hist. Nat.*, Paris, 4:469.
Type Locality. Japan.
Type Specimen. None designated.

Global Distribution

Before right whales in the North Pacific were heavily exploited by commercial whalers, concentrations were found in the Gulf of Alaska, eastern Aleutian Islands, south-central Bering Sea, Sea of Okhotsk, and Sea of Japan (Braham and Rice 1984; Cummings 1985).

Alaska Distribution

The current distribution of right whales in Alaska waters (map 88) is poorly known. The Gulf of Alaska, the eastern Aleutians, and the Bering Sea were historic summer feeding areas, and some of the last areas to suffer from overexploitation of this once abundant species (Gilmore 1978).

Beginning in 1996, small aggregations of right whales began to be seen annually in the Bering Sea. A group of three to four right whales was sighted in western Bristol Bay in July 1996 and may have included a juvenile animal (Goddard and Rugh 1998). During July 1997, a group of four to five individuals was encountered in Bristol Bay (Tynan 1999). During dedicated surveys in July 1998, July 1999, and July 2000, five, six, and thirteen right whales were again found in the same general region of the southeastern Bering Sea (LeDuc et al. 2001). Two right whales were observed in the central Bering Sea in July 1999 (Moore et al. 2000). In 2002 there were probably seven sightings of right whales, including the first confirmed sighting of a calf in decades. In 2004, more right whales were found, comprising a minimum of seventeen individuals, including two probable calves (Angliss and Outlaw 2007).

There are fewer recent sightings of right whales in the Gulf of Alaska than in the Bering Sea (Brownell et al. 2001), although little survey effort has been expended. Waite et al. (2003) summarized sightings from 1959 to 1997. Seven sightings of right whales were reported, but only one sighting of four right whales at the mouth of Yakutat Bay in 1979 could be positively confirmed. Sightings of a single right whale off eastern Kodiak Island occurred in July 1998 and in August 2004 and 2005 (Angliss and Outlaw 2007). Acoustic monitoring at seven sites in the Gulf of Alaska detected right whale calls at only two: one off eastern Kodiak and the other in deep water south of the Alaska Peninsula.

Habitat

Remnant populations of right whales occur mostly in temperate and subpolar waters. These whales are zooplankton specialists.

Status

Right whales were given worldwide protection in 1935, but have not shown a notable increase

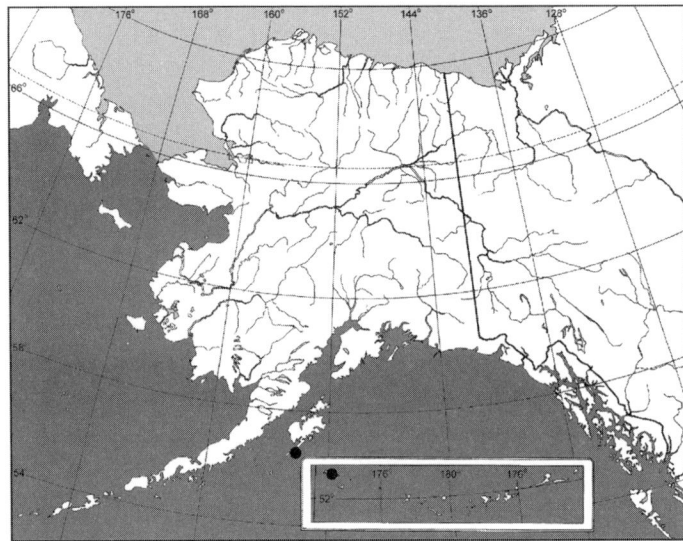

MAP 88. Distribution of North Pacific right whale, *Eubalaena japonica*.

■ = Projected Range (after Angliss and Outlaw 2007)
● = Specimen Record (n = 3): Attu Island (1 USNM); South of Kodiak Island (1 USNM); Alaska, offshore waters (1 UWBM).

in numbers since the cessation of whaling (Potter and Birchler 1999c). Wada (1973) estimated a total population of one hundred to two hundred in the North Pacific, and Tynan et al. (2001) speculated that the eastern North Pacific stock probably totaled only tens of animals. Angliss and Outlaw (2007) concluded that a reliable estimate of either abundance or trend is currently not available.

The North Pacific right whale is listed under Appendix I of CITES, as Endangered in the ESA, as Endangered by IUCN, and Endangered by COSEWIC. The National Marine Fisheries Service was petitioned on 4 October 2000 to designate critical habitat in the southeast portion of the Bering Sea near Bristol Bay. On 14 June 2005, the NMFS was ordered to designate critical habitat for the North Pacific populations. In 2006, NMFS issued a final rule designating two areas as northern right whale critical habitat, one in the Gulf of Alaska and one in the Bering Sea (71 FR 38277, 6 July 2006).

Fossils

The oldest fossil balaenid (*Morenocetus parvus*) is from the Early Miocene of South America. The Balaenidae (represented by the living right whales and bowhead) is the oldest extant family of the Mysticeti (Fordyce 2002).

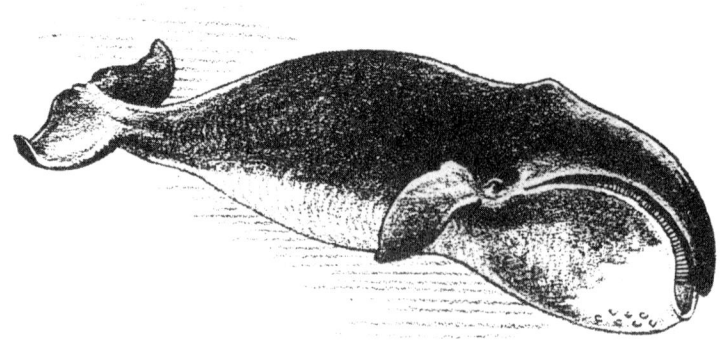

Bowhead, *Balaena mysticetus* (B. Hines)

Family BALAENOPTERIDAE Gray, 1864—rorquals

The balaenopterids contain seven to nine extant species in two genera (Deméré et al. 2005; Mead and Brownell 2005); five species are found in Alaska waters.

Fossils of Balaenopterids range from Middle Miocene to Pleistocene in North America, and Miocene in eastern Asia (Rice 1984).

Members of this family, also known as rorquals, have a sleek body, a small, posteriorly positioned dorsal fin, and characteristic pleats on the underside of the mouth. Rorquals range in size from the 2-m-long dwarf minke whale to the 30-m-long, 160-ton blue whale, the world's largest living animal. The humpback whale has distinctive long flippers (one-third its body length), in contrast to all other short-flippered rorquals. Rorquals are highly specialized "gulp" feeders, capable of opening their mouths very wide to engulf swarms of krill or schools of small fish (Rice 1984). Most rorquals migrate from their summer, high-latitude feeding grounds to lower-latitude waters in winter to mate and calve (Clapham 1999b).

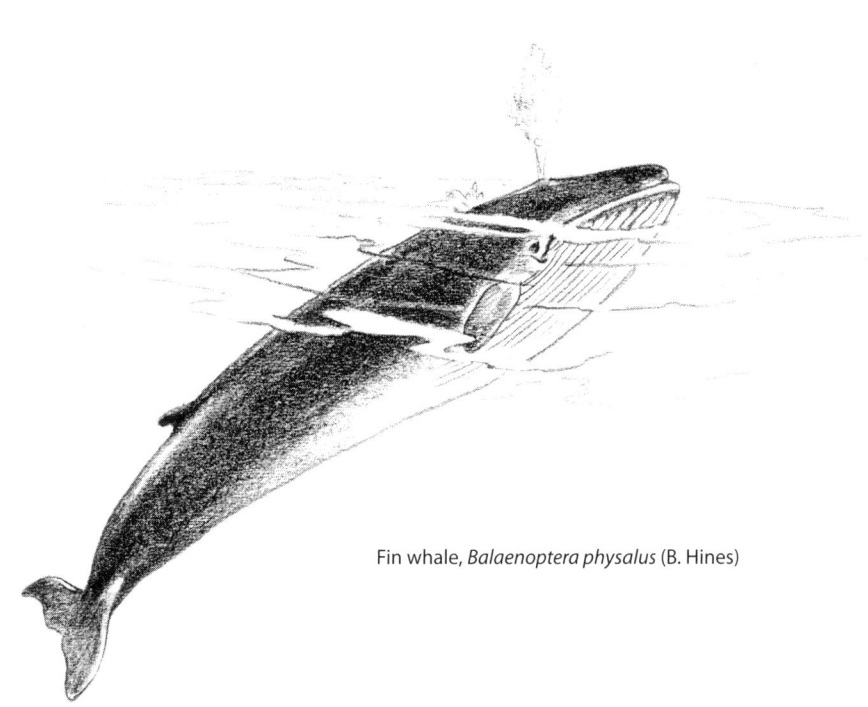

Fin whale, *Balaenoptera physalus* (B. Hines)

Common Minke Whale

Balaenoptera acutorostrata
Lacépède, 1804

Other Common Names

Northern minke whale, lesser rorqual, piked whale, little piked whale, sharpheaded finner (see Stewart and Leatherwood 1985).

Systematics

The taxonomy of the minke whale has long been problematic (Stewart and Leatherwood 1985). The northern minke whale, *B. acutorostrata*, is now considered a separate species from *B. bonaerensis*, the antarctic minke whale of the Southern Hemisphere, by various authors (Brownell et al. 2000; IWC 2001; Mead and Brownell 2005; Rice 1998). Two subspecies have been recognized in the Northern Hemisphere, with the North Pacific race occurring in Alaska waters. A provisional third subspecies, a distinctive, diminutive form of *B. acutorostrata*, occurs mainly in lower latitudes of the Southern Hemisphere in areas of sympatry with the Antarctic species (Kato and Fujise 2000; Rice 1998). Minke whales in the eastern North Pacific may be a stock distinct from those in the Okhotsk Sea, Sea of Japan, East China Sea, and Yellow Sea in the western North Pacific (B. S. Stewart 1999).

Data from mtDNA and allozymes supported the division of the minke whale complex into four distinct genetic units: North Pacific, North Atlantic, "normal" Southern Hemisphere, and sympatric "dwarf" Southern Hemisphere (Baker and Palumbi 1997).

Balaenoptera acutorostrata scammoni

Original Description. 1986. *Balaenoptera acutorostrata scammoni* Deméré, *Marine Mammal Science* 2:277–298.

Type Locality. Admiralty Inlet, Puget Sound, Washington.

Type Specimen. USNM 12177.

Range. North Pacific Ocean.

Remarks. The long-used name *B. a. davidsoni* Scammon, 1872, for this North Pacific race was found to be preoccupied (Deméré 1986).

Global Distribution

The minke whale is found from the polar ice edge to the tropics in all the world's oceans. In the North Pacific in summer, minke whales range from the Bering and Chukchi seas south to the East China Sea, 30° N in the central

Pacific, and the coast of central Baja California. In winter their range probably extends from the East China Sea and central California south to nearly the equator (Rice 1998).

Alaska Distribution

In summer, minke whales of the North Pacific move northward through nearshore waters to the edge of pack ice in the Bering Sea and, infrequently, the Chukchi Sea as far east as Barrow (Leatherwood and Reeves 1983) (map 89). In autumn, they evidently migrate south through deeper waters offshore (Dorsey et al. 1990).

Habitat

Minke whales are found in both deep waters far offshore and shallow coastal waters, often in and near ice. Recent surveys (Moore et al. 2002) found minke whale abundance similar in the central-eastern Bering Sea and the southeastern Bering Sea. They occurred throughout the area surveyed, but most minke whales in the central-eastern Bering Sea were seen in waters 100–200 m deep (Moore et al. 2000); sightings in the southeastern Bering Sea occurred along the north side of the Alaska Peninsula and were associated with the 100-m contour near the Pribilof Islands (Moore et al. 2002).

In the North Pacific they feed on euphasiids, copepods, sand lance, and herring (B. S. Stewart 1999).

Status

The entire Northern Hemisphere population is about 125,000 animals (Martin 1990), but no estimates have been made of the number of minke whales in the North Pacific alone. Survey estimates on the numbers of minke whales in the Bering Sea during 1999–2000 (in Angliss and Outlaw 2007) indicated about eight hundred minke whales in the central-eastern Bering Sea and about one thousand in the southeastern Bering Sea.

Small numbers of minke whales were harvested by commercial whalers near Akutan, Alaska, and near British Columbia in the early 1900s (B. S. Stewart 1999). All commercial hunting was stopped in 1986. Native hunters from St. Lawrence Island still occasionally take minke whales for subsistence purposes (Stewart and Leatherwood 1985). The minke whale is listed in Appendix I of CITES, and Near Threatened by IUCN (2007).

Fossils

The fossil record of Balaenopteridae extends from the Late Miocene (Berta and Sumich

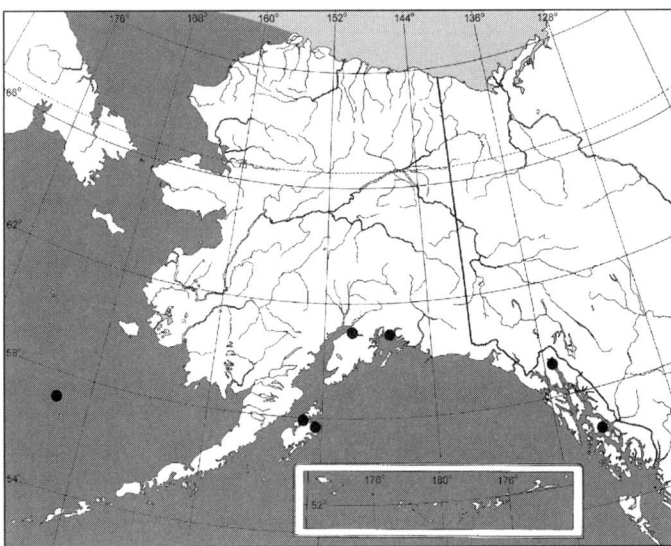

MAP 89. Distribution of common minke whale, *Balaenoptera acutorostrata*.

■ = Projected Range (after Angliss and Outlaw 2007)
● = Specimen Record (n = 8): 60.93333, −149.16667, Gulf of Alaska, Seward Quad., no specific locality recorded (1 UAM); 57.91667, −153.25, Gulf of Alaska, Kodiak Quad., Kodiak Island, Viekoda Bay (1 UAM); 57.68333, −152.45, Gulf of Alaska, Kodiak Quad., Kodiak Island, Middle Bay (1 UAM); Cook Inlet, Turnagain Arm (1 AMNH); Duncan Canal, Pearl Island (1 USNM); Prince William Sound, Glacier Island (1 USNM); Lynn Canal, Davidson Glacier, Glacier Point (1 USNM); Bering Sea, Pribilof Islands, St. Paul Island (1 USNM).

1999). A fossil record of *B. acutorostrata* from the Pliocene of Japan has not been confirmed (Deméré et al. 2005).

Sei Whale

Balaenoptera borealis
Lesson, 1828

Other Common Names

Coalfish whale, pollack whale, Rudolphi's rorqual, sardine whale, Japan finner.

Systematics

Tomilin (1946) recognized two subspecies: one in the Northern Hemisphere (*B. b. borealis*) and the other in the Southern Hemisphere (*B. b. schlegellii*). Masaki (1977) suggested the occurrence of three distinct subpopulations of sei whales for the eastern, central, and western North Pacific.

Mitochondrial and nuclear gene sequences reflected shallow evolutionary divergence among conspecific populations of sei whales (Hoelzel 1994). Allozyme analyses by Wada and Numachi (1991) indicated that the sei and Bryde's whale (*B. brydei*) of tropical and sub-tropical oceans are sister species that separated less than three hundred thousand years ago.

Balaenoptera borealis borealis

Original Description. 1828. *Balaenoptera borealis* Lesson, *Histoire naturelle . . . des mammifères*

et des oiseaux découverts depuis 1788, cétacés, p. 342.

Type Locality. Schleswig-Holstein, Lubeck Bay, near Gromitz, Germany.

Type Specimen. Skeleton in Berlin Museum (MNHU) according to Hershkovitz (1966), missing or lost according to Deméré et al. (2005).

Range. Northern Hemisphere.

Global Distribution

Sei whales are pelagic in temperate regions of the world and do not appear to be associated with coastal features (Carretta et al. 2007). Sei whales migrate seasonally between northern latitudes to feed and southern latitudes to breed (Rice 1998). During the summer, sei whales occur from California to the Gulf of Alaska (Wynne 1993). During the winter, this species migrates as far south as 20° N latitude (Potter and Birchler 1999d).

Alaska Distribution

Sei whales migrate northward to feeding grounds in the Gulf of Alaska and north of the Aleutian Islands into the southeastern corner of the southwestern basin of the Bering Sea (Rice 1998) (map 90).

Habitat

Sei whales are pelagic throughout the year, skim feeding on a wide variety of crustaceans, small fish, and squid (Wynne 1993).

Status

Sei whales were commercially harvested until 1977 (Wynne 1993). The species is listed in

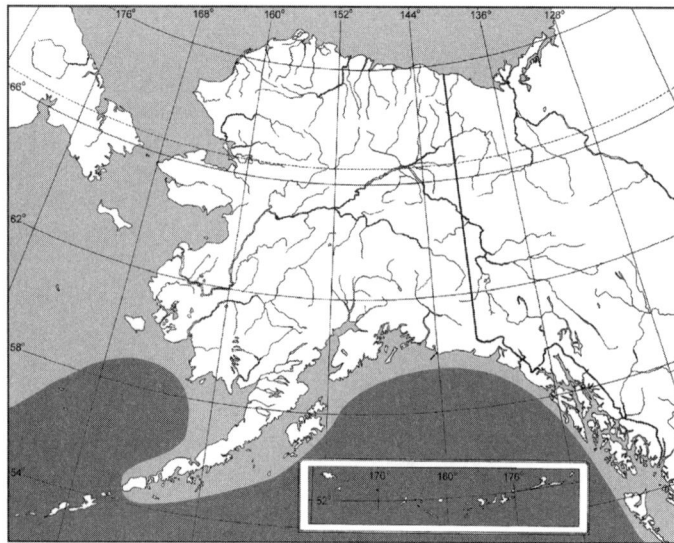

MAP 90. Distribution of sei whale, *Balaenoptera borealis*.

■ = Projected Range (after Angliss and Outlaw 2007)
Specimen Record (n = 0)

Appendix 1 of CITES, and as Endangered by IUCN, U.S. ESA, and COSEWIC. The North Pacific population was roughly estimated at thirteen thousand whales in 1974, down from about sixty-three thousand in 1963 (Martin 1990). There are no current data on trends in sei whale abundance in the eastern North Pacific waters (Carretta et al. 2007).

Fossils

A fossil record reported from the Pleistocene of Japan needs confirmation (Deméré et al. 2005).

Blue Whale

Balaenoptera musculus
(Linnaeus, 1758)

Other Common Names

Blue rorqual, great northern rorqual, Sibbald's rorqual, sulphur-bottom.

Systematics

For a period of time some authors separated the blue whale into the monotypic genus *Sibbaldus*, but this has not been widely accepted (Rice 1998). Three subspecies of *B. musculus* are currently recognized (Rice 1998); one occurs in the waters of the North Pacific.

Although only one management stock for blue whales in the North Pacific is currently recognized by the International Whaling

Commission, as many as five separate populations possibly occur there (Carretta et al. 2007).

Balaenoptera musculus musculus

Original Description. 1758. [*Balaena*] *musculus* Linnaeus, *Syst. Nat.*, 10th ed., 1:76.
Type Locality. Firth of Forth, Scotland.
Type Specimen. None formally designated.
Range. Northern Hemisphere.

Global Distribution

Blue whales are found throughout every ocean of the world, from the equator to polar regions (Martin 1990). A specific breeding ground for blue whales in any ocean remains unknown. Regular sightings of mothers and calves in the Gulf of California, Mexico, in late winter and spring suggests that a portion of the northeast Pacific stock could be breeding there (Sears 2002).

Alaska Distribution

In the eastern North Pacific, blue whales move in the summer months into the immediate offshore waters around the Gulf of Alaska and along the Aleutian Islands (map 91). Some individuals venture into the Bering Sea, rarely as far north as the Chukchi Sea (Rice 1978a; Yochem and Leatherwood 1985), but this far northern population may be separate from the whales summering off California (Rice 1998). Where the northern population winters is unknown (Potter and Birchler 1999a).

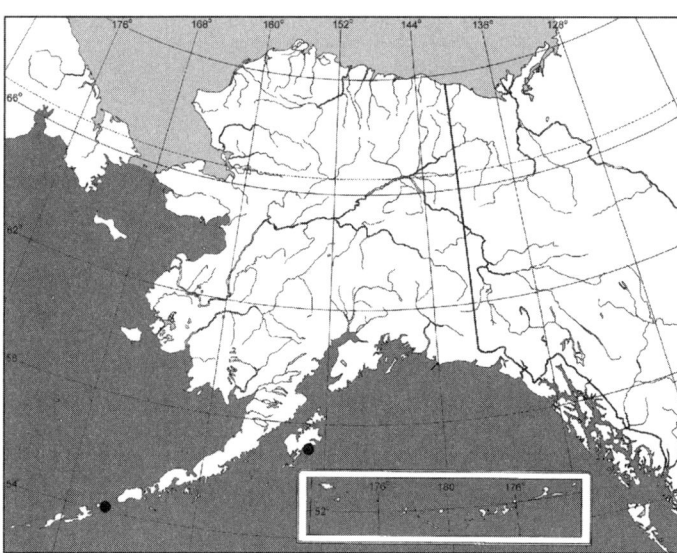

MAP 91. Distribution of blue whale, *Balaenoptera musculus*.

One of several blue whales sighted 100–150 nautical miles SE of Prince William Sound in July, 2004 (NOAA photo).

■ = Projected Range (after Angliss and Outlaw 2007)
● = Specimen Record (n = 5): Kodiak Archipelago, Sitkalidak Island, Port Hobron (1 USNM); Aleutian Islands, Krenitzin Island, Rootok Island (1 USNM); Alaska, unknown locality (1 USNM); Southeast Alaska, possibly Wrangell or Shakan (2 UWBM).

Although blue whales have been protected since 1966, they have remained absent or uncommon in areas of historical abundance, including the Gulf of Alaska, the eastern Aleutians, and the far western Aleutians (Yochem and Leatherwood 1985; Zimmerman 1994b). The sighting and nonlethal skin and blubber sampling of several blue whales about 100 nautical miles southeast of Prince William Sound in July 2004 is the first confirmed report of this species in Alaska waters in three decades (Joling 2004).

Habitat

Blue whales occur primarily along the edge of the continental shelves and along ice fronts, but also venture into deep oceanic zones and shallow inshore regions (Rice 1978a).

Status

The blue whale is listed in Appendix I of CITES, and as Endangered under the U.S. ESA and by IUCN. A recent population estimate of blue whales in the Northern Hemisphere was about three thousand animals, down from about eight thousand prior to the peak years of exploitation in the 1930s (Martin 1990; Sears 2002). The North Pacific population has been roughly estimated at twelve to seventeen hundred animals (Zimmerman 1994b).

Fossils

A blue whale fossil reported from the Pliocene of Japan was considered questionable by Deméré et al. (2005).

Fin Whale

Balaenoptera physalus
(Linnaeus, 1758)

Other Common Names

Finback whale, common rorqual, herring whale, razorback whale.

Systematics

The fin whale is closely related to the other balaenopterids, particularly the blue whale, from which it may have diverged between 3.5 and 5 million years ago. Several hybrids of blue and fin whale crosses have been described (Aguilar 2002; Berube and Aguilar 1998). Two poorly defined subspecies, one each for the Northern and Southern hemispheres, are recognized (Rice 1998). A third subspecies, the pygmy fin whale *B. p. patachonica*, has been proposed but still lacks sufficient evidence for recognition (Deméré et al. 2005). In the North Pacific, some mixing between eastern and western stocks occurs in the Bering Sea in July and August (Angliss and Outlaw 2007).

Genetic evidence from mitochondrial and nuclear genes suggested a lack of deep evolutionary separation among conspecific populations of fin whales and several other wide-ranging cetacean species (Hoelzel 1994).

Balaenoptera physalus physalus
Original Description. 1758. [*Balaena*] *physalus* Linnaeus, *Syst. Nat.*, 10th ed., 1:75.

CETACEA

243

Type Locality. Spitzbergen Seas.
Type Specimen. None formally designated.
Range. North Pacific and North Atlantic oceans.

Global Distribution

Fin whales are found throughout the world's oceans, but tend to be less common in tropical waters (Leatherwood and Reeves 1983; Rice 1998). They are highly pelagic and rarely seen in inshore coastal waters. They migrate between lower latitudes in winter and higher latitudes in summer (Leatherwood and Reeves 1983).

Alaska Distribution

The summer range of *B. physalus* in the North Pacific extends along the Aleutian Islands and the Gulf of Alaska into the Bering and Chukchi seas (Rice 1998) (map 92). Some individuals remain in North Pacific waters as far north as the Aleutians throughout the year (Rearden 1981).

Surveys in 1999 and 2000 found fin whale abundance estimates were nearly five times higher in the central-eastern Bering Sea than in the southeastern Bering Sea (Moore et al. 2002), and most sightings in the central-eastern Bering Sea occurred in a zone of particularly high productivity along the shelf break (Moore et al. 2000).

Habitat

The fin whale is generally pelagic but may use deep coastal waters. They feed primarily in summer on a wide variety of small schooling fish and invertebrates (Wynne 1993).

Status

Balaenoptera physalus was legally protected in 1976. It is listed in Appendix I of CITES, Endangered under the ESA and by IUCN, and of Special Concern by COSEWIC. North Pacific fin whales today number about twenty thousand animals, down from over fifty thousand animals prior to their overexploitation between the late 1950s and the early 1960s (Gambell 1985; Martin 1990). A recent minimum estimate of the fin whale population in Alaska waters west of the Kenai Peninsula was 5,703 animals (Angliss and Outlaw 2007). These authors concluded that reliable information on trends in abundance for the Northeast Pacific stock of fin whales is not available.

Fossils

There is no fossil record for this species (Deméré et al. 2005).

Humpback Whale

Megaptera novaeangliae
(Borowski, 1781)

Other Common Names

Hump whale, hunchbacked whale.

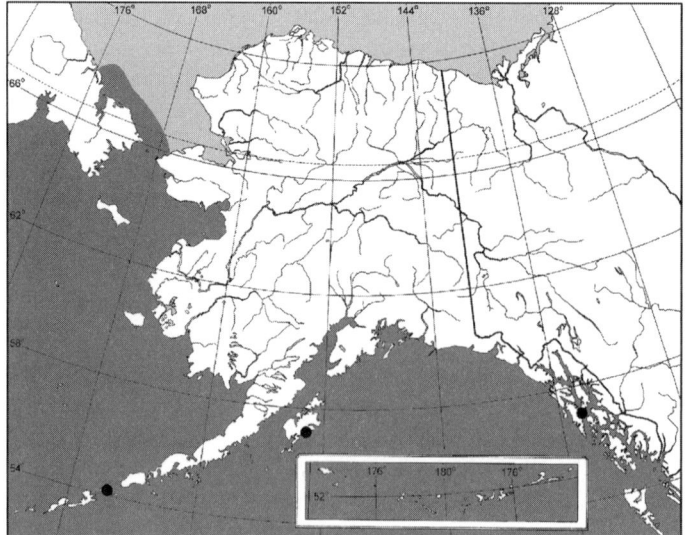

MAP 92. Distribution of fin whale, *Balaenoptera physalus*.

■ = Projected Range (after Angliss and Outlaw 2007)
● = Specimen Record (n = 15): Aleutian Islands, vicinity Akutan Island (7 USNM, fetuses); Kodiak Archipelago, Sitkalidak Island, Port Hobron (1 USNM); Alaska, unknown locality (2 USNM); Southeast Alaska, Admiralty Island, Tyee (2 AMNH); Alaska, no locality (1 AMNH); Alaska, on Pacific Coast (2 UWBM).

Systematics

Megaptera nodosa Bonnaterre, 1789 was used for the humpback whale until Kellogg (1932) demonstrated that Borowski's *novaeangliae* had priority. The genus *Megaptera* is monotypic and no subspecies are currently recognized (Mead and Brownell 2005).

Baker et al. (1994, 1998) attributed the subdivision of North Pacific humpback whales into two separate mtDNA stocks largely to maternal fidelity to distinct migratory destinations. Analyses of nuclear gene sequences found some evidence for male-mediated gene flow between stocks (Palumbi and Baker 1994).

Megaptera novaeangliae

Original Description. 1781. *Balaena novae angliae* Borowski, *Gemein. Naturgesch. Thier.*, 2(1):21.

Type Locality. Coast of New England, United States.

Type Specimen. None formally designated.

Global Distribution

Humpback whales are worldwide in distribution, with discrete Northern Hemisphere and Southern Hemisphere populations. In the North Pacific, they migrate seasonally between northern feeding grounds and three discrete southern wintering grounds (Rice 1998). The historic feeding range of humpback whales in the North Pacific encompassed coastal and inland waters around the Pacific Rim from Point Conception, California, north to the Gulf of Alaska and the Bering Sea, and west along the Aleutian Islands to the Kamchatka Peninsula and into the Sea of Okhotsk (Johnson and Wolman 1984; Nemoto 1957; Tomilin 1957).

Alaska Distribution

In the North Pacific, the summer feeding grounds extend from California to the southern Chukchi Sea (Winn and Reichley 1985; Rice 1998) (map 93). From spring through autumn, the largest concentrations are found in the Bering Sea and the eastern Aleutians, Prince William Sound, and in southeast Alaska (Nagorsen 1990; Rearden 1981). Most humpback whales that summer in Alaska waters winter primarily in the Hawaiian Island chain where they mate and calve (Clapham 1999c). The population status and migratory destinations of humpback whales that feed in the central Aleutian Islands have not been determined; however, marking studies have demonstrated a connection between the eastern Bering Sea and Japanese waters (Nishiwaki 1966).

Habitat

The humpback whale is commonly found in coastal or shelf waters, except during migration when it frequently travels across deep water (Clapham and Mead 1999). In summer, it commonly feeds on schooling fish and krill (Clapham 1999c).

Status

Due to overexploitation, North Pacific humpbacked whales decreased from an estimated

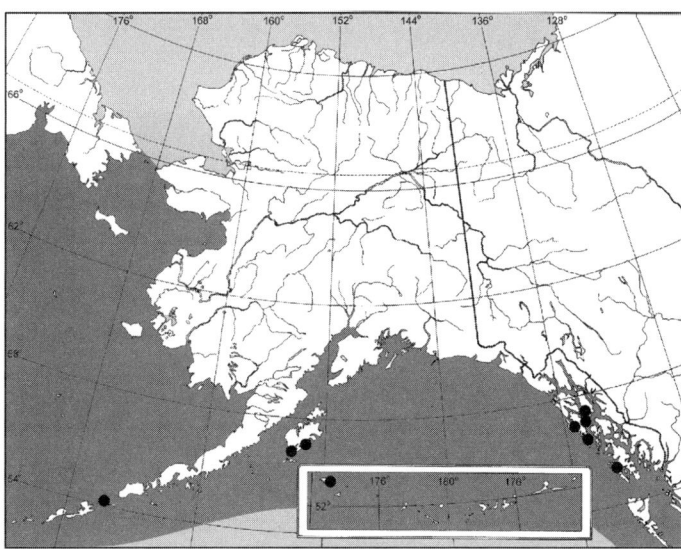

MAP 93. Distribution of humpback whale, *Megaptera novaeangliae*.

■ = Projected Range (after Angliss and Outlaw 2007)

● = Specimen Record (n = 23)

original population of fifteen thousand to about two thousand animals by the late 1980s (Martin 1990). North Pacific whales were given full protection in 1966 and appear to be making a strong recovery (Clapham 2002). Calambokidis et al. (1997) estimated the North Pacific population at six thousand to eight thousand animals. The humpback whale is listed in Appendix I of CITES, under the ESA as Endangered, by IUCN as Vulnerable, and by COSEWIC as Threatened.

Fossils

Fossils ascribed to *Megaptera* suggest their divergence from the *Balaenoptera* lineage by the Late Miocene (Fordyce 2002). Fossils of the species have been reported from the Pleistocene of Florida and Canada, and the Holocene of Japan (Deméré et al. 2005).

Humpback whale, *Megaptera novaeangliae* (B. Hines)

Family ESCHRICHTIIDAE Ellerman and Morrison-Scott, 1951—gray whale

The gray whale, *Eschrichtius robustus*, is the sole member of this family (Mead and Brownell 2005; Rice 1998).

Gray whales became extinct in the North Atlantic in early historical times but survive in two allo-patric populations in the North Pacific (Rice 1998). Fossils of Late Pleistocene age are known from western North America (Rice 1984).

Measuring up to 14 m in length, gray whales have a mottled gray body, a narrow, V-shaped head, short and yellow-colored baleen, and no dorsal fin. They are bottom "suction" feeders. Most of the eastern North Pacific population spends the summer in the shallow waters of the Bering Sea and farther north, and the rest of the year migrating (the longest known migration of any mammal) along the coast south to the lagoons of western Mexico where this species winters and females calve (Reeves et al. 2002).

Gray whale, *Eschrichtius robustus* (B. Hines)

Gray Whale

Eschrichtius robustus
(Lilljeborg, 1861)

Other Common Names

California gray whale.

Systematics

Molecular analyses (Árnason and Gullberg 1996; Árnason et al. 2004; Hasegawa et al. 1997) position the gray whale within the balaenopterids, while analyses based on morphology suggest a closer link with the balaenids and the pygmy right whale, *Caperea marginata* (Jones and Swartz 2002). This species was included in a phylogenetic analysis of baleen and toothed whales by Milinkovitch (1997). A combined analysis of partial and full cytochrome *b* gene sequences supported the hypothesis of a sister relationship between the gray whale and other baleen whales (fin, humpback, and bowhead). Other molecular studies (in Rice 1998) have failed to resolve the evolutionary relationships between the gray whale and species of rorquals, Balaenopteridae. Hall (1981) considered that the name *E. gibbosus* had priority over *E. robustus* (see Barnes and McLeod 1984). No subspecies are currently recognized (Mead and Brownell 2005).

Two stocks have been recognized in the North Pacific: the Eastern North Pacific stock, which lives along the west coast of North America, and the Western North Pacific stock, which lives along the eastern coast of Asia (Rice et al. 1984).

Eschrichtius robustus

Original Description. 1861. *Balaenoptera robusta* Lilljeborg, *Förh. Skand. Naturf. Ottende Møde*, Kjöbenhavn, 1860, 8:602 [1861].

Type Locality. Graso, Roslagen, Upland, Sweden.

Type Specimen. Partial skeleton (subfossil) in University Museum of Uppsala, Sweden.

Global Distribution

The gray whale is now restricted to the North Pacific Ocean. Two populations are recognized: one in the western Pacific, from southern China to the Sea of Okhotsk, and the other, eastern, from Mexico to the Bering, Chukchi, and western Beaufort seas in the north (Martin 1990; Rice 1998, 1999; Wolman 1985). North Atlantic populations were hunted to extinction by the 1700s (Nagorsen 1990).

Alaska Distribution

Gray whales migrate annually along the coasts of southeast Alaska between their summer feeding grounds in arctic waters and their wintering grounds in Mexico (map 94). In April and May, they enter the Bering Sea, primarily through Unimak Pass and continue moving along the coast of Bristol Bay, passing Nunivak Island and then St. Lawrence Island by May or June. They then disperse to feed primarily in shallow waters of the northern and western Bering Sea and Chukchi Sea. Since the mid-1990s, Stafford et al. (2007) reported an

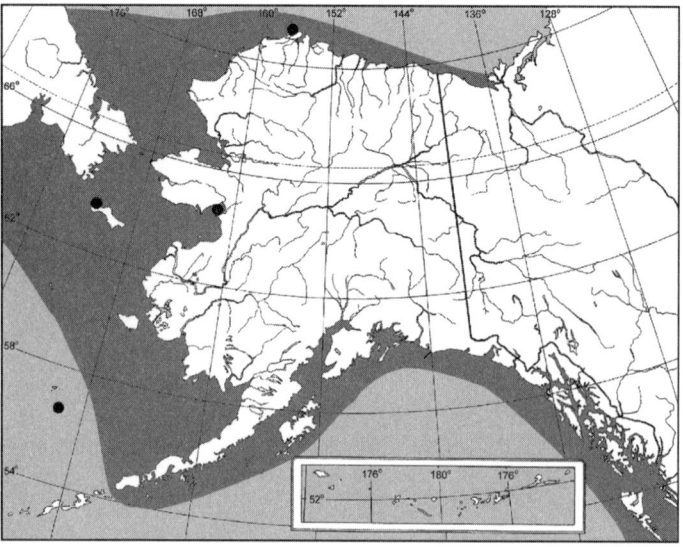

MAP 94. Distribution of gray whale, *Eschrichtius robustus*.

■ = Projected Range (after Angliss and Outlaw 2007)
● = Specimen Record (n = 6): Bering Sea, Norton Bay, silt flats between Inglutalik and Ungalik rivers (1 AMNH); Barrow (1 MVZ, 2 LACM); 56,57546, −169.61467, Bering Sea, Pribilof Islands, St. George Island (1 PSM); Bering Sea, St. Lawrence Island (1 UWBM).

increase in gray whales near Barrow, Alaska. The main southward migration begins in mid-October, passing through Unimak Pass between late October and early January (Frost 1994).

Habitat

Gray whales are the most coastal of the great whales, and are usually found in shallow coastal waters over the continental shelf (Jones and Swartz 2002). Unlike other baleen whales, the gray whale habitually forages on the ocean floor (Rice 1999).

Status

The gray whale is listed in Appendix I of CITES, and at Lower Risk (Northeast Pacific stock) by IUCN. Since their protection, the eastern North Pacific population of gray whales has recovered to pre-exploitation abundance and in 1998 were estimated at 26,600 animals (Rugh et al. 1999). This whale was delisted from the United States Endangered Species List in 1994 (Federal Register 59:31094) and is thought to be increasing (Angliss and Outlaw 2007).

Fossils

Repenning (1983) reported a Late Pleistocene fossil of gray whale from the Teshekpuk Lake Gubik Formation in northern Alaska. A fossil of this species was also reported from Late Pleistocene deposits in California (Barnes and McLeod 1984).

Family DELPHINIDAE Gray, 1821—dolphins

The Delphinids are the largest and most diverse family of marine mammals, comprising about thirty-four species in seventeen genera (Mead and Brownell 2005); eight species in seven genera have been found in Alaska waters. There is considerable debate among systematists on the classification of this group, particularly with regard to subfamilies and genera (Reeves et al. 2002). These debates were summarized by Rice (1998).

The fossil history of this family ranges from Late Miocene to Pleistocene in North America, Early Pliocene to Pleistocene in Europe, and Pliocene to Pleistocene in eastern Asia (Rice 1984).

Members of the dolphin family vary greatly in appearance, size, dentition, food habits, social behavior, and habitat preferences (Reeves et al. 2002). Most share some features in common, including numerous conical teeth, compression and fusion of neck vertebrae, and various anatomical specializations associated with their echolocation systems (Wilson and Ruff 1999). Dolphins have a prominent dorsal fin, a beak, and a bulbous forehead. Adult size ranges in length from about 1.5 m to 9.5 m in male killer whales (Rice 1984). Dolphins are shallow divers that surface frequently to breathe. Many species vocalize extensively and are highly gregarious, occurring in schools of hundreds or even thousands of animals (Rice 1984).

Killer whale, *Orcinus orca* (O. MacDonald)

Short-Finned Pilot Whale

Globicephala macrorhynchus
Gray, 1846

Other Common Names

Blackfish, Pacific pilot whale, short-finned blackfish.

Systematics

The genus *Globicephala* has a complicated fossil history (Hershkovitz 1966). Two species are currently recognized, although *G. macrorhynchus* sometimes is included with *G. melaena* (now *G. melas*), the long-finned pilot whale (see van Bree 1971). In the North Pacific, a separate species, *G. scammoni*, the Pacific pilot whale, was formerly recognized (van Bree 1971). No subspecies are currently recognized (Mead and Brownell 2005; Rice 1998).

Northern and southern forms of *G. macrorhynchus* have been noted near Japan (Kasuya et al. 1988) that appear to be genetically distinct (Wada 1988), and each is associated with different oceanic current systems (Reeves et al. 2002).

Globicephala macrorhynchus

Original Description. 1846. *Globicephalus macrorhynchus* Gray, *Zool. Voy. H.M.S. "Erebus" and "Terror"*, 1(Mammalia):33.

Type Locality. South Seas.

Type Specimen. Not known to exist.

Global Distribution

Short-finned pilot whales occur worldwide, but generally in tropical and warm-temperate waters. In the Pacific, this species ranges from the Gulf of Alaska to Guatemala (Martin 1990).

Alaska Distribution

The distribution of this tropical and temperate species regularly extends into the cold waters of the Gulf of Alaska from southeast Alaska northwestward to the Alaska Peninsula (Leatherwood and Dahlheim 1978; Orr 1951; Reilly 1978) (map 95). Leatherwood et al. (1987) suggested that movements of this whale into the more northern waters of the eastern North Pacific appear to be related to periodic incursions of warmer water.

Habitat

Short-finned pilot whales are usually found seaward, off the outer edges of the continental shelf in warm temperate and tropical waters (Abend 1999; Stacey and Baird 1993). Home (1980) reported this species spending long periods in shallow inshore areas in southeast Alaska.

Status

Globicephalus macrorhynchus is considered a rare but regular visitor to Canadian waters off the British Columbia coast (Stacey and Baird 1993); its status farther north in Alaska waters is unclear. This species virtually disappeared from the West Coast of the United States between 1984 and 1992 following a strong El

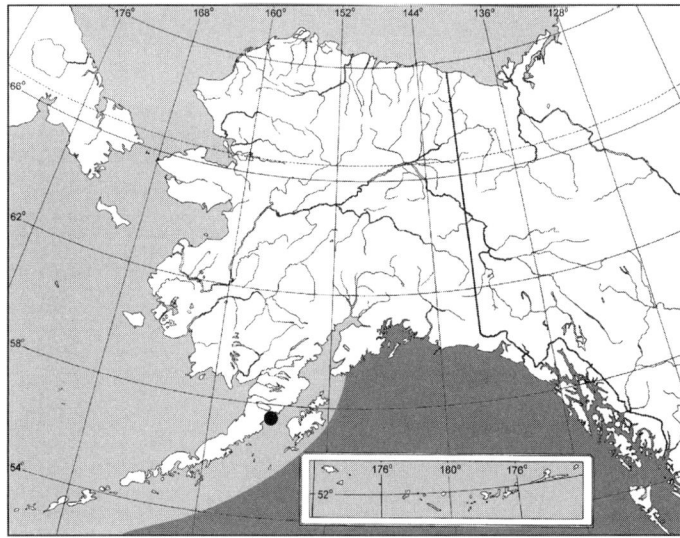

MAP 95. Distribution of short-finned pilot whale, *Globicephala macrorhynchus*.

■ = Projected Range (after Angliss and Outlaw 2007)
● = Specimen Record (n = 1): Alaska Peninsula, Kanatak area (1 CAS). [A fetus listed from St. George Island (USNM 218768) and 4 skulls said to be from Eschscholtz Bay (AMNH 181367–181370) may be in error and are not denoted on this map.]

Niño event in 1982–1983. Since then sightings (or mortalities) have remained rare (Carretta et al. 2005). The species is listed in Appendix II of CITES, at Lower Risk by IUCN, and Not at Risk by COSEWIC.

Fossils

Fossils of the extant delphinid groups appeared in the Middle to Late Miocene, and fossils of the genus *Globicephala* have been uncovered that date from the Pleistocene (Olson and Reilly 2002). Skulls of *G. melas* that dated to the eighth to twelfth centuries have been recovered in Japan, where today only *G. macrorhynchus* occurs (Olson and Reilly 2002).

Risso's Dolphin

Grampus griseus
G. Cuvier, 1812

Other Common Names

Grampus, gray grampus, grey dolphin, mottled grampus, white-headed grampus.

Systematics

Ellerman and Morrison-Scott (1951), Schevill (1954), and Hershkovitz (1961) clarified the appropriate use of the generic name *Grampus* for this monotypic species.

Grampus griseus

Original Description. 1812. *Delphinus griseus* G. Cuvier, *Ann. Mus. Nat. Hist.*, Paris 19:14.

Type Locality. Brest, France.
Type Specimen. Stuffed skin and skull in MNHN.

Global Distribution

Risso's dolphins are found in tropical and temperate waters throughout the world's oceans (Martin 1990).

Alaska Distribution

The northern limit of this species in the cooler northern waters of the eastern North Pacific Ocean probably does not exceed the northern Gulf of Alaska (map 96), where it is a rare visitor during the summer months (Braham 1983; Kruse et al. 1999; Rice 1998; Shelden 1999). Shults et al. (1982) necropsied a Risso's dolphin found beached on Middleton Island.

Habitat

Risso's dolphins inhabit deep offshore waters, occurring close to shore only where the continental shelf is narrow (Martin 1990; Shelden 1999). They apparently prefer steep shelf-edge habitats between about 400 and 1,000 m in depth (Baird 2002a).

Status

Grampus griseus is considered relatively abundant at lower latitudes (Martin 1990). It is listed in Appendix II of CITES, as Data Deficient by IUCN, and Not at Risk by COSEWIC.

Fossils

No known fossils exist.

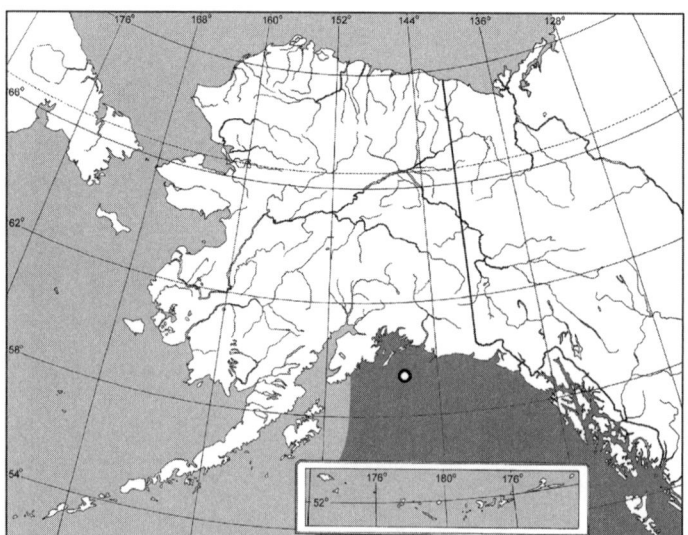

MAP 96. Distribution of Risso's dolphin, *Grampus griseus*.

■ = Projected Range (after Angliss and Outlaw 2007)
O = Specimen Record (n = 0; found on beach on Middleton Island, deposition of specimen unknown; see Shults et al. 1982).

Pacific White-Sided Dolphin

Lagenorhynchus obliquidens
(Gill, 1865)

Other Common Names

Pacific white-striped dolphin.

Systematics

The genus *Lagenorhynchus* traditionally includes six species of cold-water dolphins (Rice 1998). This hypothesis was supported in a recent analysis of cranial morphology and vertebral formulae (Miyazaki and Shikano 1997). Analysis of the cytochrome *b* gene among delphinids, however, indicated that *Lagenorhynchus* is polyphyletic (LeDuc et al. 1999). Although the Pacific white-sided dolphin and the dusky dolphin are traditionally placed in the genus *Lagenorhynchus*, a recent molecular analysis (May-Collado and Agnarsson 2006) indicates that they are more closely related to the southern dolphins of the genus *Cephalorhynchus*. Some authors (e.g., Honacki et al. 1982) have considered *L. obliquidens* a junior synonym of the dusky dolphin, *L. obscurus*, from the Southern Hemisphere, a premise not supported by recent analyses of molecular data (Cipriano 1997; May-Collado and Agnarsson 2006).

The North Pacific population is divided into two stocks; one is in the northeastern Pacific, and the other in the northwestern Pacific. An area with low population density separates the two stocks along the southern Aleutian Islands (IUCN 1991).

No subspecies are recognized, but distinguishable populations have been reported from the northeast Pacific and in Japanese waters (van Waerebeek and Würsig 2002; Walker et al. 1986).

Lagenorhynchus obliquidens

Original Description. 1865. *Lagenorhynchus obliquidens* Gill, *Proc. Acad. Nat. Sci. Phil.* 17:177.
Type Locality. Near San Francisco, California.
Type and Co-Type Specimens. USNM 1961, 1962, 1963.

Global Distribution

Pacific white-sided dolphins are found in the North Pacific Ocean north of 20° N (Brownell et al. 1999), but primarily between 38° N and 47° N (Hobbs and Jones 1993). It is common on the high seas, along the continental margins, and enters the inshore coastal areas of Alaska, British Columbia, and Washington (Dahlheim and Towell 1994; Ferrero and Walker 1996).

Alaska Distribution

Pacific white-sided dolphins occur in Alaska waters as far north as the southern Bering Sea (rarely), the Gulf of Alaska, and the Aleutian Islands (Angliss and Outlaw 2007; Kajimura and Loughlin 1988; Leatherwood et al. 1984; Walker et al. 1986) (map 97).

Habitat

Pacific white-sided dolphins are found in cold-temperate waters mainly on the continental

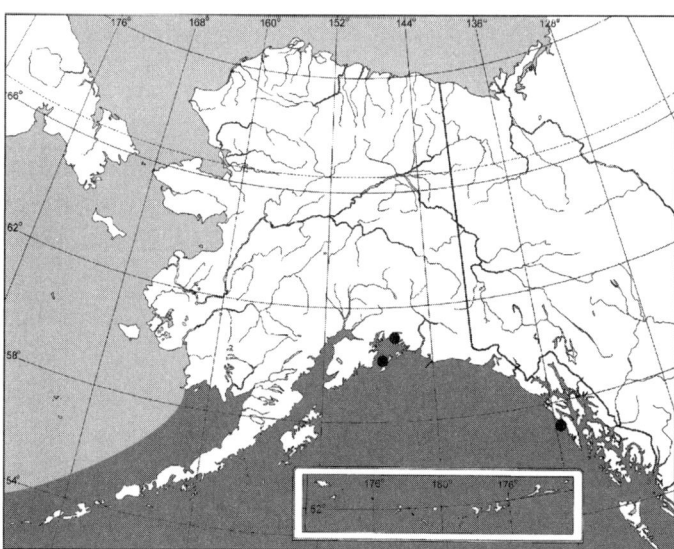

MAP 97. Distribution of Pacific white-sided dolphin, *Lagenorhynchus obliquidens*.

■ = Projected Range (after Angliss and Outlaw 2007)
● = Specimen Record (n = 3): Prince William Sound, Swan Point near Valdez (1 AMNH); Gulf of Alaska, Montague Strait (1 USNM); "Sitka Island"(1 USNM) = Kruzof or Baranof Island (Orth, 1971).

CETACEA

shelf, continental slope, and farther offshore; however, they enter protected, but deep, inshore marine waters (Brownell et al. 1999; Martin 1990).

Status

The Pacific white-sided dolphin may be the most abundant dolphin in the temperate northeastern Pacific. Brownell (1999) reported that about nine hundred thousand to one million Pacific white-sided dolphins currently inhabit the North Pacific. No reliable information on trends in abundance for the North Pacific population is currently available (Angliss and Outlaw 2007).

Residents of southeast Alaska have noted a large influx of Pacific white-sided dolphins into the region since the 1990s (MacDonald and Cook 2007). Surveys in the Gulf of Alaska in 1997 found one group of 164 dolphins off Dixon Entrance, and similar surveys in Bristol Bay in 1999 reported a group of 18 dolphins off Port Moller (Angliss and Outlaw 2007). This species is listed in Appendix II of CITES, as Lower Risk by IUCN, and Not at Risk by COSEWIC.

Fossils

A Late Pleistocene skull belonging to the genus *Lagenorhynchus* was reported from Southern California (Barnes 1976).

Northern Right Whale Dolphin

Lissodelphis borealis
(Peale, 1848)

Other Common Names

Northern right whale porpoise, Pacific right whale dolphin.

Systematics

The genus *Lissodelphis* has a complicated taxonomic history (see Hershkovitz 1966; Jefferson and Newcomer 1993). Although Honacki et al. (1982) and Mead and Brownell (2005) speculate that *L. borealis* may be a subspecies of *L. peronii*, the southern right whale dolphin, most authorities still accept them as separate species. No subspecies are recognized (Rice 1998) or stocks identified (Mitchell 1975).

Lissodelphis borealis

Original Description. 1848. *Delphinapterus borealis* Peale, *U.S. Exploring Expeditions*, 8 (Mammalogy and Ornithology):35.

Type Locality. North Pacific, 46°6'50" N, 134°5' W (10° west of Astoria, Oregon).

Type Specimen. None designated.

Global Distribution

The northern right whale dolphin, a North Pacific endemic, normally ranges from 30° to 50° N in the eastern Pacific and 35° to 51° N in the western Pacific (Jefferson et al. 1994),

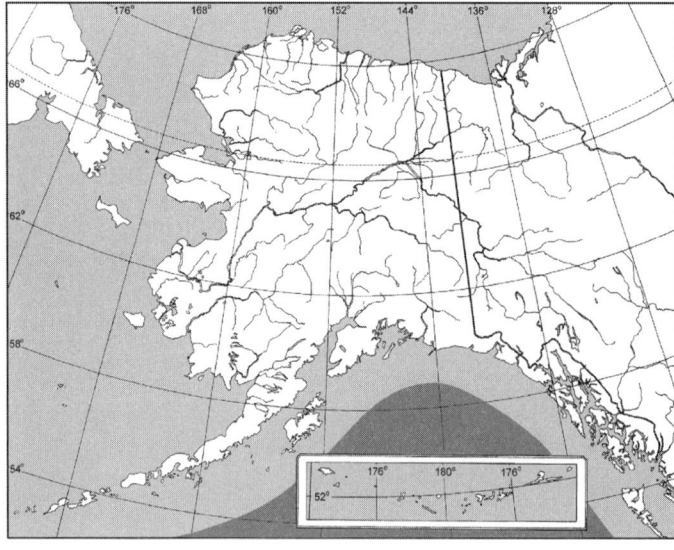

MAP 98. Distribution of northern right whale dolphin, *Lissodelphis borealis*.

■ = Projected Range (after Angliss and Outlaw 2007)
Specimen Record (n = 0)

and only rarely enters subarctic or the coldest temperate waters (Leatherwood and Reeves 1983). The eastern and western Pacific populations may be disjunct (Leatherwood and Reeves 1983).

Alaska Distribution

The occurrence of this species in Alaska waters is based on a limited number of sightings as far north as 59° N in the Gulf of Alaska and just south of the Aleutian Islands in the central Pacific (Kajimura and Loughlin 1988) (map 98). Scammon's (1874) report of this species in the Bering Sea lacks confirmation.

Habitat

This species inhabits cool, deep offshore waters (Jefferson et al. 1994).

Status

The northern right whale dolphin is probably extralimital to Alaska waters. The species is listed in Appendix II of CITES.

Fossils

No known fossils of this species exist.

Killer Whale

Orcinus orca
(Linnaeus, 1758)

Other Common Names

Blackfish, orca.

Systematics

Most authors recognize one species in the genus. Berzin and Vladimirov (1982, 1983) described a second species, *O. glacialis*, from the Antarctic, but Rice (1998) and others have suggested that further studies are needed to ascertain whether these small killer whales deserve recognition. Currently no subspecies are recognized for this broad-ranging species, for which a worldwide review of specimens is needed to document geographical variation (Dahlheim and Heyning 1999).

Three non-associating group types—resident, transient, and offshore—occur off the coasts of Washington, British Columbia, and southern Alaska (Dahlheim et al. 1997). The "resident" and "transient" types are believed to differ in several aspects of morphology, ecology, and behavior. Studies on mtDNA restriction patterns provide evidence that the "resident" and "transient" types are genetically distinct (Hoelzel 1991; Hoelzel and Dover 1991; Hoelzel et al. 1998; Stevens et al. 1989). Less is known about the "offshore" type killer whales that are encountered primarily off the coasts of California, Oregon, and British Columbia and, rarely, in southeast Alaska (Dahlheim et al. 1997; Ford et al. 1994).

Orcinus orca

Original Description. 1758. [*Delphinus*] *orca* Linnaeus, *Syst. Nat.*, 10th ed., 1:77.
Type Locality. European seas.

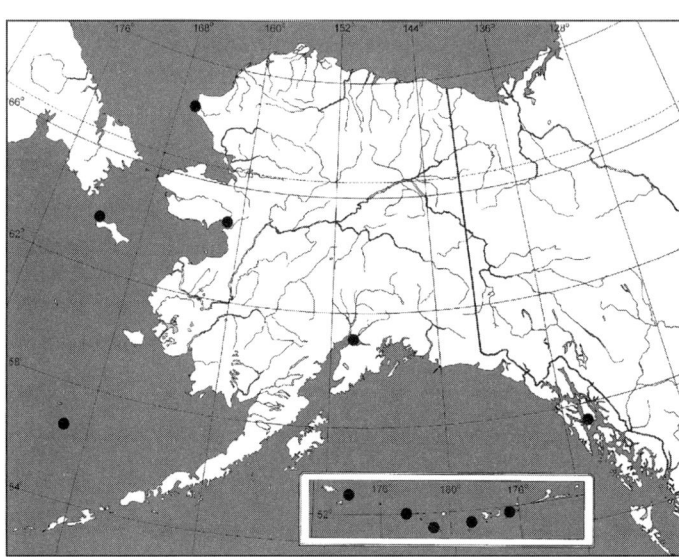

MAP 99. Distribution of killer whale, *Orcinus orca*.

■ = Projected Range (after Angliss and Outlaw 2007)
● = Specimen Record (n = 11)

Type Specimen. None designated.

Global Distribution

Killer whales are found in all the world's oceans and seas, from equatorial regions to the polar pack-ice zones (Rice 1998). They may be the most widely distributed mammal in the world (Heyning 1999).

Alaska Distribution

Killer whales occur throughout the marine waters of Alaska and are most common over the waters of the continental shelf from southeast Alaska through the Aleutian Islands and northward throughout the Bering Sea and into the Chukchi and Beaufort seas during ice-free seasons (Dahlheim and Heyning 1999) (map 99).

Habitat

A top-level marine carnivore, killer whales appear to be equally at home in coastal or oceanic waters and inhabit a variety of marine environments (Heyning and Dahlheim 1988; Martin 1990).

Status

Although densities may vary, killer whales are considered an abundant whale with a world-wide estimate of at least one hundred thousand animals (Klinowska 1991). Pod size in Alaska water ranges from one to one hundred animals, with only 1 percent of these groups containing twenty or more individuals (Braham and Dahlheim 1982). The species is listed in Appendix II of CITES, and as Lower Risk by IUCN. COSEWIC lists the resident type of killer whale in British Columbia waters as endangered or threatened, the transient type as threatened, and the offshore type of special concern.

Fossils

Fossil teeth attributed to *O. orca* or a closely related species have been reported from Pliocene-age deposits in Japan (Matsumoto 1937) and Italy (Sarra 1933).

False Killer Whale

Pseudorca crassidens
(Owen, 1846)

Other Common Names

None.

Systematics

Placement of this genus along with *Orcinus* in the subfamily Orcininae has been contentious (Stacey et al. 1994). The two species do not appear to be closely related based on genetic similarity (Baird 2002b; May-Collado and Agnarsson 2006). There are no subspecies currently recognized (Rice 1998).

Pseudorca crassidens

Original Description. 1862. *Phocaena crassidens* Owen, *Hist. Brit. Foss. Mamm. Birds*, p. 516, fig. 213.

Type Locality. Lincolnshire Fens, near Stanford, England (subfossil skull).

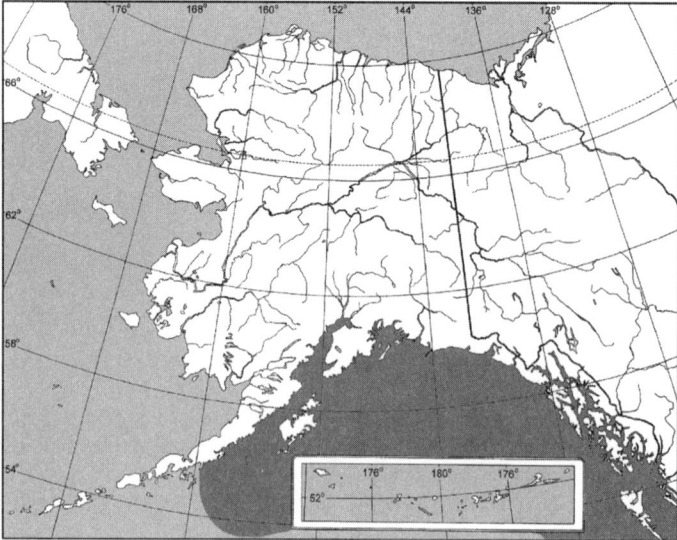

MAP 100. Distribution of false killer whale, *Pseudorca crassidens*.

■ = Projected Range (after Angliss and Outlaw 2007)
Specimen Record (n = 0)

RECENT MAMMALS OF ALASKA

256

Type Specimen. Originally in the Museum of Stanford, then possibly in London's College of Surgeons, followed by Cambridge University Museum, before said to be lost (Flower, 1884, *Cat. Osteol. Spec. Roy. Coll. Surgeons London*, 2(Mammalia): 573).

Global Distribution

The false killer whale is found worldwide in tropical and temperate waters. In the northeastern Pacific, the most northerly extensions of this gregarious species appear related to spring-summer warming of the waters (Leatherwood and Reeves 1983).

Alaska Distribution

Inclusion of this small cetacean in Alaska waters is based on reports from Prince William Sound (Leatherwood and Reeves 1983; Leatherwood et al. 1988), and on the recent sighting (and positive identification with photographs) of a lone animal near Juneau in May 2003 (NOAA-Alaska Region News Release, 2 July) (map 100).

Habitat

This schooling species is thought to prefer warmer offshore waters, migrating from north to south with the seasonal changes in water temperature.

Status

The false killer whale is a rare visitor to Alaska waters (Leatherwood and Reeves 1983; Leatherwood et al. 1988). The current number of false killer whales is unknown but they are never abundant. They are listed in Appendix II of CITES and as Lower Risk by IUCN.

Fossils

This species was first discovered as a subfossil in an English fen in 1846 and thought to be extinct until a school of them stranded on the coast of Germany fifteen years later (Rice 1998).

Pantropical Spotted Dolphin

Stenella attenuata
(Gray, 1846)

Other Common Names

None.

Systematics

Taxonomic revisions and synonymies are discussed by Perrin et al. (1987) and Perrin (2001). Perrin (1975) distinguished Hawaiian, offshore, and coastal subspecies; however, Rice (1998) indicated that a review of subspecies throughout most of this species range has not been attempted because of the limited availability of specimens. Douglas et al. (1984) suggested that inshore and offshore stocks warrant taxonomic separation, which Perrin (2001) recognized as subspecies, *S. a. attenuata* and *S. a. graffmani*.

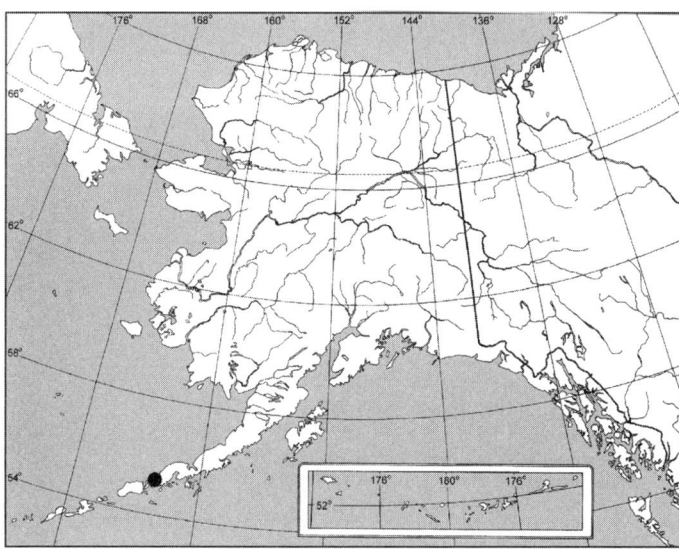

MAP 101. Distribution of pantropical spotted dolphin, *Stenella attenuata*.

● = Specimen Record (n = 1): Alaska Peninsula, Cold Bay (USNM 550771).

Stenella attenuata

Original Description. 1846. *Steno attenuatus* Gray, *Zool.Voy. H.M.S. "Erebus" and "Terror",* 1:44.

Type Locality. None given, unknown.

Type Specimen. Not known to exist.

Global Distribution

This species is found in warm oceans worldwide.

Alaska Distribution

The occurrence of this species in Alaska waters is based on a stranded animal from Cold Bay, Alaska Peninsula, preserved as a skull (USNM 550771) in 1983 (map 101).

Habitat

This species occurs nearshore and offshore in tropical to warm temperate waters (Reeves et al. 2002).

Status

Alaska vagrant. The pantropical spotted dolphin is considered one of the most abundant cetaceans worldwide (Reeves et al. 2002). It is listed in Appendix II of CITES, and as Lower Risk (Conservation Dependent) by IUCN.

Fossils

No fossils are known (Perrin 2001).

Family MONODONTIDAE Gray, 1821—beluga and narwhal

Only two living species, the beluga and the narwhal, are in the family Monodontidae (Mead and Brownell 2005). Both occur in Alaska waters, although the status of the narwhal is that of a rare visitor. While the close systematic relationship between the two species has long been accepted by most authorities, there has been a history of debate related to classification at the super- and subfamily level (Rice 1998).

The only known fossil of a monodontid is from the Late Miocene of Baja California (Rice 1998).

Monodontids are medium-sized whales with prominent foreheads and almost no beak (Reeves et al. 2002). As adults, belugas are all white in color and narwhals are mottled black and white. Unlike belugas, narwhals have no functional teeth. Male (and some female) narwhals develop a long (up to 3 m), spiraling tusk (occasionally two) from the left side of the upper jaw. The purpose of the tusk has been much debated; most recently, Nweeia (2007) provided evidence that it functions primarily as an acute sensory organ. Both species are gregarious inhabitants of arctic waters and pack ice. Narwhals are the more pelagic and ice-loving of the two, living in deep sounds and channels. Belugas prefer shallow estuaries and even ascend major rivers on occasion.

Beluga, *Delphinapterus leucas* (W. D. Berry)

Beluga

Delphinapterus leucas
(Pallas, 1776)

Other Common Names
Belukha, white whale.

Systematics
Kasuya (1973) placed *Delphinapterus* and *Orcaella* together in the Delphinapteridae, but this arrangement has not been followed by other authors (Brodie 1989). Some Russian authors split the belugas into three species or subspecies (Rice 1998). Although there is some variation in body size between stocks, a single species is generally accepted and no subspecies are currently recognized (R. Stewart 1999).

O'Corry-Crowe et al. (1997) examined mtDNA sequences from the control region for 324 belugas. They suggested a rapid radiation of these whales into several genetically distinctive groups occurred following the Pleistocene. These groups can be identified by summering concentrations. The Cook Inlet population is divergent from all others and may have been isolated since the last glacial period (Murray and Fay 1979; O'Corry-Crowe et al. 1997, 2002).

Delphinapterus leucas
Original Description. 1776. *Delphinus leucas* Pallas, *Reise durch verschiedene Provinzen des Russischen Reichs*, 3(book 1):85, footnote.

Type Locality. Mouth Obi [Ob] River, northeastern Siberia.
Type Specimen. None designated.

Global Distribution
Belugas inhabit seasonally ice-covered arctic and subarctic waters around the world, ranging from the Bering Sea, through the Canadian Arctic, in Hudson Bay, around Greenland, and north across the Arctic coast of Russia (R. Stewart 1999). The southernmost belugas occupy areas in the St. Lawrence River estuary, White Seas, Sea of Okhotsk, northern Gulf of Alaska, and James Bay (Stewart and Stewart 1989).

Alaska Distribution
In Alaska, belugas have a discontinuous distribution from Yakutat Bay and Cook Inlet in the Gulf of Alaska to the Alaska-Yukon border in the Beaufort Sea (map 102).

Belugas of the Bering Sea population range throughout the Bering, Chukchi, and Beaufort seas. Major concentrations occur in Bristol Bay, Yukon River and Norton Sound, Kotzebue Sound, and waters in and adjacent to Kasegaluk Lagoon (Hazard 1988). Belugas from Bristol Bay may overlap in range with whales that summer in more northern Alaska, Canada, or Russia waters (Hazard 1988).

Belugas inhabit Cook Inlet during all seasons and, infrequently, from as far east in the Gulf of Alaska as Yakutat to as far west as Shelikof Strait near Kodiak Island (Hazard 1988). During spring and summer months, these whales are typically concentrated near

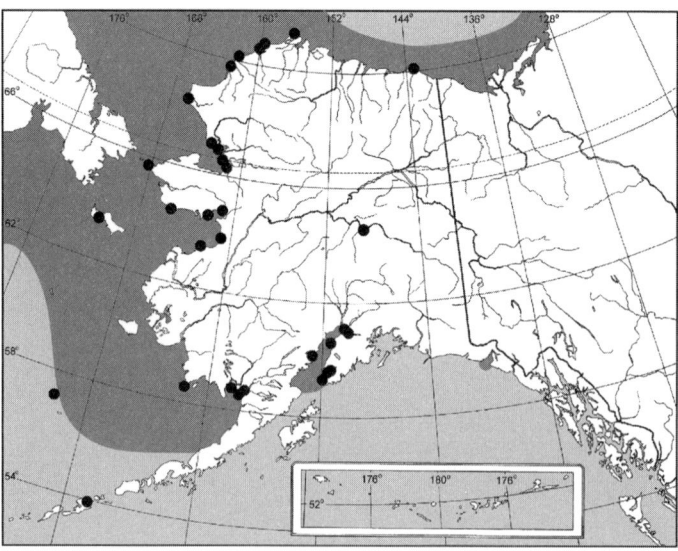

MAP 102. Distribution of beluga, *Delphinapterus leucas*.

■ = Projected Range (after Angliss and Outlaw 2007)
● = Specimen Record (n = 346)

river mouths in northern Cook Inlet (Rugh et al. 2000). A small number of beluga whales (under twenty animals and considered part of the Cook Inlet stock) occur seasonally, perhaps year-round, in Yakutat Bay (O'Corry-Crowe et al. 2006).

Belugas are known to ascend large rivers such as the Beluga and Susitna rivers in Cook Inlet and the Naknek and Kvichak rivers in Bristol Bay. They have been sighted on the Yukon River at Tanana, Rampart, and Fort Yukon (four animals in 1993) (Lowry 1994). In 2006, a dead beluga was found on the Tanana River approximately 6.5 km above Nenana (UAM 85670). Such movements may be less common than they were a century ago (Lowry 1994).

Habitat

Belugas are usually found in offshore ice-covered areas in winter and spring and in coastal waters and estuaries in summer and autumn. These whales regularly ascend major rivers and seem to be unaffected by changes in salinity (Lowry 1994).

Status

Belugas were formerly abundant throughout the arctic and subarctic. The current total world population has been estimated at close to one hundred thousand animals (IWC 1992), of which about twenty-five thousand are found in the Bering Sea (Wynne 1993). Many of the sixteen stocks provisionally recognized by the IWC Scientific Committee have been seriously reduced by overexploitation. The apparently isolated Cook Inlet stock, one of five recognized within Alaska waters (O'Corry-Crowe et al. 1997, 2002), experienced a sharp decline between 1994 and 1998 and then remained stable through 2004 at about 360 whales (NMFS 2005). An estimate in 2005 suggested further decline to 278 animals (NMFS 2006). The species is listed in Appendix II of CITES. IUCN (2007) lists the species as Vulnerable and the Cook Inlet population as Critically Endangered. Cook Inlet belugas were listed on 17 October 2008 under the ESA as Endangered.

Fossils

Beluga fossils of Pleistocene age have been reported from deposits in eastern Canada and Vermont (Harington 1981). Repenning (1983) reported Late Pleistocene fossils from the Teshekpuk Lake Gubik Formation in northern Alaska. Numerous beluga bones were recovered from Late Holocene archaeological sites in the Cape Denbigh region in western Alaska (Giddings 1964; Savinetsky et al. 2004).

Narwhal

Monodon monoceros
Linnaeus, 1758

Other Common Names

Narwhale, unicorn whale.

Systematics

All recent authorities have included the narwhal and the beluga (*Delphinapterus leucas*) as a single family, Monodontidae, in the superfamily Delphinoidea, except for Kasuya (1973). He placed the beluga in a separate family, Delphinapteridae, along with the Irrawaddy dolphin (*Orcaella brevirostris*). Some authors have recognized separate subfamilies for *Monodon* and *Delphinapterus* (Miller and Kellogg 1955). *Monodon monoceros* is monotypic and no subspecies are recognized (Rice 1998). Four geographically distinct stocks have been suggested (Martin 1990).

MtDNA analysis by Palsbøll et al. (1997) was used to determine population structure and seasonal movements of narwhals in the northwestern Atlantic.

Monodon monoceros

Original Description. 1758. [*Monodon*] *monoceros* Linnaeus, *Syst. Nat.*, 10th ed., 1:75.

Type Locality. Arctic seas.

Type Specimen. None designated.

Global Distribution

The narwhal has a discontinuous High Arctic distribution, occurring regularly from the central Canadian Arctic and west Greenland, eastward to central Russia (Hay and Mansfield 1989). They are considered infrequent or rare in eastern Siberia, Alaska, and the western Canadian Arctic (Martin 1990; Reeves and Tracey 1980). All populations are migratory (Reeves 1999).

Alaska Distribution

Narwhals are occasionally found in Alaska waters (Reeves 1978b). Documented records and observations (Bee and Hall 1956; Geist et al. 1960; Huey 1952) extend from the mouth of the Colville River in the western Beaufort Sea, to Nelson Lagoon on the north side of the Alaska Peninsula (map 103).

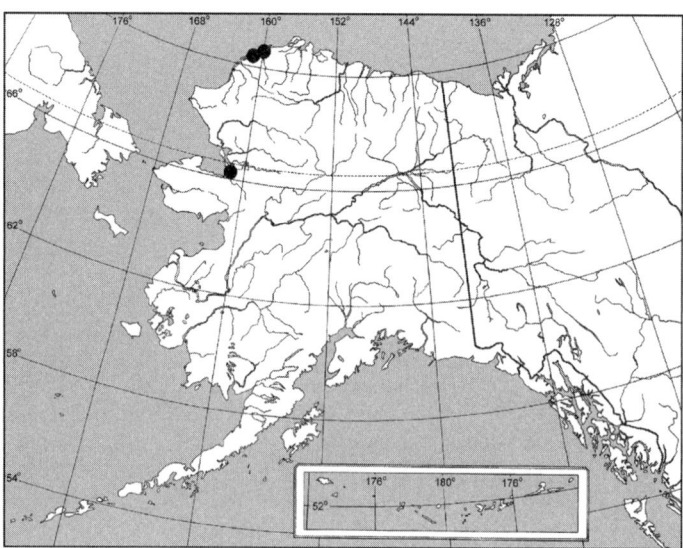

MAP 103. Distribution of narwhal, *Monodon monoceros*.

● = Specimen Record (n = 3): Chukchi Sea, Kotzebue Sound, Spafarief Bay, Kiwalik (1 UAM); Chukchi Sea, old village site halfway between Wainwright and Icy Cape (1 PSM); 70.3815, −160.0145, Chukchi Sea, Wainwright (1 UWBM).

Habitat

In contrast to the beluga, the narwhal is a deep-water species that only occasionally enters shallow bays and estuaries (Martin 1990). This species is rarely found far from ice, and has been known to penetrate deep into the polar pack ice in summer (Rutilevskii 1958).

Status

Monodon monoceros is considered locally abundant, especially in the Western Hemisphere (Leatherwood and Reeves 1983). The species is listed in Appendix II of CITES, and as Data Deficient by IUCN. It is an infrequent visitor to Alaska waters.

Fossils

Monodontids occupied temperate waters as far south as Baja California during the Late Miocene and Pliocene (Berta and Sumich 1999, and citations therein). Narwhal fossils have been found in an Early Pleistocene deposit in England, and in a Late Wisconsin–Early Holocene deposit in the Gulf of St. Lawrence (Reeves and Tracey 1980).

Narwhal, *Monodon monoceros* (B. Hines)

Family PHOCOENIDAE Gray, 1825—porpoises

The porpoise family consists of six species in three genera (Rice 1998); two species of two genera occur in Alaska waters.

Although some authorities have divided the phocoenids into two subfamilies and four genera (Barnes 1984, 1985), this remains an unresolved issue and an arrangement not corroborated by genetic studies (Rosel et al. 1995).

The geologic range of phocoenids is Late Miocene to Early Pliocene in western North America, and Pleistocene in eastern Asia (Rice 1984).

The porpoises are small cetaceans (up to 2 m in *Phocoenoides dalli*) with relatively small flippers and no prominent beak. All but one species possess a dorsal fin. Their spade-shaped teeth form a cutting edge, and are rudimentary in *Phocoenoides* (Rice 1984). Most species inhabit shallow, nearshore waters and are gregarious.

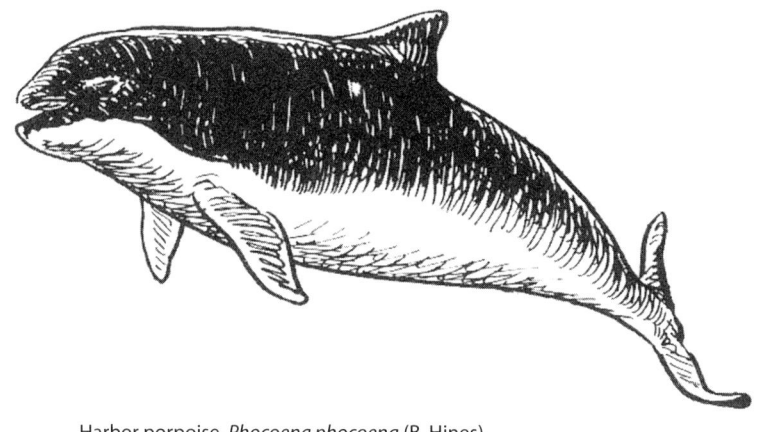

Harbor porpoise, *Phocoena phocoena* (B. Hines)

Harbor Porpoise

Phocoena phocoena
(Linnaeus, 1758)

Other Common Names
Common porpoise.

Systematics
Based on skull morphology, Miyazaki et al. (1987) and Amano and Miyazaki (1992) recognized two subspecies, one in the Atlantic and one in the Pacific. They noted that western Pacific animals differ sufficiently from those in the eastern Pacific to warrant subspecific separation (as yet unnamed according to Rice 1998). Unless further studies validate the naming of a separate subspecies for the western North Pacific population (*sensu* Rice 1998), only one subspecies occurs in Alaska waters.

Two distinct mtDNA clades exist along the west coast of North America (Rosel 1992). One is present in California, Washington, British Columbia, and Alaska (no samples were available from Oregon), while the other is found only in California and Washington (Angliss and Outlaw 2007).

Phocoena phocoena vomerina
Original Description. 1865. *Phocaena vomerina* Gill, *Proc. Acad. Nat. Sci. Philadelphia*, 17:178.

Type Locality. Puget Sound, Washington.

Type Specimen. USNM 4149.

Range. North Pacific Ocean.

Global Distribution
Harbor porpoises are restricted to temperate and subarctic waters in the Northern Hemisphere (Martin 1990; Read 1999). In the North Pacific, they occur from Japan and the Bering Sea, and to a lesser extent the Chukchi Sea, south to California.

Alaska Distribution
Two apparently separate populations occur in Alaska waters. In the western North Pacific, harbor porpoises are found from Japan north to the base of the Kamchatka Peninsula, including the Commander and Near islands in the western Aleutians. Members of this population are considered vagrants north through the Bering Strait as far as Wrangel Island in the western Chukchi Sea (Rice 1998).

In the eastern North Pacific, the species ranges along the coast from Southern California northward through the Gulf of Alaska to Unimak Island in the eastern Aleutians, and into the Bering Sea to the Pribilof Islands and southeastern Bristol Bay (map 104). Harbor porpoises are seen seasonally farther north along the Alaska coast to Point Barrow and the Mackenzie River in Canada (Read 1999; Rice 1998; Suydam and George 1992).

Habitat
Harbor porpoises occur most frequently in waters less than 100 m in depth (Angliss and

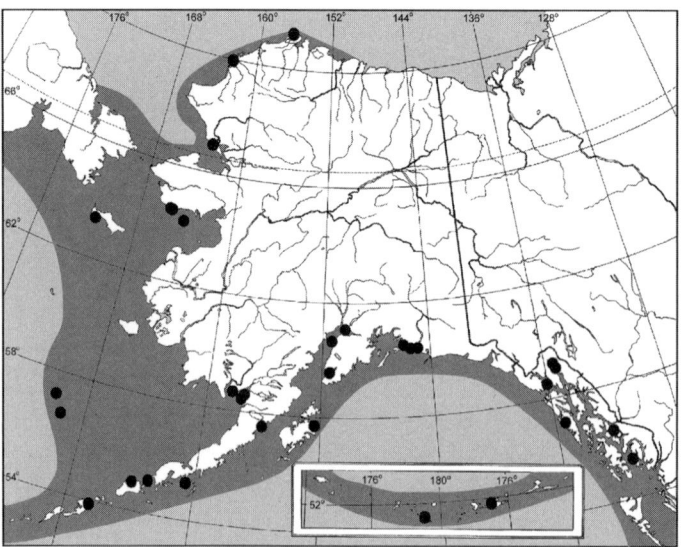

MAP 104. Distribution of harbor porpoise, *Phocoena phocoena*.

■ = Projected Range (after Angliss and Outlaw 2007)
● = Specimen Record (n = 65)

Outlaw 2007) in bays and estuaries, areas of coastal upwelling, and in tidal races (Martin 1990).

Status

The world population is unknown, but there are strong indications that harbor porpoise numbers are declining and their ranges are constricting (Reeves and Leatherwood 1994; Rosel 1997). Areas of high density occur in Glacier Bay and Yakutat Bay (Angliss and Outlaw 2007). The species is currently listed in Appendix II of CITES, as Vulnerable by IUCN, and of Partial Status of Special Concern by COSEWIC.

Fossils

Phocoenids are a rather old group of small toothed whales, with the earliest fossils, all from the margins of the Pacific, dating from the Late Miocene (Gaskin 1999). Fossils of *P. phocoena* have been reported from Pliocene deposits in Russia (Kirpichnikov 1951).

Dall's Porpoise

Phocoenoides dalli
(True, 1885)

Other Common Names

Whitefin porpoise, whiteflank porpoise, whitesided porpoise, spray porpoise.

Systematics

Two morphologically distinct forms of Dall's porpoise are known: the broadly distributed and more oceanic *dalli* type and the more restricted, more coastal *truei* type (Houck and Jefferson 1999; McMillan and Bermingham 1996). These two forms have been considered species (Andrews 1911), subspecies (Kuroda 1954; Mead and Brownell 2005; Rice 1998), and color morphs of the same species, *P. dalli* (Houck 1976). Most authors do not recognize subspecies, although Rice (1998) considered the evidence of geographical variation in the color-phase ratio presented by Morejohn (1979) and others sufficient to permit the recognition of two subspecies. In a recent analysis of mtDNA data, Escorza-Treviño et al. (2004) found shared haplotypes between the two morphotypes suggestive of a lack of reproductive isolation between them, and leading the authors to conclude that the two forms are conspecific.

As many as eight stocks have been proposed, each centered on what are thought to be major calving grounds (IWC 1992).

Phocoenoides dalli

Original Description. 1885. *Phocaena dalli* True, *Proc. U.S. Nat. Mus.*, 8:95, June 19.
Type Locality. Strait west of Adakh [= Adak], Aleutian Islands, Alaska.
Type Specimen. USNM 21762.

Global Distribution

Dall's porpoises are found only in the North Pacific Ocean, usually near the continental

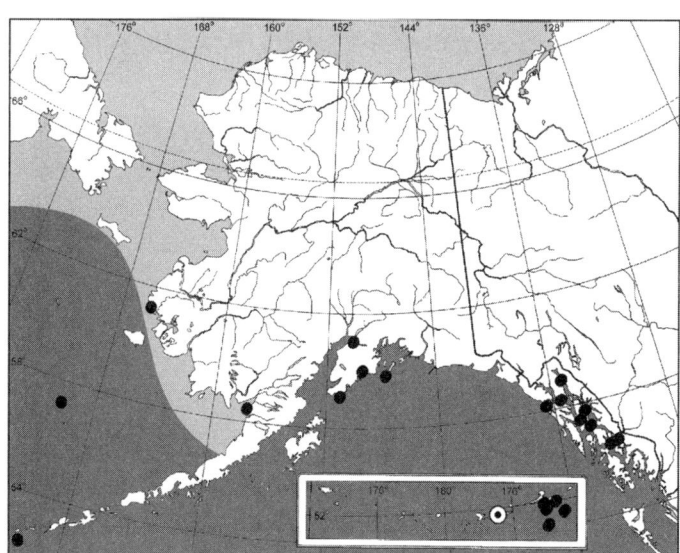

MAP 105. Distribution of Dall's porpoise, *Phocoenoides dalli*.

■ = Projected Range (after Angliss and Outlaw 2007)
● = Specimen Record (n = 28)

shelf and slope and near the coastal shores in deeper waters (Martin 1990). They seasonally range from Japan and the southern Bering Sea south along the northern Gulf of Alaska to Baja California (Nagorsen 1990).

Alaska Distribution

Reports of Dall's porpoise (predominately *dalli* type) are infrequent north of 62° N in the Bering Sea (Nishiwaki 1967) (map 105), but there have been occasional sightings as far north as the Chukchi Sea (Sleptsov 1961). They are especially abundant in the southern Bering Sea, and at least in summer remain as far north as the Pribilof Islands (Leatherwood and Reeves 1983). The only apparent distribution gaps in Alaska waters are upper Cook Inlet and the shallow eastern flats of the Bering Sea. Although present in Alaska waters during all months of the year, there may be movements of populations out of Prince William Sound and areas in the Gulf of Alaska and Bering Sea in winter (Angliss and Outlaw 2007).

Habitat

Dall's porpoises frequent cold water, where the surface temperature is between 3° C and 20° C. They appear to prefer the continental shelf and slope. When close to shore, they are found in areas with deep-water channels (Martin 1990).

Status

Dall's porpoise is one of the most common species of whales in the North Pacific Ocean (Martin 1990), although some populations may be increasingly threatened (McMillan and Bermingham 1996). The Alaska population is estimated at 83,400 animals; no reliable information is available on population trends (Angliss and Outlaw 2007). *Phocoenoides dalli* is listed in Appendix II of CITES, as Lower Risk by IUCN, and Not at Risk by COSEWIC.

Fossils

Although the family Phocoenidae is known from the Late Miocene, there is no fossil record for *P. dalli* (Barnes 1976).

Dall's porpoise, *Phocoenoides dalli* (S. O. MacDonald)

Family PHYSETERIDAE Gray, 1821—sperm whales

This family is represented by three species. The pygmy sperm whales are often assigned to their own family, Kogiidae (e.g., Rice 1998). These are the sole survivors of a rich and widespread array of some twenty genera of physeterids in the Miocene and Pliocene (Rice 1998).

The sperm whale has a worldwide distribution but is rarely seen in waters less than 200 m deep. Females, calves, and young males remain in tropical or temperate waters while older males migrate as far north as the Bering Sea in summer (Wynne 1993). The pygmy sperm whale, *Kogia breviceps*, is a worldwide inhabitant of tropical to temperate waters and has only recently been documented in Alaska waters.

Sperm whale, *Physeter catodon* (B. Hines)

Pygmy Sperm Whale

Kogia breviceps
(Blainville, 1838)

Other Common Names
None.

Systematics
The pygmy sperm whale, *Kogia breviceps*, is a monotypic species, which until recently was considered conspecific with the similar-looking and broadly sympatric dwarf sperm whale, *K. sima* (Rice 1998). The genus *Kogia* is placed in the family Kogiidae by some authors (e.g., Rice 1998). Available data are insufficient to delineate possible stock boundaries (Carretta et al. 2007).

Kogia breviceps
Original Description. 1838. *Physeter breviceps* Blainville, *Ann. Franc. Etr. Anat. Phys.*, 2:337.
Type Locality. Cape of Good Hope, South Africa.
Type Specimen. Skull only at MNHN

Global Distribution
The pygmy sperm whale is found in tropical and temperate waters around the world (Rice 1998). It has been documented in the eastern Pacific as far north as Washington State (Rice 1998). Knowledge of its distribution is based primarily on strandings and reports are often equivocal due to the difficulty of distinguishing it from the dwarf sperm whale (Leatherwood and Reeves 1983).

Alaska Distribution
The discovery of a moderately decomposed pygmy sperm whale at Boilers Beach near Yakutat in July of 2003 is the first and only record of this species in Alaska waters (map 106).

Habitat
The pygmy sperm whale is evidently an open-ocean species that lives mostly beyond the edge of the continental shelf in tropical and temperate waters.

Status
Never the targets of large-scale commercial whaling, neither species of *Kogia* is considered abundant. The frequency of strandings suggested that they are uncommon to fairly common in local waters (Caldwell and Caldwell 1989; Reeves et al. 2002). The pygmy sperm whale is listed in Appendix II of CITES, as Lower Risk by IUCN, and Not at Risk by COSEWIC.

Fossils
Fossil forms only distantly related to extant *Kogia* spp. have rarely been described from fragments of teeth, cranium, and lower jaws of Late Miocene to Early Pliocene age (McAlpine 2002).

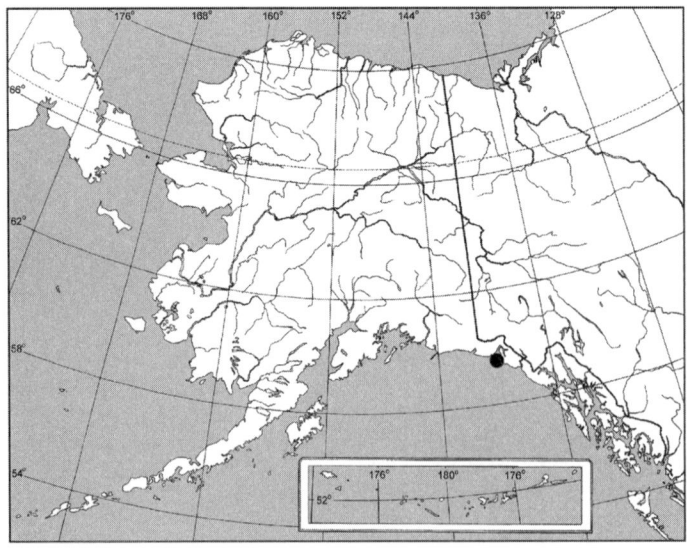

MAP 106. Distribution of pygmy sperm whale, *Kogia breviceps*.

● = Specimen Record (n = 1): 59.511667, –139.849611, Yakutat Quad., Boilers Beach, Ocean Cape area, near Yakutat (1 UAM).

Sperm Whale

Physeter catodon
Linnaeus, 1758

Other Common Names

Giant sperm whale, pot whale.

Systematics

Analysis of DNA sequence data by Milinkovitch (1997) suggested a close relationship between *P. catodon*, a toothed whale (Odontoceti), and baleen whales (Mysticeti). Subsequent morphological and genetic studies, however, indicated their inclusion among a monophyletic Odontoceti (Whitehead 2002). There is ongoing disagreement over whether the correct specific name for the sperm whale is *macrocephalus* or *catodon* (e.g., Holthuis 1987; Schevill 1986). Mead and Brownell (2005) argued that *P. catodon* has line priority and, according to the diagnosis of Linnaeus, is the only name applicable.

Although a number of separate breeding stocks have been identified (Nagorsen 1990), geographic variation is slight (Berzin 1971; Ivanova 1955; Machin 1974), and no subspecies are recognized (Rice 1998). Lyrholm et al. (1996) and Lyrholm and Gyllensten (1998) found shallow but significant mtDNA variation in a global sampling of sperm whales.

Physeter catodon

Original Description. 1758. [*Physeter*] *catodon* Linnaeus, *Syst. Nat.*, 10th ed., 1:76.

Type Locality. "Oceano Europaeo," restricted by Husson and Holthius (1974) to Berkhey, province of Zuid-Holland, The Netherlands.

Type Specimen. Neotype of the Berkhey specimen (RMNH 5828) was selected by Husson and Holthius (1974).

Global Distribution

The sperm whale ranges throughout all the deep oceans of the world from the equator to the edges of the polar pack ice (Rice 1989).

Alaska Distribution

During summer in the eastern North Pacific, male sperm whales are thought to move north to feed in the Gulf of Alaska, Bering Sea, and waters around the Aleutian Islands, while females and young remain in tropical and temperate waters year-round (Angliss and Outlaw 2007) (map 107).

Habitat

Sperm whales are found in deep waters and feed almost exclusively on squid and octopi (Potter and Birchler 1999b).

Status

The number of sperm whales in the North Pacific was reported to be 1,260,000 prior to exploitation, which was reduced to 930,000 whales by the late 1970s (Rice 1989). There is no recent or reliable estimate of abundance for the North Pacific stock, and the number of sperm whales occurring within Alaska waters is unknown (Angliss and Outlaw 2007). However, on the basis of total abundance,

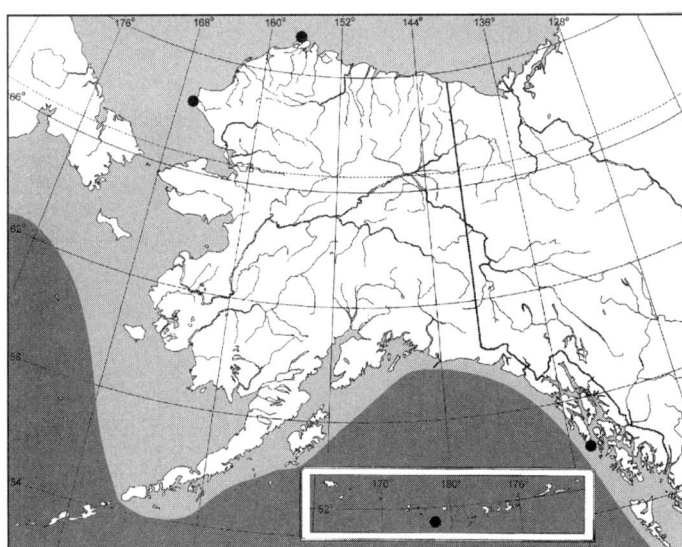

MAP 107. Distribution of sperm whale, *Physeter catodon*.

■ = Projected Range (after Angliss and Outlaw 2007)

● = Specimen Record (n = 5): Aleutian Islands, Amchitka Island (1 USNM); Baranof Island, Port Armstrong (1 UWBM); 68.34083, −166.83333, Chukchi Sea, near Pt. Hope (1 LACM); 71.3875, −156.48111, Pt. Barrow (1 LACM); Alaska, unknown locality (1 UWBM).

current distribution, and regulatory measures that are currently in place, H. Braham (in Angliss and Outlaw 2007) postulated that it is unlikely that the North Pacific stock is in danger of extinction or threatened in the foreseeable future. The sperm whale is listed in Appendix I of CITES, as Vulnerable by IUCN, as Endangered under the ESA, and Not at Risk by COSEWIC.

Fossils

The fossil record of the Physeteridae extends to the Miocene. Fossil physeterids have been found in South America, eastern and western North America, western Europe, the Mediterranean region, Australia, New Zealand, and Japan (Hirota and Barnes 1995).

Family ZIPHIIDAE Gray, 1865—beaked whales

The ziphiids includes twenty-one living species in six genera (Mead and Brownell 2005). Most members of this family are poorly known, with only a few studied in detail. Several are known only from a few stranded specimens. As a result, ziphiid taxonomy remains tentative and uncertain (Reeves et al. 2002).

Fossil remains of ziphiids are known from Early Miocene deposits in South America, Europe, and Africa; and Middle to Late Miocene and Pliocene deposits in North America. As many as fourteen extinct genera have been described (Rice 1984).

Shared characteristics of this diverse group include a beak of variable length, a pair of grooves on the throat that converge anteriorly, and the presence of "flipper pockets," shallow depressions in the body wall to tuck away the flippers and perhaps reduce drag (Reeves et al. 2002). The various species of beaked whales are usually distinguished from one another in the size and shape of the teeth and their position in the lower jaw and, more recently, in molecular analyses. Beaked whales range in size from the pygmy beaked whale (1.6 m long) to Baird's beaked whale (4.5 m). The latter species, along with Stejneger's beaked whale and Cuvier's beaked whale, are the only species of ziphiids known to occur in Alaska waters.

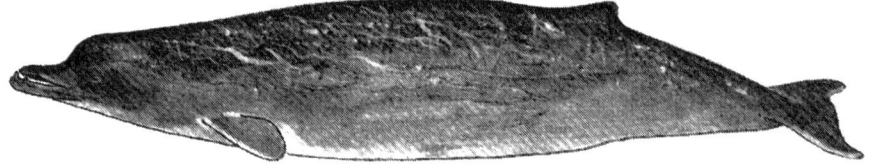

Baird's beaked whale, *Berardius bairdii* (Pike 1956)

Baird's Beaked Whale

Berardius bairdii
Stejneger, 1883

Other Common Names

North Pacific bottlenose whale, northern giant bottlenose whale, northern four-toothed whale.

Systematics

Two species of *Berardius* are currently recognized (Rice 1998). The North Pacific population may be a northern form of the Arnoux's beaked whale, *B. arnuxii*, of the Southern Hemisphere (Balcomb 1989; Dalebout et al. 1998; Davies 1963). Separate populations may inhabit the Sea of Japan, the Sea of Okhotsk and the western North Pacific near Japan (Kasuya 1986). No subspecies are currently recognized.

Berardius bairdii

Original Description. 1883. *Berardius bairdii* Stejneger, *Proc. U.S. Nat. Mus.*, 6:75, June 30.

Type Locality. Eastern shore of Bering Island, Bering Sea, North Pacific Ocean.

Type Specimen. USNM 20992.

Global Distribution

Baird's beaked whale is confined to the North Pacific Ocean from the southern Bering Sea south to Japan and Southern California (Martin 1990; Rearden 1981), and the central North Pacific along seamounts north of the Hawaiian Islands (Mead 1999a).

Alaska Distribution

This species occurs in Alaska waters from the southern Bering Sea to the Aleutian Islands and the Gulf of Alaska (Balcomb 1989) (map 108). Balcomb (1989) suggested a hiatus in distribution occurs in the eastern Gulf of Alaska. Kasuya and Ohsumi (1984) reported numerous sightings from the middle of the Gulf of Alaska westward. There are stranding records in the eastern Bering Sea from St. Matthew Island (Hanna 1920) and the Pribilof Islands (True 1910). Other strandings in Alaska waters were reported by Reeves and Mitchell (1993). Tomilin (1957) reported this species moving among ice floes as far north as 63° N on the west side of the Bering Sea. The northernmost record may be the reported sighting of two individuals in the southern Chukchi Seas near Cape Uelen in early September 1948 (Sleptsov 1961). Few winter in the northern part of their range (Tomilin 1957).

Habitat

This species is more or less restricted to waters off the continental shelf and to areas where seamounts and escarpments occur (Balcomb 1989; Martin 1990). They feed primarily on squid and other deep-water species (Wynne 1993).

Status

Reliable estimates of abundance for Alaska are unavailable (Hill and DeMaster 1999).

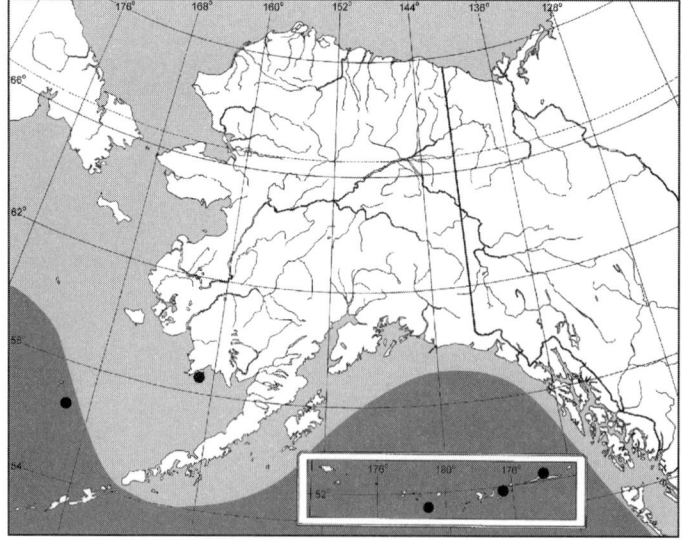

MAP 108. Distribution of Baird's beaked whale, *Berardius bairdii*.

■ = Projected Range (Angliss and Outlaw 2007)

● = Specimen Record (n = 7): 58.5, −161.75, Bristol Bay, Hagemeister Island (1 UAM); 51.86333, −176.74139, Bering Sea, Adak, Andreanof Islands, Adak Island, Shagak Bay (1 UAM); Aleutian Islands, Unalaska Island (2 USNM); Aleutian Islands, Amchitka Island, Crown Reefer Point (1 USNM); Pribilof Islands, St. George Island (2 USNM).

Berardius bairdii is listed in Appendix I of CITES, and as Lower Risk by IUCN.

Fossils

Ziphioid whales first appear in the fossil record in the Middle Miocene (Barnes et al. 1985). Fossils of ziphiids are known from the Miocene and Pliocene of Europe, North and South America, Japan, and Australia (Berta and Sumich 1999). *Berardius* may have arisen early in ziphiid evolution (Moore 1968).

Stejneger's Beaked Whale

Mesoplodon stejnegeri
True, 1885

Other Common Names

Bering Sea beaked whale, sabre-toothed beaked whale.

Systematics

Mesoplodon contains at least thirteen described species, making it the most speciose genus of living cetacean (Rice 1998). While there has been some confusion over the taxonomic status of some of the Pacific species, few authors disagree over the validity and distinctiveness of *M. stejnegeri* (Mead 1989). No subspecies or stocks are recognized (Angliss and Outlaw 2007).

Mesoplodon stejnegeri

Original Description. 1885. *Mesoplodon Stejnegeri* True, *Proc. Nat. Mus.*, 8:585, November 21.

Type Locality. Bering Island, Bering Sea, North Pacific Ocean.

Type Specimen. USNM 21112.

Global Distribution

This poorly known species is apparently confined to the subarctic and cold temperate waters of the North Pacific Ocean from the Bering Sea to Japan and central California (Mead 1989; Rice 1998).

Alaska Distribution

Reports from strandings and sightings in Alaska waters suggest that Stejneger's beaked whales range from the Gulf of Alaska and Aleutian Islands northward into the Bering Sea as far as the Pribilof Islands and Bristol Bay (Jellison 1953; Loughlin et al. 1982; Mead 1989; Rice 1978b) (map 109).

Habitat

Mesoplodon stejnegeri is most likely a deep-water whale; however, the habits and habitat requirements of this species are poorly understood. Mead (1999b) suggested that this species frequents the Aleutian Basin and the Aleutian Trench rather than the shallow waters of the northern or eastern Bering Sea.

Status

Reliable estimates of size and trends for Alaska populations are currently unavailable (Angliss and Outlaw 2007; Hill and DeMaster 1999).

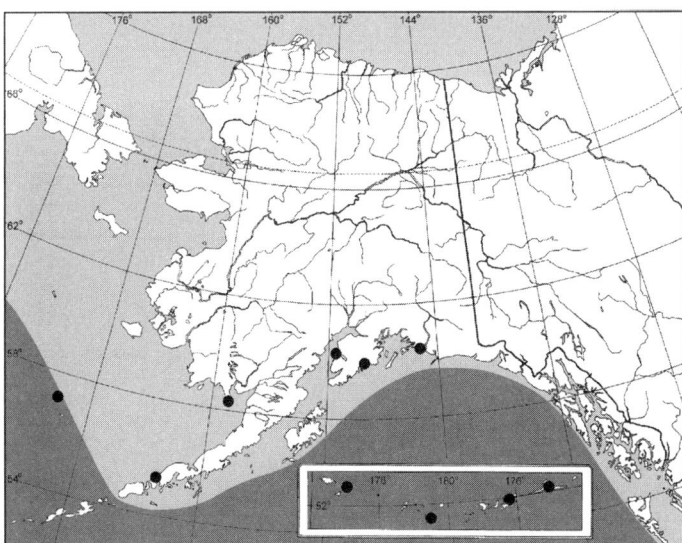

MAP 109. Distribution of Stejneger's beaked whale, *Mesoplodon stejnegeri*.

■ = Projected Range (after Angliss and Outlaw 2007)
● = Specimen Record (n = 22)

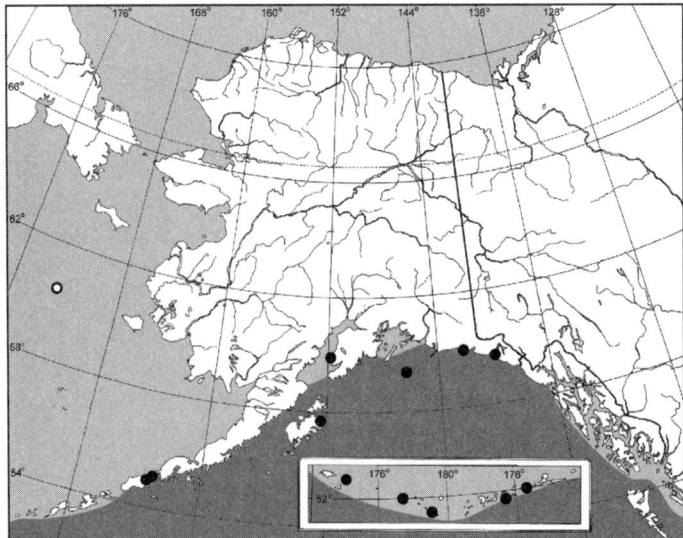

MAP 110. Distribution of Cuvier's beaked whale, *Ziphius cavirostris.*

■ = Projected Range (after Angliss and Outlaw 2007)
● = Specimen Record (n = 20)
○ = Marginal Record. Bering Sea, south side St. Matthew Island (1 USNM).

Mesoplodon stejnegeri is listed in Appendix II of CITES, as Data Deficient by IUCN, and Not at Risk by COSEWIC.

Fossils
No fossils of *Mesoplodon* have been identified (Loughlin and Perez 1985).

Cuvier's Beaked Whale

Ziphius cavirostris
G. Cuvier, 1823

Other Common Names
Goosebeak whale.

Systematics
There has been no study of geographic variation throughout this species' range and no subspecies are recognized (Rice 1998).

Ziphius cavirostris
Original Description. 1823. *Ziphius cavirostris* G. Cuvier, *Recherches sur les ossemens fossiles*, 2nd ed., 5(1):350.
Type Locality. Near Fos, Bouches-du-Rhone, France.
Type Specimen. Fossilized skull at MNHN.

Global Distribution
Cuvier's beaked whales occur in all the world's temperate and tropical waters (Heyning 1989; Rice 1998). No migrations are known (Martin 1990).

Alaska Distribution
In the Pacific, the Cuvier's beaked whale ranges north into the Gulf of Alaska and the Aleutian Islands (Mitchell 1968b; Rice 1998) (map 110). Foster and Hare (1990) noted the stranding of at least sixteen individuals of this species from Alaska since 1967 at latitudes as high as 59° N in the Bering Sea.

Habitat
Cuvier's beaked whales are restricted to deep waters (Martin 1990).

Status
Reliable estimates of size and trends for the goose-beaked whale population in Alaska are currently unavailable (Hill and DeMaster 1999), but are assumed stable (Wynne 1993). *Ziphius cavirostris* is listed in Appendix II of CITES, as Data Deficient by IUCN, and Not at Risk by COSEWIC.

Fossils
No fossils of this species have been identified.

Literature Cited

Abbott, R. J., and C. Brochmann. 2003. History and evolution of the arctic flora: In the footsteps of Eric Hultén. *Molecular Ecology* 12:299–313.

Abend, A. 1999. Short-finned pilot whale, *Globicephala macrorhynchus*. Pp. 285–286, in *The Smithsonian book of North American mammals* (D. E. Wilson and S. Ruff, eds.). Smithsonian Institution Press, Washington, D.C., in association with the American Society of Mammalogists.

Abramov, A. V. 1999. A taxonomic review of the genus *Mustela* (Mammalia, Carnivora). *Zoosystematica Rossica* 8:357–364.

ACIA. 2005. *Arctic Climate Impact Assessment: Scientific Report.* Cambridge University Press.

Ackerman, R. E. 1996. Cave 1, Lime Hills. Pp. 470–477, in *American beginnings: The prehistory and palaeoecology of Beringia* (F. H. West, ed.). University of Chicago Press, Chicago and London.

Ackerman, R. E., D. Georgina, J. Mastrogiuseppe, and A. Ruter. 1997. The archaeology of Lime Hills Cave, southwestern Alaska: A multidisciplinary approach. Beringian Paleoenvironments Workshop (September 20–23, Florissant, Colorado). Program and Abstracts.

ADFG (Alaska Department of Fish and Game). 1973. Alaska's wildlife and habitat. Anchorage, Alaska. 144 pp. + maps.

———. 1978. Alaska's wildlife and habitat, volume II. Anchorage, Alaska.

———. 1995. Tugidak Island critical habitat area management plan. ADFG, Divisions of Habitat and Restoration and Wildlife Conservation, Anchorage, AK.

———. 2004. Furbearer management report of survey-inventory activities 1 July 2000–30 June 2003 (C. Brown, ed.). Juneau, Alaska.

———. 2005. Muskox management report of survey-inventory activities 1 July 2002–30 June 2004 (C. Brown, ed.). Juneau, Alaska.

———. 2007. Alaska's comprehensive wildlife conservation strategy. http://www.sf.adfg.state.ak.us/statewide/ngplan (accessed 8 July 2007).

ADNR (Alaska Department of Natural Resources, Division of Parks and Outdoor Recreation). 2008. Shuyak Island Natural History. http://www.ksp.kodiakislanddesign.com/shuyak_nat_history.htm (accessed 27 February 2008).

Agler, B. A., S. J. Kendall, P. E. Seiser, and J. R. Lindell. 1995. Estimates of marine bird and sea otter abundance in southeast Alaska during summer 1994. Migratory Bird Management, U. S. Fish and Wildlife Service, Anchorage, Alaska.

Aguilar, A. 2002. Fin whale, *Balaenoptera physalus*. Pp. 435–442, in *Encyclopedia of marine mammals* (W. F. Perrin, B. Würsig, and J. G. M. Thewissen, eds.). Academic Press, San Diego and London.

Alekseev, A. I. 1987. *The odyssey of a Russian scientist: I. G. Voznesenskii in Alaska, California and Siberia 1839–1849* (W. C. Follette, trans.; R. A. Pierce, ed.). Limestone Press, Kingston, Ontario.

Alexander, L. F. 1996. A morphometric analysis of geographic variation within *Sorex monticolus* (Insectivora: Soricidae). The University of Kansas Natural History Museum, Miscellaneous Publication 88:1–54.

Allen, G. M. 1933. The least weasel: A circumboreal species. *Journal of Mammalogy* 14:316–319.

Allen, J. A. 1898. Revision of the chickarees, or North American red squirrels (subgenus *Tamiasciurus*). *Bulletin of the American Museum of Natural History* 10:249–298.

———. 1902. List of mammals collected in Alaska by the Andrew J. Stone Expedition of 1901. *Bulletin of the American Museum of Natural History* 19 (Art. 18):215–230.

———. 1903. Mammals collected in Alaska and Northern British Columbia by the Andrew J. Stone Expedition of 1902. *Bulletin of the American Museum of Natural History* 19 (Art. 21):521–567.

———. 1904. Mammals collected in Alaska by the Andrew J. Stone Expedition of 1903. *Bulletin of the American Museum of Natural History* 20 (Art. 24):273–292.

———. 1912. The probable recent extinction of the muskox in Alaska. *Science* 36:720–722.

———. 1913. Ontogenetic and other variations in muskoxen, with a systematic review of the muskox group, recent and extinct. *Memoirs American Museum of Natural History, new series* 1:101–226.

Amano, M., and N. Miyazaki. 1992. Geographic variation in skulls of the harbor porpoise, *Phocoena phocoena. Mammalia* 56:133–144.

AMNWF (Alaska Maritime National Wildlife Refuge). 2006. Alien/invasive mammals of the Aleutian Islands. http://alaskamaritime.fws.gov/wildlife-wildlands/wildlife/nonnative (accessed 14 March 2006).

Amstrup, S. C. 2003. Polar bear, *Ursus maritimus*. Pp. 587–610, in *Wild mammals of North America: Biology, management, and conservation* (G. A. Feldhamer, B. C. Thomson, and J. A. Chapman, eds). John Hopkins University Press, Baltimore.

Amstrup, S. C., and D. P. DeMaster. 1988. Polar Bear, *Ursus maritimus*. Pp. 39–45, in *Selected marine mammals of Alaska: Species accounts with research and management recommendations* (J. W. Lentfer, ed.). Marine Mammal Commission, Washington, D.C.

Amstrup, S. C., and C. Gardner. 1991. Research on polar bears in northern Alaska, 1985–88. Pp. 43–53, in *Polar bears: Proceedings of the tenth working meeting of the IUCN/SSC Polar Bear Specialist Group* (S. C. Amstrup and Ø. Wiig, eds.). IUCN, Gland, Switzerland.

———. 1994. Polar bear maternity denning in the Beaufort Sea. *Journal of Wildlife Management* 58:1–10.

Amstrup, S. C., T. L. McDonald, and I. Stirling. 2001. Polar bears in the Beaufort Sea: A 30-year mark-recapture case history. *Journal of Agricultural, Biological, and Environmental Statistics* 2:221–234.

Anderson, A. E., and O. C. Wallmo. 1984. *Odocoileus hemionus. Mammalian Species* 219:1–9.

Anderson, C. G. 1999. Arctic fox, *Alopex lagopus*. Pp. 146–148, in *The Smithsonian book of North American mammals* (D. E. Wilson and S. Ruff, eds.). Smithsonian Institution Press, Washington, D.C., in association with the American Society of Mammalogists.

Anderson, E. 1968. Fauna of the Little Box Elder Cave, Converse County, Wyoming. University of Colorado Studies Ser. *Earth Science* 6:1–59.

———. 1970. Quaternary evolution of the genus *Martes* (Carnivora, Mustelidae). *Acta Zoologica Fennica* 130:1–132.

———. 1973. Ferret from the Pleistocene of central Alaska. *Journal of Mammalogy* 54:778–779.

———. 1977. Pleistocene Mustelidae (Mammalia, Carnivora) from Fairbanks, Alaska. *Bulletin of the Museum of Comparative Zoology* 148:1–21.

Anderson, H. L. 1974. Natural history and systematics of the tundra hare (*Lepus othus* Merriam) in western Alaska. Unpublished thesis. University of Alaska Fairbanks.

———. 1978. Range of the tundra hare. *Murrelet* 59:72–74.

Anderson, P. K. 1995. Competition, predation, and the evolution and extinction of Steller's sea cow, *Hydrodamalis gigas. Marine Mammal Science* 11:391–394.

Anderson, P. K., and D. P. Domning. 2002. Steller's sea cow *Hydrodamalis gigas*. Pp. 1178–1181, in *Encyclopedia of marine mammals* (W. F. Perrin, B. Würsig, and J. G. M. Thewissen, eds.). Academic Press, San Diego and London.

Anderson, R. M. 1946. Catalogue of Canadian Recent mammals. *Bulletin of the National Museum of Canada* 102:1–238.

Anderson, S. 1960. *The baculum in microtine rodents*. University of Kansas Publications, Museum of Natural History 12:181–216.

Andrews, R. C. 1911. A new porpoise from Japan. *Bulletin of the American Museum of Natural History* 30:31–52.

Angliss, R. P., and R. B. Outlaw. 2007. Alaska marine mammal stock assessments, 2006. U. S. Dep. Commer., NOAA Tech. Memo. NMFSAFSC-168.

Antell, S. 1987. Systematics and zoogeography of mammals in Southeast Alaska. Unpublished dissertation. Washington State University, Pullman.

Arbogast, B. S. 1999. Mitochondrial DNA phylogeography of the New World flying squirrels (*Glaucomys*): Implications for Pleistocene biogeography. *Journal of Mammalogy* 80:142–155.

———. 2007. A brief history of the New World flying squirrels: Phylogeny, biogeography, and conservation genetics. *Journal of Mammalogy* 88:840–849.

Armstrong, M. 2007. Rare fur seal found in bay now back home. http://homernews.com/stories/103120 07/news_1_002.shtml (accessed 19 Dec. 2007).

Árnason, Ú., K. Bodin, A. Gullberg, C. Ledje, and S. Mouchaty. 1995. A molecular view of pinniped relationships with particular emphasis on the true seals. *Journal of Molecular Evolution* 40:78–85.

Árnason, Ú., and A. Gullberg. 1994. Relationship of baleen whales established by cytochrome *b* gene sequence comparison. *Nature* 367:726–728.

———. 1996. Cytochrome *b* nucleotide sequences and the identification of five primary lineages of extant cetaceans. *Molecular Biology and Evolution* 13:407–417.

Árnason, Ú., A. Gullberg, and A. Janke. 2004. Mitogenomic analyses provide new insights into cetacean origin and evolution. *Gene* 333:27–34.

Averianov, A. O. 1998. Late Pleistocene hares (*Lepus*) of the Russian plain. *Illinois State Museum Scientific Papers* 27:41–68.

Avise, J. 2000. *Phylogeography: The history and formation of species*. Harvard University Press, Cambridge.

Bailey, A. M. 1920. Smithsonian Institution Archives, Record Unit 7176, Box 5, Folder 1. U. S. Fish and Wildife Service 1860–1961.

Bailey, A. M., and R. W. Hendee. 1926. Notes on the mammals of northwestern Alaska. *Journal of Mammalogy* 7:9–28.

Bailey, E. P. 1976. Breeding bird distribution and abundance in the Barren Islands, Alaska. *Murrelet* 57:2–12.

———. 1978. Breeding seabird distribution and abundance in the Shumagin Islands, Alaska. *Murrelet* 59:82–91.

———. 1992. Red foxes, *Vulpes vulpes*, as biological control agents for introduced arctic foxes, *Alopex lagopus*, on Alaskan islands. *Canadian Field-Naturalist* 106:200–205.

———. 1993. Introduction of foxes to Alaskan islands: History, effects on avifauna, and eradication. *U.S. Fish and Wildlife Service, Technical Report Series* 193:1–53.

Bailey, E. P., and N. H. Faust. 1981. Summer distribution and abundance of marine birds and mammals between Mitrofania and Sutwik islands south of the Alaska Peninsula. *Murrelet* 62:34–42.

Bailey, V. 1900. Revision of American voles of the genus *Microtus*. *North American Fauna* 17.

Baird, R. W. 2002a. Risso's Dolphin, *Grampus griseus*. Pp. 1037–1039, in *Encyclopedia of marine mammals* (W. F. Perrin, B. Würsig, and J. G. M. Thewissen, eds.). Academic Press, San Diego and London.

———. 2002b. False killer whale, *Pseudorca crassidens*. Pp. 411–412, in *Encyclopedia of marine mammals* (W. F. Perrin, B. Würsig, and J. G. M. Thewissen, eds.). Academic Press, San Diego and London.

Baker, A. J., J. L. Eger, R. L. Peterson, and T. H. Manning. 1983. Geographic variation and taxonomy of arctic hares. *Acta Zoologica Fennica* 174:45–48.

Baker, A. R., T. R. Loughlin, V. Burkanov, C. W. Matson, R. G. Trujillo, D. G. Calkins, J. K. Wickliffe, and J. W. Bickha. 2005. Variation of mitochondrial control region sequences of Steller sea lions: The three-stock hypothesis. *Journal of Mammalogy* 86:1075–1084.

Baker, C. S., L. Meddrano-Gonzalez, J. Calmabokidis, A. Perry, F. Pichler, H. Rosenbaum, J. M. Straley, J. Urbán-Ramirez, M. Yamaguchi, and O. Von Ziegesar. 1998. Population structure of nuclear and mitochondrial DNA variation among humpback whales in the North Pacific. *Molecular Ecology* 7:695–707.

Baker, C. S., and S. R. Palumbi. 1997. The genetic structure of whale populations: Implications for management. Pp. 117–146, in *Molecular genetics of marine mammals* (A. E. Dizon, S. J. Chivers, and W. F. Perrin, eds.). Special Publication Number 3, The Society for Marine Mammalogy. Lawrence, Kansas.

Baker, C. S., R. W. Slade, J. L. Bannister, R. B. Alernethy, M. T. Weinrich, J. Lien, J. Urban, P. Corkeron, J. Calmabokidis, O. Vasquez, and S. R. Palumbi. 1994. Hierarchical structure of mitochondrial DNA gene flow among humpback whales, *Megaptera novaeangliae,* world-wide. *Molecular Ecology* 3:313–327.

Baker, R. J., L. C. Bradley, J. W. Dragoo, M. D. Engstrom, R. S. Hoffmann, C. A. Jones, F. Reid, D. W. Rice, and C. Jones. 2003. Revised checklist of North American mammals north of Mexico, 2003. Occasional Papers, Museum of Texas Tech University, Lubbock.

Bakker, V. J., and K. Hastings. 2002. Den trees used by northern flying squirrels (*Glaucomys sabrinus*) in southeastern Alaska. *Canadian Journal of Zoology* 80:1623–1633.

Balcomb, K. C., III. 1989. Baird's beaked whale, *Beradius bairdii* Stejneger, 1883: Arnoux's beaked whale, *Berardius arnuzii* Duvernoy, 1851. Pp. 261–288, in *Handbook of marine mammals. Volume 4: River dolphins and the larger toothed whales* (S. H. Ridgway and R. J. Harrison, eds.). Academic Press, London.

Ballard, W. B. 1977. Status and management of the mountain goat in Alaska. Pp. 15–23, in *Proceedings of the First International Mountain Goat Symposium* (W. Samuel and W. G. Macgregor, eds.). Queen's Printer, Victoria, British Columbia.

Ballard, W. B., L. A. Ayres, P. R. Krausman, D. J Reed, and S. G. Fancy. 1997. Ecology of wolves in relation to a migratory caribou herd in northwest Alaska. *Wildlife Monograph* 135:1-47.

Banci, V., and A. S. Arestad. 1990. Home range and habitat use of wolverines *Gulo gulo* in Yukon, Canada. *Holarctic Ecology* 13:195–200.

Banfield, A. W. F. 1961. A revision of the reindeer and caribou, genus *Rangifer. National Museum of Canada, Bulletin No. 177, Biological Series No. 66.*

———. 1974. *The mammals of Canada.* The University of Toronto Press, Toronto.

Bangs, E. E. 1979. The effects of tree crushing on small mammal populations in Southcentral Alaska. Unpublished thesis. University of Nevada, Reno.

Baranova, G. I., A. A. Gureev, and P. P. Strelkov. 1981. Catalogue of type specimens in the collection of the Zoological Institute of the USSR, mammals (Mammalia). Part 1. shrews (Insectivora), bats (Chiroptera), hares (Lagomorpha). Nauka, Leningrad.

Barbour, R. W., and W. H. Davis. 1969. *Bats of America.* University Press of Kentucky, Lexington.

Barnes, B. M. 1989. Freeze avoidance in a mammal: Body temperatures below 0 degree C in an Arctic hibernator. *Science* 244:1593–1595.

Barnes, I., P. Matheus, B. Shapiro, D. Jensen, and A. Cooper. 2002. Dynamics of Pleistocene population extinctions in Beringian brown bears. *Science* 295:2267–2270.

Barnes, L. G. 1976. Outline of eastern North Pacific fossil cetacean assemblages. *Systematic Zoology* 23:321–343.

———. 1984. Fossil odontocetes (Mammalia: Cetacea) from the Almejas Formation, Isla Cedros, Mexico. PaleoBios, Museum of Paleontology, University of California, Berkeley 42:1-46.

———. 1985. Evolution, taxonomy and antitropical distributions of the porpoises (Phocoenidae, Mammalia). Marine Mammal Science 1:149-165.

Barnes, L. G., D. P. Downing, and C. E. Ray. 1985. Status of studies on fossil marine mammals. *Marine Mammal Science* 1:15–53.

Barnes, L. G., and S. A. McLeod. 1984. The fossil record and phyletic relationships of gray whales. Pp. 4–32, in *The gray whale, Eschrichtius robustus* (M. L. Jones, S. L. Swartz, and S. Leatherwood, eds.). Academic Press, Toronto.

Barten, N. L. 2004a. Unit 1C furbearer management report. Pp. 24–31, in *Furbearer management report of survey and inventory activities* 1 July 2000–30 June 2003 (C. Brown, ed.). Alaska Department of Fish and Game. Project 7.0. Juneau, Alaska.

————. 2004b. Unit 1C moose management report. Pp. 22–44, in *Moose management report of survey and inventory activities* 1 July 2001–30 June 2003 (C. Brown, ed.). Alaska Department of Fish and Game. Project 1.0. Juneau, Alaska.

————. 2004c. Unit 5 moose management report. Pp. 68–89, in *Moose management report of survey and inventory activities* 1 July 2001–30 June 2003 (C. Brown, ed.). Alaska Department of Fish and Game. Project 1.0. Juneau, Alaska.

Barton, C. M., G. A. Clark, D. R. Yesner, and G. A. Pearson (Eds.). 2004. *The settlement of the American continents: A multidisciplinary approach to human biogeography.* University of Arizona Press, Tucson.

Batzli, G. O., and H. Henttonen. 1993. Home range and social organization of the singing vole (*Microtus miurus*). *Journal of Mammalogy* 74:868–878.

Baum, D. A., S. D. Smith, and S. S. Donovan. 2005. The tree-thinking challenge. *Science* 310:979–980.

Bayha, K., and J. Kormendy (Eds.). 1990. Proceedings of sea otter symposium, 17–19 April 1999. U. S. Fish and Wildlife Service, *Biological Report* 90:1–485.

Bechstein, J. M. 1799. Thomas Pennant's Allgemeine Uebersicht der Vierfüssigen Thiere. *Comptoir's, Weimar, Germany* 1:1–318.

Bee, J. W., and E. R. Hall. 1956. Mammals of northern Alaska on the arctic slope. University of Kansas Museum of Natural History, Miscellaneous Publication 8:1–309.

Beebe, B. F. 1980. Pleistocene peccary, *Platygonus compressus* Le Conte, from Yukon Territory, Canada. *Canadian Journal of Earth Sciences* 17:1204–1209.

Beebe, B. F., and T. J. Hulland. 1988. Mandibular and dental abnormalities of two Pleistocene American lions (*Panthera leo atrox*) from Yukon Territory. *Canadian Journal of Veterinary Research* 52:468–472.

Beier, P. 1999. Cougar, *Puma concolor.* Pp. 226–227, in *The Smithsonian book of North American mammals* (D. E. Wilson and S. Ruff, eds.). Smithsonian Institution Press, Washington, D.C., in association with the American Society of Mammalogists.

Bekoff, M. 1977. *Canis latrans. Mammalian Species* 79:1–9.

————. 1999. Coyote, *Canis latrans.* Pp. 139–141, in *The Smithsonian book of North American mammals* (D. E. Wilson and S. Ruff, eds.). Smithsonian Institution Press, Washington, D.C., in association with the American Society of Mammalogists.

Belcher, R. L., and T. E. Lee, Jr. 2002. *Arctocephalus townsendi. Mammalian Species* 700:1–5.

Belk, M. C., C. L. Pritchett, and H. D. Smith. 1990. Patterns of microhabitat use by *Sorex monticolus* in summer. *The Great Basin Naturalist* 50:387–389.

Bell, C. J., E. L. Lundelius, A. D. Barnosky, R. W. Graham, E. H. Lindsay, K. R. Ruez, Jr., H. A. Semken, Jr., S. D. Webb, and R. J. Zakrzewski. 2004. The Blancan, Irvingtonian, and Rancholabrean mammal ages. Pp. 232–314, in *Late Cretaceous and Cenozoic mammals of North America: biostratigraphy and geochronology* (M. O. Woodburne, ed.). Columbia University Press, New York.

Beneski, J. T., and D. W. Stinson. 1987. *Sorex palustris. Mammalian Species* 296:1–6.

Bergerud, A. T. 2000. Caribou. Pp. 658–693, in *Ecology and management of large mammals in North America* (S. Demarais and P. R. Krausman, eds.). Prentice-Hall, Inc., Upper Saddle River, NJ.

Bernardi, G., S. R. Fain, J. P. Gallo-Reynoso, A. L. Figueroa-Carranza, and B. J. Le Boeuf. 1998. Genetic variability in Guadalupe fur seals. *Journal of Heredity* 89:301–305.

Berta, A., and T. A. Deméré. 1986. *Callorhinus gilmorei* n. sp., (Carnivora: Otariidae) from the San Diego Formation (Blancan) and its implication for otariid phylogeny. *Transactions of the San Diego Society of Natural History* 21:111–126.

Berta, A., and G. S. Morgan. 1985. A new sea otter (Carnivora: Mustelidae) from the late Miocene and early Pliocene (Hemphillian) of North America. *Journal of Paleontology* 59:809–819.

Berta, A., and J. L. Sumich. 1999. *Marine mammals: Evolutionary biology.* Academic Press, London.

Berta, A., and A. R. Wyss. 1994. Pinniped phylogeny. *Proceedings of the San Diego Society of Natural History* 29:33–56.

Berube, M., and A. Aguilar. 1998. A new hybrid between a blue whale, *Balaenoptera musculus,* and a fin whale, *B. physalus:* Frequency and implications of hybridization. *Marine Mammal Science* 14:82–98.

Berzin, A. A. 1971. Kashalot [The sperm whale]. Izdat. Pishchevaya Promyshlennost, Moscow. English translation, 1972, Israel Program for Scientific Translations, Jerusalem.

Berzin, A. A., and N. V. Doroshenko. 1981. Right whales of the Okhotsk Sea. *Reports of the International Whaling Commission* 31:451–455.

Berzin, A. A., and V. L. Vladimirov. 1982. Novyi vid kosatok iz Antarktiki. *Priroda* 71(6):31.

———. 1983. Novyi vid kosatki (Cetacea, Delphinidae) iz vod Antarktiki. *Zoologicheskii Zhurnal* 62:287–295.

Best, T. L. 1999. Alaskan hare, *Lepus othus*. Pp. 702–704, in *The Smithsonian book of North American mammals* (D. E. Wilson and S. Ruff, eds.). Smithsonian Institution Press, Washington, D.C., in association with the American Society of Mammalogists.

Best, T. L., and T. H. Henry. 1994. *Lepus othus*. *Mammalian Species* 458:1–5.

Bickham, J. W., J. C. Patton, and T. R. Loughlin. 1996. High variability for control-region sequences in a marine mammal: Implications for conservation and biogeography of Steller sea lions (*Eumetopias jubatus*). *Journal of Mammalogy* 77:95–108.

Bidlack, A. L. 2000. Phylogeography and population genetics of northern flying squirrels (*Glaucomys sabrinus*) in Southeast Alaska. Unpublished thesis. University of Alaska Fairbanks.

Bidlack, A. L., and J. A. Cook. 2001. Reduced genetic variation in insular northern flying squirrels (*Glaucomys sabrinus*) along the North Pacific Coast. *Animal Conservation* 4:283–290.

———. 2002. A nuclear perspective on endemism in northern flying squirrels (*Glaucomys sabrinus*) of the Alexander Archipelago. *Conservation Genetics* 3:247–259.

Black, L. T. 1984. *Atka: An ethnohistory of the western Aleutians*. Limestone Press, Kingston, Ontario.

Bobrinskii, N. A., B. A. Kuznetsov, and A. P. Kuzyakin. 1965. [*Guide to the mammals of the U. S. S. R.*] Second edition. Proveshchenie, Moscow (in Russian).

Bodkin, J. L, K. A. Kloecker, G. G. Esslinger, D. H. Monson, J. D. DeGroot, and J. Doherty. 2002. Sea otter studies in Glacier Bay National Park and Preserve. 2001 Annual Report. U. S. Geological Survey, Biological Resources Division, Anchorage, AK. http://www.absc.usgs.gov/research/sea_otters/popchange.htm (accessed 10 February 2006).

Boland, J. L. 2007. Distribution of bats in southeast Alaska and selection of day-roosts in trees by Keen's myotis on Prince of Wales Island, southeast Alaska. Unpublished thesis. Oregon State University, Corvallis.

Boness, D. J. 1999a. Ribbon seal, *Phoca fasciata*. Pp. 203–204, in *The Smithsonian book of North American mammals* (D. E. Wilson and S. Ruff, eds.). Smithsonian Institution Press, Washington, D.C., in association with the American Society of Mammalogists.

———. 1999b. Hooded seal, *Cystophora cristata*. Pp. 215–216, in *The Smithsonian book of North American mammals* (D. E. Wilson and S. Ruff, eds.). Smithsonian Institution Press, Washington, D.C., in association with the American Society of Mammalogists.

Bonner, N. 1994. *Seals and sea lions of the world*. Facts on File, Inc., New York.

Bonnichsen, R., R. W. Graham, J. S. Geppert, et al. 1986. False Cougar and Shield Trap Caves, Pryor Mountains, Montana. *National Geographic Research* 2:276–290.

Bork, A. M., C. M. Strobeck, F. C. Yeh, R. J. Hudson, and R. K. Salmon. 1991. Genetic relationship of wood bison and plains bison based on restriction fragment length polymorphisms. *Canadian Journal of Zoology* 69:43–48.

Boudreau, T. A. 2004. Unit 19 bison management report. Pp. 16–27, in *Bison management report of survey and inventory activities* 1 July 2001–30 June 2003 (C. Brown, ed.). Alaska Department of Fish and Game. Project 9.0. Juneau, Alaska.

Bowers, P. M. 1978. Geology and archaeology of the Carlo Creek site, and Early Holocene campsite in the central Alaska Range. *American Quaternary Association*, Abstracts of the Fifth Biennial Meeting (September 2–4, Edmonton, Alberta).

Bowyer, R. T., and D. M. Leslie, Jr. 1992. *Ovis dalli*. *Mammalian Species* 393:1–7.

Bowyer, R. T., V. van Ballenberghe, and J. G. Kie. 1998. Timing and synchrony of parturition in Alaskan moose: Long-term versus proximal effects of climate. *Journal of Mammalogy* 79:1332–1344.

Boyce, M. S. 1978. Climatic variability and body size variation in the muskrats (*Ondatra zibethicus*) of North America. *Oecologia* 36:1–19.

Boyeskorov, G. 1999. New data on moose (*Alces*, Artiodactyla) Systematics. *Säugetierkundliche Mitteilungen* 44:3–13.

Braham, H. W. 1983. Northern records of Risso's dolphin, *Grampus griseus*, in the northeast Pacific. *Canadian Field-Naturalist* 97:89–90.

Braham, H. W., and M. E. Dahlheim. 1982. Killer whales in Alaska documented in the Platforms of Opportunity Program. *Report of the International Whaling Commission* 32:643–646.

Braham, H. W., F. E. Durham, G. H. Jarrell, and S. Leatherwood. 1980. Ingutuk: A morphological variant of the bowhead whale, *Balaena mysticetus*. *Marine Fisheries Review* 42:70–73.

Braham, H. W., R. D. Everitt, and D. J. Rugh. 1980. Northern sea lion population decline in the eastern Aleutian Islands. *Journal of Wildlife Management* 44:25–33.

Braham, H. W., and D. W. Rice. 1984. The right whale, *Balaena glacialis*. *Marine Fisheries Review* 46(4):38–44.

Brandt, J. F. 1861. Rapport sur un memoire qui, en tratant l'osteologie Comparée de la Rhytine, constitue la second partie de mes Symbolae Sirenologicae. *Mélanges Biol. Bull. Acad. Sci. St.-Pétersbourg* 4:75–77.

Brechbill, R. A. 1977. Status of the Norway rat. Pp. 261–267, in *The environment of Amchitka Island, Alaska* (L. Merritt and R. G. Fuller, eds.). Number TID-26712, Technical Information Center, Energy Research and Development Administration.

Brewster, W. G., and S. H. Fritts. 1995. Taxonomy and genetics of the gray wolf in western North America: A review. Pp. 353–373, in *Ecology and conservation of wolves in a changing world* (L. N. Carbyn, S. H. Fritts, and D. R. Seip, eds.). Canadian Circumpolar Institute, University of Alberta, Edmonton.

Brink, C. H., and F. C. Dean. 1966. Spruce seed as a food of red squirrels and flying squirrels in interior Alaska. *Journal of Wildlife Management* 30:503–512.

Broadbooks, H. E. 1965. Ecology and distribution of the pikas of Washington and Alaska. *American Midland Naturalist* 73:299–335.

Brodie, P. F. 1989. The white whale, *Delphinapterus leucas* (Pallas, 1776). Pp. 119–144, in *Handbook of marine mammals. Volume 4: River dolphins and the larger toothed whales* (S. H. Ridgway and R. J. Harrison, eds.). Academic Press, London.

Brooks, D. R., and E. P. Hoberg. 2006. Systematics and emerging infectious diseases: From management to solution. *Journal of Parasitology* 92:426–429.

Broughton, J. M. 1999. *Resource depression and intensification during the late Holocene, San Francisco Bay: Evidence from the Emeryville shellmound vertebrate fauna*. University of California Press, Berkeley.

Brownell, R. L., Jr. 1999. Pacific white-sided dolphin, *Lagenorhynchus obliquidens*. Pp. 279–280, in *The Smithsonian book of North American mammals* (D. E. Wilson and S. Ruff, eds.). Smithsonian Institution Press, Washington, D.C., in association with the American Society of Mammalogists.

Brownell, R. L., Jr., P. J. Clapham, T. Miyashita, and T. Kasuya. 2001. Conservation status of North Pacific right whales. *Journal of Cetacean Research and Management* (Special Issue) 2:269–286.

Brownell, R. L., Jr., W. F. Perrin, L. A. Pastene, P. J. Palsbøll, J. G. Mead, A. N. Zerbini, T. Kasuya, and D. D. Tormosov. 2000. Worldwide taxonomic status and geographic distribution of minke whales (*Balaenoptera acutorostrata* and *B. bonarensis*). International Whaling Commission Meeting Document SC/52/O27, 1–13.

Brownell, R. L., Jr., W. A. Walker, and K. A. Forney. 1999. Pacific white-sided dolphin, *Lagenorhynchus obliquidens* Gill, 1865. Pp. 57–84, in *Handbook of marine mammals. Volume 6: The second book of dolphins and the porpoises* (S. H. Ridgway and R. Harrison, eds.). Academic Press, London.

Brubaker, L. B., P. M., anderson, M. E. Edwards, and A. V. Lozhkin. 2005. Beringia as a glacial refugium for boreal trees and shrubs: New perspectives from mapped pollen data. *Journal of Biogeography* 32:833–848.

Brunet, M., et al. 2002. A new hominid from the Upper Miocene of Chad, Central Africa. *Nature* 418:145–151.

———. 2005. New material of the earliest hominid from the Upper Miocene of Chad. *Nature* 434:752–755.

Brunner, S. 2002. Geographic variation in skull morphology of adult Steller sea lions (*Emetopias jubatus*). *Marine Mammal Science* 18:206–222.

Bryant, H. N. 1987. Wolverine from the Pleistocene of the Yukon: Evolutionary trends and the taxonomy of *Gulo* (Carnivora: Mustelidae). *Canadian Journal of Earth Sciences* 24:654–663.

Bryant, H. N., A. P. Russell, and W. D. Fitch. 1993. Phylogenetic relationships within the extant Mustelidae (Carnivora): Appraisal of cladistic status of the Simpsonian subfamilies. *Zool. Journal of the Linnaean Society* 108:301–334.

Bryant, J. P., F. S. Chapin III, and D. R. Klein. 1983. Carbon/nutrient balance of boreal plants in relation to vertebrate herbivory. *Oikos* 40:357–368.

Bryant, L. D., and C. Maser. 1982. Classification and distribution. Pp. 1–59, in *Elk of North America: Ecology and management* (J. W. Thomas and D. E. Toweill, eds.). Stackpole Books, Harrisburg, PA.

Buck, C. L., and B. M. Barnes. 1999. Temperatures of hibernacula and changes in body composition of arctic ground squirrel over winter. *Journal of Mammalogy* 80:1264–1276.

Buckley, J. L., and W. L. Libby. 1957. Research and reports on aerial interpretation of terrestrial bioenvironments and faunal populations. Arctic Aeromedical Laboratory, Fairbanks, Alaska, Technical Report 57–32.

Burg, T. M., A. W. Trites, and M. J. Smith. 1999. Mitochondrial and microsatellite DNA analyses of harbour seal population structure in the northeast Pacific Ocean. *Canadian Journal of Zoology* 77:930–943.

Burger, J., W. Rosendahl, O. Loreille, H. Memmer, T. Eriksson, A. Götherström, J. Hiller, M. J. Collins, T. Wess, and K. W. Alt. 2004. Molecular phylogeny of the extinct cave lion *Panthera leo spelaea*. *Molecular Phylogenetics and Evolution* 30:841–849.

Burkanov, V. N., A. R. Semenov, S. A. Mashagin, and Y. V. Kitayev. 1988. Data on the abundance of the ice forms of seals in the Karaginski Gulf of the Bering Sea in 1986–1987. Voyage of exploration on marine mammals of North Pacific in 1986–1987. *VNIRO*, M.:71–80.

Burke, A., and J. Cinq-Mars. 1998. Paleoethological reconstruction and taphonomy of *Equus lambei* from the Bluefish Caves, Yukon Territory, Canada. *Arctic* 51:105–115.

Burles, D. W., A. G. Edie, and P. M. Bartier. 2004. Native land mammals and amphibian of Haida Gwaii with management implications for Gwaii Haanas National Park Reserve and Haida Heritage Site. Parks Canada Technical Reports in Ecosystem Science No. 40.

Burnett, C. D. 1983. Geographic and secondary sexual variation in the morphology of *Eptesicus fuscus*. *Annals of the Carnegie Museum of Natural History* 52:139–162.

Burns, J. J. 1970. Remarks on the distribution and natural history of pagophilic pinnipeds in the Bering and Chukchi seas. *Journal of Mammalogy* 51:445–454.

———. 1978. Ice seals. Pp. 193–205, in *Marine mammals of eastern North Pacific and Arctic waters* (D. Haley, ed.). Pacific Search Press, Seattle.

———. 1981a. Ribbon seal *Phoca fasciata* Zimmermann, 1783. Pp. 89–109, in *Handbook of marine mammals, Vol. 2: seals* (S. H. Ridgway and R. J. Harrison, eds.). Academic Press, London.

———. 1981b. Bearded seal *Erignathus barbatus*, Erxleben 1777. Pp. 145–170, in *Handbook of marine mammals, Vol. 2: Seals* (S. H. Ridgway and R. J. Harrison, eds.). Academic Press, London.

———. 1994a. Spotted seal. Alaska Department of Fish and Game Wildlife Notebook Series. http://www.adfg.state.ak.us/pubs/notebook/marine/spt-seal.php (accessed 26 March 2007).

———. 1994b. Ribbon seal. Alaska Department of Fish and Game Wildlife Notebook Series. http://www.adfg.state.ak.us/pubs/notebook/marine/rib-seal.php (accessed 26 March 2007).

———. 1994c. Walrus. Alaska Department of Fish and Game Wildlife Notebook Series. http://www.adfg.state.ak.us/pubs/notebook/marine/walrus.php (accessed 26 March 2007).

Burns, J. J., and F. H. Fay. 1970. Comparative morphology of the skull of the ribbon seal, *Histriophoca fasciata*, with remarks on systematics of Phocidae. *Journal of Zoology* (London) 161:363–394.

Burns, J. J., F. H. Fay, and G. A. Fedoseev. 1984. Craniological analysis of harbor and spotted seals of the North Pacific Region. Pp. 5–16, in *Soviet-American cooperative research on marine mammals. Pinnipeds* (F. H. Fay and G. A. Fedoseev, eds.). National Oceanic and Atmospheric Administration, Technical Report, National Marine Fisheries Service 1:1–104.

Burns, J. J., and A. Gavin. 1980. Recent records of hooded seals, *Cystophora cristata* Erxleben, from the western Beaufort Sea. *Arctic* 33:326–329.

Burns, J. J., J. J. Montague, and C. J. Cowles (Eds.). 1993. *The bowhead whale.* Society of Marine Mammalogy, Special Publication No. 2.

Burris, O. E., and D. E. McKnight. 1973. Game transplants in Alaska. *Alaska Department of Fish and Game, Wildlife Technical Bulletin* 4:1–57.

Buskirk, S. 1983. The ecology of marten in southcentral Alaska. Unpublished dissertation, University of Alaska Fairbanks.

Buskirk, S., and P. S. Gipson. 1980. Zoogeography of arctic foxes (*Alopex lagopus*) on the Aleutian Islands. Pp. 38–54, in *Worldwide Furbearer Conference Proceedings* (J. A. Chapman and D. P. Pursley, eds.). Frostburg, MD.

Buskirk, S. W., and S. O. MacDonald. 1984. Seasonal food habits of marten in south-central Alaska. *Canadian Journal of Zoology* 62:944–950.

Buskirk, S. W., and R. A. Powell. 1994. Habitat ecology of fishers and American martens. Pp. 283–296, in *Martens, sables, and fishers: Biology and conservation* (S. W. Buskirk, A. S. Harestad, M. G. Raphael, and R. A. Powell, eds.). Cornell University Press, Ithaca and London.

Butler, L. B. 2004a. Unit 9 & 10 furbearer management report. Pp. 117–125, in *Furbearer management report of survey and inventory activities* 1 July 2000–30 June 2003 (C. Brown, ed.). Alaska Department of Fish and Game. Project 7.0. Juneau, Alaska.

————. 2004b. Unit 9 moose management report. Pp. 113–120, in *Moose management report of survey and inventory activities* 1 July 2001–30 June 2003 (C. Brown, ed.). Alaska Department of Fish and Game. Project 1.0. Juneau, Alaska.

————. 2005. Unit 10 caribou management report. Pp. 57–60, in *Caribou management report of survey and inventory activities* 1 July 2002–30 June 2004 (C. Brown, ed.). Alaska Department of Fish and Game. Juneau, Alaska.

Byrd, G. V., E. P. Bailey, and W. Stahl. 1997. Restoration of island populations of Black Oystercatchers and Pigeon Guillemots by removing foxes. *Colonial Waterbirds* 20:253–260.

Byrd, G. V., and N. Norvell. 1993. Status of the Pribilof shrew based on summer distribution and habitat use. *Northwestern Naturalist* 74:49–54.

Byrd, G. V., and J. C. Williams. 2002. Pre-fox removal surveys at Avatanak Island, Alaska. U. S. Fish and Wildlife Service Report. AMNWR 02/01.

Byun, S. A., B. K. Koop, and T. E. Reimchen. 1997. North American black bear mtDNA phylogeography: Implications for morphology and the Haida Gwaii glacial refugium controversy. *Evolution* 51:1647–1653.

Cahalane, V. H. 1947. *Mammals of North America.* The Macmillan Co., New York.

————. 1959. A biological survey of Katmai National Monument. *Smithsonian Miscellaneous Collections* 138:1–246.

Calambokidis, J., G. H. Steiger, J. M. Straley, T. J. Quinn II, L. M. Herman, S. Cerchio, D. R. Salden, M. Yamaguchi, F. Sato, J. Urbán R., J. Jacobsen, O. von Ziegesar, K. C. Balcomb, C. M. Gabriele, M. E. Dahlheim, N. Higashi, S. Uchida, J. K. B. Ford, Y. Miyamura, P. L. Guevara P., S. A. Mizroch, L. Schlender, and K. Rasmussen. 1997. Population abundance and structure of humpback whales in the North Pacific basin. Final report to Southwest Fisheries Science Center, La Jolla, California.

Caldwell, D. K., and M. C. Caldwell. 1989. Pygmy sperm whale *Kogia breviceps* (de Blainville, 1838): Dwarf sperm whale *Kogia simus* Owen, 1866. Pp. 235–260, in *Handbook of marine mammals. Volume 4: River dolphins and the larger toothed whales* (S. H. Ridgway and R. Harrison, eds.). Academic Press, London.

Cardini, A. 2003. The geometry of the marmot (Rodentia: Sciuridae) mandible: Phylogeny and patterns of morphological evolution. *Systematic Biology* 52:186–205.

Cardini, A., R. S. Hoffmann, and R. W. Thorington, Jr. 2005. Morphological evolution in marmots (Rodentia, Sciuridae): Size and shape of the dorsal and lateral surfaces of the cranium. *Journal of Zoological Systematics and Evolutionary Research* 43:258–268.

Carleton, M. D. 1980. Phylogenetic relationships in neotomine-peromyscine rodents (Muroidea) and a reappraisal of the dichotomy within New World Cricetinae. Miscellaneous Publications of the Museum of Zoology, University of Michigan 157:1–146.

————. 1984. Introduction to rodents. Pp. 255–265, in *Orders and families of Recent mammals of the world* (S. Anderson and J. K. Jones, Jr., eds.). John Wiley and Sons, New York.

Carleton, M.D., and G.G. Musser. 1984. Muroid rodents. Pp. 289–379, in *Orders and families of Recent mammals of the world* (S. Anderson and J.K. Jones, Jr., eds.). John Wiley and Sons, New York .

―――――. 2005. Order Rodentia. Pp. 745–752, in *Mammal species of the world: A taxonomic and geographic reference* (D.E. Wilson and D.M. Reeder, eds.). Third Edition. Johns Hopkins University Press, Baltimore.

Carr, S.M., and S.A. Hicks. 1997. Are there two species of marten in North America? Pp. 15–28, in Martes: *Taxonomy ecology techniques and management* (G. Proulx, H.N. Bryant, and P.M. Woodard, eds.). The Provincial Museum of Alberta, Edmonton.

Carr, S.M., and E.A. Perry. 1997. Intra- and interfamilial systematic relationships of phocid seals as indicated by mitochondrial DNA sequences. Pp. 277–290, in *Molecular genetics of marine mammals*. Special Publication Number 3 (A.E. Dizon, S.J. Chivers, and W.F. Perrin, eds.). The Society for Marine Mammalogy, Lawrence, KS.

Carrara, P.E., T.A. Ager, and J.F. Baichtal. 2007. Possible refugia in the Alexander Archipelago of southeastern Alaska during the late Wisconsin glaciation. *Canadian Journal of Earth Sciences* 44:229–244.

Carraway, L.N. 1990. A morphologic and morphometric analysis of the "*Sorex vagrans* species complex" in the Pacific Coast region. Special Papers of the Museum, Texas Tech University 32:1–76.

Carretta, J.V., K.A. Forney, M.M. Muto, J. Barlow, J. Baker, B. Hanson, and M. Lowry. 2005. U.S. Pacific Marine Mammal Stock Assessments: 2004. U.S. Department of Commerce, NOAA Technical Memorandum NMFS-SWFSC-375.

―――――. 2007. U.S. Pacific Marine Mammal Stock Assessments: 2006. U.S. Department of Commerce, NOAA Technical Memorandum NMFS-SWFSC-398.

Carroll, G. 1994. Bowhead whale. Alaska Department of Fish and Game Wildlife Notebook Series. http://www.adfg.state.ak.us/pubs/notebook/marine/bowhead.php (accessed 26 March 2007).

―――――. 2004a. Unit 26A furbearer management report. Pp. 352–359, in *Furbearer management report of survey and inventory activities* 1 July 2000–30 June 2003 (C. Brown, ed.). Alaska Department of Fish and Game. Project 7.0. Juneau, Alaska.

―――――. 2004b. Unit 26A moose management report. Pp. 597–612, in *Moose management report of survey and inventory activities* 1 July 2001–30 June 2003 (C. Brown, ed.). Alaska Department of Fish and Game. Project 1.0. Juneau, Alaska.

Castrodale, L. 2006. Bats and rabies in Alaska—2006 update. *State of Alaska Epidemiology Bulletin No. 20* (September 25, 2006).

Ceballos-G, G. 1999. Northern river otter, *Lontra canadensis*. Pp. 179–180, in *The Smithsonian book of North American mammals* (D.E. Wilson and S. Ruff, eds.). Smithsonian Institution Press, Washington, D.C., in association with the American Society of Mammalogists.

Chapin F.S., III, M.W. Oswood, K. Van Cleve, L.A. Viereck, and D.L. Verbyla. 2006. *Alaska's changing boreal forest*. Oxford University Press, New York.

Chapman, J.A., K.R. Dixon, W. Lopez-Forment, and D.E. Wilson. 1983. The New World jackrabbits and hares (genus *Lepus*). 1. Taxonomic history and population status. *Acta Zoologica Fennica* 174:49–51.

Chapman, P.M. 2005. Toxic effects of contaminants in polar marine environments. *Environmental Science and Technology* 399:200A–207A.

Chapskii, K.K. 1940. Distribution of the walrus in the Laptev and East Siberian seas. *Problemy Severa* 1940:80–94.

―――――. 1955. Opyt peresmotra sistemy i diagnostiki tyulenei podsemeistva Phocinae [An attempt at a revision of the systematics and diagnostics of seals of the subfamily Phocinae]. *Trudy Zoologicheskogo Instituta Akademii Nauka SSSR* 17:160–199.

Chernyavskyi, F.B. 1962. On the systematic relationships and history of the snow sheep of the Old and New Worlds (translated from Russian). *Moscow Society of Naturalists Bulletin, Biology Section* 47:17–26.

————. 1984. *Mlekopitayushchie krainego severa Sibiri* (Mammals of the Siberian far north). Nauka, Moscow.

Chesemore, D. L. 1968a. Distribution and movements of white foxes in northern and western Alaska. *Canadian Journal of Zoology* 46:849–854.

————. 1968b. Occurrence of moose near Barrow, Alaska. *Journal of Mammalogy* 49:528–529.

Childs, H. E., Jr. 1969. Birds and mammals of the Pitmegea River region, Cape Sabine, northwestern Alaska. *Biological Papers of the University of Alaska No. 10.*

Chumbley, K., J. Sease, M. Strick, and R. Towell. 1997. Field studies of Steller sea lions (*Eumetopias jubatus*) at Marmot Island, Alaska, 1979 through 1994. U. S. Department of Commerce, NOAA Technical Memorandum NMFS-AFSC-77.

Cinq-Mars, J., and R. E. Morlan. 1999. Bluefish Caves and Old Crow Basin: A new rapport. Pp. 200–212, in *Ice-Age people of North America: Environment, origins, and adaptations* (R. Bonnichsen and K. L. Turnmire, eds.). Oregon State University Press for the Center for the Study of the First Americans, Corvallis.

Cipriano, F. 1997. Antitropical distribution and speciation in dolphins of the genus *Lagenorhynchus*: A preliminary analysis. Pp. 305–316, in *Molecular genetics of marine mammals* (A. E. Dizon, S. J. Chivers, and W. F. Perrin, eds.). Society for Marine Mammalogy, Lawrence, KS.

Clapham, P. J. 1999a. Family Balaenidae. P. 241, in *The Smithsonian book of North American mammals* (D. E. Wilson and S. Ruff, eds.). Smithsonian Institution Press, Washington, D.C., in association with the American Society of Mammalogists.

————. 1999b. Family Balaenopteridae. P. 246, in *The Smithsonian book of North American mammals* (D. E. Wilson and S. Ruff, eds.). Smithsonian Institution Press, Washington, D.C., in association with the American Society of Mammalogists.

————. 1999c. Humpback whale, *Megaptera novaeangliae*. Pp. 256–258, in *The Smithsonian book of North American mammals* (D. E. Wilson and S. Ruff, eds.). Smithsonian Institution Press, Washington, D.C., in association with the American Society of Mammalogists.

————. 2002. Humpback whale *Megaptera novaeangliae*. Pp. 589–592, in *Encyclopedia of marine mammals* (W. F. Perrin, B. Würsig, and J. G. M. Thewissen, eds.). Academic Press, San Diego and London.

Clapham, P. J., and J. G. Mead. 1999. *Megaptera novaeangliae. Mammalian Species* 604:1–9.

Clark, T. W., E. Anderson, C. Douglas, and M. Strickland. 1987. *Martes americana. Mammalian Species* 289:1–8.

Clark, W. K. 1958. The land mammals of the Kodiak Islands. *Journal of Mammalogy* 39:574–577.

Cleator, H. J., and I. Stirling. 1999. Bearded seal, *Erignathus barbatus*. Pp. 213–215, in *The Smithsonian book of North American mammals* (D. E. Wilson and S. Ruff, eds.). Smithsonian Institution Press, Washington, D.C., in association with the American Society of Mammalogists.

Clough, G. C. 1976. Mammals of the Yukon-Charley River area, Alaska. Pp. 204–232, in *The environment of the Yukon-Charley Rivers area, Alaska* (S. B. Young, ed.). The Center for Northern Studies No. 9, Wolcott, VT.

Coltrane, J. 2004. Unit 13D and 14 mountain goat management report. Pp. 146–164, in *Mountain goat management report of survey and inventory activities* 1 July 2001–30 June 2003 (C. Brown, ed.). Alaska Department of Fish and Game. Project 12.0. Juneau, Alaska.

Conroy, C. J. 1998. Molecular phylogenetics of arvicoline rodents. Unpublished dissertation. University of Alaska Fairbanks.

Conroy, C. J., and J. A. Cook. 1998. *Microtus longicaudus* (Merriam 1888), long-tailed vole. Pp. 93–95, in *North American rodents: Status survey and conservation action plan* (D. J. Hafner, E. Yensen, and G. L. Kirkland, Jr., eds.). IUCN/SSC Rodent Specialist Group. Gland, Switzerland and Cambridge, UK.

————. 1999. *Microtus xanthognathus. Mammalian Species* 627:1–5.

————. 2000a. Phylogeography of a post-glacial colonizer: *Microtus longicaudus* (Rodentia: Muridae). *Molecular Ecology* 9:165–175.

————. 2000b. Molecular systematics of a Holarctic rodent (*Microtus*: Muridae). *Journal of Mammalogy* 81:344–359.

Cook, J. A. 1998a. *Spermophilus parryii* (Richardson 1825), arctic ground squirrel. Pp. 49–50, in *North American rodents: Status survey and conservation action plan* (D. J. Hafner, E. Yensen, and G. L. Kirkland, Jr., eds.). IUCN/SSC Rodent Specialist Group. Gland, Switzerland and Cambridge, UK.

————. 1998b. *Dicrostonyx exsul* G. M. Allen 1919, St. Lawrence Island collared lemming and *Dicrostonyx unalascensis* Merriam 1900, Unalaska collared lemming. Pp. 87–89, in *North American rodents: Status survey and conservation action plan* (D. J. Hafner, E. Yensen, and G. L. Kirkland, Jr., eds.). IUCN/SSC Rodent Specialist Group. Gland, Switzerland and Cambridge, UK.

————. 1998c. *Microtus abbreviatus* Miller 1899, insular vole. Pp. 89–90, in *North American rodents: Status survey and conservation action plan* (D. J. Hafner, E. Yensen, and G. L. Kirkland, Jr., eds.). IUCN/SSC Rodent Specialist Group. Gland, Switzerland and Cambridge, UK.

————. 1998d. *Marmota caligata* (Eschscholtz 1829), hoary marmot. Pp. 39–40, in *North American rodents: Status survey and conservation action plan* (D. J. Hafner, E. Yensen, and G. L. Kirkland, Jr., eds.). IUCN/SSC Rodent Specialist Group. Gland, Switzerland and Cambridge, UK.

Cook, J. A., A. L. Bidlack, C. J. Conroy, J. R. Demboski, M. A. Fleming, A. M. Runck, K. D. Stone, and S. O. MacDonald. 2001. A phylogeographic perspective on endemism in the Alexander Archipelago of southeast Alaska. *Biological Conservation* 97:215–227.

Cook, J. A., C. J. Conroy, and J. D. Herriges, Jr. 1997. Northern record of the water shrew, *Sorex palustris,* in Alaska. *Canadian Field-Naturalist* 111:638–640.

Cook, J. A., N. Dawson, and S. O. MacDonald. 2006. Management of highly fragmented systems: The north temperate Alexander Archipelago. *Biological Conservation* 133:1–15.

Cook, J. A., N. Dawson, S. O. MacDonald, and A. M. Runck. 2004. Mammal diversity: inventories of Alaska National Parks stimulate new perspectives. *Alaska Park Science* 3(2):23–27.

Cook, J. A., K. Galbreath, E. Waltari, D. McDonald, and B. J. Hayward. 2000. Mammal inventory of western Alaska Range, 1999. University of Alaska Museum report to Bureau of Land Management.

Cook, J. A., E. P. Hoberg, A. Koehler, H. Henttonen, L. Wickström, V. Haukisalmi, K. Galbreath, F. Chernyavski, N. Dokuchaev, A. Lahsuhtkin, S. O. MacDonald, A. Hope, E. Waltari, A. Runck, A. Veitch, R. Popko, E. Jenkins, S. Kutz, and R. Eckerlin. 2005. Beringia: Intercontinental exchange and diversification of high latitude mammals and their parasites during the Pliocene and Quaternary. *Mammal Study* 30: S33–S44.

Cook, J. A., and G. L. Kirkland, Jr. 1998. *Clethrionomys gapperi* (Vigors 1830), southern red-backed vole; Gapper's red-backed vole. P. 87, in *North American rodents: Status survey and conservation action plan* (D. J. Hafner, E. Yensen, and G. L. Kirkland, Jr., eds.). IUCN/SSC Rodent Specialist Group. Gland, Switzerland and Cambridge, UK.

Cook, J. A., and D. R. Klein. 1999. Insular vole, *Microtus abbreviatus.* Pp. 623–624, in *The Smithsonian book of North American mammals* (D. E. Wilson and S. Ruff, eds.). Smithsonian Institution Press, Washington, D.C., in association with the American Society of Mammalogists.

Cook, J. A., and S. O. MacDonald. 2001. Should endemism be a focus of conservation efforts along the North Pacific Coast of North America? *Biological Conservation* 97:207–213.

————. 2002. Mammal inventory of Alaska's National Parks and Preserves: Denali National Park and Preserve. National Park Service Alaska Region, Inventory and Monitoring Program Annual Report 2002.

————. 2003. Mammal inventory of Alaska's National Parks and Preserves: Wrangell–St. Elias National Park and Preserve. National Park Service Alaska Region, Inventory and Monitoring Program Annual Report 2001–2002.

————. 2004a. Mammal inventory of Alaska's National Parks and Preserves, Arctic Network: Bering Land Bridge NP, Cape Krusenstern NM, Kobuk Valley NP, Noatak NP, Gates of the Arctic NP&P. National Park Service Alaska Region, Inventory and Monitoring Program Final Report.

————. 2004b. Mammal inventory of Alaska's National Parks and Preserves: Lake Clark National Park and Preserve. National Park Service Alaska Region, Inventory and Monitoring Program Annual Report 2003.

─────. 2004c. Mammal inventory of Alaska's National Parks and Preserves: Kenai Fjords National Park. National Park Service Alaska Region, Inventory and Monitoring Program Annual Report 2003.

─────. 2006. Mammal inventory of Alaska's National Parks and Preserves: Southwest Alaska Network—Kenai Fjords National Park, Lake Clark National Park and Preserve, and Katmai National Park and Preserve. National Park Service Alaska Region, Inventory and Monitoring Program Final Report 2005.

Cook, J. A., A. M. Runck, and C. J. Conroy. 2004. Historical biogeography at the crossroads of the northern continents: Molecular phylogenetics of red-backed voles (Rodentia: Arvicolinae). *Molecular Phylogenetics and Evolution* 30:767–777.

Corbet, G. B. 1978. *The mammals of the Palaearctic region: A taxonomic review.* British Museum (Natural History), London.

Corbet, G. B., and J. E. Hill. 1980. *A world list of mammalian species.* British Museum (Natural History), London.

─────. 1986. *A world list of mammalian species.* Second edition. British Museum (Natural History), London.

COSEWIC (Committee on the Status of Endangered Wildlife in Canada). 2007. http://www.cosewic.gc.ca (accessed 26 Sept. 2007).

Cotton, C. L., and K. L. Parker. 2000. Winter habitat and nest trees used by northern flying squirrels in subboreal forests. *Journal of Mammalogy* 81:1071–1086.

Cowan, I. M. 1936. Distribution and variation in deer (Genus *Odocoileus*) of the Pacific coastal region of North America. *California Fish and Game* 22:155–246.

─────. 1940. Distribution and variation in the native sheep of North America. *American Midland Naturalist* 24:505–580.

─────. 1956. What and where are the mule and black-tailed deer? Pp. 334–359, in *The deer of North America* (W. P. Taylor, ed.). The Stackpole Company, Harrisburg, PA, and The Wildlife Management Institute, Washington, D.C.

─────. 1989. Birds and mammals of Haida Gwaii. Pp. 175–186, in *The outer shores* (G. G. E. Scudder and N. Gessler, eds.). Queen Charlotte Islands Museum Press, Skidgate, British Columbia.

Cowan, I. M., and C. J. Guiguet. 1965. The mammals of British Columbia. British Columbia Provincial Museum Handbook No. 11, 2nd ed. Province of British Columbia, Department of Recreation and Conervation.

Cowan, I. M., and W. McCrory. 1970. Variation in the mountain goat, *Oreamnos americanus* (Blainville). *Journal of Mammalogy* 51:60–73.

Cranford, J. A. 1999. Western jumping mouse, *Zapus princeps.* Pp. 668–669, in *The Smithsonian book of North American mammals* (D. E. Wilson and S. Ruff, eds.). Smithsonian Institution Press, Washington, D.C., in association with the American Society of Mammalogists.

Crockford, S. J., and S. G. Frederick. 2007. Sea ice expansion in the Bering Sea during the Neoglacial: Evidence from archaeozoology. *The Holocene* 17:699–706.

Cronin, M. A. 1991. Mitochondrial and nuclear genetic relationships of deer (*Odocoileus* spp.) in western North America. *Canadian Journal of Zoology* 69:1270–1279.

─────. 1992. Intraspecific variation in mitochondrial DNA of North American cervids. *Journal of Mammalogy* 73:70–82.

Cronin, M. A., S. C. Amstrup, G. W. Garner, and E. R. Vyse. 1991. Interspecific and intraspecific mitochondrial DNA variation in North American bears (*Ursus*). *Canadian Journal of Zoology* 69:2985–2992.

Cronin, M. A., J. Bodkin, B. Ballachey, J. Estes, and J.C. Patton. 1996. Mitochondrial-DNA variation among subspecies and populations of sea otters (*Enhydra lutris*). *Journal of Mammalogy* 77:546–557.

Cronin, M. A., S. Hills, E. W. Born, and J. C. Patton. 1994. Mitochondrial DNA variation in Atlantic and Pacific walruses. *Canadian Journal of Zoology* 72:1035–1043.

Cronin, M. A., W. J. Spearman, W. Buchholz, S. Miller, L. Comerci, and L. Jack. 2002. Microsatellite DNA and mitochondrial DNA variation in Alaskan sea otters. Alaska Fisheries Technical Report.

Crossen, K. J., R. W. Graham, D. W. Veltre, and D. Yesner. 2003. A Pribilof Island cave: Late Quaternary mammal bone assemblages from St. Paul Island, Bering Sea, Alaska. Abstract, Annual Geological Society of America Conference NW Division 424 (Geological Society of America, Seattle).

Crowley, D. W. 2003. Unit 6 wolf management report. Pp. 52–57, in *Wolf management report of survey and inventory activities* 1 July 1999–30 June 2002 (C. Healy, ed.). Alaska Department of Fish and Game. Juneau, Alaska.

————. 2004a. Unit 6 furbearer management report. Pp. 80–89, in *Furbearer management report of survey and inventory activities* 1 July 2000–30 June 2003 (C. Brown, ed.). Alaska Department of Fish and Game. Project 7.0. Juneau, Alaska.

————. 2004b. Unit 6 moose management report. Pp. 90–106, in *Moose management report of survey and inventory activities* 1 July 2001–30 June 2003 (C. Brown, ed.). Alaska Department of Fish and Game. Project 1.0. Juneau, Alaska.

————. 2004c. Unit 6 mountain goat management report. Pp. 82–105, in *Mountain goat management report of survey and inventory activities* 1 July 2001–30 June 2003 (C. Brown, ed.). Alaska Department of Fish and Game. Project 12.0. Juneau, Alaska.

————. 2005a. Unit 6 black bear management report. Pp. 129–151, in *Black bear management report of survey and inventory activities* 1 July 2001–30 June 2005 (C. Brown, ed.). Alaska Department of Fish and Game. Project 17.0. Juneau, Alaska.

————. 2005b. Unit 6 deer management report. Pp. 93–108, in *Deer management report of survey and inventory activities* 1 July 2002–30 June 2004 (C. Brown, ed.). Alaska Department of Fish and Game. Juneau, Alaska.

Cummings, W. C. 1985. Right whales. Pp. 275–304, in *Handbook of marine mammals. Volume 3: The sirenians and baleen whales* (S. H. Ridgway and R. Harrison, eds.). Academic Press, London.

Currier, M. J. 1983. *Felis concolor. Mammalian Species* 200:1–7.

Dahlheim, M. E., D. Ellifrit, and J. Swenson. 1997. *Killer whales of Southeast Alaska: A catalogue of photoidentified individuals.* Day Moon Press, Seattle.

Dahlheim, M. E., and J. E. Heyning. 1999. Killer whale, *Orcinus orca* (Linnaeus, 1758). Pp. 281–322, in *Handbook of marine mammals. Volume 6: The second book of dolphins and the porpoises* (S. H. Ridgway and R. Harrison, eds.). Academic Press, London.

Dahlheim, M. E., and R. G. Towell. 1994. Occurrence and distribution of Pacific white-sided dolphins (*Lagenorhynchus obliquidens*) in southeastern Alaska, with notes on an attack by killer whales (*Orcinus orca*). *Marine Mammal Science* 10:458–464.

Dalebout, M. L., A. van Helden, K. van Waerebeek, and C. S. Baker. 1998. Molecular genetic identification of Southern Hemisphere beaked whales. *Molecular Ecology* 7:687–694.

Dall, W. H. 1870. *Alaska and its resources.* Lee and Shepard, Boston.

Dalquest, W. W., and M. C. Ramage. 1946. Notes on the long-legged bat (*Myotis volans*) at Old Fort Tejon and vicinity, California. *Journal of Mammalogy* 27:60–63.

Darimont, C. T., P. C. Paquet, T. E. Reimchen, and V. Crichton. 2005. Range expansion by moose into coastal temperate rainforests of British Columbia, Canada. *Diversity and Distributions* 11:235–239.

Darwin, C. 1859. *On the origin of species by means of natural selection, or the preservation of favoured races in the struggle for life.* Murray, London.

Dau, J. 2004a. Unit 23 furbearer management report. Pp. 315–324, in *Furbearer management report of survey and inventory activities* 1 July 2000–30 June 2003 (C. Brown, ed.). Alaska Department of Fish and Game. Project 7.0. Juneau, Alaska.

————. 2004b. Unit 23 moose management report. Pages 523–541, in *Moose management report of survey and inventory activities* 1 July 2001–30 June 2003 (C. Brown, ed.). Alaska Department of Fish and Game. Project 1.0. Juneau, Alaska.

————. 2005. Unit 23 and Subunit 26A Dall sheep management report. Pp. 148–159, in *Dall sheep management report of survey and inventory activities* 1 July 2001–30 June 2004 (C. Brown, ed.). Alaska Department of Fish and Game. Project 6.0. Juneau, Alaska.

Daugherty, C. H., A. Cree, J. M. Hay, and M. B. Thompson. 1990. Neglected taxonomy and continuing extinctions of tuatara (*Sphenodon*). *Nature* 347:177–179.

Davies, J. L. 1963. The antitropical factor in cetacean speciation. *Evolution* 17:107–116.

Davis, B. L. 2001. Sea mammal hunting and the Neoglacial: An archaeofaunal study of environmental change and subsistence technology at Margaret Bay, Unalaska. Pp. 71–85, in *Archaeology in the Aleut Zone of Alaska* (D. E. Dumond, ed.). University of Oregon Anthropological Papers 58, Eugene.

Davis, C. S., I. Delisle, I. Sterling, D. B. Siniff, and C. Strobeck. 2004. A phylogeny of the extant Phocidae inferred from complete mitochondrial DNA coding regions. *Molecular Phylogenetics and Evolution* 33:363–377.

Davis, J., and W. Z. Lidicker, Jr. 1975. The taxonomic status of the southern sea otter. *Proceedings of the California Academy of Sciences* 40:429–437.

Davis, M. 1950. Hybrids of the polar and Kodiak bear. *Journal of Mammalogy* 31:449–450.

Dawson, N. G. 2008. Vista Norteña: Tracking historical diversification and contemporary structure in high latitude mesocarnivores. Unpublished dissertation. University of New Mexico, Albuquerque.

Dawson, W. 2007. Laurence Irving: An appreciation. *Physiological and Biochemical Zoology* 80:9–24.

de Muizon, C. 1982. Phocid phylogeny and dispersal. *Annals of the South African Museum* 89:175–213.

Dean, F. C., and D. L. Chesemore. 1974. Studies of birds and mammals in the Baird and Schwatka mountains, Alaska. *Biological Papers of the University of Alaska*, No. 15.

Degerbøl, M. 1935. Report of the mammals collected by the fifth Thule Expedition to Arctic North America; Zoology. I. Mammals. *Report of the Fifth Thule Expedition, 1921–1922*, 2:1–67.

Del Frate, G. G. 2004. Unit 16A moose management report. Pp. 224–232, in *Moose management report of survey and inventory activities* 1 July 2001–30 June 2003 (C. Brown, ed.). Alaska Department of Fish and Game. Grants W-27–3 and 4, Project 1.0. Juneau, Alaska.

DeLong, R. L. 1978. Northern elephant seal. Pp. 207–211, in *Marine mammals of eastern North Pacific and Arctic waters* (D. Haley, ed.). Pacific Search Press, Seattle.

DeMaster, D. P., and I. Stirling. 1981. *Ursus maritimus. Mammalian Species* 145:1–7.

Demboski, J. 1999. Molecular systematics and biogeography of long-tailed shrews (Insectivora: *Sorex*) and northern flying squirrels (Rodentia: *Glaucomys*). Unpublished dissertation. University of Alaska Fairbanks.

Demboski, J. R., and J. A. Cook. 2001. Phylogeography of the dusky shrew, *Sorex monticolus* (Insectivora, Soricidae): Insight into deep and shallow history in northwestern North America. *Molecular Ecology* 10:1227–1240.

———. 2003. Phylogenetic diversification within the *Sorex cinereus* group (Soricidae). *Journal of Mammalogy* 84:144–158.

Demboski, J. R., J. A. Cook, and G. L. Kirkland, Jr. 1998. *Glaucomys sabrinus* (Shaw 1801), northern flying squirrel. Pp. 37–39, in *North American rodents: Status survey and conservation action plan* (D. J. Hafner, E. Yensen, and G. L. Kirkland, Jr., eds.). IUCN/SSC Rodent Specialist Group. Gland, Switzerland and Cambridge, UK.

Demboski, J. R., B. K. Jacobsen, and J. A. Cook. 1998. Implications of cytochrome *b* sequence variation for biogeography and conservation of the northern flying squirrels (*Glaucomys sabrinus)* of the Alexander Archipelago, Alaska. *Canadian Journal of Zoology* 76:1771–1777.

Demboski, J. R., K. D. Stone, and J. A. Cook. 1999. Further perspectives on the Haida Gwaii glacial refugium controversy. *Evolution* 53:2008–2012.

Deméré, T. A. 1986. The fossil whale, *Balaenoptera davidsonii* (Cope 1872), with a review of other Neogene species of Balaenoptera (Cetacea: Mysticeti). *Marine Mammal Science* 2:277–298.

———. 1994. The family Odobenidae: A phylogenetic analysis of living and fossil taxa. *Proceedings of the San Diego Society of Natural History* 29:99–123.

Deméré, T. A., A. Berta, and M. R. McGowen. 2005. The taxonomic and evolutionary history of fossil and modern balaenopteroid mysticetes. *Journal of Mammalian Evolution* 12(1/2):99–143.

Dice, L. R. 1921. Notes on the mammals of interior Alaska. *Journal of Mammalogy* 2:20–28.

Diersing, V. E. 1980. Systematics and evolution of the pygmy shrews (subgenus *Microsorex*) of North America. *Journal of Mammalogy* 61:76–101.

————. 1984. Lagomorphs. Pp. 241–254, in *Orders and families of Recent mammals of the world* (S. Anderson and J. K. Jones, Jr., eds.). John Wiley and Sons, New York.

Dillon, L. S. 1961. Historical subspeciation in the North American marten. *Systematic Zoology* 10:49–64.

Dixon, E. J. 1984. Context and environment in taphonomic analysis: Examples from Alaska's Porcupine River caves. *Quaternary Research* 22:201–215.

————. 1999. *Bones, boats, and bison: Archaeology and the first colonization of western North America.* University of New Mexico Press, Albuquerque.

Dixon, E. J., D.C. Plaskett, and R. M. Thorson. 1979. *Report of the 1979 archaeological and geological reconnaissance of cave deposits, Porcupine River, Alaska.* University of Alaska.

Dixon, J. S. 1938. Birds and mammals of Mount McKinley National Park. National Park Service, Fauna Series No. 3.

Dixon, K. R., J. A. Chapman, G. R. Willner, D. E. Wilson, and W. Lopez-Forment. 1983. The New World jackrabbits and hares (genus *Lepus*). 2. Numerical taxonomic analysis. *Acta Zoologica Fennica* 174:53–56.

Dobson, G. E. 1889. Description of a new species of water-shrew from Unalaska Island. *Annals and Magazine of Natural History* (6th Series) 4(13):372–374.

Dobzhansky, T. 1973. Nothing in biology makes sense except in the light of evolution. *The American Biology Teacher* 35:125–129.

Dokuchaev, N. E. 1997. A new species of shrew (Soricidae, Insectivora) from Alaska. *Journal of Mammalogy* 78:811–817.

Domning, D. P. 1977. An ecological model for late Tertiary sirenian evolution in the North Pacific Ocean. *Systematic Zoology* 25:352–362.

————. 1978. Sirenian evolution in the North Pacific Ocean. *University of California Publications in Geological Science* 118:1–176.

————. 1994. A phylogenetic analysis of the Sirenia. *Proceedings of the San Diego Society of Natural History* 29:177–190.

————. 1996. Bibliography and index of the Sirenia and Desmostylia. *Smithsonian Contributions to Paleobiology* 80:1–611.

Domning, D. P., J. Thomason, and D. G. Corbett. 2007. Steller's sea cow in the Aleutian Islands. *Marine Mammal Science* 23:976–983.

Dorsey, E. M., S. J. Stern, A. R. Hoelzel, and J. Jacobsen. 1990. Minke whale (*Balaenoptera acutorostrata*) from the west of North America: Individual recognition and small scale site fidelity. *Report of the International Whaling Commission* (Special Issue 12):357–368.

Douglas, M. E., G. D. Schnell, and D. J. Hough. 1984. Differentiation between inshore and offshore spotted dolphins in the tropical Pacific Ocean. *Journal of Mammalogy* 65:375–387.

Douglass, R. J. 1984. Ecological distribution of small mammals in the DeLong Mountains of northwestern Alaska. *Arctic* 37:148–154.

Doyle, A. T. 1990. Use of riparian and upland habitats by small mammals. *Journal of Mammalogy* 71:14–23.

Dragoo, J. W., and R. L. Honeycutt. 1997. Systematics of mustelid-like carnivores. *Journal of Mammalogy* 78:426–443.

Drew, R. E., J. G. Hallett, K. B. Aubry, K. W. Cullings, S. M. Koepf, and W. J. Zielinski. 2003. Conservation genetics of the fisher (*Martes pennanti*) based on mitochondrial DNA sequencing. *Molecular Ecology* 12:51–62.

DuBois, S. D. 2004. Unit 20D bison management report. Pp. 28–71, in *Bison management report of survey and inventory activities* 1 July 2001–30 June 2003 (C. Brown, ed.). Alaska Department of Fish and Game. Project 9.0. Juneau, Alaska.

Dufresne, F. 1942. Mammals and birds of Alaska. *Fish and Wildlife Service Circular* 3:1–37.

————. 1946. *Alaska's animals and fishes.* A. S. Barnes and Co., New York.

Durall, S. 1993. Phylogenetic analysis of northern hair seals based on nucleotide sequences of the mitochondrial cytochrome *b* gene. Unpublished thesis. University of Alaska Fairbanks.

Durner, G. M., and S. C. Amstrup. 1995. Movements of a polar bear from northern Alaska to northern Greenland. *Arctic* 48:338–341.

Dyke, A. S., J. Hooper, C. R. Harington, and J. M. Savelle. 1999. The Late Wisconsinan and Holocene record of walrus (*Odobenus rosmarus*) from North America: A review with new data from Arctic and Atlantic Canada. *Arctic* 52:160–181.

Ebbert, S. E., and G. V. Byrd. 2002. Eradications of invasive species to restore natural biological diversity on Alaska Maritime National Wildlife Refuge. Pp. 102–109, in *Turning the tide: The eradication of invasive species* (C. R. Veitch, and M. N. Clout, eds.). IUCN SSC Invasive Species Specialist Group. IUCN, Gland, Switzerland, and Cambridge, UK.

Eckert, J. 2007. Historische Aspekte der Echinococcose: Einer uralten, aber noch immer relevanten Zoonose. *Schweizer Archiv fuer Tierheilkunde* 149:5–14.

Eddingsaas, A. A. 2001. The effects of Pleistocene glacial processes on Beringia: The evolutionary history of the arctic ground squirrel (*Spermophilus parryii*). Unpublished thesis. Idaho State University, Pocatello.

Eddingsaas, A. A., B. K. Jacobsen, E. P. Lessa, and J. A. Cook. 2004. Evolutionary history of the arctic ground squirrel (*Spermophilus parryii*) in Nearctic Beringia. *Journal of Mammalogy* 85:601–610.

Eger, J. L. 1990. Patterns of geographic variation in the skull of Nearctic ermine (*Mustela erminea*). *Canadian Journal of Zoology* 68:1241–1249.

Eide, S., and S. Miller. 1994. Brown bear. Alaska Department of Fish and Game Wildlife Notebook Series. http://www.adfg.state.ak.us/pubs/notebook/biggame/brnbear.php (accessed 26 March 2007).

Eley, T. J., Jr. 1994. Ringed seal. Alaska Department of Fish and Game Wildlife Notebook Series. http://www.adfg.state.ak.us/pubs/notebook/marine/rin-seal.php (accessed 26 March 2007).

Elias, S. A., S. K. Short, C. H. Nelson, and H. H. Birks. 1996. Life and times of the Bering land bridge. *Nature* 382:60–63.

Elkins, W. A., and U. C. Nelson. 1954. Wildlife introductions and transplants in Alaska. Proceedings, 5th Alaska Science Conference.

Ellerman, J. R., and T. C. S. Morrison-Scott. 1951. Checklist of Palaearctic and Indian mammals 1758 to 1946. Trustees of the British Museum (Natural History), London.

———. 1966. Checklist of Palaearctic and Indian mammals 1758 to 1946. Second edition. British Museum of Natural History, London.

Elliott, H. W. 1881. A monograph of the Pribilov Islands group, or the seal-islands of Alaska. U. S. Government Printing Office, Washington, D.C.

———. 1896. Report upon the present condition of the fur-seal rookeries of the Pribilov Islands of Alaska. House Documents 54th Congress, Document No. 175. U. S. Government Printing Office, Washington, D.C.

Engstrom, M. D. 1999. Collared lemmings, *Dicrostonyx*. Pp. 658–659, in *The Smithsonian book of North American mammals* (D. E. Wilson and S. Ruff, eds.). Smithsonian Institution Press, Washington, D.C., in association with the American Society of Mammalogists.

Erbajeva, M. A., J. I. Mead, and S. L. Swift. 2003. Evolution and development of Asian and North American ochotonids. Poster abstract, Third International Mammoth Conference, 24–29 May 2003, Dawson City, Yukon Territory.

Erdbrink, D. P. 1953. A review of fossil and Recent bears of the Old World. Krukkerij Jan De Lange, Deventer 2:321–597.

Eschricht, D. F., and J. Reinhardt. 1866. On the Greenland right-whale (*Balaena mysticetus*, Linn.); with especial reference to its geographical distribution and migrations in times past and present, and to its external and internal characteristics. Pp. 1–150, in *Recent memoirs on the Cetacea* (W. H. Flower, ed.). Published for the Ray Society by Robert Hardwicke, London. Translated from the Danish by J. Reinhardt.

Escorza-Treviño, S., L. A. Pastene, and A. E. Dizon. 2004. Molecular analyses of the *truei* and *dalli* morphotypes of Dall's porpoise (*Phocoenoides dalli*). *Journal of Mammalogy* 85:347–355.

Estes, J. A. 1980. *Enhydra lutris*. *Mammalian Species* 133:1–8.

———. 1999. Sea otter, *Enhydra lutris*. Pp. 180–182, in *The Smithsonian book of North American mammals* (D. E. Wilson and S. Ruff, eds.). Smithsonian Institution Press, Washington, D.C., in association with the American Society of Mammalogists.

Ewer, R. F. 1973. *The carnivores*. Cornell University Press, Ithaca, New York.

Eyerdam, W. J. 1936. Mammal remains from an Aleut Stone Age village. *Journal of Mammalogy* 17:61.

Fall, J. A., D. B. Andersen, L. Brown, M. Coffing, G. Jennings, C. Mishler, A. Page, C. J. Utermohle, and V. Vanek. 1993. Noncommercial harvests and uses of wild resources in Sand Point, Alaska, 1992. Technical Paper No. 226. Alaska Department of Fish and Game, Division of Subsistence, Juneau, Alaska.

Farmer, C. J., D. K. Person, and R. T. Bowyer. 2006. Risk factors and mortality of black-tailed deer in a managed forest landscape. *Journal of Wildlife Management* 70:1403–1415.

Fay, F. H. 1982. Ecology and biology of the Pacific walrus, *Odobenus rosmarus divergens,* Illiger. *North American Fauna* 74.

———. 1985. *Odobenus rosmarus. Mammalian Species* 238:1–7.

———. 1995. An earlier Pacific record of a hooded seal. *Marine Mammal Science* 11:415.

———. 1999. Walrus, *Odobenus rosmarus.* Pp. 193–194, in *The Smithsonian book of North American mammals* (D. E. Wilson and S. Ruff, eds.). Smithsonian Institution Press, Washington, D.C., in association with the American Society of Mammalogists.

Fay, F. H., and J. L. Sease. 1985. Preliminary status of selected small mammals. Final report to U. S. Fish and Wildlife Service. University of Alaska Fairbanks, Institute of Marine Science.

Fedje, D. W., Q. Mackie, E. J. Dixon, and T. H. Heaton. 2004. Late Wisconsin environments and archaeological visibility on the northern Northwest Coast. Pages 97–138, in *Entering America: Northeast Asia and Beringia before the last glacial maximum* (D. B. Madsen, ed.). University of Utah Press, Salt Lake City.

Fedorov, V. B., and A. V. Goropashnaya. 1999. The importance of ice ages in diversification of arctic collared lemmings (*Dicrostonyx*): Evidence from the mitochondrial cytochrome *b* region. *Hereditas* 130:301–307.

Fedorov, V. B., K. Fredga, and G. H. Jarrell. 1999. Mitochondrial DNA variation and the evolutionary history of chromosome races of collared lemmings (*Dicrostonyx*) in the Eurasian Arctic. *Journal of Evolutionary Biology* 12:134–145.

Fedorov, V. B., A. V. Goropashnaya, M. Jaarola, and J. A. Cook. 2003. Phylogeography of lemmings (*Lemmus*): No evidence for postglacial colonization of Arctic from the Beringian refugium. *Molecular Ecology* 12:725–731.

Fedorov, V. B., A. V. Goropashnaya, G. H. Jarrell, and K. Fredga. 1999. Phylogeographic structure and mitochondrial DNA variation in true lemmings (*Lemmus*) from the Eurasian Arctic. *Biological Journal of the Linnean Society* 66:357–371.

Fedorov, V. B., and N. C. Stenseth. 2002. Multiple glacial refugia in the North American Arctic: Inference from phylogeography of the collared lemming (*Dicrostonyx groenlandicus*). Proceedings Royal Society London B 269:2071–2077.

Fedoseev, G. A. 1984. Morfoekologicheskie razlichiya v populyatsiyakh largi (*Phoca larga*) i kry-latki (*Histriophoca fasciata*) Beringova Morya [Morphoecological differences in the populations of harbor seal and ribbon seal in the Bering Sea]. Pp. 108–120, in *Morskie mlekopitayushchie Dalnego Vostoka* (A. S. Perlov, ed.). Tikhookeanskii Nauchno-Issledovatel'skii Institut Rybnogo Khozyaistva i Okeanografii, Vladivostok. (Russian with English abstract.)

Fedoseev, G. A., Y. V. Raslivalov, and G. G. Bobrova. 1988. Distribution abundance of the ice forms of pinnipeds on the ice of the Bering Sea in April and May 1987. Voyage of exploration on marine mammals of North Pacific in 1986–1987, *VNIRP*, M:44–70.

Fedyk, S. 1970. Chromosomes of *Microtus* (*Stenocranius*) *gregalis* major (Ognev, 1923) and phylogenetic connections between sub-arctic representatives of the genus *Microtus* Schrank, 1798. *Acta Theriologica* 15:143–152.

Fenton, M. B. 1999. Little Brown Bat, *Myotis lucifugus.* Pp. 94–95, in *The Smithsonian book of North American mammals* (D. E. Wilson and S. Ruff, eds.). Smithsonian Institution Press, Washington, D.C., in association with the American Society of Mammalogists.

Fenton, M. B., and R. M. R. Barclay. 1980. *Myotis lucifugus. Mammalian Species* 142:1–8.

Ferrero, R. C., and W. A. Walker. 1996. Age, growth and reproductive patterns of the Pacific white-sided dolphin (*Lagenorhynchus obliquidens*) taken in high seas driftnets in the central North Pacific Ocean. *Canadian Journal of Zoology* 74:1673–1687.

Fetter, A. W., and J. I. Everitt. 1981. Determination of the gross and microscopic structures of selected tissues and organs of the bowhead whale, *Balaena mysticetus,* with emphasis on bone, blubber and the lymphoimmune and cardiovascular systems. Pp. 51–88, in *Tissue structural studies and other investigations on the biology of endangered whales in the Beaufort Sea* (T. Albert, ed.). Final Report for the period April 1, 1980 through June 30, 1981. Prepared for U. S. Department of Interior, Bureau of Land Management, Alaska OCS Office, Anchorage.

Fichter, L. S. 1972. The North American beavers of the genus *Castor,* post incisor dentition: A multivariate study. Unpublished dissertation. University of Michigan, Ann Arbor.

Filardi, C. E., and R. G. Moyle. 2005. Single origin of a pan-Pacific bird group and upstream colonization of Australasia. *Nature* 438:216–219.

Findley, J. S. 1955. Speciation of the wandering shrew. University of Kansas Publications, *Museum of Natural History* 9:1–68.

Fiscus, C. H. 1978. Northern fur seal. Pp. 153–159, in *Marine mammals of eastern North Pacific and Arctic waters* (D. Haley, ed.). Pacific Search Press, Seattle.

Fitch, J. H., and K. A. Shump. 1979. *Myotis keenii. Mammalian Species* 121:1–3.

Fleming, M. F., and J. A. Cook. 2002. Phylogeography of endemic ermine (*Mustela erminea*) in Southeast Alaska. *Molecular Ecology* 11:795–808.

Flerov, C. C. 1933. Review of the Palaearctic reindeer or caribou. *Journal of Mammalogy* 14:328–338.

Flint, R. F. 1957. *Glacial and Pleistocene geology.* John Wiley and Sons, New York.

Flux, J. E. C. 1983. Introduction to taxonomic problems in hares. *Acta Zoologica Fennica* 174:7–10.

Flynn, R. W., and G. Blundell. 1992. Ecology of martens in southeast Alaska. Alaska Department of Fish and Game, Federal Aid in Wildlife Restoration Research Progress Report, Project W-23-5, Study 7.16, December.

Ford, J. K. B., G. M. Ellis, and K. C. Balcomb III. 1994. *Killer whales: The natural history and geneology of Orcinus orca in British Columbia and Washington State.* University of British Columbia Press, Vancouver.

Fordyce, R. E. 2002. Fossil record. Pp. 453–471, in *Encyclopedia of marine mammals* (W. F. Perrin, B. Würsig, and J. G. M. Thewissen, eds.). Academic Press, San Diego and London.

Fordyce, R. E., and L. G. Barnes. 1994. The evolutionary history of whales and dolphins. *Annual Review of Earth and Planetary Science* 22:419–455.

Forney, K. A., J. Barlow, M. M. Muto, M. Lowry, J. Baker, G. Cameron, J. Mobley, C. Stinchcomb, and J. V. Carretta. 2000. U. S. Pacific marine mammal stock assessments: 2000. U. S. Department of Commerce, NOAA Technical Memorandum NMFS-SWFSC-300.

Forsten, A., and P. M. Youngman. 1982. *Hydrodamalis gigas. Mammalian Species* 165:1–3.

Foster, J. B. 1965. Evolution of the mammals of Haida Gwaii, British Columbia. Occasional Papers of the British Columbia Museum 14. Victoria, British Columbia.

Foster, N. R., and M. P. Hare. 1990. Cephalopod remains from a Cuvier's beaked whale (*Ziphius cavirostris*) stranded in Kodiak, Alaska. *Northwestern Naturalist* 71:49–51.

Fox, J. L., C. A. Smith, and J. W. Schoen. 1989. Relation between mountain goats and their habitat in southeast Alaska. General Technical Report PNW-GTR-246. U. S. Forest Service, Washington, D.C.

Frances, J. 2008. Spatial genetic structure and demographic history of wolverine in North America with an emphasis on northern peripheral populations. Unpublished thesis. University of New Mexico, Albuquerque.

Frank, F. 1985. Zur evolution und systematik der kleinen wiesel (*Mustela nivalis* Linnaeus, 1766). *Zeitschrift für Säugetierkunde* 50:208–225.

Franzmann, A. W. 1981. *Alces alces. Mammalian Species* 154:1–7.

———. 1988. A review of Alaskan wildlife translocations. Pp. 210–229, in *Translocations of wild animals* (L. Nielsen and R. D. Brown, eds.). Wisconsin Humane Society and Caesar Kleberg Wildlife Research Institute, Milwaukee, WI, and Kingsville, TX.

———. 2000. Moose. Pp. 578–600, in *Ecology and management of large mammals in North America* (S. Demarais and P. R. Krausman, eds.). Prentice-Hall, Inc., Upper Saddle River, NJ.

Fritts, E. I. 2006. Wildlife and humans at risk: A needs assessment and strategic action plan for returning Alaska to its rat-free state. Draft Plan 10/17/06. http://www.adfg.state.ak.us/special/invasive/invasive.php (accessed 15 March 2007).

Frost, K. J. 1994. Gray whale. Alaska Department of Fish and Game Wildlife Notebook Series. http://www.adfg.state.ak.us/pubs/notebook/marine/gray.php (accessed 26 March 2007).

Frost, K. J., and L. F. Lowry. 1980. Feeding of ribbon seals (*Phoca fasciata*) in the Bering Sea in spring. *Canadian Journal of Zoology* 58:1601–1607.

———. 1981. Ringed, Baikal and Caspian seals. Pp. 29–53, in *Handbook of marine mammals. Volume 2: Seals* (S. H. Ridgway and R. J. Harrison, eds.). Academic Press, London.

Fuller, T. K. 1981. Small mammal populations on the Kenai Peninsula, Alaska. *Northwest Science* 55:298–303.

Galbreath, K. 2002. Genetic consequences of Ice Ages for a Holarctic rodent: Phylogeography and post-glacial colonization of the tundra vole, *Microtus oeconomus,* in Beringia. Unpublished thesis. University of Alaska Fairbanks.

Galbreath, K., and J. Cook. 2004. Genetic consequences of Pleistocene glaciations for the tundra vole (*Microtus oeconomus*) in Beringia. *Molecular Ecology* 13:135–148.

Galindo, C., and C. J. Krebs. 1985. Habitat use by singing voles and tundra voles in the southern Yukon. *Oecologia* 66:430–436.

Gallo-Reynoso, J. P. 1994. Factors affecting the population status of the Guadalupe fur seal, *Arctocephalus townsendi* (Merriam, 1897), at Isla de Guadalupe, Baja California, Mexico. Unpublished dissertation. University of California, Santa Cruz.

Gambell, R. 1985. Fin whale, *Balaenoptera physalus* (Linnaeus, 1758). Pp. 171–192, in *Handbook of marine mammals. Volume 3: The sirenians and baleen whales* (S. H. Ridgway and R. Harrison, eds.). Academic Press, London.

Gannon, W. L., R. S. Sikes, and Animal Care and Use Committee of the American Society of Mammalogists. 2007. Guidelines of the American Society of Mammalogists for the use of wild mammals in research. *Journal of Mammalogy* 88:809–823.

García-Perea, R. 1992. New data on the systematics of lynxes. *Cat News* 16:15–16.

Gard, L. M. Jr., G. E. Lewis, and F.C. Whitmore, Jr. 1972. Steller's sea cow in Pleistocene interglacial beach deposits on Amchitka, Aleutian Islands. *Geological Society of America Bulletin* 83:867–870.

Gardner, A. L. 1974. Mammals of the Noatak River Valley. Pp. 224–250, in *The environment of the Noatak River Basin, Alaska* (S.B. Young, ed.). Alaska Contributions, Center for Northern Studies, Number 1. Wolcott, VT.

Gardner, A. L., and C. B. Robbins. 1998. Generic names of northern and southern fur seals (Mammalia: Otariidae). *Marine Mammal Science* 14:544–551.

Gardner, C. L. 1985. The ecology of wolverines in southcentral Alaska. Unpublished thesis. University of Alaska Fairbanks.

Gardner, C. L., W. B. Ballard, and R. H. Jessup. 1986. Long distance movement by an adult wolverine. *Journal of Mammalogy* 67:603.

Gaskin, D. E. 1999. Harbor porpoise, *Phocoena phocoena.* Pp. 295–296, in *The Smithsonian book of North American mammals* (D. E. Wilson and S. Ruff, eds.). Smithsonian Institution Press, Washington, D.C., in association with the American Society of Mammalogists.

Geffen, E., A. Mecure, D. J. Girman, D. W. MacDonald, and R. K. Wayne. 1992. Phylogenetic relationships of the fox-like canids. *Journal of Zoology* (London) 228:27–39.

Geist, O. W. 1934. Brown bear seen on St. Lawrence Island. *Journal of Mammalogy* 15:316–317.

Geist, O. W., J.L. Buckley, and R. H. Manville. 1960. Alaska records of the narwhal. *Journal of Mammalogy* 41:250–253.

Geist, V. 1991. Phantom subspecies: The wood bison *Bison bison* "*athabascae*", Rhoads 1897, is not a valid taxon, but an ecotype. *Arctic* 44:283–300.

———. 1998. *Deer of the world: their evolution, behavior, and ecology.* Stackpole Books, Mechanicsburg, PA.

Gentry, R. L., and D. E. Withrow. 1978. Steller sea lion. Pp. 167–171, in *Marine mammals of eastern North Pacific and Arctic waters* (D. Haley, ed.). Pacific Search Press, Seattle.

George, S. B. 1988. Systematics, historical biogeography, and evolution of the genus *Sorex*. *Journal of Mammalogy* 69:443–461.

―――. 1999a. Montane shrew, *Sorex monticolus*. Pp. 31–33, in *The Smithsonian book of North American mammals* (D. E. Wilson and S. Ruff, eds.). Smithsonian Institution Press, Washington, D.C., in association with the American Society of Mammalogists.

―――. 1999b. Eastern heather vole, *Phenacomys ungava*. Pp. 618–619, in *The Smithsonian book of North American mammals* (D. E. Wilson and S. Ruff, eds.). Smithsonian Institution Press, Washington, D.C., in association with the American Society of Mammalogists.

Georgina, D. 2001. The small mammals of Lime Hills Cave I, Alaska. Pp. 23–31, in *People and Wildlife in Northern North America: Essays in Honor of R. Dale Guthrie* (S. G. Gerlach and M. S. Murray, eds.). BAR International Series 944.

Giannico, G. R., and D. Nagorsen. 1989. Geographic and sexual variation in the skull of Pacific coast marten (*Martes americana*). *Canadian Journal of Zoology* 67:1386–1393.

Gibson, D. D., and G. V. Byrd. 2007. Birds of the Aleutian Islands, Alaska. Nuttall Ornithological Club and AOU Monographs.

Giddings, J. L. 1964. *The archeology of Cape Denbigh*. Brown University Press, Providence, RI.

Gillespie, R. G., and G. K. Roderick. 2002. Arthropods on islands: Colonization, speciation, and conservation. *Annual Review of Entomology* 47:595–632.

Gilmore, R. M. 1933. Notes on the Unalaska collared lemming. *Journal of Mammalogy* 14:257–258.

―――. 1978. Right whale. Pp. 62–69, in *Marine mammals of eastern North Pacific and arctic waters* (D. Haley, ed.). Pacific Search Press, Seattle.

Gipson, P. S. 1983. Evaluation and control implications of behavior of feral dogs in interior Alaska. *Vertebrate Pest Control Management Material* 4:285–294.

Goddard, P. C., and D. J. Rugh. 1998. A group of right whales seen in the Bering Sea in July 1996. *Marine Mammal Science* 14:344–349.

Goldman, D. P., R. Giri, and S. J. O'Brien. 1989. Molecular genetic-distance estimates among the Ursidae as indicated by one- and two-dimensional protein electrophoresis. *Evolution* 43:282–295.

Goldman, E. A. 1935. New American mustelids of the genera *Martes*, *Gulo* and *Lutra*. *Proceedings of the Biological Society of Washington* 48:175–186.

―――. 1944. Classification of wolves. Pp. 387–636, in *The wolves of North America, part 2.* American Wildlife Institute, Washington, D.C.

Goodman, M., J. Czelusniak, S. Page, and C. M. Meireles. 2001. Where DNA sequences place *Homo sapiens* in a phylogenetic classification of Primates. Pp. 279–289, in *Humanity from African naissance to coming millennia* (P. V. Tobias, M. A. Raath, J. Moggi-Cecchi, and G. A. Doyle, eds.). Firenze University Press, Florence, and Witwatersrand University Press, Johannesburg.

Gorbics, C. S., and J. L. Bodkin. 2001. Stock structure of sea otters (*Enhydra lutris kenyoni*) in Alaska. *Marine Mammal Science* 17:632–647.

Gorn, T. 2004. Unit 22 furbearer management report. Pp. 304–314, in *Furbearer management report of survey and inventory activities* 1 July 2000–30 June 2003 (C. Brown, ed.). Alaska Department of Fish and Game. Project 7.0. Juneau, Alaska.

Graham, C. E., S. Ferrier, F. Huettman, C. Moritz, and A. T. Peterson. 2004. New developments in museum-based informatics and applications in biodiversity analysis. *Trends in Ecology and Evolution* 19:497–503.

Graham, R. W. 1972. Biostratigraphy and paleoecological significance of the Conard Fissure local fauna with emphasis on the genus *Blarina*. Unpublished M. S. thesis, University of Iowa, Iowa City.

―――. 1991. Variability in the size of North American Quaternary black bears (*Ursus americanus*) with the description of a fossil black bear from Bill Neff Cave, Virginia. Pp. 237–250, in *Beamers, bobwhites, and blue-points: Tributes to the career of Paul W. Parmalee* (J. R. Purdue, W. E. Kippel, and B. W. Styles, eds.). *Illinois State Museum Scientific Papers* 23:1–436.

Graham, R. W., and M. A. Graham. 1994. The Late Quaternary distribution of *Martes* in North America. Pp. 26–58, in *Martens, sables, and fishers: Biology and conservation* (S. W. Buskirk, A. S. Harestad, M. G. Raphael, and R. A. Powell, eds.). Cornell University Press, Ithaca, NY.

Grant, P. R. (Ed.). 1998. *Evolution on islands*. Oxford University Press, Oxford.

Gray, F. H. 1915. Smithsonian Institution Archives, Record Unit 7176, Box 7, Folder 3. U. S. Fish and Wildlife Service, 1860–1961.

Grayson, D. K. 1984. Time of extinction and nature of adaptation of the noble marten, *Martes noblis*. Pp. 233–240, in *Contributions in Quaternary vertebrate paleontology: A volume in memorial to John E. Guilday* (H. H. Genoways and M. R. Dawson, eds.). Special Publication, Carnegie Museum of Natural History 8:1–538.

————. 1991. Late Pleistocene mammalian extinctions in North America: Taxonomy, chronology, and explanations. *Journal of World Prehistory* 5:193–231.

————. 2005. A brief history of Great Basin pikas. Journal of Biogeography 32:2103–2111.

Griffin, B., and D. M. Johnson. 1994. American bison. Alaska Department of Fish and Game Wildlife Notebook Series. http://www.adfg.state.ak.us/pubs/notebook/biggame/bison.php (accessed 26 March 2007).

Grinnell, H. W. 1918. Notes on some bats from Alaska and British Columbia. *University of California Publications in Zoology* 17:431–433.

Gromov, I. M., and G. I. Baranova (Eds.). 1981. *Katalog mlekopitayushchikh SSSR* [Catalog of mammals of the USSR]. Nauka, Leningrad. (In Russian.)

Gromov, I. M., D. I. Bibikov, N. I. Kalabukhov, and M. N. N. Meier. 1965. Fauna SSSR, Mlekopitayushchie, tom. 3, vyp. 2 [Fauna of the U. S. S. R. Mammals, vol. 3, No. 2]. Nazemnye belich'e [Ground Squirrels]. Nauka, Moscow-Leningrad. (In Russian.)

Groves, C. P. 1981. Systematic relationships in the Bovini (Artiodactyla, Bovidae). *Zeitschrift für Zoologische Systematik und Evolutionsforschung* 4:264–278.

————. 1982. Cranial and dental characteristics in the systematics of Old World Felidae. *Carnivore* 5:28–39.

————. 2005. Order Primates. Pp. 111–184, in *Mammal species of the world: A taxonomic and geographic reference* (D. E. Wilson and D. M. Reeder, eds.). Third Edition. Johns Hopkins University Press, Baltimore.

Groves, C. P., and P. Grubb. 1987. Relationships of living deer. Pp. 21–59, in *Biology and management of the cervidae* (C. Wemmer, ed.). Smithsonian Institution Press, Washington, D.C.

Groves, P. 1997. Intraspecific variation of mitochondrial DNA of muskoxen based on control region sequences. *Canadian Journal of Zoology* 75:568–575.

Groves, P., and G. F. Shields. 1996. Phylogenetics of the Caprinae based on cytochrome *b* sequence. *Molecular Phylogenetics and Evolution* 5:467–476.

————. 1997. Cytochrome *b* sequence suggests convergent evolution of Asian takin and Arctic muskox. *Molecular Phylogenetics and Evolution* 8:363–374.

Grubb, P. 1993. Order Artiodactyla. Pp. 377–414, in *Mammal species of the world: A taxonomic and geographic reference* (D. E. Wilson and D. M. Reeder, eds.). Second Edition. Smithsonian Institution Press, Washington, D.C.

————. 2005. Order Artiodactyla. Pp. 637–722, in *Mammal species of the world: A taxonomic and geographic reference* (D. E. Wilson and D. M. Reeder, eds.). Third Edition. Johns Hopkins University Press, Baltimore.

Guilday, J. E. 1963. Pleistocene zoogeography of the lemming, *Dicrostonyx*. Evolution 17:194–197.

————. 1968. Grizzly bears from eastern North America. *The American Midland Naturalist* 79:247–250.

————. 1982. Dental variation in *Microtus xanthognathus, M. chrotorrhinus,* and *M. pennsylvanicus* (Rodentia: Mammalia). *Annals of the Carnegie Museum* 51:211–230.

Guilday, J. E., and M. S. Bender. 1960. Late Pleistocene records of the yellow-cheeked vole, *Microtus xanthognathus* (Leach). *Annals of the Carnegie Museum* 35:315–330.

Guilday, J. E., P. S. Martin, and A. D. McCrady. 1964. New Paris number 4: A late Pleistocene cave deposit in Bedford County, Pennsylvania. *Bulletin of the National Speleological Society* 26:121–194.

Guilday, J. E., P. W. Parmalee, and H. W. Hamilton. 1977. The Clark's cave bone deposit and the late Pleistocene paleoecology of the central Appalachian Mountains of Virginia. *Bulletin of the Carnegie Museum of Natural History* 2:1–87.

Gureev, A. A. 1964. Lagomorpha. Fauna SSSR, Mammals, 3 [pt. 10]:1–275. Akad. Nauk SSSR, Zool. Inst., Moscow [new ser. 87].

Guthrie, R. D. 1966. The extinct wapiti of Alaska and Yukon Territory. *Canadian Journal of Zoology* 44:47–57.

———. 1967. Fire melanism among animals. *American Midland Naturalist* 77:227–230.

———. 1968a. Paleoecology of a late Pleistocene small mammal community from interior Alaska. *Arctic* 21:223–244.

———. 1968b. Paleoecology of the large mammal community in interior Alaska during the late Pleistocene. *American Midland Naturalist* 79:346–363.

———. 1970. Bison evolution and zoogeography in North America during the Pleistocene. *The Quarterly Review of Biology* 45:1–15.

———. 1973. Mummified pika (*Ochotona*) carcass and dung pellets from Pleistocene deposits in interior Alaska. *Journal of Mammalogy* 54:970–971.

———. 1980. Bison and man in North America. *Canadian Journal of Anthropology* 1:55–73.

———. 1983. Paleoecology of the Dry Creek site and its implications for early hunters. Pp. 209–287, in *Archaeology and paleoecology of a late Pleistocene Alaskan hunting camp* (W. R. Powers, R. D. Guthrie, and J. F. Hoffecker, eds.). National Park Service, Washington, D.C.

———. 1984. Alaskan megabucks, megabulls, and megarams: The issue of Pleistocene gigantism. Pp. 483–500, in *Contributions in Quaternary vertebrate paleontology: A volume in memorial to John E. Guilday* (H. H. Genoways and M. R. Dawson, eds.). Carnegie Museum of Natural History Special Publication No. 8.

———. 1990a. *Frozen fauna of the Mammoth Steppe: The story of Blue Babe*. The University of Chicago Press, Chicago.

———. 1990b. New dates on Alaskan Quaternary dogs and wolves. *Current Research in the Pleistocene* 7:109–110.

———. 2001a. Origin and causes of the mammoth steppe: A story of cloud cover, woolly mammal tooth pits, buckles, and inside-out Beringia. *Quaternary Science Reviews* 20:549–574.

———. 2001b. Paleobehavior in Alaskan Pleistocene horses: social structure, maturation dates, uses of the landscape, and mortality patterns. Pp. 32–49, in *People and wildlife in northern North America: Essays in honor of R. Dale Guthrie* (S. G. Gerlach and M. S. Murray, eds.). BAR International Series 944.

———. 2003. Rapid body size decline in Alaskan Pleistocene horses before extinction. *Nature* 426:169–171.

———. 2004. Radiocarbon evidence of mid-Holocene mammoths stranded on an Alaskan Bering Sea island. *Nature* 429:746–749.

———. 2006. New carbon dates link climatic change with human colonization and Pleistocene extinctions. *Nature* 441:207–209.

Guthrie, R. D., and J. V. Matthews, Jr. 1971. The Cape Deceit fauna: Early Pleistocene mammalian assemblage from the Alaskan Arctic. *Quarternary Research* 1:474–510.

Guthrie, R. D., A. V. Sher, and C. R. Harington. 2001. New radiocarbon dates on Saiga antelopes (*Saiga tatarica*) from Alaska, Canada, and Siberia: their paleoecological significance. Pp. 50–57, in *People and wildlife in northern North America: Essays in honor of R. Dale Guthrie* (S. G. Gerlach and M. S. Murray, eds.). BAR International Series 944.

Hafner, D. J., K. E. Petersen, and T. L. Yates. 1981. Evolutionary relationships of jumping mice (genus *Zapus*) of the southwestern United States. *Journal of Mammalogy* 62:501–512.

Hafner, D. J., E. Yensen, and G. L. Kirkland, Jr. (Eds.). 1998. North American rodents: Status survey and conservation action plan. IUCN/SSC Rodent Specialist Group. Gland, Switzerland, and Cambridge, UK.

Hagmeier, E. M. 1956. Distribution of marten and fisher in North America. *Canadian Field-Naturalist* 70:150–168.

———. 1958. The inapplicability of the subspecies concept to the North American marten. *Systematic Zoology* 7:150–168.

———. 1959. A re-evaluation of the subspecies of fisher. *Canadian Field-Naturalist* 73:185–197.

————. 1961. Variation and relationships in North American marten. *Canadian Field-Naturalist* 75:122–137.

Halanych, K. M., J. R. Demboski, B. J. van Vuuren, D. R. Klein, and J. A. Cook. 1999. Cytochrome *b* phylogeny of North American hares and jackrabbits (*Lepus*, Lagomorpha) and the effects of saturation in outgroup taxa. *Molecular Phylogenetics and Evolution* 11:213–221.

Haley, D. 1978. Steller Sea Cow. Pp. 236–241, in *Marine mammals of Eastern North Pacific and Arctic waters* (D. Haley, ed.). Pacific Search Press, Seattle.

Hall, E. R. 1936. Mustelid mammals from the Pleistocene of North America with systematic notes on some recent members of the genera *Mustela, Taxidea,* and *Mephitis. Carnegie Institute Washington Publications* 473:41–119.

————. 1946. Zoological subspecies of man at the peace table. *Journal of Mammalogy* 27:358–364.

————. 1951. American weasels. University of Kansas Publication, *Museum of Natural History* 4:1–446.

————. 1981. *The mammals of North America.* Wiley-Interscience, New York.

Hall, E. R., and E. L. Cockrum. 1952. Comments on the taxonomy and geographic distribution of North American microtines. University of Kansas Publication, *Museum of Natural History* 5:293–312.

Hall, E. R., and R. M. Gilmore. 1932. New mammals from St. Lawrence Island, Bering Sea, Alaska. *University of California Publications in Zoology* 38:391–404.

————. 1934. *Marmota caligata broweri,* a new marmot from northern Alaska. *Canadian Field Naturalist* 48:57–59.

Hall, E. R., and K. Kelson. 1959. *The mammals of North America.* 2 volumes. Ronald Press, New York.

Hanna, G. D. 1920. Mammals of the St. Matthew Islands, Bering Sea. *Journal of Mammalogy* 1:118–122.

————. 1923. Rare mammals of the Pribilof Islands, Alaska. *Journal of Mammalogy* 4:209–215.

Hanni, K. D., D. J. Long, R. E. Jones, P. Pyle, and L. E. Morgan. 1997. Sightings and strandings of Guadalupe fur seals in Central and Northern California, 1988–1995. *Journal of Mammalogy* 78:684–690.

Harfenist, A., K. R. MacDowell, T. Golumbia, G. Schultze, and Laskeek Bay Conservation Society. 2000. Monitoring and control of raccoons on seabird colonies in Haida Gwaii (Queen Charlotte Islands). Proceedings of a Conference on the Biology and Management of Species and Habitats at Risk, Kamloops, British Columbia, 15–19 Feb.,1999. Volume One (L. M. Darling, ed.). British Columbia Ministry of Environment, Lands and Parks, Victoria, and University College of the Cariboo, Kamloops, British Columbia.

Harington, C. R. 1961. History, distribution and ecology of the muskoxen. Unpublished thesis. McGill University, Montreal, Quebec.

————. 1970. A postglacial muskox (*Ovibos moschatus*) from Grandview, Manitoba, and comments on the zoogeography of *Ovibos. National Museum of Natural Science Publications in Paleontology* 2:1–13.

————. 1971. A Pleistocene mountain goat from British Columbia with comments on the dispersal history of *Oreamnos. Canadian Journal of Earth Science* 8:1081–1093.

————. 1977. Pleistocene mammals of the Yukon Territory. Unpublished dissertation, University of Alberta, Edmonton.

————. 1978. Quaternary vertebrate faunas of Canada and Alaska and their suggested chronological sequence. *Syllogeus* 15:1–105.

————. 1980a. Pleistocene mammals from Lost Chicken Creek, Alaska, USA. *Canadian Journal of Earth Science* 17:168–198.

————. 1980b. Faunal exchanges between Siberia and North America: Evidence from Quaternary land mammal remains in Siberia, Alaska and Yukon Territory. *Canadian Journal of Anthropology* 1:45–49.

————. 1981. Whales and seals of the Champlain Sea. *Trail and Landscape* 15:32–47.

————. 1984. Mammoths, bison and time in North America. Pp. 299–309, in *Quaternary dating methods* (W. C. Mahaney, ed.). Elsevier Science Publications, Amsterdam.

————. 1989. Pleistocene vertebrate localities in the Yukon. Pp. 93–98, in *Late Cenozoic history of the interior basins of Alaska and the Yukon* (L. D. Carter, D. Thomas, and J. P. Galloway, eds.). U. S. Geological Survey Circular 1026.

————. 1990. Vertebrates of the last inter-glaciation in Canada: A review with new data. *Géographie physique et Quaternaire* 44:375–387.

————. 1995. Woolly mammoth. Yukon Beringia Interpretive Center Research Notes. http://www.beringia.com (accessed 26 June 2007).

————. 1996a. American lion. Yukon Beringia Interpretive Center Research Notes. http://www.beringia.com (accessed 3 June 2007).

————. 1996b. North American short-faced bear. Yukon Beringia Interpretive Center Research Notes. http://www.beringia.com (accessed 3 June 2007).

————. 1996c. Giant beaver. Yukon Beringia Interpretive Center Research Notes. http://www.beringia.com (accessed 10 August 2007)

————. 2003. *Annotated bibliography of Quaternary vertebrates of northern North America with radiocarbon dates.* University of Toronto Press, Toronto.

Harington, C. R., and J. Cinq-Mars. 1995. Radiocarbon dates on saiga antelope (*Saiga tatarica*) fossils from Yukon and the Northwest Territories. *Arctic* 48:1–7.

Harris, A. H. 1999. Water shrew, *Sorex palustris.* Pp. 38–39, in *The Smithsonian book of North American mammals* (D. E. Wilson and S. Ruff, eds.). Smithsonian Institution Press, Washington, D.C., in association with the American Society of Mammalogists.

Harrison, R. G., S. M. Bogdanowicz, and R. S. Hoffmann. 2003. Phylogeny and evolutionary history of the ground squirrels (Rodentia: Marmotinae). *Journal of Mammalian Evolution* 10:249–276.

Hartman, L. H., and D. S. Eastman. 1999. Distribution of introduced raccoons *Procyon lotor* on the Queen Charlotte Islands: Implications for burrow-nesting seabirds. *Biological Conservation* 88:1–13.

Hasegawa, M., J. Adachi, and M. C. Milinkovitch. 1997. Novel phylogeny of whales supported by total evidence. *Journal of Molecular Evolution 44* (Supplement 1):S117–S120.

Haskell, S. P. 2006. First record of a river otter, *Lutra canadensis,* captured on the northeastern coast of Alaska. *Canadian Field-Naturalist* 120:235–236.

Hatch, S. A., and M. A. Hatch. 1983. Populations and habitat use of marine birds in the Semidi Islands, Alaska. *Murrelet* 64:39–46.

Haukisalmi, V., H. Henttonen, J. Niemimaa, and R. L. Rausch. 2002. *Paranoplocephala etholeni* n.sp. (Cestoda: Anoplocephalidae) in *Microtus pennsylvanicus* from Alaska, with a synopsis of *Paranoplocephala*-species in Holarctic rodents. *Parasite* 9:305–314.

Hawes, M. L. 1976. Odor as a possible isolating mechanism in sympatric species of shrews (*Sorex vagrans* and *Sorex obscurus*). *Journal of Mammalogy* 57:404–406.

————. 1977. Home range, territoriality, and ecological separation in sympatric shrews, *Sorex vagrans* and *Sorex obscurus. Journal of Mammalogy* 58:354–367.

Hay, K. A., and A. W. Mansfield. 1989. Narwhal, *Monodon monoceros* Linnaeus, 1758. Pp. 145–176, in *Handbook of marine mammals. Volume 4: River dolphins and the larger toothed whales* (S. H. Ridgway and R. J. Harrison, eds.). Academic Press, London.

Haynes, G. 1991. *Mammoths, mastodonts, and elephants: Biology, behavior and the fossil record.* Cambridge University Press, Cambridge.

Hazard, K. 1988. Beluga whale, *Delphinapterus leucas.* Pp. 195–235, in *Selected marine mammals of Alaska: Species accounts with research and management recommendations* (J. W. Lentfer, ed.). Marine Mammal Commission, Washington, D.C.

Heaton, T. H. 1995. Middle Wisconsin bear and rodent remains discovered on Prince of Wales Island, Alaska. *Current Research in the Pleistocene* 12:92–94.

————. 2001. Late Pleistocene and Holocene vertebrates from the Southeast Alaskan mainland. *Journal of Vertebrate Paleontology* 21:59A–60A.

Heaton, T. H., and F. Grady. 1992. Preliminary report on the fossil bears of El Capitan Cave, Prince of Wales Island, Alaska. *Current Research in the Pleistocene* 9:97–99.

———. 1993. Fossil grizzly bears from Prince of Wales Island, Alaska, offer new insights into animal dispersal, interspecific competition, and age of deglaciation. *Current Research in the Pleistocene* 10:98–100.

———. 2003. The Late Wisconsin vertebrate history of Prince of Wales Island, southeast Alaska. Chapter 2. Pp. 17–53, in *Ice age cave faunas of North America* (B. W. Schubert, J. I. Mead, and R. W. Graham, eds.). Indiana University Press, Bloomington.

Helgen, K. M. 2005. Family Castoridae. Pp. 842–843, in *Mammal species of the world: A taxonomic and geographic reference* (D. E. Wilson and D. M. Reeder, eds.). Third Edition. Johns Hopkins University Press, Baltimore.

Heller, E. 1909. The mammals. Pp. 245–264, in *Birds and mammals of the 1907 Alexander Expedition to southeast Alaska* (J. Grinnell, ed.). *University of California Publication in Zoology* 5(2):171–264.

———. 1910. Mammals of the 1908 Alexander Alaska Expedition, with descriptions of the localities visited and notes on the flora of the Prince William Sound Region. *University of California Publications in Zoology* 5(11):321–360.

Hemmer, H. 1978. The evolutionary systematics of living Felidae: Present status and current problems. *Carnivore* 1:71–79.

Hennings, D., and R. S. Hoffmann. 1977. A review of the taxonomy of the *Sorex vagrans* species complex from western North America. *Occasional Papers of the Museum of Natural History, University of Kansas* 68:1–35.

Henttonen, H., and V. A. Peiponen. 1982. Polarrötelmaus, *Clethrionomys rutilus*. Pp. 165–176, in *Handbuch der Säugetiere Europas* (J. Niethammer and F. Krapp, eds.). Akademische Verlagsgesellschaft (Wiesbaden) 2/I:1–649.

Heptner, V. G. 1965. Eshchë raz o stellerovoi korove. *Priroda* 54:91–94.

Heptner, V. G., K. K. Chapskii, V. A. Arsen'ev, and V. E. Sokolov (Eds.). 1976. Pinnipeds and toothed whales. *Mammals of the Soviet Union.* Vysshaya Shkola, Moscow 2:1–718.

Heptner, V. G., A. A. Nasimovic, and A. G. Bannikov. 1988. Mammals of the Soviet Union: Artiodactyla and Perissodactyla [A translation of Heptner et al. 1961, Mlekopitayushchie Sovetskovo Soyuza: Parnokopytnye i neparnokopytnye]. Smithsonian Institution Libraries and National Science Foundation, Washington, D.C. 1:1–1147.

Heptner, V. G., and N. P. Naumov. 1974. *Die Säugetiere der Sowjetunion. Volume 2. Seekühe und Raubtiere.* Jena, Fisher Verlag.

Hershkovitz, P. 1961. On the nomenclature of certain whales. *Fieldiana Zoology* 39:547–565.

———. 1966. Catalog of living whales. *U. S. National Museum Bulletin* 246:1–259.

Heyning, J. E. 1989. Cuvier's beaked whale, *Ziphius cavirostris* G. Curier, 1823. Pp. 289–308, in *Handbook of marine mammals. Volume 4: River dolphins and the larger toothed whales* (S. H. Ridgway and R. J. Harrison, eds.). Academic Press, London.

———. 1999. Killer whale, *Orcinus orca.* Pp. 287–289, in *The Smithsonian book of North American mammals* (D. E. Wilson and S. Ruff, eds.). Smithsonian Institution Press, Washington, D.C., in association with the American Society of Mammalogists.

Heyning, J. E., and M. E. Dahlheim. 1988. *Orcinus orca. Mammalian Species* 304:1–9.

Heyning, J. E. 2002. Cuvier's beaked whale, *Ziphius cavirostris.* Pp. 305–307, in *Encyclopedia of marine mammals* (W. F. Perrin, B. Würsig, and J. G. M. Thewissen, eds.). Academic Press, London.

Higdon, J. W., O. R. P. Bininda-Emonds, R. M. D. Beck, and S. H. Ferguson. 2007. Phylogeny and divergence of the pinnipeds (Carnivora: Mammalia) assessed using a multigene dataset. BMC Evolutionary Biology 7:216, doi:10.1186/1471-2148-7-216.

Hill, J. E. 1942. *Citellus parryii* from the Pleistocene of Alaska. *Geological Society of America Bulletin* 53:1842.

Hill, P. S., and D. P. DeMaster. 1999. Alaska marine mammal stock assessments 1999. National Marine Mammal Laboratory, Seattle.

Hirons, A., D. M. Schell, and B. P. Finney. 2001. Temporal records of d13C and d15N in North Pacific pinnipeds: Inferences regarding environmental change and diet. *Oecologia* 129:591–601.

Hirota, K., and L. G. Barnes. 1995. A new species of middle Miocene sperm whale of the genus *Scaldicetus* (Cetacea; Physeteridae) from Shiga-mura, Japan. *Island Arc* 3:453–472.

Hjeljord, O. 1973. Mountain goat forage and habitat preference in Alaska. *Journal of Wildlife Management* 37:353–362.

Hobbs, R. C., and L. L. Jones. 1993. Impacts of high seas driftnet fisheries on marine mammal populations in the North Pacific. *International North Pacific Fisheries Commission Bulletin* 53:409–434.

Hoberg, E. P., S. J. Kutz, K. E. Galbreath, and J. A. Cook. 2003. Arctic biodiversity: From discovery to faunal baselines-revealing the history of a dynamic system. *Journal of Parasitology* 89 (Suppl.):S84-S95.

Hoefs, M. 1989. Thinhorn sheep (*Ovis dalli*), distribution, abundance and management. Pp. 105–137, in *Symposium on wild sheep of the world*, Prague, Czechoslovakia.

———. 2001. Mule, *Odocoileus hemionus,* and white-tailed, *O. virginianus,* deer in the Yukon. *Canadian Field-Naturalist* 115:296–300.

Hoekstra, P. F., T. M. O'Hara, S. M. Backus, C. Hanns, and D. C. G. Muir. 2005. Concentrations of persistent organochlorine contaminants in bowhead whale tissues and other biota from northern Alaska: Implications for human exposure from a subsistence diet. *Environmental Research* 98:329–340.

Hoelzel, A. R. 1991. Analysis of regional mitochondrial DNA variation in the killer whale; implications for cetacean conservation. *Report of the International Whaling Commission, Special Issue* 13:225–233.

———. 1994. Genetics and ecology of whales and dolphins. *Annual Review of Ecology and Systematics* 25:377–399.

Hoelzel, A. R., M. E. Dahlheim, and S. J. Stern. 1998. Low genetic variation among killer whales (*Orcinus orca*) in the Eastern North Pacific, and genetic differentiation between foraging specialists. *Journal of Heredity* 89:121–128.

Hoelzel, A. R., and G. A. Dover. 1991. Genetic differentiation between sympatric killer whale populations. *Heredity* 66:191–195.

Hoffmann, R. S. 1964. (Review of Gromov et al., 1963, and Heptner et al., 1961). *Journal of Mammalogy* 45:153–156.

———. 1981. Different voles for different holes: Environmental restrictions on refugial survival of mammals. Pp. 25–45, in *Evolution today* (G. G. E. Scudder and J. L. Reveal, eds.). Proceedings of the Second International Congress of Systematic and Evolutionary Biology 2:1–486.

———. 1999. Alaska marmot, *Marmota broweri.* Pp. 393–395, in *The Smithsonian book of North American mammals* (D. E. Wilson and S. Ruff, eds.). Smithsonian Institution Press, Washington, D.C., in association with the American Society of Mammalogists.

Hoffmann, R. S., C. G. Anderson, R. W. Thorington, Jr., and L. R. Heaney. 1993. Order Sciuridae. Pp. 419–465, in *Mammal species of the world: A taxonomic and geographic reference.* Second Edition. (D. E. Wilson and D. M. Reeder, eds.). Smithsonian Institution Press, Washington. D.C.

Hoffmann, R. S., J. W. Koeppl, and C. F. Nadler. 1979. The relationships of the amphiberingian marmots (Mammalia: Sciuridae). University of Kansas Museum of Natural History, *Occasional Paper* 83:1–56.

Hoffmann, R. S., and R. S. Peterson. 1967. Systematics and zoogeography of *Sorex* in the Bering Strait area. *Systematic Zoology* 16:127–136.

Hoffmann, R. S., and A. T. Smith. 2005. Order Lagomorpha. Pp. 185–211, in *Mammal species of the world: A taxonomic and geographic reference* (D. E. Wilson and D. M. Reeder, eds.). Third Edition. Johns Hopkins University Press, Baltimore.

Hoffmann, R. S., and R. D. Taber. 1967. Origin and history of Holarctic tundra ecosystems, with special reference to their vertebrate faunas. Pp. 143–170, in *Arctic and alpine environments* (H. E. Wright, Jr. and W. H. Osburn, eds.). Indiana University Press, Bloomington.

Hogan, K., M. C. Hedin, H. S. Koh, S. K. Davis, and I. F. Greenbaum. 1993. Systematic and taxonomic implications of karyotypic, electrophoretic, and mitochondrial-DNA variation in *Peromyscus* from the Pacific Northwest. *Journal of Mammalogy* 74:819–831.

Holden, M. E., and G. G. Musser. 2005. Family Dipodidae. Pp. 871–893, in *Mammal species of the world: A taxonomic and geographic reference* (D. E. Wilson and D. M. Reeder, eds.). Third Edition. Johns Hopkins University Press, Baltimore.

Hollister, N. 1911. A Systematic Synopsis of the Muskrats. *North American Fauna* 32.

Holmes, C. E. 1996. Broken Mammoth. Pp. 312–318, in *American beginnings: The prehistory and palaeoecology of Beringia* (F. H. West, ed.). University of Chicago Press, Chicago and London.

Holthuis, L. B. 1987. The scientific name of the sperm whale. *Marine Mammal Science* 3:87–90.

Home, W. S. 1973. The mammals of Glacier Bay. Unpublished National Park Service Report. (Manuscript not dated; 1973 surmised by B. Page, pers. comm.)

Home, W. S. 1980. Pacific pilot whales: Repeated, localized sightings in southeast Alaska. *Wasmann Journal of Biology* 38:18–20.

Honacki, J. H., K. E. Kinman, and J. W. Koeppl (Eds.). 1982. *Mammal species of the world: A taxonomic and geographic reference*. Allen Press, Inc. and The Association of Systematics Collections, Lawrence, KS.

Hopkins, D. M. 1967. *The Bering land bridge*. Stanford University Press, Stanford, CA.

Hopkins, D. M., J. V. J. Matthews, C. E. Schweger, and S. B. Young. 1982. *Paleoecology of Beringia*. Academic Press, New York.

Houck, W. J. 1976. The taxonomic status of the porpoise genus *Phocoenoides*. *Reports of the Advisory Committee on Marine Resources Research* 114:1–13.

Houck, W. J., and T. A. Jefferson. 1999. Dall's porpoise, *Phocoenoides dalli* (True, 1885). Pp. 443–472, in *Handbook of marine mammals. Volume 6: The second book of dolphins and the porpoises* (S. H. Ridgway and R. Harrison, eds.). Academic Press, London.

Howell, A. H. 1915. Revision of the American marmots. *North American Fauna* 37.

———. 1918. Revision of the American flying squirrels. *North American Fauna* 44.

———. 1924. Revision of the American Pikas (Genus Ochotona). *North American Fauna* 47.

———. 1926. Voles of the Genus Phenacomys. *North American Fauna* 48.

———. 1927. Revision of the American Lemming Mice (Genus Synaptomys). *North American Fauna* 50.

———. 1934. Description of a new race of flying squirrel from Alaska. *Journal of Mammalogy* 15:64.

———. 1938. Revision of the North American ground squirrels, with a classification of the North American Sciuridae. *North American Fauna* 56.

———. 1940. Brown bear killed on St. Lawrence Island. *Journal of Mammalogy* 21:216.

Huchon, D., O. Madsen, M. J. J. B. Sibbald, K. Ament, M. Stanhope, F. Catzeflis, W. W. de Jong, and E. J. P. Douzery. 2002. Rodent phylogeny and a timescale for the evolution of Glires: Evidence from an extensive taxon sampling using three nuclear genes. *Molecular Biology and Evolution* 19:1053–1065.

Huey, L. M. 1952. An Alaskan record of the narwhal. *Journal of Mammalogy* 33:496.

Hultén, E. 1937. Outline of the history of arctic and boreal biota during the Quaternary Period: Stockholm, Sweden, Bokförlags Aktiebolaget Thule.

Hundertmark, K. J., R. T. Bowyer, G. F. Shields, and C. C. Schwartz. 2003. Mitochondrial phylogeography of moose (*Alces alces*) in North America. *Journal of Mammalogy* 84:718–728.

Hundertmark, K. J., R. T. Bowyer, G. F. Shields, C. C. Schwartz, and M. H. Smith. 2006. Colonization history and taxonomy of moose *Alces alces* in southeastern Alaska inferred from mtDNA variation. *Wildlife Biology* 12:331–338.

Hundertmark, K. J., G. F. Shields, I. G. Udina, R. T. Bowyer, A. A. Danilkin, and C. C. Schwartz. 2002. Mitochrondial phylogeography of moose (*Alces alces*): Late Pleistocene divergence and population expansion. *Molecular Phylogenetics and Evolution* 22:375–387.

Huntington, J., and L. Elliott. 2002. *On the edge of nowhere*. Epicenter Press, Kenmore, WA.

Husson, A. M., and L. B. Holthuis. 1974. *Physeter macrocephalus* Linnaeus, 1758, the valid name for the sperm whale. *Zoologische Mededelingen* 48:205–217.

Hutterer, R. 2005. Order Soricomorpha. Pp. 220–311, in *Mammal species of the world: A taxonomic and geographic reference* (D. E. Wilson and D. M. Reeder, eds.). Third Edition. Johns Hopkins University Press, Baltimore.

ICZN (International Commission on Zoological Nomenclature). 1956. Opinion 417. Rejection for nomenclatorial purposes of volume 3 (Zoologie) of the work by Lorenz Oken entitled Okens Lehrbuch der Naturgeschichte published in 1815–1816. *Opinions and Declarations Rendered by the International Commission on Zoological Nomenclature*, 14:1–42.

————. 1960. Opinion 581. Determination of the generic names for the fallow deer of Europe and the Virginian deer of America (Class Mammalia). *Bulletin of Zoological Nomenclature* 17:267–275.

Irving, L. 1939. Respiration in diving mammals. *Physiological Reviews* 19:112–134.

Irving, W. N., A. V. Jopling, and I. Kritsch-Armstrong. 1989. Studies of bone technology and taphonomy, Old Crow Basin, Yukon Territory. Pp. 347–379, in *Bone modification* (R. Bonnichsen and M. H. Sorg, eds.). Center for the Study of the First Americans, University of Maine, Orono.

IUCN (World Conservation Union). 1991. Dolphins, porpoises and whales of the world: IUCN Red Data Book (M. Klinowska, compiler). IUCN, Gland, Switzerland, and Cambridge, UK.

————. 2007. 2007 IUCN Red List of threatened species. http://www.iucnredlist.org (accessed 5 February 2008).

Ivanova, E. I. 1955. Kharakteristika proportsii tela kashalota (*Physeter catodon* L.). *Trudy Instituta Okeanologii Akademii Nauk SSSR* 18:100–112.

IWC (International Whaling Commission). 1992. Report of the Scientific Committee. *Reports of the International Whaling Commission* 42:178–234.

————. 2001. Report of the Scientific Committee. *Journal of Cetacean Resources Management 3* (Supplement).

Jackson, H. H. T. 1926. An unrecognized water shrew from Wisconsin. *Journal of Mammalogy* 7:58.

————. 1928. A taxonomic review of the long-tailed shrews, genera *Sorex* and *Microsorex*. *North American Fauna* 51.

————. 1951. Classification of the races of coyotes. Part II. Pp. 227–341, in *The clever coyote* (S. P. Young and H. H. T. Jackson, eds.). The Stackpole Company, Harrisburg.

Janis, C. M., and K. M. Scott. 1987. The interrelationships of higher ruminant families with special emphasis on the members of the Cervoidea. *American Museum Novitates* 2893:1–85.

Jarrell, G. H. 1981. Cytogenetic and morphological investigation of variability in the bowhead whale, *Balaena mysticetus*. Pp. 213–231, in *Tissue structural studies and other investigations on the biology of endangered whales in the Beaufort Sea* (T. Albert, ed.). Final Report for the period April 1, 1980 through June 30, 1981. Prepared for U. S. Department of Interior, Bureau of Land Management, Alaska OCS Office, Anchorage.

Jarrell, G. H., and K. Fredga. 1993. How many kinds of lemmings? A taxonomic overview. Pp. 45–57, in *The biology of lemmings* (N. C. Stenseth and R. A. Ims, eds.). Academic Press, London.

Jefferson, T. A., and M. W. Newcomer. 1993. *Lissodelphis borealis*. *Mammalian Species* 425:1–6.

Jefferson, T. A., M. W. Newcomer, S. Leatherwood, and K. V. Waerebeek. 1994. Right whale dolphins, *Lissodelphis borealis* (Peale, 1848) and *Lissodelphis peronii* (Lacépède, 1804). Pp. 335–362, in *Handbook of marine mammals. Volume 5: The first book of dolphins* (S.H. Ridgway and R. Harrison, eds.). Academic Press, London.

Jellison, W. M. 1953. A beaked whale, *Mesoplodon* sp., from the Pribilofs. *Journal of Mammalogy* 34:249–251.

Jenkins, E. J., A. M. Veitch, S. J. Kutz, E. P. Hoberg, and L. Polley. 2006. Climate change and the epidemiology of protostrongylid nematodes in northern ecosystems: *Parelaphostrongylus odocoilei* and *Protostrongylus stilesi* in Dall's sheep (*Ovis dalli dalli*). *Parasitology* 132:387–401.

Jenkins, S. H., and P. E. Busher. 1979. *Castor canadensis*. *Mammalian Species* 120:1–8.

Jernsletten, J. L., and K. Klokov. 2002. Sustainable reindeer husbandry. Arctic Council 2000–2002. University of Tromsø. http://www.reindeer-husbandry.uit.no (accessed 29 April 2007).

Jingfors, K. Y., and D. R. Klein. 1982. Productivity in recently established muskox populations in Alaska. *Journal of Wildlife Management* 46:1092–1096.

Johannessen, O. M., L. Bengtsson, M. W. Miles, S. I. Kuzmina, V. A. Semenov, G. V. Alexseev, A. P. Nagurnyi, V. F. Zakharov, L. P. Bobylev, L. H. Pettersson, K. Hasselmann, and H. P. Cattle. 2004. Arctic climate change: Observed and modeled temperature and sea-ice variability. *Tellus* 56A:328–341.

Johnson, J. H., and A. A. Wolman. 1984. The humpback whale, *Megaptera novaeangliae*. *Marine Fisheries Review* 46:30–37.

Johnson, K. P. 2003. Island biogeography theory and evolution: Genetic divergence and speciation of island taxa. *Comments on Theoretical Biology* 8:339–356.

Johnson, L. 1994. Black bear. Alaska Department of Fish and Game Wildlife Notebook Series. http://www.adfg.state.ak.us/pubs/notebook/biggame/blkbear.php (accessed 26 March 2007).

Johnson, W. E., J. A. Godoy, F. Palomares, M. Delibes, M. Fernandes, E. Revilla, and S. J. O'Brien. 2004. Phylogenetic and phylogeographic analysis of Iberian lynx populations. *Journal of Heredity* 95:19–28.

Johnson-Murray, J. L. 1977. Myology of the gliding membranes of some Petauristine rodents (genera: *Glaucomys, Pteromys, Petinomys,* and *Petaurista*). *Journal of Mammalogy* 58:374–384.

Jolicoeur, P. 1959. Multivariate geographical variation in the wolf *Canis lupus* L. *Evolution* 13:283–299.

Joling, D. 2004. Scientists spot rare blue whales in Alaska. *Fairbanks Daily News-Miner*, July 28.

Jones, G. S. 1981. The systematics and biology of the genus *Zapus* (Mammalia, Rodentia, Zapodidae). Unpublished dissertation. Indiana State University.

Jones, M. L., and S. L. Swartz. 2002. Gray whale, *Eschrichtius robustus*. Pp. 524–536, in *Encyclopedia of marine mammals* (W. F. Perrin, B. Würsig, and J. G. M. Thewissen, eds.). Academic Press, London.

Jones, R. E. 1967. A *Hydrodamalis* skull fragment from Monterey Bay, California. *Journal of Mammalogy* 48:143–144.

Jopling, A. V., W. N. Irving, and B. F. Beebe. 1981. Stratigraphic, sedimentological and faunal evidence for the occurrence of pre-Sangamonian artefacts in northern Yukon. *Arctic* 34:3–33.

Judd, S. R., and S. P. Cross. 1980. Chromosomal variation in *Microtus longicaudus* (Merriam). *Murrelet* 61:2–5.

Jung, T. S., and P. J. Merchant. 2005. First confirmation of cougar, *Puma concolor,* in the Yukon. *Canadian Field-Naturalist* 119:580–581.

Junge, J. A., and R. S. Hoffmann. 1981. An annotated key to the long-tailed shrews (genus *Sorex*) of the United States and Canada, with notes on Middle-American shrews. University of Kansas, Museum Natural History, Occasional Paper 94:1–48.

Junge, J. A., R. S. Hoffmann, and R. W. Darby. 1983. Relationships with the Holarctic *Sorex arcticus–Sorex tundrensis* species complex. *Acta Theriologica* 28:339–350.

Kajimura, H., and T. R. Loughlin. 1988. Marine mammals in the oceanic food web of the eastern subarctic Pacific. *Bulletin of the Ocean Research Institute*, University of Tokyo 26:187–223.

Kaliszewska, Z. A., J. Seger, V. J. Rowntree, S. G. Barco, R. Benegas, P. B. Best, M. W. Brown, R. L. Brownell, Jr., A. Carribero, R. Harcourt, A. R. Knowlton, K. Marshall-Tilas, N. J. Patenaude, M. Rivarola, C. M. Schaeff, M. Sironi, W. A. Smith, and T. K. Yamada. 2005. Population histories of right whales (Cetacea: *Eubalaena*) inferred from mitochondrial sequence diversities and divergences of their whale lice (Amphipoda: *Cyamus*). *Molecular Ecology* 14:3439–3456.

Karlstrom, T. N. V., and G. E. Ball. 1969. *The Kodiak Island refugium: Its geology, flora, fauna, and history*. Ryerson Press, Toronto.

Kasuya, T. 1973. Systematic consideration of Recent toothed whales based on the morphology of the tympano-periotic bone. *Scientific Reports of the Whales Research Institute* 25:1–103.

———. 1986. Distribution and behavior of Baird's beaked whales off the Pacific coast of Japan. *Scientific Reports of the Whales Research Institute* 37:61–83.

Kasuya, T., T. Miyashita, and F. Kasamatsu. 1988. Segregation of two forms of short-finned pilot whales off the Pacific coast of Japan. *Scientific Reports of the Whales Research Institute* 39:77–90.

Kasuya, T., and S. Ohsumi. 1984. Further analysis of the Baird's beaked whale stock in the western North Pacific. *Reports of the International Whaling Commission* 34:587–595.

Kato, H., and Y. Fujise. 2000. Dwarf minke whales; morphology, growth and life history with some analyses on morphometric variation among the different forms and regions. International Whaling Commission Meeting Document SC/52/OS3, 1–30.

Kawamichi, T. 1981. Vocalizations of *Ochotona* as a taxonomic character. Pp. 324–339, in *Proceedings of the World Lagomorph Conference held in Guelph, Ontario, August 1979* (K. Myers and C. D. MacInnes, eds.). University of Guelph, Guelph, Ontario.

Kellnhauser, J. T. 1983. The acceptance of *Lontra* Gray for the New World river otters. *Canadian Journal of Zoology* 61:278–279.

Kellogg, R. 1922. Pinnipeds from Miocene and Pleistocene deposits of California. University of California Publications, *Bulletin of the Department of Geological Sciences* 13:23–132.

————. 1932. New names for mammals proposed by Borowski in 1780 and 1781. *Proceedings of the Biological Society of Washington* 45:147–148.

————. 1936. Mammals from a native village site on Kodiak Island. *Proceedings of the Biological Society of Washington* 49:37–38.

Kelly, B. P. 1988a. Ribbon seal *Phoca fasciata.* Pp. 95–106, in *Selected marine mammals of Alaska* (J.W. Lentfer, ed.). Marine Mammal Commission, Washington, D.C.

————. 1988b. Ringed seal. Pp. 57–75, in *Selected marine mammals of Alaska* (J. W. Lentfer, ed.). Marine Mammal Commission, Washington, D.C.

————. 2001. Climate change and ice breeding pinnipeds. Pp. 43–55, in *"Fingerprints" of climate change: Adapted behaviour and shifting species ranges* (G. R. Walther, C. A. Burga, and P. J. Edwards, eds.). Kluwer Academic/Plenum Publishers, New York and London.

————. 2004. Walruses, seals, and climate change. Abstracts of the 34th Annual International Arctic Workshop, Institute of Arctic and Alpine Research, Boulder, Colorado, 11–13 March. http://www.colorado.edu/INSTAAR/AW2004/get_abstr.html?id=27 (accessed 5 February 2008).

Kenney, R. D. 2002. North Atlantic, North Pacific, and southern right whales: *Eubalaena glacialis, E. japonica,* and *E. australis.* Pp. 806–813, in *Encyclopedia of marine mammals* (W. F. Perrin, B. Würsig, and J. G. M. Thewissen, eds.). Academic Press, San Diego and London.

Kenyon, K. W. 1964. Wildlife and historical notes on Simeonof Island, Alaska. *The Murrelet* 45:1–8.

————. 1969. The sea otter in the eastern Pacific Ocean. *North American Fauna* 68.

Kenyon, K. W., and D. L. Spencer. 1960. Sea otter population and transplant studies in Alaska, 1959. U. S. Fish and Wildlife Service, Special Scientific Report, Wildlife No. 48.

Kessel, B., S. O. MacDonald, D. D. Gibson, B. A. Cooper, and B. A. Anderson. 1982. Birds and non-game mammals baseline studies. Susitna Hydroelectric Project Environmental Studies Phase I Final Report to the Alaska Power Authority.

King, C. M. 1983. *Mustela erminea. Mammalian Species* 195:1–8.

King, J. E. 1966. Relationships of the hooded and elephant seals (genera *Cystophora* and *Mirounga*). *Journal of Zoology* (London), 148:385–398.

————. 1983. *Seals of the world.* Second edition. Cornell University Press, Ithaca, NY.

Kinkhart, E., and K. Pitcher. 1994. Harbor Seal. Alaska Department of Fish and Game Wildlife Notebook Series. http://www.adfg.state.ak.us/pubs/notebook/marine/harseal.php (accessed 26 March 2007).

Kirchhoff, M. D., and D. N. Larsen. 1998. Dietary overlap between native Sitka black-tailed deer and introduced elk in Southeast Alaska. *Journal of Wildlife Management* 62:236–242.

Kirpichnikov, A. A. 1951. O del'finakh iz otlozhenii aspheronskogo yarusa. *Dokl. Akad. Nauk SSSR* 79:1021–1024.

Klein, D. R. 1959. St. Matthew Island reindeer-range study. U. S. Fish and Wildlife Service Special Scientific Report, Wildlife 43. Washington, D.C.

————. 1965. Postglacial distribution patterns of mammals in the southern coastal regions of Alaska. *Journal of the Arctic Institute of North America* 18:7–20.

————. 1968. The introduction, increase, and crash of reindeer on St. Matthew Island. *Journal of Wildlife Management* 32:350–367.

————. 1987. Vegetation recovery patterns following overgrazing by reindeer on St. Matthew Island. *Journal of Range Management* 40:336–338.

————. 1988. The establishment of muskox populations by translocation. Pp. 298–318, in *Translocation of wild animals* (L. Nielsen and R. D. Brown, eds.). Wisconsin Humane Society and Caesar Kleberg Wildlife Research Institute, Milwaukee, WI, and Kingsville, TX.

————. 1995. Tundra or arctic hares. P. 359, in *Our living resources, a report to the nation on the distribution, abundance, and health of U. S. plants, animals, and ecosystems* (E. T. LaRoe, G. S Farris,

C. E. Puckett, P. D. Doran, and M. J. Mac, eds.). U.S. Department of the Interior, National Biological Service, Washington, D.C.

―――. 1996. The introduction, increase and demise of wolves on Coronation Island, Alaska. Pp. 275–280, in *Ecology and conservation of wolves in a changing world* (L. N. Carbyn, S. H. Fritts, and D. R. Seip, eds.). Canadian Circumpolar Institute, University of Alberta, Edmonton.

―――. 2000. The muskox. Pp. 545–558, in *Ecology and management of large mammals in North America* (S. Demarais and P. R. Krausman, eds.). Prentice-Hall, Inc., Upper Saddle River, NJ.

Klingener, D. 1984. Gliroid and dipodoid rodents. Pp. 381–388, in *Orders and families of Recent mammals of the world* (S. Anderson and J. K. Jones, Jr., eds.). John Wiley and Sons, New York.

Klinowska, M. 1991. *Dolphins, porpoises and whales of the world*. The IUCN Red Data Book. IUCN, Gland, Switzerland and Cambridge, UK.

Koehler, A. V. A. 2006. Systematics, phylogeography, distribution, and life cycle of *Soboliphyme baturini*. Unpublished thesis. University of New Mexico, Albuquerque.

Koenigswald, W. V., and L. D. Martin. 1984. Revision of the fossil and recent Lemminae (Rodentia, Mammalia). Pp. 122–137, in *Papers in vertebrate paleontology honoring Robert Warren Wilson* (R. M. Mengel, ed.). Special Publication of Carnegie Museum of Natural History 9:1–186.

Koopman, K. F. 1984. Bats. Pp. 145–186, in *Orders and families of Recent mammals of the world* (S. Anderson and J. K. Jones, Jr., eds.). John Wiley and Sons, New York.

Korobitsyna, K. V., C. F. Nadler, N. N. Vorontsov, and R. S. Hoffmann. 1974. Chromosomes of the Siberian snow sheep, *Ovis nivicola*, and implications concerning the origin of the amphiberingian wild sheep (Subgenus *Pachyceros*). *Quaternary Research* 4:235–245.

Kovacs, K. M., and D. M. Lavigne. 1986. *Cystophora cristata*. *Mammalian Species* 258:1–9.

Kratochvíl, J. 1982. Karyotyp und System der Familie Felidae (Carnivora, Mammalia). *Folia Zoologica* 31:289–304.

Krause, J., P. H. Dear, J. L. Pollack, M. Slatkin, H. Spriggs, I. Barnes, A. M. Lister, I. Ebersberger, S. Pääbo, and M. Hofreiter. 2006. Multiplex amplification of the mammoth mitochondrial genome and the evolution of Elephantidae. *Nature* 439:724–727.

Kruse, S., D. K. Caldwell, and M. C. Caldwell. 1999. Risso's dolphin, *Grampus griseus* (G. Cuvier, 1812). Pp. 183–212, in *Handbook of marine mammals. Volume 6: The second book of dolphins and the porpoises* (S. H. Ridgway and R. Harrison, eds.). Academic Press, London.

Krutzsch, P. H. 1954. North American jumping mice (genus *Zapus*). University of Kansas Publication, *Museum of Natural History* 7:349–472.

Kucheruk, V. V. 1990. Areal. [Range]. Pp. 34–84, in *Seraya krysa: Sistematika, ekologiya, reguliatsiya chislennosti* [Norway rat: systematics, ecology, population control] (V. E. Sokolov and E. V. Karasjova, eds.). Nauka, Moscow. (In Russian.)

Kunz, T. H. 1982. *Lasionycteris noctivagans*. *Mammalian Species* 172:1–5.

Kuroda, N. 1954. On the affinity of the Dall's and True's porpoises. *Bulletin of the Yamashina Ornithology Research Institute* 5:44–46.

Kurta, A., and R. H. Baker. 1990. *Eptesicus fuscus*. *Mammalian Species* 356:1–10.

Kurtén, B. 1963. Notes on some Pleistocene mammal migrations from the Palaearctic to the Nearctic. *Eiszeitalter und Gegenwart* 14:96–103.

―――. 1964. The evolution of the polar bear, *Ursus maritimus* Phipps. *Acta Zoologica Fennica* 108:1–30.

―――. 1968. *Pleistocene mammals of Europe*. Weidenfield and Nicolson, London.

―――. 1971. *The age of mammals*. Weidenfeld and Nicolson, London.

―――. 1985. The Pleistocene lion of Beringia. *Annales Zoologici Fennici* 22:117–121.

Kurtén, B., and E. Anderson. 1980. *Pleistocene mammals of North America*. Columbia University Press, New York.

Kurtén, B., and R. L. Rausch. 1959. Biometric comparisons between North American and European mammals. I. A comparison between Alaskan and Fennoscandian wolverine (*Gulo gulo* Linnaeus). *Acta Arctica* 11:1–21.

Kutz, S. J., E. P. Hoberg, J. Nagy, L. Polley, and B. Elkin. 2004. "Emerging" parasitic infections in Arctic ungulates. *Integrative and Comparative Biology* 44:109–118.

Kutz, S. J., E. P. Hoberg, L. Polley, and E. J. Jenkins. 2005. Global warming is changing the dynamics of Arctic host-parasite systems. *Proceedings of the Royal Society London*, B. 272: 2571–2576.

Kwiecinski, G. G. 1998. *Marmota monax. Mammalian Species* 591:1–8.

Kyle, C. J., and C. Strobeck. 2001. Genetic structure of North American wolverine (*Gulo gulo*) populations. *Molecular Ecology* 10:337–347.

Laakkonen, J., A. Smith, K. Hildebrandt, J. Niemimaa, and H. Henttonen. 2005. Significant morphological, but little molecular differences between *Trypanosoma* of rodents from Alaska. *Journal of Parasitology* 91:201–203.

Laing, H. M., and R. M. Anderson. 1929. Notes on mammals of upper Chitina River region, Alaska. Pp. 96–107, in *Birds and mammals of the Mount Logan Expedition, 1925* (H. M. Laing, P. A. Taverner, and R. M. Anderson, eds.). Annual Report, National Museum of Canada Bulletin No. 56.

Lamont, M. M., J. T. Vida, J. T. Harvey, S. Jeffries, R. Brown, H. H. Huber, R. DeLong, and W. K. Thomas. 1996. Genetic substructure of the Pacific harbor seal (*Phoca vitulina richardsi*) off Washington, Oregon, and California. *Marine Mammal Science* 12:402–413.

Lance, E. W. 1995. Phylogeographic variation and the island syndrome in Holarctic tundra voles (*Microtus oeconomus*). Unpublished thesis. University of Alaska Fairbanks.

―――. 2002. Montague Island marmot: A conservation assessment. General Technical Report PNW-GTR-541.U. S. Department of Agriculture, Forest Service, Pacific Northwest Research Station, Portland, Oregon.

Lance, E. W., and J. A. Cook. 1998a. Biogeography of tundra voles (*Microtus oeconomus*) of Beringia and the southern coast of Alaska. *Journal of Mammalogy* 79:53–65.

―――. 1998b. *Microtus oeconomus* (Pallas 1776), tundra vole. Pp. 97–99, in *North American rodents: Status survey and conservation action plan* (D. J. Hafner, E. Yensen, and G. L. Kirkland, Jr., eds.). IUCN/SSC Rodent Specialist Group. Gland, Switzerland, and Cambridge, UK.

Larivière, S. 1999. *Mustela vison. Mammalian Species* 608:1–9.

―――. 2001. *Ursus americanus. Mammalian Species* 647:1–11.

Larivière, S., and M. Pasitschniak-Arts. 1996. *Vulpes vulpes. Mammalian Species* 537:1–11.

Larivière, S., and L. R. Walton. 1998. *Lontra canadensis. Mammalian Species* 587:1–8.

Larson, K. R. 2002. BLM summer fieldwork report. Mammal collection, University of Alaska Museum.

Larter, N. C., and C. C. Gates. 1991. Diet and habitat selection of wood bison in relation to seasonal changes in forage quantity and quality. *Canadian Journal of Zoology* 69:2677–2685.

Lavrov, L. S., and V. N. Orlov. 1973. [Karyotypes and taxonomy of modern beavers (*Castor*, Castoridae, Mammalia)]. *Zoologicheskii Zhurnal* 52:734–742. (In Russian.)

Leatherwood, J. S., and M. E. Dahlheim. 1978. Worldwide distribution of pilot whales and killer whales. Naval Ocean Systems Center Technical Report 295.

Leatherwood, S., and R. R. Reeves. 1983. *The Sierra Club Handbook of whales and dolphins*. Sierra Club Books, San Francisco.

Leatherwood, S., R. R. Reeves, A. E. Bowles, B. S. Stewart, and K. R. Goodrich. 1984. Distribution, seasonal movements and abundance of Pacific white-sided dolphins in the eastern North Pacific. *Scientific Reports of the Whales Research Institute* 35:129–157.

Leatherwood, S., R. R. Reeves, W. F. Perrin, and W. E. Evans. 1988. *Whales, dolphins, and porpoises of the Eastern North Pacific and adjacent Arctic waters: A guide to their identification*. Dover Publications, Toronto.

Leatherwood, S., B. S. Stewart, and P. A. Folkens. 1987. Cetaceans of the Channel Islands National Marine Sanctuary. National Oceanic and Atmospheric Administration, Channel Islands National Marine Sanctuary.

LeDuc, R. G., W. F. Perrin, and A. E. Dizon. 1999. Phylogenetic relationships among the delphinid cetaceans based on full cytochrome *b* sequences. *Marine Mammal Science* 15:619–648.

LeDuc, R. G., W. L. Perryman, J. W. Gilpatrick, Jr., J. Hyde, C. Stinchcomb, J. V. Carretta, and R. L. Brownell, Jr. 2001. A note on recent surveys for right whales in the southeastern Bering Sea. *Journal of Cetacean Resource Management* (Special Issue 2):287–289.

Lenart, E. A. 2005. Units 26B and 26C muskox management report. Pp. 49–68, in *Muskox management report of survey and inventory activities* 1 July 2002–30 June 2004 (C. Brown, ed.). Project 16.0. Alaska Department of Fish and Game. Juneau.

Lensink, C. J. 1954. Occurrence of *Microtus xanthognathus* in Alaska. *Journal of Mammalogy* 35:259–260.

————. 1960. Status and distribution of sea otters in Alaska. *Journal of Mammalogy* 41:172–182.

————. 1962. The history and status of sea otters in Alaska. Unpublished dissertation. Purdue University, Lafayette, IN..

Lensink, C. J., R. O. Skoog, and J. L. Buckley. 1955. Food habits of marten in interior Alaska and their significance. *Journal of Wildlife Management* 19:364–368.

Lent, P. C. 1978. Muskox. Pp. 135–147, in *Big game of North America: Ecology and management* (J. L. Schmidt and D. L. Gilbert, eds.). Wildlife Management Institute and Stackpole Books, Harrisburg, PA.

————. 1988. *Ovibos moschatus. Mammalian Species* 302:1–9.

————. 1998. Alaska's indigenous muskoxen: A history. *Rangifer* 18:133–144.

Lentfer, J. W. 1974. Agreement on conservation of polar bears. *Polar Record* 17:327–330.

————. 1994. Polar bear. Alaska Department of Fish and Game Wildlife Notebook Series. http://www.adfg.state.ak.us/pubs/notebook/marine/polarbea.php (accessed 26 March 2007).

Leonard, J. A., R. K. Wayne, and A. Cooper. 2000. Population genetics of Ice Age brown bears. *Proceedings of the National Academy of Sciences* 97:1651–1654.

Lessa, E. P., J. A. Cook, and J. L. Patton. 2003. Genetic footprints of demographic expansion in North America, but not Amazonia, following the Late Pleistocene. *Proceedings of the National Academy of Sciences* 100:10331–10334.

Libby, W. L. 1958. Records of the pika in the Tanana Hills, Alaska. *Journal of Mammalogy* 39:448–449.

Lidicker, W. Z., Jr., and A. Yang. 1986. Morphology of the penis in the taiga vole (*Microtus xanthognathus*). *Journal of Mammalogy* 67:497–502.

Lister, A. M., and P. Bahn. 1994. *Mammoths*. Macmillian, New York.

Lister, A. M., and A. V. Sher. 2001. The origin and evolution of the woolly mammoth. *Science* 294:1094–1097.

Lloyd, D. S., R. B. Smith, and K. A. Sunberg. 1987. Introduction of European wild boar to Marmot Island, Alaska. *Murrelet* 68:57–58.

Loehr, J., K. Worley, A. Grapputo, J. Carey, A. Veitch, and D. W. Coltman. 2006. Evidence for cryptic glacial refugia from North American mountain sheep mitochondrial DNA. *Journal of Evolutionary Biology* 19:419–430.

Long, C. A. 1974. *Microsorex hoyi* and *Microsorex thompsoni*. *Mammalian Species* 33:1–4.

————. 1999. Pygmy shrew, *Sorex hoyi*. Pp. 25–27, in *The Smithsonian book of North American mammals* (D. E. Wilson and S. Ruff, eds.). Smithsonian Institution Press, Washington, D.C., in association with the American Society of Mammalogists.

Loring, J. A. 1902. *Notes on mammals and birds observed in southern Alaska in 1901*. New York Zoological Society, Sixth Annual Report. Pp. 145–159.

Lotze, J. H., and S. Anderson. 1979. *Procyon lotor. Mammalian Species* 119:1–8.

Loughlin, T. R., C. H. Fiscus, A. M. Johnson, and D. J. Rugh. 1982. Observations of *Mesoplodon stejnegeri* (Ziphiidae) in the central Aleutian Islands, Alaska. *Journal of Mammalogy* 63:697–700.

Loughlin, T. R., and R. V. Miller. 1989. Growth of the northern fur seal colony on Bogoslof Island, Alaska. *Arctic* 42:368–372.

Loughlin, T. R., and M. A. Perez. 1985. *Mesoplodon stejnegeri. Mammalian Species* 250:1–6.

Loughlin, T. R., M. A. Perez, and R. L. Merrick. 1987. *Eumetopias jubatus. Mammalian Species* 283:1–7.

Lowell, R. E. 2004a. Unit 1B moose management report. Pp. 9–21, in *Moose management report of survey and inventory activities* 1 July 2001–30 June 2003 (C. Brown, ed.). Alaska Department of Fish and Game. Project 1.0. Juneau, Alaska.

————. 2004b. Unit 3 moose management report. Pp. 58–67, in *Moose management report of survey and inventory activities* 1 July 2001–30 June 2003 (C. Brown, ed.). Alaska Department of Fish and Game. Project 1.0. Juneau, Alaska.

————. 2004c. Unit 3 elk management report. Pp. 1–8, in *Elk management report of survey and inventory activities* 1 July 2001–30 June 2003 (C. Brown, ed.). Alaska Department of Fish and Game Project 13.0. Juneau, Alaska.

Lowry, L. F. 1994. Beluga whale. Alaska Department of Fish and Game Wildlife Notebook Series. http://www.adfg.state.ak.us/pubs/notebook/marine/beluga.php (accessed 26 March 2007).

Lowry, L. F., V. N. Burkanov, K. J. Frost, M. A. Simpkins, A. Springer, D. P. DeMaster, and R. Suydam. 2000. Habitat use and habitat selection by spotted seals (*Phoca larga*) in the Bering Sea. *Canadian Journal of Zoology* 78:1959–1971.

Lowry, L. F., and K. J. Frost. 1984. Foods and feeding of bowhead whales in western and northern Alaska. *Scientific Reports of the Whales Research Institute* 35:1–16.

Lowry, L. F., K. J. Frost, and J. J. Burns. 1980. Feeding of bearded seals in the Bering and Chukchi seas and trophic interactions with pacific walruses. *Arctic* 33:330–342.

Lowry, L. F., K. J. Frost, D. J. Calkins, G. L. Swartzman, and S. Hills. 1982. Feeding habits, food requirements, and status of Bering Sea marine mammals. Council Document 19 North Pacific Fishery Management Council, Anchorage, Alaska.

Lowry, L. F., K. J. Frost, R. Davis, D. P. DeMaster, and R. S. Suydam. 1998. Movements and behavior of satellite-tagged spotted seals (*Phoca largha*) in the Bering and Chukchi Seas. *Polar Biology* 19:221–230.

Lowry, L. F., K. J. Frost, and T. R. Loughlin. 1989. Importance of walleye pollock in the diets of marine mammals in the Gulf of Alaska and Bering Sea, and implications for fishery management. Pp. 710–726, in *Proceedings of the International Symposium on the Biology and Management of Walleye Pollock*, November, 1988. Anchorage, Alaska.

Lucid, M. K. 2003. Phylogeography of Keen's mouse (*Peromyscus keeni*) in a naturally fragmented landscape. Unpublished thesis. Idaho State University, Pocatello.

Lucid, M. K., and J. A. Cook. 2004. Phylogeography of Keen's mouse (*Peromyscus keeni*) in a naturally fragmented landscape. *Journal of Mammalogy* 85:1149–1159.

————. 2007. Cytochrome *b* haplotypes suggest an undescribed *Peromyscus* species from the Yukon. *Canadian Journal of Zoology* 85:916–919.

Ludt, C. J., W. Schroeder, O. Rottmann, and R. Kuehn. 2004. Mitochondrial DNA phylogeography of red deer (*Cervus elaphus*). *Molecular Phylogenetics and Evolution* 31:1064–1083.

Lukáčová, L., J. Zima, and V. T. Volobouev. 1996. Karyotypic variation in *Sorex tundrensis* (Soricidae, Insectivora). *Hereditas* 125:233–238.

Lydekker, R. 1898. *Wild oxen, sheep, and goats of all lands living and extinct*. Rowland Ward, London.

Lynch, A. J., D. W. Duszynski, and J. A. Cook. 2007. Species of Coccidia (Apicomplexa: Eimeriidae) infecting pikas from Alaska, U. S. A. and Northeastern Siberia, Russia. *Journal of Parasitology* 93:1230–1234.

Lyrholm, T., and U. Gyllensten. 1998. Global matrilineal population structure in sperm whales as indicated by mitochondrial DNA sequences. *Proceedings of the Royal Society London* B 265:1679–1684.

Lyrholm, T., O. Leimer, and U. Gyllensten. 1996. Low diversity and biased substitution patterns in the mitochondrial DNA control region of sperm whales: Implications for estimates of time since common ancestry. *Molecular Biology and Evolution* 13:1318–1326.

MacArthur, R. H., and E. O. Wilson. 1963. An equilibrium theory of insular zoogeography. *Evolution* 17: 373–387.

————. 1967. *The theory of island biogeography*. Princeton University Press, Princeton, NJ.

MacDonald, S. O. 1980. Habitats of small mammals and birds: Evaluating the effects of agricultural development in the Delta Junction area, Alaska. Unpublished report for the State of Alaska, Department of Natural Resources, Division of Lands and Water Management, Fairbanks.

MacDonald, S. O., and J. A. Cook. 1996. The land mammal fauna of southeast Alaska. *Canadian Field-Naturalist* 110:571–599.

————. 1998. *Castor canadensis* Kuhl 1820, beaver; American beaver. Pp. 59–60, in *North American rodents: Status survey and conservation action plan* (D. J. Hafner, E. Yensen, and G. L. Kirkland, Jr., eds.). IUCN/SSC Rodent Specialist Group. Gland, Switzerland, and Cambridge, UK.

————. 2001. Mammal inventory of Alaska's National Parks and Preserves: Yukon-Charley Rivers National Preserve. National Park Service Alaska Region, Inventory and Monitoring Program Annual Report 2001.

————. 2002. Mammal inventory of Alaska's National Parks and Preserves, Northwest Network: Bering Land Bridge National Preserve, Cape Krusenstern National Monument, Kobuk Valley National Park, Noatak National Preserve. National Park Service Alaska Region, Inventory and Monitoring Program Annual Report 2001.

————. 2003. Mammal inventory of Alaska's National Parks and Preserves: Gates of the Arctic National Park and Preserve. National Park Service Alaska Region, Inventory and Monitoring Program Annual Report 2002.

————. 2007. Mammals and amphibians of Southeast Alaska. *The Museum of Southwestern Biology, Special Publication* 8:1–191.

MacDonald, S. O., J. A. Cook, G. L. Kirkland, Jr., and E. Yensen. 1998. *Microtus pennsylvanicus* (Ord 1815), meadow vole. Pp. 99–101, in *North American rodents: Status survey and conservation action plan* (D. J. Hafner, E. Yensen, and G. L. Kirkland, Jr., eds.). IUCN/SSC Rodent Specialist Group. Gland, Switzerland, and Cambridge, UK.

MacDonald, S. O., and C. Elliot. 1984. Distribution of the water shrew (*Sorex palustris*) in Alaska. *Murrelet* 65:45.

MacDonald, S. O., and C. Jones. 1987. *Ochotona collaris. Mammalian Species* 281:1–4.

MacDonald, S. O., A. M. Runck, and J. A. Cook. 2004. The heather vole (genus *Phenacomys*) in Alaska. *Canadian Field-Naturalist* 118:438–440.

MacDonald, S. O., E. Waltari, R. A. Nofchissey, Y. E. Sawyer, G. D. Ebel, and J. A. Cook. In press. First records of deermice (*Peromyscus maniculatus*) in the Copper River Basin, southcentral Alaska. *Northwestern Naturalist*.

MacFadden, B. J., and H. Galiano. 1981. Late Hemphillian cat (Mammalia: Felidae) from the Bone Valley formation of central Florida. *Journal of Paleontology* 55:218–226.

Machin, D. 1974. A multivariate study of the external measurements of the sperm whale (*Physeter catadon*). *Journal of Zoology, London* 172:267–288.

Macpherson, A. 1965. The origin of diversity in mammals of the Canadian Arctic tundra. *Systematic Zoology* 14:153–173.

Madsen, D. B. (Ed.). 2004. *Entering America: Northeast Asia and Beringia before the Last Glacial Maximum.* University of Utah Press, Salt Lake City.

Magoun, A. J. 1985. Population characteristics, ecology and management of wolverines in Northwestern Alaska. Unpublished thesis. University of Alaska Fairbanks.

Magoun, A. J., and J. P. Copeland. 1998. Characteristics of wolverine reproductive den sites. *Journal of Wildlife Management* 62:1313–1320.

Magoun, A. J., and P. Valkenburg. 1977. The river otter (*Lutra canadensis*) on the north slope of the Brooks Range, Alaska. *Canadian Field-Naturalist* 91:303–305.

Majluf, P. 1999. California sea lion, *Zalophus californianus*. Pp. 201–202, in *The Smithsonian book of North American mammals* (D. E. Wilson and S. Ruff, eds.). Smithsonian Institution Press, Washington, D.C., in association with the American Society of Mammalogists.

Major, H. L., I. L. Jones, G. V. Byrd, and J. C. Williams. 2006. Assessing the effects of introduced Norway rats (*Rattus norvegicus*) on survival and productivity of Least Auklets (*Aethia pusilla*). *Auk* 123:681–694.

Maldonado, J. E., F. O. Davila, B. S. Stewart, E. Geffen, and R. K Wayne. 1995. Intraspecific genetic differentiation in California sea lions (*Zalophus californianus*) from southern California and the Gulf of California. *Marine Mammal Science* 11:46–58.

Maniscalco, J. M., K. Wynne, K. W. Pitcher, M. B. Hanson, S. R. Melin, and S. Atkinson. 2004. The occurrence of California sea lions (*Zalophus californianus*) in Alaska. *Aquatic Mammals* 30:427–433.

Manning, T. H. 1971. Geographical variation in the polar bear, *Ursus maritimus* Phipps. *Canadian Wildlife Service Report Series* 13:1–27.

Manville, R. H., and S. P. Young. 1965. Distribution of Alaskan mammals. U. S. Fish and Wildlife Service Circular 211.

Marshall, J. T., Jr., and R. D. Sage. 1981. Taxonomy of the house mouse. Symposia of the Zoological Society of London 47:15–25.

Martin, A. R. 1990. *The illustrated encyclopedia of whales and dolphins*. Portland House, New York.

Martin, L. D. 1989. Fossil history of the terrestrial Carnivora. Pp. 536–568, in *Carnivore behavior, ecology, and evolution* (J. L. Gittleman, ed.). Cornell University Press, Ithaca, New York.

Martin, L. D., W. von Koenigswald, and J. D. Stewart. 1986. Pleistocene *Phenacomys* from Kansas with remarks on other fossil records. *Transactions of the Nebraska Academy of Science* 14:35–39.

Masaki, Y. 1977. The separation of the stock units of sei whales in the North Pacific. *Reports of the International Whaling Commission* (Special Issue 1): 71–79.

Masuda, R., and M. C. Yoshida. 1994. A molecular phylogeny of the Family Mustelidae (Mammalia, Carnivora), based on comparison of mitochondrial cytochrome *b* nucleotide sequences. *Zoological Science* 11:605–612.

Matheus, P., J. Burns, J. Weinstock, and M. Hofreiter. 2004. Pleistocene brown bears in the mid-continent of North America. *Science* 306 (5699):1150.

Matheus, P. E. 1994. Pleistocene mammals. Pp. 54–69, in *Prehistoric Alaska. Alaska Geographic* 21(4):1–112.

———. 1997. Paleoecology and ecomorphology of the giant short-faced bear in Eastern Beringia. Unpublished dissertation. University of Alaska Fairbanks.

Matsumoto, H. 1937. A new species of orca from the Basal Calabrian at Naganuma, Minato Town, Province of Kazusa, Japan. *Zoology Magazine* (Japan) 49:191–193.

Mattern, M. Y., and D. A. McLennan. 2000. Phylogeny and speciation of felids. *Cladistics* 16:232–253.

Matthee, C. A., and S. K. Davis. 2001. Molecular insights into the evolution of the Family Bovidae: a nuclear DNA perspective. *Molecular Biology and Evolution* 18:1220–1230.

Matyushkin, E. N. 1979. Rysi Golarktiki [Lynx of the Holarctic]. Pp. 76–162, in *Mlekopitayushchie: Issledovaniya po faune Sovetskogo Soyuza* [Mammals: Investigations on the fauna of the Soviet Union] (O. L. Rossolimo, ed.). Sbornik Trudov Zoologicheskogo Muzeya MGU 13:1–279. (In Russian.)

May, R. M. 1993. Global change: The need and concern for collecting and preserving. Pp. 35–42, in *International symposium and first world congress on the preservation and conservation of natural history collections*. (C. L. Rose, S. L. Williams and J. Gisbert, eds.). Congress Book, Vol. 3, Dirección General de Bellas Artes y Archivos, Ministerio de Cultura, Madrid.

May-Collado, L., and I. Agnarsson. 2006. Cytochrome *b* and Bayesian inference of whale phylogeny. *Molecular Phylogenetics and Evolution* 38, 344–354.

Mayr, E. 1969. *Principles of systematic biology*. McGraw-Hill, New York.

McAllister, J. A., and R. S. Hoffmann. 1988. *Phenacomys intermedius. Mammalian Species* 305:1–8.

McAlpine, D. F. 2002. Pygmy and dwarf sperm whales, *Kogia breviceps* and *K. sima*. Pp. 1007–1009, in *Encyclopedia of marine mammals* (W. F. Perrin, B. Würsig, and J. G. M. Thewissen, eds.). Academic Press, San Diego and London.

McDonald, H. G., C. R. Harington, and G. De Iuliis. 2000. The ground sloth *Megalonyx* from Pleistocene deposits of the Old Crow Basin, Yukon, Canada. *Arctic* 53:213–220.

McDonald, J. N. 1981. *North American bison: Their classification and evolution*. University of California Press, Berkeley.

McDonald, J. N., and C. E. Ray. 1989. The autochthonous North American musk oxen *Bootherium, Symbos,* and *Gidleya* (Mammalia: Artiodactyla: Bovidae). Smithsonian Contributions to Paleobiology 66.

McDonough, T. J. 2004. Unit 15B moose management report. Pp. 209–215, in *Moose management report of survey and inventory activities* 1 July 2001–30 June 2003 (C. Brown, ed.). Alaska Department of Fish and Game. Project 1.0. Juneau, Alaska.

McDonough, T. J., and E. Rexstad. 2005. Short-term demographic response of the red-backed vole to spruce beetle infestations in Alaska. *Journal of Wildlife Management* 69:246–254.

McGowan, C., L. A. Howes, and W. S. Davidson. 1999. Genetic analysis of an endangered pine marten (*Martes americana*) population from Newfoundland using randomly amplified polymorphic DNA markers. *Canadian Journal of Zoology* 77:661–666.

McKenna, M. C., and S. K. Bell. 1997. *Classification of mammals above the species level*. Columbia University Press, New York.

McLaren, I. A. 1966. Taxonomy of harbor seals of the western North Pacific and evolution of certain other hair seals. *Journal of Mammalogy* 47:466–473.

McLaughlin, C. A. 1984. Protrogomorph, Sciuromorph, Castorimorph, Myomorph (Geomyoid, Anomaluroid, Pedetoid, and Ctenodactyloid) rodents. Pp. 267–288, in *Orders and families of Recent mammals of the world* (S. Anderson and J. K. Jones, Jr., eds.). John Wiley and Sons, New York.

McMillan, W. O., and E. Bermingham. 1996. The phylogeographic pattern of mitochondrial DNA variation in the Dall's porpoise *Phocoenoides dalli*. *Molecular Ecology* 5:47–61.

Mead, J. G. 1989. Beaked whales of the genus *Mesoplodon*. Pp. 349–430, in *Handbook of marine mammals. Volume 4: River dolphins and the larger toothed whales* (S. H. Ridgway and R. J. Harrison, eds.). Academic Press, London.

———. 1999a. Baird's beaked whale, *Berardius bairdii*. Pp. 304–305, in *The Smithsonian book of North American mammals* (D. E. Wilson and S. Ruff, eds.). Smithsonian Institution Press, Washington, D.C., in association with the American Society of Mammalogists.

———. 1999b. Stejneger's beaked whale, *Mesoplodon stejnegeri*. Pp. 314–315, in *The Smithsonian book of North American mammals* (D. E. Wilson and S. Ruff, eds.). Smithsonian Institution Press, Washington, D.C., in association with the American Society of Mammalogists.

Mead, J. G., and R. L. Brownell, Jr. 2005. Order Cetacea. Pp. 723–743, in *Mammal species of the world: A taxonomic and geographic reference* (D. E. Wilson and D. M. Reeder, eds.). Third Edition. Johns Hopkins University Press, Baltimore.

Mead, J. I. 1987. Quaternary records of pika, *Ochotona*, in North America. *Boreas* 16:165–171.

Mead, J. I., and F. Grady. 1996. *Ochotona* (Lagomorpha) from Late Quaternary cave deposits in eastern North America. *Quaternary Research* 45:93–101.

Meagher, M. 1986. *Bison bison*. *Mammalian Species* 266:1–8.

Mech, L. D. 1970. *The wolf: The ecology and behavior of an endangered species*. Natural History Press (Doubleday), New York.

———. 1974. *Canis lupus*. *Mammalian Species* 37:1–6.

———. 1999. Gray wolf, *Canis lupus*. Pp. 141–143, in *The Smithsonian book of North American mammals* (D. E. Wilson and S. Ruff, eds.). Smithsonian Institution Press, Washington, D.C., in association with the American Society of Mammalogists.

Merriam, C. H. 1890. Descriptions of twenty-six new species of North American mammals. *North American Fauna* 4.

———. 1895. Synopsis of the American shrews of the genus *Sorex*. *North American Fauna* 10:57–98.

———. 1896. Synopsis of the weasels of North America. *North American Fauna* 11.

———. 1900. Papers from the Harriman Alaska Expedition. I. Descriptions of twenty-six new mammals from Alaska and British North America. *Proceedings of the Washington Academy of Science* 2:13–30.

———. 1918. Review of the grizzly and big brown bears of North America (genus *Ursus*) with description of a new genus, *Vetularctos*. *North American Fauna* 41.

Merriam, H., J. Schoen, and D. Hardy. 1994. Sitka black-tailed deer. Alaska Department of Fish and Game Wildlife Notebook Series. http://www.adfg.state.ak.us/pubs/notebook/biggame/bt_deer.php (accessed 26 March 2007).

Merritt, J. F. 1981. *Clethrionomys gapperi*. *Mammalian Species* 146:1–9.

Meylan, A., and J. Hausser. 1991. The karyotype of the North American *Sorex tundrensis* (Mammalia; Insectivora). Pp. 125–129, in *The cytogenetics of the Sorex araneus group and related topics*. Mémoires de la Société Vaudoise des Sciences Naturelles 19:1–151.

Milinkovitch, M. C. 1997. The phylogeny of whales: A molecular approach. Pp. 317–338, in *Molecular genetics of marine mammals* (A. E. Dizon, S. J. Chivers, and W. F. Perrin, eds.). Special Publication Number 3, The Society for Marine Mammalogy.

Miller, G. S., Jr. 1896. Genera and subgenera of voles and lemmings. *North American Fauna* 12.

———. 1897. Revision of the North American bats of the family Vespertilionidae. *North American Fauna* 13.

———. 1924. List of North American Recent mammals, 1923. *U.S. National Museum Bulletin* 128:1–673.

Miller, G. S., Jr., and G. M. Allen. 1928. The American bats of the genera *Myotis* and *Pizonyx*. *Bulletin of the U. S. National Museum* 144:1–218.

Miller, G. S., Jr., and R. Kellogg. 1955. List of North American Recent mammals. *U. S. National Museum Bulletin* 205:1–954.

Miller, R. R. 1990. New records of postglacial walrus and a review of Quaternary marine mammals in New Brunswick. *Atlantic Geology* 26:97–107.

Miller, W. E. 1971. Pleistocene vertebrates of the Los Angeles Basin and vicinity (exclusive of Rancho Le Brea). *Scientific Bulletin, Los Angeles County Museum of Natural History* 10:1–124.

Millien, V., K. Lyons, L. Olson, F. Smith, A. Wilson, and Y. Yom-Tov. 2006. Ecotypic variation in the context of global climate change: Revisiting the rules. *Ecology Letters* 9:853–869.

Milyutin, A. I. 1990. Sistematika [Systematics]. Pp. 25–33, in *Seraya krysa: Sistematika, ekologiya, reguliatsiya chislennosti* [Norway rat: systematics, ecology, population control] (V. E. Sokolov and E. V. Karasjova, eds.). Nauka, Moscow. (In Russian.)

Mitchell, E. 1968a. The Mio-Pliocene pinniped *Imagotaria*. *Journal of the Fisheries Research Board of Canada* 25:1843–1900.

———. 1968b. Northeast Pacific stranding distribution and seasonality of Cuvier's beaked whale, *Ziphius cavirostris*. *Canadian Journal of Zoology* 46:265–279.

Mitchell, E. (Ed.). 1975. Report of the meeting on smaller cetaceans, Montreal, April 1–11, 1974. *Journal of the Fisheries Research Board of Canada* 32:889–983.

Mitchell, E., and R. H. Tedford. 1973. The Enaliarctinae, a new group of extinct aquatic Carnivora and a consideration of the origin of the Otariidae. *Bulletin of the American Museum of Natural History* 151:201–284.

Miyamoto, M. M., S. M. Tanhauser, and P. J. Laipis. 1989. Systematic relationships in the artiodactyl tribe Bovini (family Bovidae), as determined from mitochondrial DNA sequences. *Systematic Zoology* 38:342–349.

Miyazaki, N., M. Amano, and Y. Fujise. 1987. Growth and skull morphology of the harbour porpoises in the Japanese waters. *Memoirs of the National Science Museum, Tokyo* 20:137–146.

Miyazaki, N., and C. Shikano. 1997. Preliminary study on comparative skull morphology and vertebral formula among the six species of the genus, *Lagenorhynchus* (Cetacea: Delphinidae). *Mammalia* 61:573–587.

Mohr, E. 1952. Beiträge zur Kenntnis der Mähnenrobben. *Zoologische Garten* 19:98–112.

Mooney, P. 2004. Unit 4 mountain goat management report. Pp. 64–72, in *Mountain goat management report of survey and inventory activities* 1 July 2001–30 June 2003 (C. Brown, ed.). Alaska Department of Fish and Game. Project 12.0 Juneau, Alaska.

Moore, J. 1968. Relationships among the living genera of beaked whales with classifications, diagnoses and keys. *Fieldiana Zoology* 53:209–298.

Moore, S. E., and R. R. Reeves. 1993. Distribution and movement. Pp. 313–386, in *The bowhead whale* (J. J. Burns, J. J. Montague, and C. J. Cowles, eds.). Special Publication Number 2, The Society for Marine Mammalogy.

Moore, S. E., J. M. Waite, N. A. Friday, and T. Honkalehto. 2002. Distribution and comparative estimates of cetacean abundance on the central and south-eastern Bering Sea shelf with observations on bathymetric and prey associations. *Progress in Oceanography* 55:249–262.

Moore, S. E., J. M. Waite, L. L. Mazzuca, and R. C. Hobbs. 2000. Provisional estimates of mysticete whale abundance on the central Bering Sea shelf. *Journal of Cetacean Research and Management* 2:227–234.

Morejohn, G. V. 1979. The natural history of Dall's porpoise in the North Pacific Ocean. Pp. 45–83, in *Behavior of marine animals: Current perspectives in research. Volume 3: Cetaceans.* (H. W. Winn and B. L. Olla, eds.). Plenum Press, New York.

Morkill, A. 2005. Wildlife at risk: Impacts of shipwrecks on the Alaska Maritime National Wildlife Refuge. Abstract, Aleutian Life Forum 2005, 16–20 August 2005, Unalaska, Alaska. http://www.faculty.uaf.edu/ffrsb/outreach/ALF2005/anne.htm (accessed 27 February 2008).

Morlan, R. E. 1983. Counts and estimates of taxonomic abundance in faunal remains: Microtine rodents from Bluefish Cave I. *Canadian Journal of Archaeology* 7:61–76.

—————. 1984. Biostratigraphy and biogeography of Quaternary microtine rodents from northern Yukon Territory. Pp. 184–199, in *Contributions in Quaternary vertebrate paleontology: A volume in memorial to John E. Guilday* (H. H. Genoways and M. R. Dawson, eds.). Special Publication, Carnegie Museum of Natural History 8:1–538.

—————. 1989. Paleoecological implications of Late Pleistocene and Holocene microtine rodents from the Bluefish Caves, northern Yukon Territory. *Canadian Journal of Earth Sciences* 26:149–156.

—————. 2003. Some primitive mammoth teeth from Old Crow Loc. 47, Northern Yukon. Poster abstract, Third International Mammoth Conference, 24–29 May 2003, Dawson City, Yukon Territory.

Mouchaty, S., J. A. Cook, and G. F. Shields. 1995. Phylogenetic analysis of northern hair seals based on nucleotide sequences of the mitochondrial cytochrome *b* gene. *Journal of Mammalogy* 76:1178–1185.

Mowrey, R. A., G. A. Laursen, and T. A. Moore. 1981. Hypogeous fungi and small mammal mycophagy in Alaska taiga. *Proceedings of the Alaska Science Conference* 32:120–121.

Mowrey, R. A., and J. C. Zasada. 1984. Den tree use and movements of northern flying squirrels in interior Alaska and implications for forest management. Pp. 351–356, in *Proceedings: Fish and wildlife relationships in old-growth forests* (W. R. Meehan, T. R. Merrell, and T. A. Hanley, eds.). American Institute of Fish Research and Biology, Juneau, Alaska.

Mullican, T. R., and L. N. Carraway. 1990. Shrew remains from Moonshiner and Middle Butte Caves. *Journal of Mammalogy* 71:351–356.

Murdoch, J., and A. M. Sergeant. 1885. Part IV, Natural History. Pp. 89–200, in *Report of the International Polar Expedition to Point Barrow, Alaska, in response to the resolution of the House of Representatives on December 11, 1884* (P. H. Ray and J. Murdoch, eds.). Government Printing Office, Washington, D.C.

Murie, A. 1944. The wolves of Mount McKinley. *United States National Park Service Fauna Series* 5:1–238.

—————. 1961. *A naturalist in Alaska.* Devin-Adair Co., New York.

Murie, M. E. 1978. *Two in the far north.* Alaska Northwest Publishing Company, Anchorage.

Murie, O. J. 1935. Alaska-Yukon caribou. *North American Fauna* 54.

—————. 1936. Notes on the mammals of St. Lawrence Island, Alaska. Pp. 337–346, in *Archaeological excavations at Kukulik, St. Lawrence Island, Alaska.* Vol. II (O. W. Geist and F. G. Rainey, eds.). Miscellaneous Publications of the University of Alaska Fairbanks.

—————. 1959. Fauna of the Aleutian Islands and Alaska Peninsula. *North American Fauna* 61.

Murray, N. K., and F. H. Fay. 1979. The white whales or belukhas, *Delphinapterus leucas,* of Cook Inlet, Alaska. Draft prepared for June 1979 meeting of the Sub-committee on Small Cetaceans of the Scientific Committee of the International Whaling Commission. College of Environmental Sciences, University of Alaska Fairbanks.

Murrell, B. P., L. A. Durden, and J. A. Cook. 2003. Host associations of the tick, *Ixodes angustus* (Acari: Ixodidae), on Alaskan mammals. *Journal of Medical Entomology* 40:682–685.

Musser, G. G., and M. D. Carleton. 1993. Family Muridae. Pp. 501–756, in *Mammal species of the world: A taxonomic and geographic reference* (D. E. Wilson and D. M. Reeder, eds.). Second Edition. Smithsonian Institution Press, Washington, D.C.

—————. 2005. Superfamily Muroidea. Pp. 894–1531, in *Mammal species of the world: A taxonomic and geographic reference* (D. E. Wilson and D. M. Reeder, eds.). Third Edition. Johns Hopkins University Press, Baltimore.

Nadler, C. F., and R. S. Hoffmann. 1977. Patterns of evolution and migration in the arctic ground squirrel, *Spermophilus parryii* (Richardson). *Canadian Journal of Zoology* 55:748–758.

Nadler, C. F., E. A. Lyapunova, R. S. Hoffmann, N. N. Vorontsov, C. F. Nadler, Jr., and I. I. Fomichova. 1974. Evolution in ground squirrels. I. Transferrins in Holarctic populations of *Spermophilus. Comparative Biochemistry and Physiology* 47A:663–681.

Nadler, C. F., V. R. Rausch, E. A. Lyapunova, R. S. Hoffmann, and N. N. Vorontsov. 1976. Chromosomal banding patterns of the Holarctic rodents, *Clethrionomys rutilus* and *Microtus oeconomus. Zeitschrift für Säugetierkunde* 41:137–146.

Nadler, C. F., N. M. Zhurkevich, R. S. Hoffmann, A. I. Kozlovskii, L. Deutsch, and C. F. Nadler, Jr. 1978. Biochemical relationships of the Holarctic vole general (*Clethrionomys, Microtus,* and *Arvicola* (Rodentia: Arvicolinae)). *Canadian Journal of Zoology* 56:1564–1575.

Nagorsen, D. W. 1985. A morphometric study of geographic variation in the snowshoe hare (*Lepus americanus*). *Canadian Journal of Zoology* 63:567–579.

————. 1990. The mammals of British Columbia: A taxonomic catalogue. Memoir No. 4, Royal British Columbia Museum.

————. 1996. *Opossums, shrews, and moles of British Columbia*. Royal British Columbia Museum Handbook Volume 2. University of British Columbia Press, Vancouver.

————. 2005. Rodents and lagomorphs of British Columbia. Royal British Columbia Museum Handbook on the mammals of British Columbia, Volume 4. Royal British Columbia Museum, Victoria.

Nagorsen, D. W., and R. M. Brigham. 1993. *Bats of British Columbia*. Royal British Columbia Museum Handbook. University of British Columbia Press, Vancouver.

Nagorsen, D. W., R. W. Campbell, and G. R. Giannico. 1991. Winter food habits of marten, *Martes americana,* in Haida Gwaii. *Canadian Field-Naturalist* 105:55–59.

Nagorsen, D. W., and D. M. Jones. 1981. First records of the tundra shrew (*Sorex tundrensis*) in British Columbia. *Canadian Field-Naturalist* 95:93–94.

Nagorsen, D. W., and G. Keddie. 2000. Late Pleistocene mountain goats (*Oreamnos americanus*) from Vancouver Island: Biogeographic implications. *Journal of Mammalogy* 81:666–675.

Nagorsen, D. W., G. Keddie, and R. J. Hebda. 1995. Early Holocene black bears, *Ursus americanus,* from Vancouver Island. *Canadian Field-Naturalist* 109:11–18.

Nagy, T. R., and B. A. Gower. 1999. Northern collared lemming, *Dicrostonyx groenlandicus.* Pp. 659–660, in *The Smithsonian book of North American mammals* (D. E. Wilson and S. Ruff, eds.). Smithsonian Institution Press, Washington, D.C., in association with the American Society of Mammalogists.

Naske, C. M., and H. E. Slotnick. 1987. *Alaska: A history of the 49th state*. Second edition. University of Oklahoma Press, Norman and London.

NatureServe. 2007. NatureServe Explorer: An online encyclopedia of life [web application]. Version 6.2 NatureServe, Arlington, Virginia. http://www. natureserve.org/explorer (accessed 8 August 2007).

Nelson, E. W. 1887. Report upon natural history collections made in Alaska between the years 1877 and 1881. Arctic Series of Publications, U. S. Signal Service, III.

Nelson, E. W., and F. W. True. 1887. Mammals of northern Alaska, pt. 2. Pp. 227–293, in *Report upon natural history collections made in Alaska between the years 1877 and 1881*. Arctic Series of Publications, U.S. Signal Service, III.

Nemoto, T. 1957. Foods of baleen whales in the northern Pacific. *Science Report of the Whales Research Institute* 12:33–89.

Nichols, L. 1978. Dall's sheep. Pp. 173–189, in *Big game of North America: Ecology and management* (J. L. Schmidt and D. L. Gilbert, eds.). Wildlife Management Institute and Stackpole Books, Harrisburg, PA.

Nishiwaki, M. 1966. Distribution and migration of the large cetaceans in the North Pacific as shown by Japanese whaling results. Pp. 171–191, in *Whales, dolphins and porpoises* (K. S. Norris, ed.). University of California Press, Berkeley.

————. 1967. Distribution and migration of marine mammals in the North Pacific area. *Bulletin of the Ocean Research Institute* 1:1–64.

Niu, Y., F. Wei, M. Li, X. Liu, and Z. Feng. 2004. Phylogeny of pikas (Lagomorpha, *Ochotona*) inferred from mitochondrial cytochrome *b* sequences. *Folia Zool.* 53:141–155.

NMFS (National Marine Fisheries Service). 1993. Final conservation plan for the northern fur seal (*Callorhinus ursinus*). Prepared by the National Marine Mammal Laboratory/Alaska Fisheries Science Center, Seattle, and the Office of Protected Resources/National Marine Fisheries Service, Silver Spring, Maryland.

————. 2005. Draft conservation plan for the Cook Inlet beluga whale (*Delphinapterus leucas*). National Oceanic and Atmospheric Administration, 16 March 2005.

————. 2006. 2005 Cook Inlet beluga whale population estimate completed. National Oceanic and Atmospheric Administration National Marine Fisheries Service Alaska Region News Release, 20 January 2006.

Noss, R. F., and A. Y. Cooperrider. 1994. *Saving nature's legacy.* Island Press, Washington, D.C.

Nowacki, G., P. Spencer, M. Fleming, T. Brock, and T. Jorgenson. 2001. Ecoregions of Alaska. U. S. Geological Survey Open-File Report 02–297 (map).

Nowak, R. M. 1973. North American Quaternary *Canis.* Unpublished dissertation. University of Kansas, Lawrence.

————. 1976. *The cougar in the United States and Canada.* U. S. Fish and Wildlife Service, Washington, D.C. and New York Zoological Society.

————. 1983. A perspective on the taxonomy of wolves in North America. Pp. 10–19, in *Wolves in Canada and Alaska: Their status, biology, and management* (L. N. Carbyn, ed.). Canadian Wildlife Service Report, Series 45.

————. 1991. *Walker's mammals of the world.* Fifth Edition. Volume I. Johns Hopkins University Press, Baltimore and London.

————. 1995. Another look at wolf taxonomy. Pp. 375–397, in *Ecology and conservation of wolves in a changing world* (L. N. Carbyn, S. H. Fritts, and D. R. Seip, eds.). Canadian Circumpolar Institute Occasional Publication 35.

NRC (National Research Council). 2003. *The decline of the Steller sea lion in Alaskan waters: Untangling food webs and fishing nets.* National Academy Press, Washington D.C.

Nweeia, M. T. 2007. The narwhal tusk, potential as an arctic water probe and weather station: Collaborative and integrative studies of science and Inuit traditional knowledge. Abstracts of the ARCUS 19th Annual Meeting and Arctic Forum, Washington, D.C., 23–24 May 2007.

O'Corry-Crowe, G. M., A. E. Dizon, R. S. Suydam, and L. F. Lowry. 2002. Molecular genetics studies of population structure and movement patterns in a migratory species: The beluga whale, *Delphinapterus leucas,* in the western nearctic. P. 464, in *Molecular and cell biology of marine mammals* (C. J. Pfeiffer, ed.). Kreiger Publishing Company, Malabar, FL.

O'Corry-Crowe, G., W. Lucey, C. Bonin, E. Henniger, and R. Hobbs. 2006. The ecology, status and stock identity of beluga whales, *Delphinapterus leucas,* in Yakutat Bay, Alaska. Report to the U. S. Marine Mammal Commission NMFS-YSB-YTT, February 2006.

O'Corry-Crowe, G. M., R. S. Suydam, A. Rosenberg, K. J. Frost, and A. E. Dizon. 1997. Phylogeography, population structure and dispersal patterns of the beluga whale, *Delphinapterus leucas,* in the western Nearctic revealed by mitochondrial DNA. Molecular Ecology 6:955–970.

O'Corry-Crowe, G. M., and R. L. Westlake. 1997. Molecular investigations of spotted seals (*Phoca largha*) and harbor seals (*P. vitulina*) and their relationship in areas of sympatry. Pp. 291–304, in *Molecular genetics of marine mammals* (A. E. Dizon, S. J. Chivers, and W. F. Perrin, eds.). Special Publication Number 3. The Society for Marine Mammalogy.

O'Farrell, T. P. 1965. The rabbits of Middleton Island, Alaska. *Journal of Mammalogy* 46:525–527.

Ognev, S. I. 1950. Mammals of U. S. S. R. and adjacent countries. Vol. 7, Rodents (continued). (Translated from Russian, 1964, by Israel Program for Scientific Translations, Jerusalem.) National Science Foundation, Washington, D.C.

————. 1963. Mammals of the USSR and adjacent countries: Rodents (continued). (Mammals of eastern Europe and northern Asia) [A translation of S. I. Ognev, 1947, Zveri SSSR i prilezhashchikh stran: Gryzuny (prodolzhenie). (Zveri vostochnoi Evropy i severnoi Azii)]. Israel Program for Scientific Translations, Jerusalem 6:1–508.

————. 1964. Mammals of the USSR and adjacent countries: Vol. 7, Rodents (continued). (Translated from Russian by Israel Program for Scientific Translations, Jerusalem.) National Science Foundation, Washington, D.C.

Okhotina, M. V. 1984. Otriad Insectivora Bowdich, 1821-nasekomoiadyne. Pp. 31–72, in *Nazemnye mlekopitaiushchie dal'nego vostoka SSSR.* Opredelitel' (V. G. Krivosheev, ed.). Nauka, Moscow.

Olson, P. A., and S. B. Reilly. 2002. Pilot whales, *Globicephala melas* and *G. macrorhynchus.* Pp. 898–903, in *Encyclopedia of marine mammals* (W. F. Perrin, B. Würsig, and J. G. M. Thewissen, eds.). Academic Press, London.

O'Neill, D. 2004. *The last giant of Beringia: The mystery of the Bering land bridge*. Westview Press, Boulder, CO.

O'Neill, M. B., D. W. Nagorsen, and R. J. Baker. 2005. Mitochondrial DNA variation in water shrews (*Sorex palustris, Sorex bendirii*) from western North America: Implications for taxonomy and pylogeography. *Canadian Journal of Zoology* 83:1469–1475.

Orr, R. T. 1939. Extension of the range of *Sorex tundrensis*. *Journal of Mammalogy* 20:251.

———. 1951. Cetacean records from the Pacific coast of North America. *Wasmann Journal of Biology* 9:147–148.

Orth, D. J. 1971. Dictionary of Alaska place names. *U.S. Geological Survey Professional Paper 567*, Second Printing.

Osborn, H. F. 1942. *Proboscidea*. Volume II, pp. 805–1676. American Museum Press, New York.

Osgood, W. H. 1900. Mammals of the Yukon Territory. Pp. 21–46, in *Results of a biological reconnaissance of the Yukon River region* (W. H. Osgood and L. B. Bishop, eds.). North American Fauna 19.

———. 1901. Natural History of the Queen Charlotte Islands, British Columbia, and Natural History of the Cook Inlet Region, Alaska. *North American Fauna* 21.

———. 1903. Two new spermophiles from Alaska. *Proceedings of the Biological Society of Washington* 16:25–28.

———. 1904. A biological reconnaissance of the base of the Alaska Peninsula. *North American Fauna* 24.

———. 1909. Biological investigations in Alaska and the Yukon Territory. *North American Fauna* 30.

Ozowa T., S. Hayashi, and V. M. Mikhelson. 1997. Phylogenetic position of mammoth and Steller's sea cow within Tethyteria demonstrated by mitochondrial DNA sequences. *Journal of Molecular Evolution* 44:406–413.

Paetkau, D., S. C. Amstrup, E. W. Born, W. Calvert, A. E. Derocher, G. W. Garner, F. Messier, I. Stirling, M. K. Taylor, Ø. Wiig, and C. Strobeck. 1999. Genetic structure of the world's polar bear populations. *Molecular Ecology* 8:1571–1584.

Paetkau, D., G. F. Shields, and C. Strobeck. 1998. Gene flow between insular, coastal and interior populations of brown bears in Alaska. *Molecular Ecology* 7:1283–1292.

Pallas, P. S. 1811 [1831]. Zoographia Rosso-Asiatica, sistens omnium Animalium in extensor Imperio Rossico et adjacentibus maribus observatorum recensionem, domicillia, mores et descriptiones, anatomen atque icons plurimorum. *Petropoli, in officina Caes. acadamiae scientiarum.* 3 vol. 1:1–568.

Palsbøll, P. J., M. P. Heide-Jørgensen, and R. Dietz. 1997. Population structure and seasonal movements of narwhals, *Monodon monoceros,* determined from mtDNA analysis. *Heredity* 78:284–292.

Palumbi, S. R., and C. S. Baker. 1994. Contrasting population structure from nuclear intron sequences and mtDNA of humpback whales. *Molecular Biology and Evolution* 11:426–435.

Paradiso, J. L., and R. H. Manville. 1961. Taxonomic notes on the tundra vole (*Microtus oeconomus*) in Alaska. *Proceedings of the Biological Society of Washington* 74:77–92.

Parker, D. I. 1996. Forest ecology and distribution of bats in Alaska. Unpublished thesis. University of Alaska Fairbanks.

Parker, D. I., and J. A. Cook. 1996. Keen's long-eared bat, *Myotis keenii,* confirmed in southeast Alaska. *Canadian Field-Naturalist* 110:611–614.

Parker, D. I., J. A. Cook, and S. W. Lewis. 1996. Effects of timber harvest on bat activity in southeast Alaska's temperate rainforests. Pp. 277–292, in *Bats and forest symposium*, October 19–21, 1995 (R. M. R. Barclay and R. M. Brigham, eds.). Working Paper 23/1996. Victoria, British Columbia.

Parker, D. I., B. E. Lawhead, and J. A. Cook. 1997. Distributional limits of bats in Alaska. *Arctic* 50:256–265.

Pasitschniak-Arts, M. 1993. *Ursus arctos. Mammalian Species* 439:1–10.

Pasitschniak-Arts, M., and S. Larivière. 1995. *Gulo gulo. Mammalian Species* 499:1–10.

Peacock, E. 2004. Population, genetics, and behavioral studies of black bear *Ursus americanus* in southeast Alaska. Unpublished dissertation. University of Nevada, Reno.

Pearson, T. 1981. Geographic and intraspecific cranial variation in North American arctic ground squirrels. Unpublished thesis. University of Kansas, Lawrence.

Peirce, K. N. 2003. A small mammal inventory on Fort Richardson, Alaska. Final report submitted to U.S. Army Environmental Resources Department, Fort Richardson, Alaska.

Peirce, K. N., and J. M. Peirce. 2000a. Range extensions for the Alaska tiny shrew and pygmy shrew in southwestern Alaska. *Northwestern Naturalist* 81:67–68.

———. 2000b. A range extension for the meadow jumping mouse, *Zapus hudsonius,* in southwestern Alaska. *Canadian Field-Naturalist* 114:311.

Perrin, W. F. 1975. Variation of spotted and spinner porpoise (genus *Stenella*) in the eastern Pacific and Hawaii. *Bulletin of the Scripps Institution of Oceanography* 21:1–206.

———. 2001. *Stenella attenuata. Mammalian Species* 683:1–8.

Perrin, W. F., E. D. Mitchell, J. G. Mead, D. K. Caldwell, M. C. Caldwell, P. J. H. van Bree, and W. H. Dawbin. 1987. Revision of the spotted dolphins, *Stenella* spp. *Marine Mammal Science* 3:99–170.

Perry, E. 1999. Ringed seal, *Phoca hispida.* Pp. 206–207, in *The Smithsonian book of North American mammals* (D. E. Wilson and S. Ruff, eds.). Smithsonian Institution Press, Washington, D.C., in association with the American Society of Mammalogists.

Perry, E. A., S. M. Carr, S. E. Bartlett, and W. S. Davidson. 1995. A phylogenetic perspective on the evolution of reproductive behavior in pagophilic seals of the northwest Atlantic as indicated by mitochondrial DNA sequences. *Journal of Mammalogy* 76:22–31.

Person, D. K., M. Kirchhoff, V. Van Ballenberghe, G. C. Iverson, and E. Grossman. 1996. The Alexander Archipelago wolf: A conservation assessment. U. S. Forest Service, Pacific Northwest Research Station, General Technical Report PNW-GTR-384.

Person, D. K., and A. L. Russell. 2008. Correlates of mortality in an exploited wolf population. *Journal of Wildlife Management* 72:1540–1549.

Persons, K. 2004. Unit 22 moose management report. Pp. 496–522, in *Moose management report of survey and inventory activities* 1 July 2001–30 June 2003 (C. Brown, ed.). Alaska Department of Fish and Game. Project 1.0. Juneau, Alaska.

Peterson, R. L. 1955. *North American moose.* University of Toronto Press, Toronto.

———. 1966. *The mammals of eastern Canada.* Oxford University Press, Toronto.

Peterson, R. O., J. D. Woolington, and T. N. Bailey. 1984. Wolves of the Kenai Peninsula, Alaska. *Wildlife Monograph* 88:1–52.

Peterson, R. S. 1967. The land mammals of Unalaska Island: Present status and zoogeography. *Journal of Mammalogy* 48:119–129.

Péwé, T. L. 1975. *Quaternary geology of Alaska.* Geological Survey Professional Paper, U. S. Government Printing Office, Washington, D.C.

Péwé, T. L., and D. M. Hopkins. 1967. Mammal remains in pre-Wisconsin age in Alaska. Pp. 266–270, in *The Bering land bridge* (D. M. Hopkins, ed.). Stanford University Press, Stanford, CA.

Pietsch, M. 1970. Vergleichende Untersuchungen an Schädeln nordamerikanischer und europäischer Bisamratten (*Ondatra zibethicus* L. 1766). *Zeitschrift für Säugetierkunde* 35:257–288.

Pike, G. C. 1956. Guide to the whales, porpoises and dolphins of the north-east Pacific and arctic waters of Canada and Alaska. Fisheries Research Board of Canada Biological Station, Nanaimo, British Columbia, Circular No. 32 (revised).

Pitcher, K. W. 1989. Studies of southeastern Alaska sea otter populations: Distribution, abundance, structure, range expansion and potential conflicts with shell fisheries. Anchorage, Alaska. Alaska Department of Fish and Game, Cooperative Agreement 14-16-0009-954 with U.S. Fish and Wildlife Service.

———. 1990. Major decline in number of harbor seals *Phoca vitulina richardsi,* on Tugidak Island, Gulf of Alaska. *Marine Mammal Science* 6:121–134.

Pitelka, F. A. 1967. Some characteristics of microtine cycles in the Arctic. Pp. 153–184, in *Arctic biology* (H. P. Hansen, ed.). Oregon State University Press, Corvallis.

Pitelka, F. A., and G. O. Batzli. 1993. Distribution, abundance and habitat use by lemmings on the north slope of Alaska. Pp. 213–236, in *The biology of lemmings* (N. C. Stenseth and R. A. Ims, eds.). Linnean Society of London.

———. 2007. Population cycles in lemmings near Barrow, Alaska: A historical review. *Acta Theriologica* 52:323–336.

Pocock, R. I. 1917. The classification of the existing Felidae. *Annals and Magazine of Natural History*, Series 8, 20:329–350.

Polly, P. D. 2003. Paleophylogeography: The tempo and geographic differentiation in marmots (*Marmota*). *Journal of Mammalogy* 84:369–384.

Polziehn, R. O., R. Beach, J. Sheraton, and C. Strobeck. 1996. Genetic relationships among North American bison populations. *Canadian Journal of Zoology* 74:738–749.

Polziehn, R. O., J. Hamr, F. F. Mallory, and C. Strobeck. 1998. Phylogenetic status of North American wapiti (*Cervus elaphus*) subspecies. *Canadian Journal of Zoology* 76:998–1010.

Popov, L. A. 1982. Status of the main ice-living seals inhabiting inland waters and coastal marine areas of the U. S. S. R. Pp. 361–381, in *Mammals in the seas, Vol. IV: Small cetaceans, seals, sirenians and otters*. FAO Fisheries Series, No. 5.

Porter, B. 2004a. Unit 1A and 2 moose management report. Pp. 1–8, in *Moose management report of survey and inventory activities* 1 July 2001–30 June 2003 (C. Brown, ed.). Alaska Department of Fish and Game. Project 1.0. Juneau, Alaska.

———. 2004b. Unit 1A mountain goat management report. Pp. 1–21, in *Mountain goat management report of survey and inventory activities* 1 July 2001–30 June 2003 (C. Brown, ed.). Alaska Department of Fish and Game. Project 12.0. Juneau, Alaska.

Porter, L. 1988. Late Pleistocene fauna of Lost Chicken Creek, Alaska. *Arctic* 41:303–313.

Potter, C. W., and B. Birchler. 1999a. Blue whale, *Balaenoptera musculus*. Pp. 251–253, in *The Smithsonian book of North American mammals* (D. E. Wilson and S. Ruff, eds.). Smithsonian Institution Press, Washington, D.C., in association with the American Society of Mammalogists.

———. 1999b. Sperm whale, *Physeter macrocephalus*. Pp. 299–301, in *The Smithsonian book of North American mammals* (D. E. Wilson and S. Ruff, eds.). Smithsonian Institution Press, Washington, D.C., in association with the American Society of Mammalogists. 750 p.

———. 1999c. Northern right whale, *Eubalaena glacialis*. Pp. 241–243, in *The Smithsonian book of North American mammals* (D. E. Wilson and S. Ruff, eds.). Smithsonian Institution Press, Washington, D.C., in association with the American Society of Mammalogists.

———. 1999d. Sei whale, *Balaenoptera borealis*. Pp. 248–249, in *The Smithsonian book of North American mammals* (D. E. Wilson and S. Ruff, eds.). Smithsonian Institution Press, Washington, D.C., in association with the American Society of Mammalogists.

Powell, R. A. 1999. Fisher, *Martes pennanti*. Pp. 167–168, in *The Smithsonian book of North American mammals* (D. E. Wilson and S. Ruff, eds.). Smithsonian Institution Press, Washington, D.C., in association with the American Society of Mammalogists.

Powell, R. R. 1981. *Martes pennanti*. *Mammalian Species* 156:1–6.

Preble, E. A. 1899. Revision of the Jumping Mice of the Genus Zapus. *North American Fauna* 15.

———. 1923. Part 1. Birds and mammals. Pp. 1–128, in *A biological survey of the Pribilof Islands, Alaska* (E. A. Preble, et al., eds.). *North American Fauna* 46.

Prestrud, P., and I. Stirling. 1994. The International Polar Bear Agreement and the current status of polar bear conservation. *Aquatic Mammals* 20:1–12.

Pruitt, W. O., Jr. 1968. Synchronous biomass fluctuations of some northern mammals. *Mammalia* 32:172–191.

Pyare, S., W. P. Smith, J. V. Nicholls, and J. A. Cook. 2002. Feeding habits of northern flying squirrels: comparisons between Southeast Alaska and beyond. *Canadian Field-Naturalist* 116:98–103.

Quakenbush, L. T. 1988. Spotted seal, *Phoca largha*. Pp. 107–124, in *Selected marine mammals of Alaska: Species accounts with research and management recommendations* (J. W. Lentfer, ed.). Marine Mammal Commission, Washington, D.C.

Quammen, D. 1996. *The song of the dodo: Island biogeography in an age of extinctions*. Scribner, New York.

Quay, W. B. 1951. Observations on mammals of the Seward Peninsula, Alaska. *Journal of Mammalogy* 32:88–99.

Rainey W. E., J. M. Lowenstein, V. M. Sarich, and D. M. Magor. 1984. Sirenian molecular systematics, including the extinct Steller's sea cow (*Hydrodamalis gigas*). *Naturwissenschaften* 71:586–588.

Raloff, J. 1993. An otter tragedy. *Science News* 143:200–202.

Ramey, R. R. 1993. Evolutionary genetics and systematics of North American mountain sheep: Implications for conservation. Unpublished dissertation. Cornell University, Ithaca, NY.

Rand, A. L. 1945. Mammals of Yukon, Canada. *Bulletin of the National Museum of Canada* 100:1–93.

————. 1948. Mammals of the eastern Rockies and western plains of Canada. *Bulletin of the National Museum of Canada* 108:1–237.

————. 1954. The ice age and mammalian speciation in North America. *Arctic* 7:31–35.

Randi, E., N. Mucci, F. Claro-Hergueta, A. Bonnet, and E. J. P. Douzery. 2001. A mitochondrial DNA control region phylogeny for the Cervinae: Speciation in *Cervus* and its implications for conservation. *Animal Conservation* 4:1–11.

Rausch, R. L. 1950. Notes on the distribution of some arctic mammals. *Journal of Mammalogy* 31:464–466.

————. 1951. Notes on the Nunamiut Eskimo and mammals of the Anaktuvuk Pass region, Brooks Range, Alaska. *Arctic* 4:147–195.

————. 1953a. On the status of some arctic mammals. *Arctic* 6:91–148.

————. 1953b. On the land mammals of St. Lawrence Island, Alaska. *Murrelet* 34:18–26.

————. 1962. Notes on the collared pika, *Ochotona collaris* (Nelson), in Alaska. *Murrelet* 42:22–24.

————. 1963. A review of the distribution of Holarctic Recent mammals. Pp. 29–43, in *Pacific Basin biogeography; a symposium* (J. L. Gressitt, ed.). Tenth Pacific Science Congress, 1961. Bishop Museum Press, Honolulu.

————. 1964. The specific status of the narrow-skulled vole (subgenus *Stenocranius* Kashchenko) in North America. *Zeitschrift für Säugetierkunde* 29:343–358.

————. 1967. New records of the pigmy shrew *Microsorex hoyi* (Baird) in Alaska. *Murrelet* 48:9–10.

————. 1969. Origin of the terrestrial mammalian fauna of the Kodiak Archipelago. Pp. 216–234, in *The Kodiak Island Refugium: Its geology, flora, fauna, and history* (T. N. V. Karlstrom and G. E. Ball, eds.). The Boreal Institute, University of Alberta, Edmonton.

————. 1994. Transberingian dispersal of cestodes in mammals. *International Journal of Parasitology* 24:1203–1212.

Rausch, R. L., F. H. Fay, and F. S. Williamson. 1990. The ecology of *Echinococcus multilocularis* (Cestoda: Taeniidae) on St. Lawrence Island, Alaska. II. Helminth populations in the definitive host. *Annales de Parasitologie Humaine et Comparée* 65:131-140.

Rausch, R. L., and V. R. Rausch. 1965. Cytogenetic evidence for the specific distinction of an Alaskan marmot, *Marmota broweri* Hall and Gilmore (Mammalia: Sciuridae). *Chromosoma* (Berlin) 16:618–623.

————. 1968. On the biology and systematic position of *Microtus abbreviatus* Miller, a vole endemic to the St. Matthew Islands, Bering Sea. *Zeitschrift für Säugetierkunde* 33:65–99.

————. 1971. The somatic chromosomes of some North American marmots (Sciuridae), with remarks on the relationships of *Marmota broweri* Hall and Gilmore. *Mammalia* 35:85–101.

————. 1972. Observations on chromosomes of *Dicrostonyx torquatus stevensoni* Nelson and chromosomal diversity of varying lemmings. *Zeitschrift für Säugetierkunde* 37:372–384.

————. 1975a. Relationships of the red-backed vole, *Clethrionomys rutilus* (Pallas) in North America: Karyotypes of the subspecies *dawsoni* and *albiventer*. *Systematic Zoology* 24:163–170.

————. 1975b. Taxonomy and zoogeography of *Lemmus* spp. (Rodentia, Arvicolinae), with notes on laboratory-reared lemmings. *Zeitschrift für Säugetierkunde* 40:8–34.

————. 1995. The taxonomic status of the shrew of St. Lawrence Island, Bering Sea (Mammalia: Soricidae). *Proceedings of the Biological Society of Washington* 108:717–728.

————. 1997. Evidence for specific independence of the shrew (Mammalia: Soricidae) of St. Paul Island (Pribilof Islands, Bering Sea). *Zeitschrift für Säugetierkunde* 62:193–202.

Rausch, V. R., and D. L. Baldwin (Eds.) 2002. *The Yukon Relief Expedition and the journal of Carl Johan Sakariassen*. University of Alaska Press, Fairbanks.

Rausch, V. R., and R. L. Rausch. 1993. Karyotypic characteristics of *Sorex tundrensis* Merriam (Mammalia: Soricidae), a Nearctic species of the *S. araneus*-group. *Proceedings of the Biological Society of Washington* 106:410–416.

Ray, C. E. 1971. Polar bear and mammoth on the Pribilof Islands. *Arctic* 24:9–19.

————. 1983. Hooded seal, *Cystophora cristata*: Supposed fossil records in North America. *Journal of Mammalogy* 64:509–512.

Read, A. J. 1999. Harbour porpoise, *Phocoena phocoena* (Linnaeus, 1758). Pp. 323–355, in *Handbook of marine mammals. Volume 6: The second book of dolphins and the porpoises* (S. H. Ridgway and R. Harrison, eds.). Academic Press, London.

Rearden, J. (Ed.). 1981. Alaska mammals. *Alaska Geographic* 8:1–184.

Reeder, W. G. 1965. Occurrence of the big brown bat in southwestern Alaska. *Journal of Mammalogy* 46:332–333.

Reeves, R. R. 1978a. Atlantic walrus (*Odobenus rosmarus rosmarus*): A literature survey and status report. U. S. Fish and Wildlife Service, *Wildlife Research Report* 10:1–41.

———. 1978b. The narwhal: The arctic's unicorn. *Alaska* 44:10–11, 88–89.

———. 1999. Narwhal, *Monodon monoceros.* Pp. 293–294, in *The Smithsonian book of North American mammals* (D. E. Wilson and S. Ruff, eds.). Smithsonian Institution Press, Washington, D.C., in association with the American Society of Mammalogists.

Reeves, R. R., and S. Leatherwood. 1985. Bowhead whale. Pp. 305–344, in *Handbook of marine mammals. Volume 3: The sirenians and baleen whales* (S. H. Ridgway and R. Harrison, eds.). Academic Press, London.

———. 1994. Dolphins, porpoises, and whales: 1994–1998 Action Plan for the Conservation of Cetaceans. IUCN, Gland, Switzerland.

Reeves, R. R., and E. Mitchell. 1993. Status of Baird's beaked whale, *Berardius bairdii. Canadian Field-Naturalist* 107:509–523.

Reeves, R. R., B. S. Stewart, P. J. Clapham, and J. A. Powell. 2002. *National Audubon Society guide to marine mammals of the world.* Chanticleer Press, New York.

Reeves, R. R., B. S. Stewart, and S. Leatherwood. 1992. *The Sierra Club handbook of seals and sirenians.* Sierra Club Books, San Francisco.

Reeves, R. R., and S. Tracey. 1980. *Monodon monoceros. Mammalian Species* 127:1–7.

Reich, M., A. Gehler, U. B. Göhlich, D. Mol, and H. van der Plicht. 2006. The rediscovery of type material of *Mammuthus primigenius* (Mammalia: Proboscidea). Pp. 155–157, in *Ancient life and modern approaches.* Abstracts of the Second International Palaeontological Congress, 17–21 June 2006 (Yang Qun, Wang Yong-dong, and E. A. Weldon, eds.). University of Science and Technology of China Press, Beijing.

Reich, M., A. Gehler, D. Mol, H. van der Plicht, and A. Lister. 2007. The rediscovery of type material of *Mammuthus primigenius* (Mammalia: Proboscidea). Pp. 190–191, in *Abstracts IV International mammoth conference, Yakutsk,* 18–22 June 2007 (G. Boeskorov, ed.). Ministry of Science and Professional Education, Republic of Sakha (Yakutia).

Reichstein, H. 1957. Schädelvariabilität europäischer Mauswiesel (*Mustela nivalis* L.) und Hermeline (*Mustela erminea* L.) in Beziehung zu Verbreitung und Geschlecht. *Zeitschrift für Säugetierkunde* 22:151–182.

Reig, S. 1997. Biogeographic and evolutionary implications of size variation in North American least weasels (*Mustela nivalis*). *Canadian Journal of Zoology* 75:2036–2049.

Reijnders, P., S. Brasseur, J. van der Toorn, P. van der Wolf, I. Boyd, J. Harwood, D. Lavigne, and L. Lowry. 1993. Seals, fur seals, sea lions, and walrus. Status Survey and Conservation Action Plan. IUCN/SSC Seal Specialist Group. IUCN, Gland, Switzerland.

Reilly, S. B. 1978. Pilot whale. Pp. 112–119, in *Marine mammals of eastern north pacific and Arctic waters* (D. Haley, ed.). Pacific Search Press, Seattle..

Repenning, C. A. 1967. Palearctic-Nearctic mammalian dispersal in the late Cenozoic. Pp. 288–311, in *The Bering land bridge* (D. M. Hopkins, ed.). Stanford University Press, Stanford, CA.

———. 1983. New evidence for the age of the Gubik Formation, Alaskan North Slope. *Quaternary Research* 19:356–372.

———. 2001 Beringian climate during intercontinental dispersal: A mouse eye view. *Quaternary Science Reviews* 20:25–40.

———. 2003. *Mimomys* in North America (Chapter 17). *Bulletin of the American Museum of Natural History* 279:469–512.

Repenning C. A., and E. M. Brouwers. 1992. Late Pliocene–Early Pleistocene ecologic changes in the Arctic Ocean Borderland. *U. S. Geological Survey Bulletin.* 2036:1–37.

Repenning, C. A., E. M. Brouwers, L. D. Carter, L. Marincovich, Jr., and T. A. Ager. 1987. The Beringian ancestry of *Phenacomys* (Rodentia: Cricetidae) and the beginning of the modern Arctic Ocean Borderland biota. *U. S. Geological Survey Bulletin* 1687:1–35.

Repenning, C. A., and F. M. Grady. 1988. The microtine rodents of the Cheetah Room fauna, Hamilton Cave, West Virginia, and the spontaneous origin of *Synaptomys. U. S. Geological Survey Bulletin* 1853:1–32.

Repenning, C. A., D. M. Hopkins, and M. Rubin. 1964. Tundra rodents in a Late Pleistocene fauna from the Tofty Placer District, central Alaska. *Arctic* 17:177–197.

Repenning, C. A., C. E. Ray, and D. Grigorescu. 1979. Pinniped biogeography. Pp. 357–369, in *Historical biogeography, plate tectonics and the changing environment* (J. Gray and A. J. Boucot, eds.). Thirty-seventh Annual Biological Colloquium, Oregon State University Press, Corvallis.

Repenning, C. A., and R. H. Tedford. 1977. Otarioid seals of the Neogene. *U. S. Geological Survey Professional Paper* 992:1–91.

Revkin, A. C. 2001. The call of the wild takes its toll on reindeer. *New York Times*, Science Section, 23 January.

Reynolds, H. V, III. 2005. Unit 20E black bear management report. Pp. 247–256, in *Black bear management report of survey and inventory activities* (C. Brown, ed.). Alaska Department of Fish and Game. Project 17.0. Juneau, Alaska.

Reynolds, H. W., R. D. Glaholt, and A. W. L. Hawley. 1982. Bison. Pp. 972–1007, in *Wild mammals of North America: Biology, management, and economics* (J. A. Chapman and G. A. Feidhammer, eds.). Johns Hopkins University Press, Baltimore.

Reynolds, P., and D. E. Ross. 1984. Population status of muskoxen in the Arctic National Wildlife Refuge, Alaska. Biological Paper, University of Alaska, Special Report No. 4:63 (abstract only).

Rhoads, S. N. 1902. Synopsis of the American martens. *Proceedings of the Academy of Natural Sciences of Philadelphia* 54:443–460.

Rhode, C. J., and W. Barker. 1953. Alaska's fish and wildlife. U. S. Department of the Interior, Fish and Wildlife Service Circular 17.

Rice, D. W. 1978a. Blue whale. Pp. 31–35, in *Marine mammals of eastern North Pacific and Arctic waters* (D. Haley, ed.). Pacific Search Press, Seattle.

———. 1978b. Beaked whales. Pp. 89–95, in *Marine mammals of eastern North Pacific and Arctic waters* (D. Haley, ed.). Pacific Search Press, Seattle.

———. 1984. Cetaceans. Pp. 447–490, in *Orders and families of Recent mammals of the world* (S. Anderson and J. K. Jones, Jr., eds.). John Wiley and Sons, New York.

———. 1989. Sperm Whale. Pp. 177–233, in *Handbook of marine mammals. Volume 4: River dolphins and the larger toothed whales* (S. H. Ridgway and R. Harrison, eds.). Academic Press, London.

———. 1998. Marine mammals of the world: Systematics and distribution. Special Publication Number 4. *The Society for Marine Mammalogy.*

———. 1999. Gray whale, *Eschrichtius robustus.* Pp. 259–261, in *The Smithsonian book of North American mammals* (D. E. Wilson and S. Ruff, eds.). Smithsonian Institution Press, Washington, D.C., in association with the American Society of Mammalogists.

Rice, D. W., A. A. Wolman, and H. W. Braham. 1984. The gray whale, *Eschrichtius robustus. Marine Fisheries Review* 46:7–14.

Richards, R. L. 1988. *Microtus xanthognathus* and *Synaptomys borealis* in the Late Pleistocene of southern Indiana. *Proceedings of the Indiana Academy of Science* 98:561–570.

Richardson, J. 1839. In *Zoology of Captain Beechey's voyage of the 'Blossom,'* List of Mammals, p. 5.

Rideout, C. B., and R. S. Hoffmann. 1975. *Oreamnos americanus. Mammalian Species* 63:1–6.

Riedman, M. L., and J. A. Estes. 1991. The sea otter (*Enhydra lutris*): Behavior, ecology, and natural history. U. S. Fish and Wildlife Service, *Biological Report* 90:1–126.

Ritchie, J. C., and L. C. Cwynar. 1982. The late-Quaternary vegetation of the north Yukon. Pp. 113–126, in *Paleoecology of Beringia* (D. M. Hopkins, J. V. Matthews Jr., C. E. Schweger, and S. B. Young, S. B., eds.). Academic Press, New York.

Roberson, K. 1986. Range extension of the Sitka black-tailed deer (*Odocoileus hemionus sitkensis*) in Alaska. *Canadian Field-Naturalist* 100:563–565.

Røed, K. H., and K. R. Whitten. 1986. Transferrin variation and evolution of Alaskan reindeer and caribou, *Rangifer tarandus* L. Rangifer (Special Issue) 1:247–251.

Roest, A. I. 1973. Subspecies of the sea otter, *Enhydra lutris.* Contributions in Science, Los Angeles County Museum 252:1–17.

Rogers, L. L. 1999. American black bear, *Ursus americanus.* Pp. 157–160, in *The Smithsonian book of North American mammals* (D. E. Wilson and S. Ruff, eds.). Smithsonian Institution Press, Washington, D.C., in association with the American Society of Mammalogists.

Rose, K. D., V. B. DeLeon, P. Missiaen, R. S. Rana, A. Sahni, L. Singh, and T. Smith. 2008. Early Eocene lagomorph (Mammalia) from Western India and the early diversification of Lagomorpha. Proceedings of the Royal Society London B 275:1203–1208.

Rosel, P. E. 1992. Genetic population structure and systematics of some small cetaceans inferred from mitochondrial DNA sequence variation. Unpublished dissertation. University of California, San Diego.

———. 1997. A review and assessment of the status of the harbor porpoise (*Phocoena phocoena*) in the North Atlantic. Pp. 209–226, in *Molecular genetics of marine mammals* (A. E. Dizon, S. J. Chivers, and W. F. Perrin, eds.). Special Publication Number 3, The Society for Marine Mammalogy.

Rosel, P. E., M. G. Haygood, and W. F. Perrin. 1995. Phylogenetic relationships among the true porpoises (Cetacea: Phocoenidae). *Molecular Phylogenetics and Evolution* 4:463–474.

Rosenbaum, H. C., R. L. Brownell, M. W. Brown, C. Schaeff, V. Portway, B. N. White, S. Malik, L. A. Pastene, N. J. Patenaude, C. S. Baker, M.Goto, P. B. Best, P. J. Clapham, P. Hamilton, M. Moore, R. Payne, V. Rowntree, C. T. Tynan, J. L. Bannister, and R. DeSalle. 2000. World-wide genetic differentiation of *Eubalaena*: Questioning the number of right whale species. *Molecular Ecology* 9:1793–1802.

Roth, E. L. 1972. Late Pleistocene mammals from Klein Cave, Kerr County, Texas. *Texas Journal of Science* 24:75–84.

Rotterman, L. M., and T. Simon-Jackson. 1988. Sea otter, *Enhydra lutris.* Pp. 237–275, in *Selected marine mammals of Alaska: Species accounts with research and management recommendations* (J. W. Lentfer, ed.). Marine Mammal Commission, Washington, D.C.

Rugh, D., D. DeMaster, A. Rooney, J. Breiwick, K. Sheldon, and S. Moore. 2003. A review of bowhead whale (*Balaena mysticetus*) stock identity. *Journal of Cetacean Resource Management* 5:267–279.

Rugh, D. J., M. M. Muto, S. E. Moore, and D. P. DeMaster. 1999. Status review of the eastern north Pacific stock of gray whales. U. S. Department of Commerce NOAA Technical Memorandum NMFS-AFSC-103.

Rugh, D. J., K. E. W. Shelden, and B. A. Mahoney. 2000. Distribution of belugas, *Delphinapterus leucas,* in Cook Inlet, Alaska, during June/July, 1993–2000. *Marine Fisheries Review* 62:6–21.

Runck, A. M. 2001. Molecular and morphological perspectives on post-glacial colonization of *Clethrionomys rutilus* and *Clethrionomys gapperi* in Southeast Alaska. Unpublished thesis. University of Alaska Fairbanks.

———. 2006. Dynamics across a broad hybrid zone between *Myodes rutilus* and *M. gapperi.* Unpublished dissertation. Idaho State University, Pocatello.

Runck, A. M., and J. A. Cook. 2005. Postglacial expansion of the southern red-backed vole (*Clethrionomys gapperi*) in North America. *Molecular Ecology* 14:1445–1456.

Russell, K. R. 1978. Mountain lion. Pp. 207–225, in *Big game of North America: Ecology and management* (J.L. Schmidt and D.L. Gilbert, eds.). Stackpole Books, Harrisburg, PA.

Rutilevskii, G. L. 1958. A narwhal in the region of drifting station North Pole 5. *Problemy Arktiki* 3:116–119 (in Russian). *American Meteorology Society Translations* 182:1–4, 1959 (in English).

Sarra, R. 1933. Denti di pesci del Cretaceo e di mammiferi del Pliocene rinvenuti in Basilicata. *Rivista Italiana di Paleontologia* 39:29–34.

Sattler, R. A. 1997. Large mammals in Lower Rampart Cave 1, Alaska: Interspecific utilization of an Eastern Beringian cave. *Geoarchaeology: An International Journal* 12:657–688.

Savage, D. E., and D. E. Russell. 1983. *Mammalian paleofaunas of the world.* Addison-Wesley Publishing Company, Reading, MA.

Savinetsky, A. B., N. K. Kiseleva, and B. F. Khassanov. 2004. Dynamics of sea mammal and bird populations of the Bering Sea region over the last several millennia. *Palaeogeography, Palaeoclimatology, Palaeoecology* 209:335–352.

Scammon, C. M. 1874. *The marine mammals of the northwestern coast of North America together with an account of the American whale-fishery.* J. H.Carmany and Company, San Francisco, CA.

Scheffer, V. B. 1947. Raccoon transplanted in Alaska. *Journal of Wildlife Management* 12:350–351.

————. 1958. *Seals, sea lions, and walruses, a review of the Pinnipedia.* Stanford University Press, Stanford, CA.

Schevill, W. E. 1986. The International Code of Zoological Nomenclature and a paradigm: The name *Physeter catadon* Linnaeus 1758. *Marine Mammal Science* 2:153–157.

Schevill, W. W. 1954. Sight records of the gray grampus, *Grampus griseus* (Cuvier). *Journal of Mammalogy* 35:123–124.

Schiller, E. L., and R. Rausch. 1956. Mammals of the Katmai National Monument, Alaska. *Arctic* 9:191–201.

Schoen, J. W. 2007. Alexander Archipelago wolf (*Canis lupus ligoni*). In *The coastal forests and mountains ecoregion of southeastern Alaska and the Tongass National Forest: A conservation assessment and resource synthesis* (J. W. Schoen and E. Dovichin, eds.). Audubon Alaska and The Nature Conservancy, Anchorage, Alaska.

Schoen, J. W., R. Flynn, and B. Clark. 2007. Marten (*Martes americana*). In *The coastal forests and mountains ecoregion of southeastern Alaska and the Tongass National Forest: A conservation assessment and resource synthesis* (J. W. Schoen and E. Dovichin, eds.). Audubon Alaska and the Nature Conservancy, Anchorage, Alaska.

Schoen, J. W., and S. Gende. 2007. Brown bear (*Ursus arctos*). In *The coastal forests and mountains ecoregion of southeastern Alaska and the Tongass National Forest: A conservation assessment and resource synthesis* (J. W. Schoen and E. Dovichin, eds.). Audubon Alaska and The Nature Conservancy, Anchorage, Alaska.

Schoen, J. W., and M. Kirchhoff. 2007. Sitka black-tailed deer (*Odocoileus hemionus sitkensis*). In *The coastal forests and mountains ecoregion of southeastern Alaska and the Tongass National Forest: A conservation assessment and resource synthesis* (J. W. Schoen and E. Dovichin, eds.). Audubon Alaska and The Nature Conservancy, Anchorage, Alaska.

Schoen, J. W., and M. D. Kirchhoff. 1981. Habitat use by mountain goats in southeast Alaska. Final Report Project W-17-R, Alaska Department of Fish and Game, Juneau.

Scholander, P. F., L. Irving, and S. W. Grinnell. 1942. On the temperature and metabolism of the seal during diving. *Journal of Cellular and Comparative Physiology* 19:67–78.

Schreiber, A., R. Wirth, M. Riffel, and H. Van Rompaey. 1989. Weasels, civets, mongooses, and their relatives: An action plan for the conservation of mustelids and viverrids. IUCN/SSC Mustelid and Viverrid Specialist Group. IUCN, Gland, Switzerland.

Schwartz, M. K., L. Scott Mills, K. S. McKelvey, L. F. Ruggiero, and F. W. Allendorf. 2002. DNA reveals high dispersal synchronizing the population dynamics of Canada lynx. *Nature* 415:520–522.

Schwarz, E. 1940. Status and affinities of the bears of Northeast Asia. *Journal of Mammalogy* 21:206–211.

Schwarz, E., and H. K. Schwarz. 1943. The wild and commensal stocks of the house mouse, *Mus musculus* Linneaus. *Journal of Mammalogy* 24:59–72.

Scribner, K. T., G. W. Garner, S. C. Amstrup, and M. A. Cronin. 1997. Population genetic studies of the polar bear (*Ursus maritimus*): A summary of available data and interpretation of results. Pp. 185–196, in *Molecular genetics of marine mammals* (A. E. Dizon, S. J. Chivers, and W. F. Perrin, eds.). Society for Marine Mammalogy Special Publication, Number 3. Allen Press, Lawrence, KS.

Sears, R. 2002. Blue whale, *Balaenoptera musculus.* Pp. 112–116, in *Encyclopedia of marine mammals* (W. F. Perrin, B. Würsig, and J. G. M. Thewissen, eds.). Academic Press, San Diego and London.

Seavoy, R.J. 2004a. Unit 18 furbearer management report. Pp. 229–240, in *Furbearer management report of survey and inventory activities* 1 July 2000–30 June 2003 (C. Brown, ed.). Alaska Department of Fish and Game. Project 7.0. Juneau, Alaska.

————. 2004b. Unit 18 moose management report. Pp. 267–292, in *Moose management report of survey and inventory activities* 1 July 2001–30 June 2003 (C. Brown, ed.). Alaska Department of Fish and Game. Project 1.0. Juneau, Alaska.

Selinger, J. 2004. Units 7 and 15 furbearer management report. Pp. 90–106, in *Furbearer management report of survey and inventory activities* 1 July 2000–30 June 2003 (C. Brown, ed.). Alaska Department of Fish and Game. Project 7.0. Juneau, Alaska.

Sellers, R. A. 2003. Unit 9 & 10 wolf management report. Pp. 66–69, in *Wolf management report of survey and inventory activities* 1 July 1999–30 June 2002 (C. Healy, ed.). Alaska Department of Fish and Game. Juneau, Alaska.

Serfass, T. L., R. P. Brooks, J. M. Novak, P. E. Johns, and O. E. Rhodes, Jr. 1998. Genetic variation among populations of river otters in North America: Considerations for reintroduction projects. *Journal of Mammalogy* 79:736–746.

Sergeant, D. E. 1976. History and present status of populations of harp and hooded seals. *Biological Conservation* 10:95–118.

Serreze, M. C., J. E. Walsh, F. S. Chapin III, T. Osterkamp, M. Dyurgerov, V. Romanovsky, W. C. Oechel, J. Morison, T. Zhang, and R. G. Barry. 2000. Observational evidence of recent change in the northern high-latitude environment. *Climatic Change* 46:159–207.

Servheen, C. 1990. The status and conservation of bears of the world. *International Conference on Bear Research and Management* 2:1–32.

Shackleton, D. 1999. *Hoofed mammals of British Columbia*. Royal British Columbia Museum Handbook Volume 3. University of British Columbia Press, Vancouver.

Shackleton, D. M., L. V. Hills, and D. A. Hutton. 1975. Aspects of variation in cranial characters of plains bison (*Bison bison bison* Linnaeus) from Elk Island National Park, Alberta. *Journal of Mammalogy* 56:871–877.

Shapiro, B., A. J. Drummond, A. Rambaut, M. C. Wilson, P. E. Matheus, A. V. Sher, O. G. Pybus, M. T. P. Gilbert, I. Barnes, J. Binladen, E. Willerslev, A. J. Hansen, G. F. Baryshnikov, J. A. Burns, S. Davydov, J. C. Driver, D. G. Froese, C. R. Harington, G. Keddie, P. Kosintsev, M. L. Kunz, L. D. Martin, R. O. Stephenson, J. Storer, R. Tedford, S. Zimov, and A. Cooper. 2004. Rise and fall of the Beringian steppe bison. *Science* 306:1561–1565.

Shaughnessy, P. D., and F. H. Fay. 1977. A review of the taxonomy and nomenclature of North Pacific harbour seals. *Journal of Zoology* (London) 182:385–419.

Shaw, J. H. 1962. The bushy-tailed woodrat in Southeast Alaska. *Journal of Mammalogy* 43:431–432.

Sheffield, S. R., and C. M. King. 1994. *Mustela nivalis. Mammalian Species* 454:1–10.

Shelden, K. E. 1999. Risso's dolphin, *Grampus griseus*. Pp. 280–282, in *The Smithsonian book of North American mammals* (D. E. Wilson and S. Ruff, eds.). Smithsonian Institution Press, Washington, D.C., in association with the American Society of Mammalogists.

Shelden, K. E. W., and D. J. Rugh. 1995. The bowhead whale (*Balaena mysticetus*): Status review. *Marine Fisheries Review* 57:1–20.

Sher, A. V. 1971. *Mammals and stratigraphy of the Pleistocene in the far north-east of the USSR and North America*. Nauka, Moscow. (In Russian.)

Shields, G. F., and T. D. Kocher. 1991. Phylogenetic relationships of North American ursids based on analysis of mitochondrial DNA. *Evolution* 45:218–221.

Shoshani, J. 1998. Understanding proboscidean evolution: a formidable task. *Trends in Ecology and Evolution* 13:480–487.

————. 2005a. Order Proboscidea. Pp. 90–91, in *Mammal species of the world: A taxonomic and geographic reference* (D. E. Wilson and D. M. Reeder, eds.). Third Edition. Johns Hopkins University Press, Baltimore.

————. 2005b. Order Sirenia. Pp. 92–93, in *Mammal species of the world: A taxonomic and geographic reference* (D. E. Wilson and D. M. Reeder, eds.). Third Edition. Johns Hopkins University Press, Baltimore.

Shults, L. M., F. H. Fay, and J. D. Hall. 1982. Helminths from Stejneger's beaked whale *Mesoplodon stejnegeri* and Risso's dolphin *Grampus griseus* in Alaska. *Proceedings of the Helminth Society of Washington* 49:146–147.

Simmons, N.B. 2005. Order chiroptera. Pp. 312–529, in *Mammal species of the world: A taxonomic and geographic reference* (D.E. Wilson and D.M. Reeder, eds.). Third Edition. Johns Hopkins University Press, Baltimore.

Simmons, N.B., and J.H. Geisler. 1998. Phylogenetic relationships of Icaronycteris, Archeonycteris, Hassianycteris, and Palaeochiropteryx to extant bat lineages, with comments on the evolution of echolocation and foraging strategies in microchiroptera. *Bulletin of the American Museum of Natural History* 235:1–182.

Simpson, C.D. 1984. Artiodactyls. Pp. 563–587, in *Orders and families of Recent mammals of the world* (S. Anderson and J.K. Jones, Jr., eds.). John Wiley and Sons, New York.

Simpson, G.G. 1945. The principles of classification and a classification of mammals. *Bulletin of the American Museum of Natural History* 85:1–350.

Simpson, M.R. 1993. *Myotis californicus. Mammalian Species* 428:1–4.

Skinner, M.F., and O.C. Kaisen. 1947. The fossil *Bison* of Alaska and preliminary revision of the genus. *Bulletin of the American Museum of Natural History* 89:123–256.

Sleptsov, M.M. 1961. Observations of small cetaceans in far eastern seas and northwest Pacific. *Trudy Instituta Morfologii Zhivotnykh* 34:136–143 (translated from Russian).

Slough, B.G., and T.S. Jung. 2007. Diversity and distribution of the terrestrial mammals of the Yukon Territory: A review. *Canadian Field-Naturalist* 121:119–127.

Small, M., K.D. Stone, and J.A. Cook. 2003. American marten (*Martes americana*) population structure across a landscape fragmented in time and space. *Molecular Ecology* 12:69–103.

Small, R.J., G.W. Pendleton, and K.W. Pitcher. 2003. Trends in abundance of Alaska harbor seals, 1983–2001. *Marine Mammal Science* 19:344–362.

Small, R.J., G.W. Pendleton, and K.M. Wynne. 1998. Harbor seal population trends in the Ketchikan, Sitka, and Kodiak Island areas of Alaska. Pp. 7–26, in *Harbor seal investigations in Alaska, 1997–1998*. Alaska Department of Fish and Game Final Report for NOAA Award NA57FX0367.

Smetzer, M.B. 2008. Polar bear killed near village in Interior, Alaska. *Fairbanks Daily News-Miner*, March 28.

Smirnov, N.A. 1929. A review of the Pinnipedia of Europe and northern Asia. *Izvestiya Otdela Prikladnoy Ikhtiologii* 9:231–268. (Translated from Russian by F.H. Fay, University of Alaska Fairbanks.)

Smirnov, N.G., and V.B. Fedorov. 2003. Holarctic collared lemmings: Traces of their spread as related to the history of the Arctic biota. *Russian Journal of Ecology* 34:332–338. (Translated from *Ekologiya* 5:370–376.)

Smith, C.A., and L. Nichols. 1984. Mountain goat transplants in Alaska: Restocking depleted herds and mitigating mining impacts. *Proceedings of the Biennial Symposia Northern Wild Sheep and Goat Council* 4:467–480.

Smith, F.A. 1997. *Neotoma cinerea. Mammalian Species* 564:1–8.

Smith, H.C., and E.J. Edmonds. 1985. The brown lemming, *Lemmus sibiricus*, in Alberta. *Canadian Field-Naturalist* 99:99–100.

Smith, M.E., and M.C. Belk. 1996. *Sorex monticolus. Mammalian Species* 528:1–5.

Smith, T.E. 1984. Status of muskoxen in Alaska. *Biological Paper of the University of Alaska Special Report* 4:15–18.

Smith, W.P., S.M. Gende, and J.V. Nichols. 2004. Ecological correlates of flying squirrel microhabitat use and density in temperate rainforests of southeastern Alaska. *Journal of Mammalogy* 84:663–674.

Smith, W.P., and J.V. Nichols. 2003. Demography of the Prince of Wales flying squirrel: An endemic of Southeastern Alaska temperate rainforest. *Journal of Mammalogy* 84:1044–1058.

Smolen, M.J., and B.L. Keller. 1987. *Microtus longicaudus. Mammalian Species* 271:1–7.

Snell, R.R., and K.M. Cunnison. 1983. Relation of geographic variation in the skull of *Microtus pennsylvanicus* to climate. *Canadian Journal of Zoology* 61:1232–1241.

Sokolov, I.I. 1973. Napravleniya evolyutsii i estestvennaya klassifikatsiya podsemeystva vydrovykh (Lutrinae, Mustelidae, Fissipedia) [Trends of evolution and the natural classification of the subfamily of otters (Lutrinae, Mustelidae, Fissipedia)]. *Byulleten' Moskovskogo Obshchestva Ispytatelei Prirody Otdel Biologicheskii* 78:45–52. (In Russian, English summary.)

Solow, A. R., D. L. Roberts, and K. M. Robbirt. 2006. On the Pleistocene extinctions of Alaskan mammoths and horses. *Proceedings of the National Academy of Sciences* 103:7351–7353.

Spencer, D. L., and J. B. Hakala. 1964. Moose and fire on the Kenai. *Proceedings of the Tall Timbers Fire Ecology Conference* 3:11–33.

Spencer, D. L., and C. J. Lensink. 1970. The muskox of Nunivak Island. *Journal of Wildlife Management* 34:1–15.

Springer, M. S., G. C. Cleven, O. Madsen, W. W. de Jong, V. G. Waddell, H. M. Amrine, and M. J. Stanhope. 1997. Endemic African mammals shake the phylogenetic tree. *Nature* 388:61–64.

Stacey, P. J., and R. W. Baird. 1993. Status of the short-finned pilot whale, *Globicephala macrorhynchus*, in Canada. *Canadian Field-Naturalist* 107:481–489.

Stacey, P. J., S. Leatherwood, and R. W. Baird. 1994. *Pseudorca crassidens*. *Mammalian Species* 456:1–6.

Stafford, K. M., S. E. Moore, M. Spillane, and S. Wiggins. 2007. Grey whale calls recorded near Barrow, Alaska, throughout the winter of 2003–04. *Arctic* 60:167–172.

Stains, H. J. 1984. Carnivores. Pp. 491–521, in *Orders and families of Recent mammals of the world* (S. Anderson and J. K. Jones, Jr., eds.). John Wiley and Sons, New York.

Steele, M. A. 1998. *Tamiasciurus hudsonicus*. *Mammalian Species* 586:1–9.

Stein, B. R. 1990. Limb mycology and phylogenetic relationships in the superfamily Dipodoidea (birch mice, jumping mice, and jerboas). *Zeitschrift für Zoologische Systematik und Evolutionsforschung* 28:299–314.

———. 1997. Annie M. Alexander: Extraordinary patron. *Journal of the History of Biology* 30:243–266.

Stejneger, L. 1883. Contributions to the history of the Commander Islands. Number 1—Notes on the natural history, including descriptions of new cetaceans. *Proceedings of the U.S. National Museum* 6:58–59.

———. 1887. How the great northern sea-cow (*Rhytina*) became exterminated. *American Naturalist* 21:1047–1054.

———. 1936. *Georg Wilhelm Steller, the pioneer of Alaskan natural history*. Harvard University Press, Cambridge, MA.

Steller, G. W. 1751. De Bestiis Marinis. *Novi Commentarii Academiae Scientiarum Imperialis Petropolitannae* 2:289–398.

———. 1899. The beasts of the sea. Pp. 179–201, in *The fur seals and fur-seal islands of the North Pacific Ocean* (D. S. Jordan, ed.). Government Printing Office. (Abridged translation of Steller, 1751, by W. and J. E. Miller.)

Stephenson, R. O., W. B. Ballard, C. A. Smith, and K. Richardson. 1995. Wolf biology and management in Alaska 1981–92. Pp. 43–54, in *Ecology and conservation of wolves in a changing world* (L. N. Carbyn, S. H. Fritts, and D. R., eds.). Canadian Circumpolar Institute, Occasional Papers 35, Edmonton, Alberta.

Stephenson, R. O., S. C. Gerlach, R. D. Guthrie, C. R. Harington, R. O. Mills, and G. Hare. 2001. Wood bison in late Holocene Alaska and adjacent Canada: Paleontological, archaeological and historical records. Pp. 125–159, in *People and Wildlife in Northern North America: Essays in Honor of R. Dale Guthrie* (S. C. Gerlach and M. S. Murray, eds.). British Archaeological Reports, International Series S944.

Steppan, S. J., M. R. Akhverdyan, E. A. Lyapunova, D. G. Fraser, N. N. Vorontsov, R. S. Hoffmann, and M. J. Braun. 1999. Molecular phylogeny of the marmots (Rodentia: Sciuridae): Tests of evolutionary and biogeographic hypotheses. *Systematic Biology* 48:715–734.

Stern, R. O., E. L. Arobio, L. L. Naylor, and W. C. Thomas. 1980. *Eskimos, reindeer and land*. Bulletin 59, Agricultural Experiment Station, University of Alaska Fairbanks.

Stevens, T. A., D. A. Duffield, E. D. Asper, K. G. Hewlett, A. Bolz, L. J. Gage, and D. Bossart. 1989. Preliminary findings of restriction fragment differences in mitochondrial DNA among killer whales (*Orcinus orca*). *Canadian Journal of Zoology* 67:2592–2595.

Stewart, B. E., and R. E. A. Stewart. 1989. *Delphinapterus leucas*. *Mammalian Species* 336:1–8.

Stewart, B. S. 1999. Minke whale, *Balaenoptera acutorostrata*. Pp. 246–247, in *The Smithsonian book of North American mammals* (D. E. Wilson and S. Ruff, eds.). Smithsonian Institution Press, Washington, D.C., in association with the American Society of Mammalogists.

Stewart, B. S., and H. R. Huber. 1993. *Mirounga angustirostris. Mammalian Species* 449:1–10.

Stewart, B. S., and S. Leatherwood. 1985. Minke whale. Pp. 91–136, in *Handbook of marine mammals. Volume 3: The sirenians and baleen whales* (S. H. Ridgway and R. Harrison, eds.). Academic Press, London.

Stewart, R. E. A. 1999. Beluga, *Delphinapterus leucas*. Pp. 291–292, in *The Smithsonian book of North American mammals* (D. E. Wilson and S. Ruff, eds.). Smithsonian Institution Press, Washington, D.C., in association with the American Society of Mammalogists.

Stone, K. D. 2000. Molecular evolution of martens (genus *Martes*). Unpublished dissertation. University of Alaska Fairbanks.

Stone, K. D., and J. A. Cook. 2000. Phylogeography of black bears (*Ursus americanus*) of the Pacific Northwest. *Canadian Journal of Zoology* 78:1218–1223.

———. 2001. Molecular evolution of the Holarctic genus *Martes. Molecular Phylogenetics and Evolution* 24:169–179.

Stone, K. D., R. Flynn, and J. Cook. 2002. Post-glacial colonization of northwestern North America by the forest associated American marten (*Martes americana*). *Molecular Ecology* 11:2049–2063.

Stone, W. S. 1900. Report on the birds and mammals obtained by the McIlhenny expedition to Pt. Barrow, Alaska. Pp. 4–49. in Proceedings of the Academy of Natural Sciences, Philadelphia.

Storer, J. E. 2003a. The Eastern Beringian vole *Microtus deceitensis* (Rodentia, Muridae, Arvicolinae) in Late Pliocene and Early Pleistocene faunas of Alaska and Yukon. *Quaternary Research* 60:84–93.

———. 2003b. Advances in the Ice Age biostratigraphy of Eastern Beringia. Abstract, Third International Mammoth Conference, 24–29 May 2003, Dawson City, Yukon Territory.

———. 2004a. A Middle Pleistocene (late Irvingtonian) mammalian fauna from Thistle Creek, Klondike Goldfields Region of Yukon Territory, Canada. *Paludicola* 4:137–150.

———. 2004b. A new species of *Mustela* (Mammalia; Carnivora; Mustelidae) from the Fort Selkirk fauna (early Pleistocene) of Yukon Territory, Canada. *Paludicola* 4:151–155.

Streator, C. P. 1895. Smithsonian Institution Archives, Record Unit 7176, Box 20, Folder 9. U.S. Fish and Wildlife Service, 1860–1961.

Strickland, M. A., C. W. Douglas, M. Novak, and N. P. Hunziger. 1982. Marten (*Martes americana*). Pp. 599–612, in *Wild mammals of North America* (J. A. Chapman and G. A. Feildhamer, eds.). The Johns Hopkins University Press, Baltimore.

Strobeck, C., R. O. Polziehn, and R. Beech. 1993. Genetic relationship between wood and plains bison assayed using mitochondrial DNA sequence. Pp. 209–221, in *Proceedings, North American Public Bison Herds Symposium*, July 27–29 (R. E. Walker, ed.). Custer State Park, Lacrosse, WI.

Suarez, A. V., and N. D. Tsutsui. 2004. The value of museum collections to research and society. *Bioscience* 54:66–74.

Suring, L. H., D.C. Crocker-Bedford, R. W. Flynn, C. L. Hale, G. C. Iverson, M. D. Kirchhoff, T. E. Schenck II, L. C. Sea, and K. Titus. 1992. A strategy for maintaining well-distributed, viable populations of wildlife associated with old growth forest in Southeast Alaska. Report of an Interagency Committee, Juneau.

Sushkin, P. P. 1925. The zoogeographical regions of Central Siberia and the adjacent parts of mountainous Asia and the historical experience of the contemporary fauna of Palaearctic Asia. *Byulleten' Moskovskogo Obshchestva Ispytatelei Prirody Otdel Biologicheskii*, Vol. 34 (in Russian).

Suydam, R. S., and J. C. George. 1992. Recent sightings of harbour porpoises, *Phocoena phocoena*, near Point Barrow, Alaska. *Canadian Field-Naturalist* 106:489–492.

———. 2004. Subsistence harvest of bowhead whales (*Balaena mysticetus*) by Alaskan Eskimos, 1974 to 2003. Unpublished report submitted to International Whaling Commission (SC/56/BRG12).

Swanson, S. A. 1996. Small mammal populations in post-fire black spruce (*Picea mariana*) seral communities in the upper Kobuk River valley, Alaska. National Park Service, Technical Report NPS/AFARBR/NRTR-96/30.

Swarth, H. S. 1911a. Birds and mammals of the 1909 Alexander Expedition. *University of California Publications in Zoology* 7:9–172.

———. 1911b. Two new species of marmots from northwestern America. *University of California Publications in Zoology* 7:201–204.

———. 1922. Birds and mammals of the Stikine River region of northern British Columbia and Southeast Alaska. *University of California Publications in Zoology* 24:125–314.

———. 1931. The lemming of Nunivak Island, Alaska. *Proceedings of the Biological Society of Washington* 44:101–104.

———. 1933. The long-tailed meadow-mouse of southeastern Alaska. *Proceedings of the Biological Society of Washington* 46:207–212.

———. 1936. Origins of the fauna of the Sitkan District, Alaska. *Proceedings of the California Academy Sciences* 223:59–78.

Szalay, F. S. 1985. Rodent and Lagomorph morphotype adaptations, origins, and relationships: Some postcranial attributes analyzed. Pp. 83–132, in *Evolutionary relationships among rodents: A multidisciplinary analysis* (W. P. Luckett and J. L. Hartenberger, eds.). Plenum, New York.

Taber, S. 1943. Perennially frozen ground in Alaska: Its origin and history. *Geological Society of America Bulletin* 54:1433–1548

Talbot, S., and G. F. Shields. 1996. A phylogeny of the bears (Ursidae) inferred from complete sequences of three mitochondrial genes. *Molecular Phylogenetics and Evolution* 5:567–575.

Tavaré, S., C. R. Marshall, O. Will, C. Soligo, and R. D. Martin. 2002. Using the fossil record to estimate the age of the last common ancestor of extant primates. *Nature* 416:726–729.

Taylor, R. H., and J. E. Brooks. 1995. A survey of Shemya rodents. Project Report. U. S. Fish and Wildlife Service Legacy Project 1244. Alaska Maritime National Wildlife Refuge, Homer.

Tener, J. S. 1965. Muskoxen in Canada: A biological and taxonomic review. *Canadian Wildlife Service Monographs* 2:1–166.

Thomas, M. G., E. Hagelberg, H. B. Jones, Z. Yang, and A. M. Lister. 2000. Molecular and morphological evidence on the phylogeny of the Elephantidae. *Proceedings of the Royal Society London* B. 267:2493–2500.

Thorington, R. W., and R. S. Hoffmann. 2005. Family Sciuridae. Pp. 754–818, in *Mammal species of the world: A taxonomic and geographic reference* (D. E. Wilson and D. M. Reeder, eds.). Third Edition. Johns Hopkins University Press, Baltimore.

Thorington, R. W., Jr., A. L. Musante, C. G. Anderson, and K. Darrow. 1996. Validity of three genera of flying squirrels: *Eoglaucomys, Glaucomys,* and *Hyloptetes. Journal of Mammalogy* 77:69–83.

Tobey, R. W. 2004a. Unit 11 Copper River Herd bison management report. Pp. 1–8, in *Bison management report of survey and inventory activities* 1 July 2001–30 June 2003 (C. Brown, ed.). Alaska Department of Fish and Game. Project 9.0. Juneau, Alaska.

———. 2004b. Unit 11 Chitina River Herd bison management report. Pp. 9–15, in *Bison management report of survey and inventory activities* 1 July 2001–30 June 2003 (C. Brown, ed.). Alaska Department of Fish and Game. Project 9.0. Juneau, Alaska.

Tomasik, E. 2003. A mitochondrial view of geographic variation in wolverine (*Gulo gulo*) of northwestern North America. Unpublished thesis. Idaho State University, Pocatello.

Tomasik, E., and J. A. Cook. 2005. Mitochondrial phylogeography and conservation genetics of wolverine (*Gulo gulo*) in Northwestern North America. *Journal of Mammalogy* 86:386–396.

Tomilin, A. G. 1946. Thermoregulation and the geographical races of cetaceans. (Termoregulyatsiya i geograficheskie rasy kitoobraznykh.) *Doklady Akademii Nauk CCCP* 54:465–472. (English and Russian.)

———. 1957. Mammals of the USSR and adjacent countries, Volume IX: Cetacea. Edited by V. G. Heptner. Nauk SSSR, Moscow. (English translation 1967 by Israel Program for Scientific Translations, Jerusalem. Published by U. S. Department of Commerce, Springfield, VA.)

Troyer, W. A., and R. J. Hensel. 1969. The brown bear of Kodiak island. Kodiak National Wildlife Refuge Report. U. S. Fish and Wildlife Service, Kodiak, Alaska.

True, F. W. 1886. An annotated list of the mammals collected by the late Charles L. McKay in the vicinity of Bristol Bay, Alaska. *Proceedings of the U. S. National Museum* 9:221–224.

—————. 1899. Mammals of the Pribilof Islands. Part 3. Pp. 345–354, in *Fur seals and fur-seal Islands of North Pacific Ocean* (D. S. Jordan et al., eds.). U. S. Treasury Department, Doc. 2017.

—————. 1910. An account of the beaked whales of the family Ziphiidae in the collection of the United States National Museum, with remarks on some specimens in other American museums. Smithsonian Institution, United States National Museum Bulletin 73.

Trujillo, R. G., T. R. Loughlin, N. J. Gemmell, J. C. Patton, and J. W. Bickham. 2004. Variation in microsatellites and mtDNA across the range of the Steller sea lion, *Eumetopias jubatus*. *Journal of Mammalogy* 85:12–20.

Tumlison, R. 1987. *Felis lynx. Mammalian Species* 269:1–8.

Turner, L. C. 1886. Contributions to the natural history of Alaska. U. S. Army Signal Service. Arctic Series Publication Number 2.

Tynan, C. 1999. Redistribution of cetaceans in the southeast Bering Sea relative to anomalous oceanographic conditions during the 1997 El Niño. Pp. 115–117, in *Proceedings of the 1998 science board symposium on the impacts of the 1997/98 El Niño event on the North Pacific Ocean and its marginal seas.* (H. J. Freeland, W. T. Peterson, and A. Tyler, eds.). PICES Scientific Report No. 10, North Pacific Marine Science Organization (PICES), Sydney, British Columbia.

Tynan, C. T., and D. P. DeMaster. 1997. Observations and predictions of Arctic climate change: Potential effects on marine mammals. *Arctic* 50:308–322.

Tynan, C. T., D. P. DeMaster, and W. T. Peterson. 2001. Endangered right whales on the southeastern Bering Sea shelf. *Science* 294:1894.

USDA (U. S. Department of Agriculture), U. S. Forest Service Alaska Region, and Alaska Department of Natural Resources–Division of Forestry. 2002. Forest insect and disease conditions in Alaska, 2002. General Technical Report R10-TP-113.

USFS (U. S. Forest Service). 1997. Tongass land management plan revision: Final environmental impact assessment. R10-MB-338b. USDA Forest Service Alaska Region, Juneau.

USFWS (U. S. Fish and Wildlife Service). 1994. Conservation plan for the sea otter in Alaska. Marine Mammals Management, U. S. Fish and Wildlife Service, Anchorage, Alaska.

—————. 2002a. Stock assessment for sea otters (*Enhydra lutris*): Southcentral Alaska stock. Marine Mammal Protection Act Stock Assessment Report. http://alaska.fws.gov/fisheries/mmm/seaotters/reports.htm (accessed 3 April 2007).

—————. 2002b. Stock assessment for sea otters (*Enhydra lutris*): Southwest Alaska stock. Marine Mammal Protection Act Stock Assessment Report. http://alaska.fws.gov/fisheries/mmm/seaotters/reports.htm (accessed 3 April 2007).

—————. 2002c. Stock assessment for sea otters (*Enhydra lutris*): Southeast Alaska stock. Marine Mammal Protection Act Stock Assessment Report. http://alaska.fws.gov/fisheries/mmm/seaotters/reports.htm (accessed 3 April 2007).

—————. 2005. Yukon Delta National Wildlife Refuge: Mammal species list. http://alaska.fws.gov/internettv/nwrtv/yukondeltatv/mammals.htm (accessed 15 March 2007).

Uspenski, S. M. 1984. Muskoxen in the USSR, some results of and perspectives on the introduction. Biological Paper of the University of Alaska, Special Report 4:12–14.

Valdez, R., and P. R. Krausman. 1999. Description, distribution, and abundance of mountain sheep in North America. Pp. 3–22, in *Mountain sheep in North America* (R. Valdez and P. R. Krausman, eds.). University of Arizona Press, Tucson.

Valkenburg, P. 1999. Caribou. Alaska Department of Fish and Game Wildlife Notebook Series. http://www.adfg.state.ak.us/pubs/notebook/biggame/caribou.php (accessed 26 March 2007).

—————. 2007. North Slope muskox mystery. *Outdoor Alaska* 15(10):7&10.

van Bree, P. J. H. 1971. On *Globicephala sieboldii* Gray, 1846, and other species of pilot whales. *Beaufortia* 19(249):79–87.

van Daele, L. J. 2003. The history of bears on the Kodiak Archipelago. Alaska Natural History Association, Anchorage.

—————. 2004. Unit 8 furbearer management report. Pp. 107–116, in *Furbearer management report of survey and inventory activities* 1 July 2000–30 June 2003 (C. Brown, ed.). Alaska Department of Fish and Game. Project 7.0. Juneau, Alaska.

————. 2005. Unit 1C deer management report. Pp. 109–125, in *Deer management report of survey and inventory activities* 1 July 2002–30 June 2004 (C. Brown, ed.). Alaska Department of Fish and Game. Juneau, Alaska.

van Daele, L. J., and J. R. Crye. 2004a. Unit 8 elk management report. Pp. 9–22, in *Elk management report of survey and inventory activities* 1 July 2001–30 June 2003 (C. Brown, ed.). Alaska Department of Fish and Game. Project 13.0. Juneau, Alaska.

————. 2004b. Unit 8 mountain goat management report. Pp. 122–137, in *Mountain goat management report of survey and inventory activities* 1 July 2001–30 June 2003 (C. Brown, ed.). Alaska Department of Fish and Game. Project 12.0. Juneau, Alaska.

van Gelder, R. G. 1978. A review of canid classification. *American Museum Novitates* 2646:1–10.

van Waerebeek, K., and B. Würsig. 2002. Pacific white-sided dolphin and dusky dolphin, *Lagenorhynchus obliquidens* and *L. obscurus.* Pp. 859–861, in *Encyclopedia of marine mammals* (W. F. Perrin, B. Würsig, and J. G. M. Thewissen, eds.). Academic Press, San Diego and London.

van Zyll de Jong, C. G. 1972. A systematic review of the Nearctic and Neotropical river otters (genus *Lutra,* Mustelidae, Carnivora). *Royal Ontario Museum, Life Sciences Contribution* 80:1–104.

————. 1975. Differentiation of the Canada lynx, *Felis* (*Lynx*) *canadensis subsolana,* in Newfoundland. *Canadian Journal of Zoology* 53:699–705.

————. 1976. A comparison between woodland and tundra forms of the common shrew (*Sorex cinereus*). *Canadian Journal of Zoology* 54:963–973.

————. 1979. Distribution and systematic relationships of long-eared *Myotis* in western Canada. *Canadian Journal of Zoology* 57:987–994.

————. 1982. Relationships of amphiberingian shrews of the *Sorex cinereus* group. *Canadian Journal of Zoology* 60:1580–1587.

————. 1983. Handbook of Canadian mammals. Part 1. Marsupials and insectivores. National Museum of Natural History.

————. 1985. Handbook of Canadian mammals. Part 2. Bats. National Museum of Natural History.

————. 1986. A systematic study of Recent bison, with particular consideration of the wood bison (*Bison bison athabascae* Rhoads, 1898). Publications in Natural Sciences 6:69. National Museums of Canada, Ottawa.

————. 1987. A phylogenetic study of the Lutrinae (Carnivora; Mustelidae) using morphological data. *Canadian Journal of Zoology* 65:2536–2544.

————. 1991. Speciation of the *Sorex cinereus* group. Pp. 65–73, in *The biology of the Soricidae* (J. S. Findley and T. L. Yates, eds.). Special Publications, Museum of Southwestern Biology 1:1–91.

————. 1999a. St. Lawrence Island shrew, *Sorex jacksoni.* P. 28, in *The Smithsonian book of North American mammals* (D. E. Wilson and S. Ruff, eds.). Smithsonian Institution Press, Washington, D.C., in association with the American Society of Mammalogists.

————. 1999b. Tundra shrew, *Sorex tundrensis.* Pp. 44–45, in *The Smithsonian book of North American mammals* (D. E. Wilson and S. Ruff, eds.). Smithsonian Institution Press, Washington, D.C., in association with the American Society of Mammalogists.

————. 1999c. Barren ground shrew, *Sorex ugyunak.* P. 45, in *The Smithsonian book of North American mammals* (D. E. Wilson and S. Ruff, eds.). Smithsonian Institution Press, Washington, D.C., in association with the American Society of Mammalogists.

van Zyll de Jong, C. G., and D. W. Nagorsen. 1994. A review of the distribution and taxonomy of *Myotis keenii* and *Myotis evotis* in British Columbia and the adjacent United States. *Canadian Journal of Zoology* 72:1069–1078.

Vartanyan, S. L., V. E. Garrut, and A. V. Sher. 1993. Holocene dwarf mammoths from Wrangel Island in the Siberian Arctic. *Nature* 382:337–340.

Veltre, D. W., D. R. Yesner, K. J. Crossen, R. W. Graham, and J. B. Coltrain. 2008. Patterns of faunal extinction and paleoclimatic change from mid-Holocene mammoth and polar bear remains, Pribilof Islands, Alaska. *Quaternary Research* 70:40–50.

Vibe, C. 1967. Arctic animals in relation to climatic fluctuations. *Meddelelser om Grønland* 170:1–227.

Vorontsov, N. N., and E. Yu. Ivanitskaya. 1973. Comparative karyology of north Palearctic pikas (Ochotona, Lagomorpha, Ochotonidae). *Zoologichesky Zhurnal* 52:584–588.

Vorontsov, N. N., and E. A. Lyapunova. 1986. Genetics and problems of Trans-Beringian connections of Holarctic mammals. Pp. 441–481, in *Beringia in the Cenozoic Era* (V. L. Kontrimavichus. ed.) (translation of Beringiya v Kainozoe, 1976). A. A. Balkema, Rotterdam.

Wada, S. 1973. The ninth memorandum on the stock assessment of whales in the North Pacific. *Report of the International Whaling Commission* 23:164–169.

————. 1988. Genetic differentiation between two forms of short-finned pilot whales off the Pacific coast of Japan. *Scientific Reports of the Whales Research Institute* 39:91–101.

Wada, S., and K. Numachi. 1991. Allozyme analyses of genetic differentiation among the populations and species of the Balaenoptera. *Reports of the International Whaling Commission Special Issue* 13:125–154.

Wahlert, J. H. 1977. Cranial foramina and relationships of *Eutypomys* (Rodentia: Eutypomyidae). *American Museum Novitates*. 2626:1–8.

Waite, J. M., K. Wynne, and D. K. Mellinger. 2003. Documented sighting of a North Pacific right whale in the Gulf of Alaska and post-sighting acoustic monitoring. *Northwestern Naturalist* 84:38–43.

Waits, L. P., S. L. Talbot, R. H. Ward, and G. F. Shields. 1998. Mitochondrial DNA phylogeography of the North American brown bear and implications for conservation. *Conservation Biology* 12:408–417.

Wake, M. H. 1988. The science educator looks at museums: Perspective of a zoologist. Pp. 305–314, in *Science learning in the informal setting* (P. G. Heltne and L. Marquardt, eds.). Chicago Academy of Sciences, Chicago.

Walker, W. A., S. Leatherwood, K. R. Goodrich, W. F. Perrin, and R. K. Stroud. 1986. Geographical variation and biology of the Pacific white-sided dolphin, *Lagenorhynchus obliquidens,* in the northeastern Pacific. Pp. 441–465, in *Research on dolphins* (M. M. Bryden and R. Harrison, eds.). Clarendon Press, Oxford.

Wall, D. A., S. K. Davis, and B. M. Read. 1992. Phylogenetic relationships in the subfamily Bovinae (Mammalia: Artiodactyla) based on ribosomal DNA. *Journal of Mammalogy* 73:262–275.

Wallmo, O. C. 1981. Mule and black-tailed deer distribution and habitats. Pp. 1–25, in *Mule and black-tailed deer of North America* (O. C. Wallmo, ed.). University of Nebraska Press, Lincoln.

Waltari, E. 2005. Comparative phylogeography of two high latitude mammals. Unpublished dissertation, Idaho State University, Pocatello.

Waltari, E., and J. A. Cook. 2005. Hares on ice: Phylogeography and historical demographics of *Lepus arcticus, L. othus,* and *L. timidus* (Mammalia: Lagomorpha). *Molecular Ecology* 14:3005–3016.

Waltari, E., E. P. Hoberg, E. P. Lessa, and J. A. Cook. 2007. Eastward ho: Phylogeographical perspectives on colonization of hosts and parasites across the Beringian nexus. *Journal of Biogeography* 34:561–574.

Wanless, R. M., A. Angel, R. J. Cuthbert, G. M. Hilton, and P. G. Ryan. 2007. Can predation by invasive mice drive seabird extinctions? *Biology Letters* 3:241–244.

Warner, R. M., and N. J. Czaplewski. 1984. *Myotis volans. Mammalian Species* 224:1–4.

Wayne, R. K., N. Lehman, and T. K. Fuller. 1995. Conservation genetics of the gray wolf. Pp. 399–407, in *Ecology and conservation of wolves in a changing world* (L. N. Carbyn, S. H. Fritts, and D. R. Seip, ed.). Canadian Circumpolar Institute, University of Alberta, Edmonton.

Weber, F. R., T. D. Hamilton, D. M. Hopkins, C. A. Repenning, and H. Haas. 1981. Canyon Creek: A late Pleistocene vertebrate locality in interior Alaska. *Quaternary Research* 16:167–180.

Weckworth, B. V. 2003. Phylogeography and population dynamics of wolves of the Pacific Northwest. Unpublished thesis. Idaho State University, Pocatello.

Weckworth B. V., S. Talbot, G. K. Sage, D. K. Person, and J. Cook. 2005. A signal for independent coastal and continental histories for North American wolves. *Molecular Ecology* 14:917–931.

Weigel, I. 1961. Das Fellmuster der wildlebenden Katzenarten und der Hauskatze in Vergleichender und Stammesgeschichter Hinsicht. *Säugetierkundliche Mitteilungen* 9:1–120.

Weigl, P. D., and D. W. Osgood. 1974. Study of the northern flying squirrel, *Glaucomys sabrinus* by temperature telemetry. *American Midland Naturalist* 92:482–486.

Weinstock, J., E. Willerslev, A. Sher, W. Tong, S. Y. W. Ho, D. Rubenstein, J. Storer, J. Burns, L. Martin, C. Bravi, A. Prieto, D. Froese, E. Scott, L. Xulong, and A. Cooper. 2005. Evolution, systematics, and phylogeography of Pleistocene horses in the new world: A molecular perspective. PloS Biology 3(8): e241.

Wells-Gosling, N., and L. R. Heaney. 1984. *Glaucomys sabrinus. Mammalian Species* 229:1–8.

Wenrich, W. 1922. Smithsonian Institution Archives, Record Unit 7176, Box 20, Folder 24. U. S. Fish and Wildlife Service, 1860–1961.

Werdelin, L. 1981. The evolution of lynxes. *Annales Zoologici Fennici* 18:37–71.

West, E. W. 1991. Status reports on selected Alaskan mammals of ecological concern. Unpublished report. Alaska Natural Heritage Program, Anchorage.

West, F. H. (Ed.). 1996. *American beginnings: The prehistory and palaeoecology of Beringia.* University of Chicago Press, Chicago.

West, S. D. 1974. Post-burn population response of the northern red-backed vole, *Clethrionomys rutilus,* in Interior Alaska. Unpublished thesis. University of Alaska Fairbanks.

———. 1979. *Habitat responses of microtine rodents to central Alaskan forest succession.* Unpublished dissertation. University of California, Berkeley.

Westlake, R. L., and G. M. O'Corry-Crowe. 2002. Macrogeographic structure and patterns of genetic diversity in harbor seals (*Phoca vitulina*) from Alaska to Japan. *Journal of Mammalogy* 83:1111–1126.

Weston, M. L. 1981. The *Ochotona alpina* complex: A statistical re-evaluation. Pp. 73–89, in *Proceedings of the World Lagomorph Conference held in Guelph, Ontario, August 1979* (K. Myers and C. D. MacInnes, eds.). University of Guelph, Ontario.

Whitaker, J. O., Jr. 1972. *Zapus hudsonius. Mammalian Species* 11:1–7.

Whitaker, J. O., Jr., and B. E. Lawhead. 1992. Foods of *Myotis lucifugus* in a maternity colony in central Alaska. *Journal of Mammalogy* 73:646–648.

Whitehead, H. 2002. Sperm whale, *Physeter macrocephalus.* Pp. 1165–1172, in *Encyclopedia of marine mammals* (W. F. Perrin, B. Würsig, and J. G. M. Thewissen, eds.). Academic Press, San Diego and London.

Whitman, J. S., W. B. Ballard, and C. L. Gardner. 1986. Home range and habitat use by wolverines in southcentral Alaska. *The Journal of Wildlife Management* 50:460–463.

Whitmore, F. C., Jr., and L. M. Gard, Jr. 1977. Steller's sea cow (*Hydrodamalis gigas*) of Late Pleistocene age from Amchitka, Aleutian Islands, Alaska. *U. S. Geological Survey Professional Paper* 1036:1–19.

Whitten, K. 1994. Dall sheep. Alaska Department of Fish and Game Wildlife Notebook Series. http://www.adfg.state.ak.us/pubs/notebook/biggame/dallshee.php (accessed 26 March 2007).

Wickström, L. M., J. Hantula, V. Haukisalmi, and H. Henttonen. 2001. Genetic and morphometric variation in the circumpolar helminth parasite *Andrya arctica* (Cestoda, Anoplocephalidae) in relation to the divergence of its lemming hosts (*Dicrostonyx* spp.). *Zoological Journal of the Linnaean Society* 131:443–457.

Wickström, L. M., V. Haukisalmi, S. Varis, J. Hantula, V. B. Fedorov, and H. Henttonen. 2003. Phylogeography of the circumpolar *Paranoplocephala arctica* species complex (Cestoda: Anoplocephalidae) parasitizing collared lemmings (*Dicrostonyx* spp.). *Molecular Ecology* 12:3359–3371.

Wickström, L. M., V. Haukisalmi, S. Varis, J. Hantula, and H. Henttonen. 2005. Molecular phylogeny and systematics of anoplocephaline cestodes in rodents and lagomorphs. *Systematic Parasitology* 62:83–99.

Wigen, R. J. 2005. History of the vertebrate fauna in Haida Gwaii. Pp. 96–115, in *Haida Gwaii: Human history and environment from the time of loon to the time of the iron people* (D. W. Fedje and R. W. Mathewes, eds.). University of British Columbia Press, Vancouver.

Wike, M. J. 1998. Mitochondrial-DNA variation among populations of *Peromyscus* from Yukon, Canada and southeastern Alaska. Unpublished thesis. Texas A&M University, College Station.

Williams, J. C., G. V. Byrd, and N. B. Konyukhov. 2003. Whiskered auklets, foxes, humans and how to right a wrong. Marine *Ornithology* 31:175–180.

Williams, J. C., and V. Tutiakoff, Jr. 2005. Aerial survey of barren-ground caribou at Adak Island, Alaska in 2005. U. S. Fish and Wildlife Service Report AMNWR 05/14. Homer.

Willner, G. R., G. A. Feldhamer, E. E. Zucker, and J. A. Chapman. 1980. *Ondatra zibethicus*. *Mammalian Species* 141:1–8.

Wilson, D. E. 1976. Cranial variation in polar bears. Pp. 447–453, in *Bears, their biology and management* (M.R. Pelton, J.W. Lentfer, and G.E. Folk, eds.). International Union for the Conservation of Nature (IUCN), New Series, 40:1–467.

Wilson, D. E., M. A. Bogan, R. L. Brownell, Jr., A. M. Burdin, and M. K. Maminov. 1991. Geographic variation in sea otters, *Enhydra lutris*. *Journal of Mammalogy* 72:22–36.

Wilson, D. E., F. R. Cole, J. D. Nichols, R. Rudran, and M. S. Foster (Eds.). 1996. *Measuring and monitoring biological diversity: Standard methods for mammals*. Smithsonian Institution Press, Washington, D.C. and London.

Wilson, D. E., and D. M. Reeder (Eds.). 2005. *Mammal species of the world: A taxonomic and geographic reference*. Third Edition. Johns Hopkins University Press, Baltimore.

Wilson, D. E., and S. Ruff (Eds.). 1999. *The Smithsonian book of North American mammals*. Smithsonian Institution Press, Washington, D.C., in association with the American Society of Mammalogists.

Wilson, G. A., and C. Strobeck. 1999. Genetic variation within and relatedness among wood and plains bison populations. *Genome* 42:483–496.

Winn, H. E., and N. E. Reichley. 1985. Humpback whale. Pp. 241–273, in *Handbook of marine mammals. Volume 3: The sirenians and baleen whales* (S. H. Ridgway and R. Harrison, eds.). Academic Press, London.

Wisely, S. A., S. W. Buskirk, G. A. Russell, K. B. Aubry, and W. J. Zielinski. 2004. Genetic diversity and structure of the fisher (*Martes pennanti*) in a peninsular and peripheral metapopulation. *Journal of Mammalogy* 85:640–648.

Wolff, J. O. 1980. The role of habitat patchiness in the population dynamics of snowshoe hares. *Ecological Monographs* 50:111–130.

————. 1999. Taiga vole, *Microtus xanthognathus*. Pp. 647–649, in *The Smithsonian book of North American mammals* (D. E. Wilson and S. Ruff, eds.). Smithsonian Institution Press, Washington, D.C., in association with the American Society of Mammalogists.

Wolff, J. O., and W. Z. Lidicker, Jr. 1980. Population ecology of the taiga vole, *Microtus xanthognathus*, in interior Alaska. *Canadian Journal of Zoology* 58:1800–1812.

Wolff, J. O., and J. C. Zasada. 1975. Red squirrel response to clearcut and shelterwood systems in Interior Alaska. U.S. Forest Service Research Note, PNW-255.

Wolman, A. A. 1985. Gray whale. Pp. 67–90, in *Handbook of marine mammals. Volume 3: The sirenians and baleen whales* (S. H. Ridgway and R. Harrison, eds.). Academic Press, London.

Wood, A. E. 1959. Eocene radiation and phylogeny of the rodents. *Evolution* 13:354–361.

Woodford, R. 2004. Mountain lions in Alaska. *Alaska Wildlife News*, February issue. http://www.wildlifenews.alaska.gov (accessed 28 March 2007).

————. 2005a. Red squirrels enliven the winter forests. *Alaska Wildlife News*, February issue. http://www.wildlifenews.alaska.gov (accessed 25 May 2007).

————. 2005b. Combating invasive rats in Alaska. *Alaska Wildlife News*, May issue. http://www.wildlifenews.alaska.gov (accessed 28 March 2007).

————. 2006. Wandering weasels, are fishers moving into Alaska? *Alaska Wildlife News*, May issue. http://www.wildlifenews.alaska.gov (accessed 28 March 2007).

Wooding, S., and R. Ward. 1997. Phylogeography and Pleistocene evolution in the North American black bear. *Molecular Biology and Evolution* 14:1096–1105.

Woods, C. A. 1973. *Erethizon dorsatum*. *Mammalian Species* 29:1–6.

————. 1984. Hystricognath rodents. Pp. 389–446, in *Orders and families of Recent mammals of the world* (S. Anderson and J. K. Jones, Jr., eds.). John Wiley and Sons, New York.

Woods, C. A., and C. W. Kilpatrick. 2005. Infraorder Hystricognathi. Pp. 1538–1600, in *Mammal species of the world: A taxonomic and geographic reference* (D. E. Wilson and D. M. Reeder, eds.). Third Edition. Johns Hopkins University Press, Baltimore.

Woolington, J. D. 2004a. Unit 17 furbearer management report. Pp. 213–228, in *Furbearer management report of survey and inventory activities* 1 July 2000–30 June 2003 (C. Brown, ed.). Alaska Department of Fish and Game. Project 7.0. Juneau, Alaska.

———. 2004b. Unit 17 moose management report. Pp. 246–266, in *Moose management report of survey and inventory activities* 1 July 2001–30 June 2003 (C. Brown, ed.). Alaska Department of Fish and Game. Project 1.0. Juneau, Alaska.

Worley, K., C. Strobeck, S. Arthur, J. Carey, H. Schwantje, A. Veitch, and D. W. Coltman. 2004. Population genetic structure of North American thinhorn sheep *Ovis dalli. Molecular Ecology* 13:2545–2556.

Wozencraft, W. C. 1989. Classification of the Recent Carnivora. Pp. 569–593, in *Carnivore behavior, ecology and evolution* (J. L. Gittleman, ed.). Cornell University Press, Ithaca, NY.

———. 1993. Order Carnivora. Pp. 279–348, in *Mammal species of the world: A taxonomic and geographic reference* (D. E. Wilson and D. M. Reeder, eds.). Second Edition. Smithsonian Institution Press, Washington, D.C.

———. 2005. Order Carnivora. Pp. 532–628, in *Mammal species of the world: A taxonomic and geographic reference* (D. E. Wilson and D. M. Reeder, eds.). Third Edition. Johns Hopkins University Press, Baltimore.

Wright, P. L. 1953. Intergradation between *Martes americana* and *Martes caurina* in western Montana. *Journal of Mammalogy* 34:70–87.

Wynne, K. 1993. *Guide to marine mammals of Alaska.* Alaska Sea Grant College Program, University of Alaska Fairbanks.

Wyss, A. R. 1988. On "retrogression" in the evolution of the Phocinae and phylogenetic affinities of the monk seals. *American Museum Novitates* 2924:1–38.

Yang, H., E. M. Golenberg, and J. Shoshani. 1996. Phylogenetic resolution within the Elephantidae using fossil DNA sequence from the American mastodon (*Mammut americanum*) as an outgroup. *Proceedings of the National Academy of Sciences* 93:1190–1194.

Yates, T. L. 1984. Insectivores, elephant shrews, tree shrews, and dermopterans. Pp. 117–144, in *Orders and families of Recent mammals of the world* (S. Anderson and J. K. Jones, Jr., eds.). John Wiley and Sons, New York.

Yates, T. L., J. N. Mills, C. A. Parmenter, et al. 2002. The ecology and evolutionary history of an emergent disease: Hantavirus pulmonary syndrome. *BioScience* 52(11):989–998.

Yesner, D. R. 2001. Human dispersal into interior Alaska: Antecedent conditions, mode of colonization, and adaptations. *Quaternary Science Reviews* 20:315–327.

Yesner, D.R., G.A. Pearson, and D.E. Stone. 2000. Additional organic artifacts from the Broken Mammoth site, Big Delta, Alaska. *Current Research in the Pleistocene* 17:87–89.

Yochem, P.K., and S. Leatherwood. 1985. Blue whale. Pp. 193–240, in *Handbook of marine mammals. Volume 3: The sirenians and baleen whales* (S.H. Ridgway and R. Harrison, eds.). Academic Press, London.

Yom-Tov, Y., and J. Yom-Tov. 2005. Global warming, Bergmann's rule and body size in the masked shrew *Sorex cinereus* Kerr in Alaska. *Journal of Animal Ecology* 74:803–808.

Yom-Tov, Y., S. Yom-Tov, D. MacDonald, and E. Yom-Tov. 2007. Population cycles and changes in body size of the lynx in Alaska. *Oecologia* 152:239–244.

Young, S.P., and E.A. Goldman. 1946. *The puma, mysterious American cat.* The American Wildlife Institute, Washington, D.C.

Youngman, P. M. 1975. Mammals of the Yukon Territory. National Museums of Canada, Ottawa, *Publications in Zoology* 10:1–192.

———. 1986. The extinct short-faced skunk *Brachyprotoma obtusata* (Mammalia, Carnivora): First records for Canada and Beringia. *Canadian Journal of Earth Sciences* 23:419–424.

———. 1993. The Pleistocene small carnivores of eastern Beringia. *Canadian Field-Naturalist* 107:139–163.

———. 1994. Beringian ferrets: Mummies, biogeography, and systematics. *Journal of Mammalogy* 75:454–461.

Youngman, P.M., and F.W. Schueler. 1991. *Martes nobilis* is a synonym of *Martes americana,* not an extinct Pleistocene-Holocene species. *Journal of Mammalogy* 72:567–577.

Yudin, B.S. 1969. Taxonomy of some shrews (Soricidae) from Palearctic and Nearctic. *Acta Theriologica* 14:21–34.

Zagorodnyuk, I.V. 1990. Kariotipicheskaya izmenchivost' i sistematika serykh polevok (Rodentia, Arvicolini). Soobshchenie 1. Vidovoi sostav i khromosomnye chisla [Karyotypic variability and systematics of the gray voles (Rodentia, Arvicolini). Communication 1. Species composition and chromosomal numbers]. *Vestnik Zoologii* 2:26–37. (In Russian.)

Zakrzewski, R.J. 1985. The fossil record. Pp. 1–51, in *Biology of New World* Microtus (R.H. Tamarin, ed.). Special Publication, American Society of Mammalogists 8.

Zazula, G.D., D.G. Froese, S.A. Elias, S. Kuzmina, and R.W. Mathewes. 2007. Arctic ground squirrels of the mammoth-steppe: Paleoecology of Late Pleistocene middens (~24 000–29 450 14C yr BP), Yukon Territory, Canada. *Quaternary Science Reviews* 26:979–1003.

Zimmerman, S.T. 1994a. Northern fur seal. Alaska Department of Fish and Game Wildlife Notebook Series. http://www.adfg.state.ak.us/pubs/notebook/marine/furseal.php (accessed 26 March 2007).

———. 1994b. Blue whale. Alaska Department of Fish and Game Wildlife Notebook Series. http://www.adfg.state.ak.us/pubs/notebook/marine/blue.php (accessed 26 March 2007).

Zimmermann, K. 1942. Zur Kenntnis von *Microtus oeconomus* (Pallas). *Archiv für Naturgeschichte, Neue Folge* 11:174–197.

Appendices

Appendix 1. Locations and Names of Alaska Quadrangle Topographic Maps (1:250,000 scale).

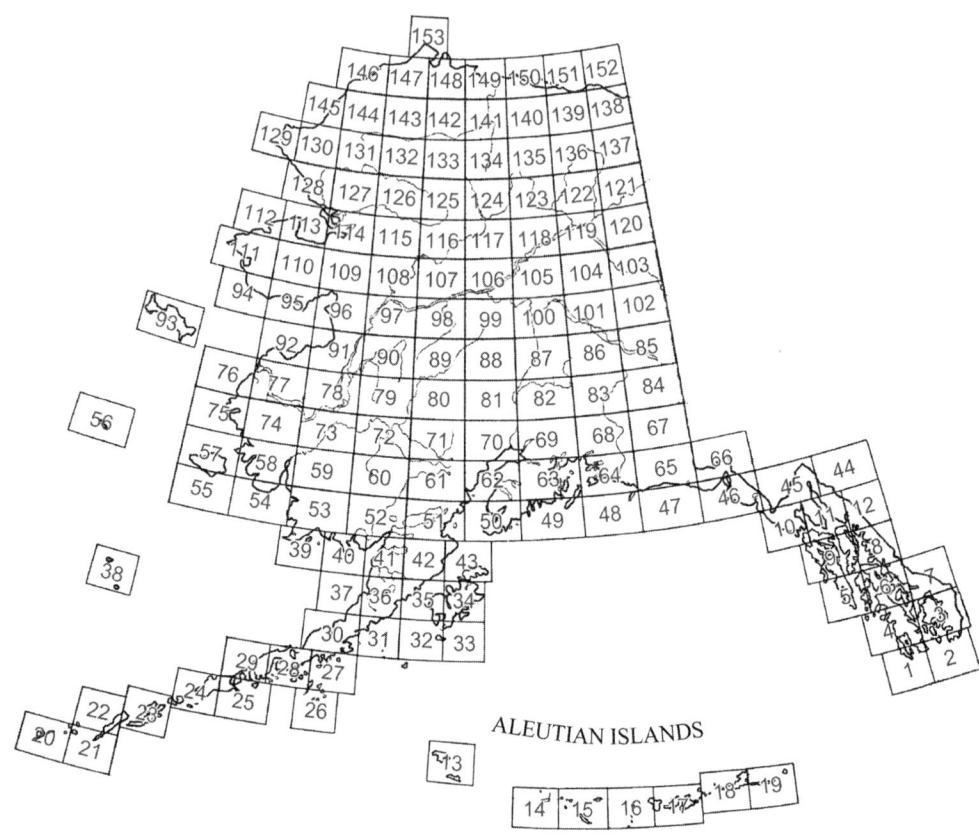

ALEUTIAN ISLANDS

Numeric

1	Dixon Entrance	24	Unimak	47	Icy Bay	70	Tyonek
2	Prince Rupert	25	False Pass	48	Middleton Island	71	Lime Hills
3	Ketchikan	26	Simeonof Island	49	Blying Sound	72	Sleetmute
4	Craig	27	Stepovak Bay	50	Seldovia	73	Russian Mission
5	Port Alexander	28	Port Moller	51	Iliamna	74	Marshall
6	Petersburg	29	Cold Bay	52	Dillingham	75	Hooper Bay
7	Bradfield Canal	30	Chignik	53	Goodnews	76	Black
8	Sumdum	31	Sutwik Island	54	Kuskokwim Bay	77	Kwiguk
9	Sitka	32	Trinity Islands	55	Cape Mendenhall	78	Holy Cross
10	Mt. Fairweather	33	Kaguyak	56	St. Matthew	79	Iditarod
11	Juneau	34	Kodiak	57	Nunivak Island	80	McGrath
12	Taku River	35	Karluk	58	Baird Inlet	81	Talkeetna
13	Attu	36	Ugashik	59	Bethel	82	Talkeetna Mts.
14	Kiska	37	Bristol Bay	60	Taylor Mts.	83	Gulkana
15	Rat Islands	38	Pribilof Islands	61	Lake Clark	84	Nabesna
16	Gareloi Islands	39	Hagemeister Island	62	Kenai	85	Tanacross
17	Adak	40	Nushugak Bay	63	Seward	86	Mt. Hayes
18	Atka	41	Naknek	64	Cordova	87	Healy
19	Seguam	42	Mt. Katmai	65	Bering Glacier	88	Mt. McKinley
20	Amukta	43	Afognak	66	Mt. St. Elias	89	Medfra
21	Samalga Island	44	Atlin	67	McCarthy	90	Ophir
22	Umnak	45	Skagway	68	Valdez	91	Unalakleet
23	Unalaska	46	Yakutat	69	Anchorage	92	St. Michael

93	St. Lawrence	109	Candle	125	Survey Pass	141	Umiat
94	Nome	110	Bendeleben	126	Ambler River	142	Ikpikpuk River
95	Solomon	111	Teller	127	Baird Mts.	143	Lookout Ridge
96	Norton Bay	112	Shishmaref	128	Noatak	144	Utukok River
97	Nulato	113	Kotzebue	129	Point Hope	145	Point Lay
98	Ruby	114	Selawik	130	De Long Mts.	146	Wainwright
99	Kantishna River	115	Shungnak	131	Misheguk Mtn.	147	Meade River
100	Fairbanks	116	Hughes	132	Howard Pass	148	Teshekpuk
101	Big Delta	117	Bettles	133	Killik River	149	Harrison Bay
102	Eagle	118	Beaver	134	Chandler Lake	150	Beechey Point
103	Charley River	119	Fort Yukon	135	Philip Smith Mts.	151	Flaxman Island
104	Circle	120	Black River	136	Arctic	152	Barter Island
105	Livengood	121	Coleen	137	Table Mtn.	153	Barrow
106	Tanana	122	Christian	138	Demarcation Point		
107	Melozitna	123	Chandalar	139	Mt. Michelson		
108	Kateel River	124	Wiseman	140	Sagavanirktok		

Alphabetic

17	Adak	52	Dillingham	67	McCarthy	112	Shishmaref
43	Afognak	1	Dixon Entrance	80	McGrath	115	Shungnak
126	Ambler River	102	Eagle	147	Meade River	26	Simeonof Island
20	Amukta	100	Fairbanks	89	Medfra	9	Sitka
69	Anchorage	25	False Pass	107	Melozitna	45	Skagway
136	Arctic	151	Flaxman Island	48	Middleton Island	72	Sleetmute
18	Atka	119	Fort Yukon	131	Misheguk Mtn.	95	Solomon
44	Atlin	16	Gareloi Islands	10	Mt. Fairweather	93	St. Lawrence
13	Attu	53	Goodnews	86	Mt. Hayes	56	St. Matthew
58	Baird Inlet	83	Gulkana	42	Mt. Katmai	92	St. Michael
127	Baird Mts.	39	Hagemeister Island	88	Mt. McKinley	27	Stepovak Bay
153	Barrow	149	Harrison Bay	139	Mt. Michelson	8	Sumdum
152	Barter Island	87	Healy	66	Mt. St. Elias	125	Survey Pass
118	Beaver	78	Holy Cross	84	Nabesna	31	Sutwik Island
150	Beechey Point	75	Hooper Bay	41	Naknek	137	Table Mtn.
110	Bendeleben	132	Howard Pass	128	Noatak	12	Taku River
65	Bering Glacier	116	Hughes	94	Nome	81	Talkeetna
59	Bethel	47	Icy Bay	96	Norton Bay	82	Talkeetna Mts.
117	Bettles	79	Iditarod	97	Nulato	85	Tanacross
101	Big Delta	142	Ikpikpuk River	57	Nunivak Island	106	Tanana
76	Black	51	Iliamna	40	Nushugak Bay	60	Taylor Mts.
120	Black River	11	Juneau	90	Ophir	111	Teller
49	Blying Sound	33	Kaguyak	6	Petersburg	148	Teshekpuk
7	Bradfield Canal	99	Kantishna River	135	Philip Smith Mts.	32	Trinity Islands
37	Bristol Bay	35	Karluk	129	Point Hope	70	Tyonek
109	Candle	108	Kateel River	145	Point Lay	36	Ugashik
55	Cape Mendenhall	62	Kenai	5	Port Alexander	141	Umiat
123	Chandalar	3	Ketchikan	28	Port Moller	22	Umnak
134	Chandler Lake	133	Killik River	38	Pribilof Islands	91	Unalakleet
103	Charley River	14	Kiska	2	Prince Rupert	23	Unalaska
30	Chignik	34	Kodiak	15	Rat Islands	24	Unimak
122	Christian	113	Kotzebue	98	Ruby	144	Utukok River
104	Circle	54	Kuskokwim Bay	73	Russian Mission	68	Valdez
29	Cold Bay	77	Kwiguk	140	Sagavanirktok	146	Wainwright
121	Coleen	61	Lake Clark	21	Samalga Island	124	Wiseman
64	Cordova	71	Lime Hills	19	Seguam	46	Yakutat
4	Craig	105	Livengood	114	Selawik		
130	De Long Mts.	143	Lookout Ridge	50	Seldovia		
138	Demarcation Point	74	Marshall	63	Seward		

Appendix 2. Regional Occurrence of Alaska Land Mammals: ● = natural occurrence;
■ = introduced

LAND MAMMAL SPECIES	ALASKA REGION						TOTAL
	SE	SC	C	SW	W	N	
RODENTIA—rodents							
Sciuridae—squirrels							
Glaucomys sabrinus, northern flying squirrel	●	●	●				3
Marmota broweri, Alaska marmot			●			●	2
Marmota caligata, hoary marmot	●	●	●	●			4
Marmota monax, woodchuck			●				1
Spermophilus parryii, arctic ground squirrel	●	●	●	●	●	●	6
Tamiasciurus hudsonicus, red squirrel	●	●	●	●	●		5
Castoridae—beavers							
Castor canadensis, American beaver	●	●	●	●	●		5
Dipodidae—jumping mice							
Zapus hudsonius, meadow jumping mouse	●	●	●	●	●		5
Zapus princeps, western jumping mouse	●						1
Cricetidae—voles, mice, rats, and lemmings							
Dicrostonyx groenlandicus, collared lemming				●	●	●	3
Lemmus trimucronatus, brown lemming	●	●	●	●	●	●	6
Microtus abbreviatus, insular vole					●		1
Microtus longicaudus, long-tailed vole	●	●	●				3
Microtus miurus, singing vole		●	●	●	●	●	5
Microtus oeconomus, root vole	●	●	●	●	●	●	6
Microtus pennsylvanicus, meadow vole	●	●	●		●		4
Microtus xanthognathus, taiga vole			●		●		2
Myodes gapperi, southern red-backed vole	●						1
Myodes rutilus, northern red-backed vole	●	●	●	●	●	●	6
Neotoma cinerea, bushy-tailed woodrat	●						1
Ondatra zibethicus, common muskrat	●	●	●	●	●		5
Peromyscus keeni, northwestern deermouse	●						1
Peromyscus maniculatus, N.A. deermouse		■		■			1
Phenacomys intermedius, western heather vole	●						1
Synaptomys borealis, northern bog lemming	●	●	●	●			4
Muridae—Old World mice and rats							
Mus musculus, house mouse	■	■					2
Rattus norvegicus, brown rat	■	■	■	■	■		5
Rattus rattus, roof rat				■			1
Erethizontidae—porcupines							
Erethizon dorsatum, North American porcupine	●	●	●	●	●	●	6
LAGOMORPHA—pikas and hares							
Ochotonidae—pikas							
Ochotona collaris, collared pika	●	●	●	●			4
Leporidae—hares and rabbits							
Lepus americanus, snowshoe hare	●	●	●	●	●	●	6
Lepus othus, Alaska hare				●	●		2
Oryctolagus cuniculus, European rabbit		■		■			2
SORICOMORPHA—shrews, moles, and solenodons							
Soricidae—shrews							
Sorex alaskanus, Glacier Bay water shrew	●						1
Sorex cinereus, cinereus shrew	●	●	●	●	●	●	6
Sorex hoyi, pygmy shrew		●	●	●	●		4
Sorex jacksoni, St. Lawrence Island shrew					●		1
Sorex monticolus, dusky shrew	●	●	●	●	●	●	6
Sorex palustris, American water shrew	●	●	●				3
Sorex pribilofensis, Pribilof Island shrew					●		1
Sorex tundrensis, tundra shrew		●	●	●	●	●	5
Sorex ugyunak, barren ground shrew					●	●	2
Sorex yukonicus, Alaska tiny shrew		●	●	●	●	●	5

LAND MAMMAL SPECIES	ALASKA REGION						TOTAL
	SE	SC	C	SW	W	N	
CHIROPTERA—bats							
Vespertilionidae—evening bats							
Lasionycteris noctivagans, silver-haired bat	●						1
Myotis californicus, California myotis	●						1
Myotis keenii, Keen's myotis	●						1
Myotis lucifugus, little brown myotis	●	●	●	●			4
Myotis volans, long-legged myotis	●						1
CARNIVORA—carnivores							
Felidae—cats							
Lynx canadensis, Canadian lynx	●	●	●	●	●	●	6
Puma concolor, cougar	●						1
Canidae—wolves and foxes							
Canis latrans, coyote	●	●	●	●	●		5
Canis lupus, wolf	●	●	●	●	●	●	6
Vulpes lagopus, arctic fox				●	●	●	3
Vulpes vulpes, red fox	●	●	●	●	●	●	6
Ursidae—bears							
Ursus americanus, American black bear	●	●	●	●	●		5
Ursus arctos, brown bear	●	●	●	●	●	●	6
Mustelidae—weasels							
Gulo gulo, wolverine	●	●	●	●	●	●	6
Lontra canadensis, North American river otter	●	●	●	●	●		5
Martes americana, American marten	●	●	●	●	●		5
Martes caurina, Pacific marten	●						1
Martes pennanti, fisher	●						1
Mustela erminea, ermine	●	●	●	●	●	●	6
Mustela nivalis, least weasel		●	●	●	●	●	5
Neovison vison, American mink	●	●	●	●			4
Procyonidae—raccoon and relatives							
Procyon lotor, raccoon	■						1
ARTIODACTYLA—even-toed ungulates							
Cervidae—deer							
Alces americanus, moose	●	●	●	●	●	●	6
Cervus canadensis, wapiti	■			■			2
Odocoileus hemionus, mule deer	●	■		■			3
Rangifer tarandus, caribou		●	●	●	●	●	5
Bovidae—bovids							
Bison bison, American bison		■	■	■			3
Oreamnos americanus, mountain goat	●	●					2
Ovibos moschatus, muskox					■	■	2
Ovis dalli, Dall's sheep		●	●	●		●	4
TOTAL NUMBER OF SPECIES	51	46	44	46	40	26	
TOTAL RESTRICTED TO REGION	13	0	1	2	3	0	
TOTAL TRANSLOCATED TO REGION	4	6	1	7	2	1	

Appendix 2 (continued)

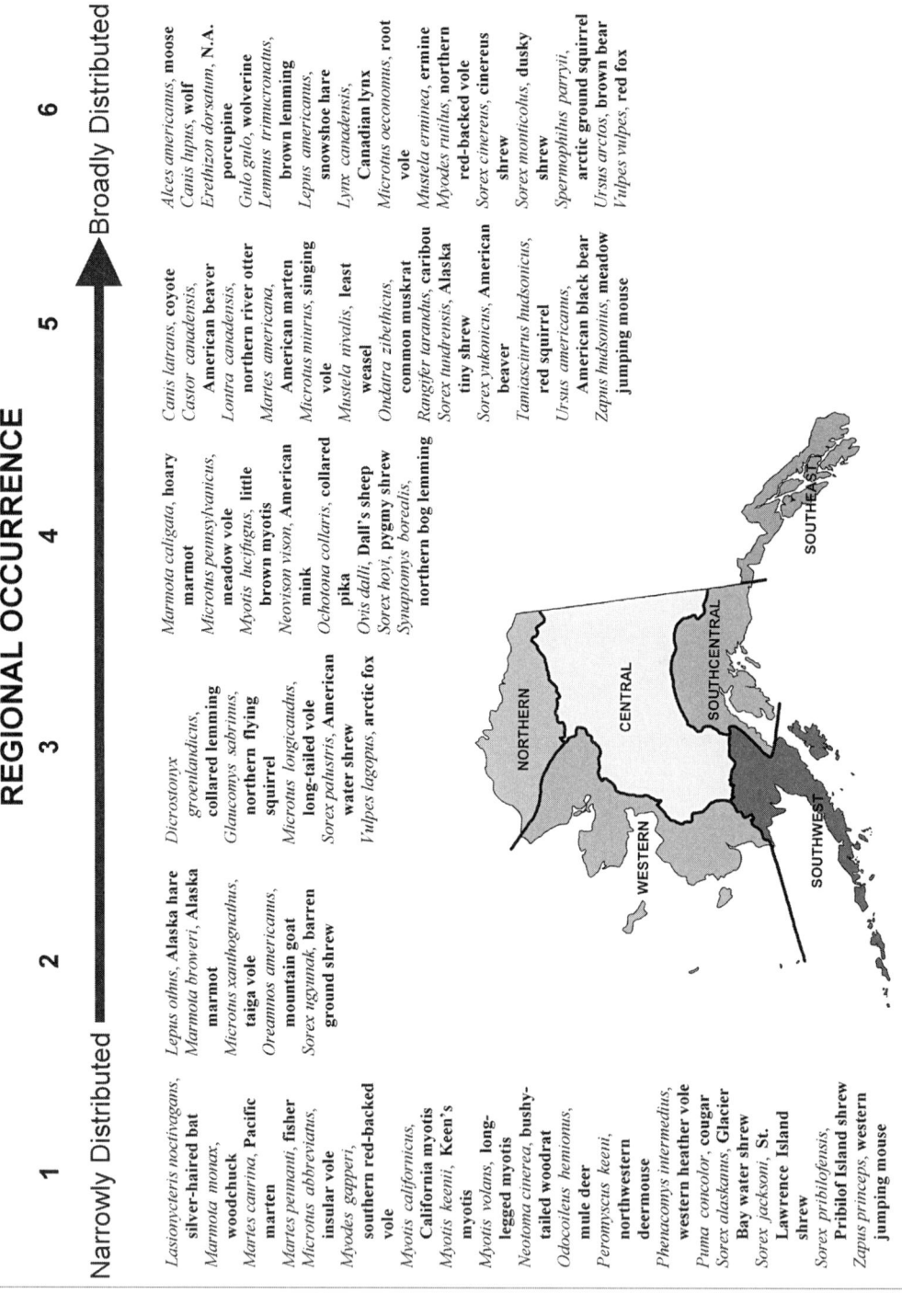

REGIONAL OCCURRENCE

Narrowly Distributed → Broadly Distributed

1	2	3	4	5	6
Lasionycteris noctivagans, **silver-haired bat** *Marmota monax*, **woodchuck** *Martes caurina*, **Pacific marten** *Martes pennanti*, **fisher** *Myodes abbreviatus*, **insular vole** *Myodes gapperi*, **southern red-backed vole** *Myotis californicus*, **California myotis** *Myotis keenii*, **Keen's myotis** *Myotis volans*, **long-legged myotis** *Neotoma cinerea*, **bushy-tailed woodrat** *Odocoileus hemionus*, **mule deer** *Peromyscus keeni*, **northwestern deermouse** *Phenacomys intermedius*, **western heather vole** *Puma concolor*, **cougar** *Sorex alaskanus*, **Glacier Bay water shrew** *Sorex jacksoni*, **St. Lawrence Island shrew** *Sorex pribilofensis*, **Pribilof Island shrew** *Zapus princeps*, **western jumping mouse**	*Lepus othus*, **Alaska hare** *Marmota broweri*, **Alaska marmot** *Microtus xanthognathus*, **taiga vole** *Oreamnos americanus*, **mountain goat** *Sorex ugyunak*, **barren ground shrew**	*Dicrostonyx groenlandicus*, **collared lemming** *Glaucomys sabrinus*, **northern flying squirrel** *Microtus longicaudus*, **long-tailed vole** *Sorex palustris*, **American water shrew** *Vulpes lagopus*, **arctic fox**	*Marmota caligata*, **hoary marmot** *Microtus pennsylvanicus*, **meadow vole** *Myotis lucifugus*, **little brown myotis** *Neovison vison*, **American mink** *Ochotona collaris*, **collared pika** *Ovis dalli*, **Dall's sheep** *Sorex hoyi*, **pygmy shrew** *Synaptomys borealis*, **northern bog lemming**	*Canis latrans*, **coyote** *Castor canadensis*, **American beaver** *Lontra canadensis*, **northern river otter** *Martes americana*, **American marten** *Microtus miurus*, **singing vole** *Mustela nivalis*, **least weasel** *Ondatra zibethicus*, **common muskrat** *Rangifer tarandus*, **caribou** *Sorex tundrensis*, **Alaska tiny shrew** *Sorex yukonicus*, **American beaver** *Tamiasciurus hudsonicus*, **red squirrel** *Ursus americanus*, **American black bear** *Zapus hudsonius*, **meadow jumping mouse**	*Alces americanus*, **moose** *Canis lupus*, **wolf** *Erethizon dorsatum*, **N.A. porcupine** *Gulo gulo*, **wolverine** *Lemmus trimucronatus*, **brown lemming** *Lepus americanus*, **snowshoe hare** *Lynx canadensis*, **Canadian lynx** *Microtus oeconomus*, **root vole** *Mustela erminea*, **ermine** *Myodes rutilus*, **northern red-backed vole** *Sorex cinereus*, **cinereus shrew** *Sorex monticolus*, **dusky shrew** *Spermophilus parryii*, **arctic ground squirrel** *Ursus arctos*, **brown bear** *Vulpes vulpes*, **red fox**

NORTHERN

WESTERN

CENTRAL

SOUTHCENTRAL

SOUTHWEST

SOUTHEAST

Appendix 3. Number of Alaska Type Specimens by Institution

Mammalian orders and families	Number of genera	Number of species	Number of subspecies	Number of Type Specimens by Institution
PROBOSCIDEA				
Elephantidae	1	1†		1 GZG
SIRENIA				
Dugongidae	1	1†		
PRIMATES				
Hominidae	1	1		
RODENTIA				
Sciuridae	4	6 (1*)	18 (9*)	12 USNM, 5 MVZ
Castoridae	1	1	2 (1*)	2 MVZ
Dipodidae	1	2	3	1 USNM, 1 AMNH
Cricetidae	9	16(1*)	59 (38*)	34 USNM, 11 MVZ, 3 NMC, 2 ANSP, 2 CAS, 2 KU, 1 AMNH, 1 MCZ
Muridae	2	3	1	
Erethizontidae	1	1	2	1 USNM, 1 AMNH
LAGOMORPHA				
Ochotonidae	1	1		1 USNM
Leporidae	2	3 (1*)	3 (2*)	3 USNM
SORICOMORPHA				
Soricidae	1	10 (4*)	12 (5*)	13 USNM, 2 MVZ, 1 AMNH, 1 UAM
CHIROPTERA				
Vespertilionidae	3	6	5	6 USNM
CARNIVORA				
Felidae	2	2	2	
Canidae	2	4	12 (6*)	9 USNM, 1 CMNH, 1 MVZ
Ursidae	1	3	11 (8*)	6 USNM, 1 AMNH, 1 MVZ
Otariidae	3	3	1	
Odobenidae	1	1	1	
Phocidae	7	8	3	1 BMNH, 1 USNM
Mustelidae	6	9	23 (12*)	11 USNM, 4 MVZ, 3 ANSP, 2 FMNH, 1 MCZ
Procynonidae	1	1		
ARTIODACTYLA				
Cervidae	4	4	9	4 USNM, 1 AMNH, 1 ROM
Bovidae	4	4	7 (2*)	2 AMNH, 1 FMNH, 1 NMC, 1 USNM
CETACEA				
Balaenidae	2	2		
Balaenopteridae	2	5	4	1 USNM
Eschrichtiidae	1	1		
Delphinidae	7	7		1 USNM, 1 MNHN
Monodontidae	2	2		
Phocoenidae	2	2	1	2 USNM
Physeteridae	2	2		1 MNHN
Ziphiidae	3	3		2 USNM, 1 MNHN
TOTALS	80	**115 (7 endemic)**	**179 (83 endemic)**	**169** (109 USNM, 26 MVZ, 8 AMNH, 5 ANSP, 4 NMC, 3 FMNH, 3 MNHN, 2 CAS, 2 KU, 2 MCZ, 1 BMNH, 1 CMNH, 1 GZG, 1 ROM, 1 UAM)

*endemic to Alaska † extinct

Appendix 4. Specimens of Alaska Mammals by Institution

SPECIES	UAM	AMNH	CAS (SU)	FMNH	KU	MCZ	MSB	MVZ	PSM	USNM	UWBM	Other*	TOTAL
SIRENIA													
Dugongidae													
Hydrodamalis gigas	1										1		2
RODENTIA													
Sciuridae													
Glaucomys sabrinus	448		3	1	1		27	15		30	1	14	540
Marmota broweri	13				6		39	5	5	8	3	2	81
Marmota caligata	38	22		2	1		55	37	5	80	4	8	252
Marmota monax	32	1	1	1			4	2	3	5	1		50
Spermophilus parryii	539	112	10	30	322	30	81	146	69	621	9	185	2154
Tamiasciurus hudsonicus	887	100	16	14	29		57	105	12	303	20	88	1631
Castoridae													
Castor canadensis	687	1	4	6		3	19	29	11	65		5	830
Dipodidae													
Zapus hudsonius	306	12	4		29		2	1		29	1	14	398
Zapus princeps	53		1				8	1					63
Cricetidae													
Dicrostonyx groenlandicus	234	55	1	8	26	69	91	275	24	257	11	107	1158
Lemmus trimucronatus	883	57	36	54	604	50	127	1556	30	454	23	631	4505
Microtus abbreviatus	78			1			13	6		30	1		129
Microtus longicaudus	997	11	20	3	453	4	37	91	9	151	5	48	1829
Microtus miurus	1613	91	5	4	417	71	64	289	48	231	2	39	2874
Microtus oeconomus	4765	164	26	24	807	262	1385	1451	89	1400	18	242	10,633
Microtus pennsylvanicus	2814	4	7	3	87		109	122	2	121	14	68	3351
Microtus xanthognathus	2110	10			4		93	104	15	52	206	12	2606
Myodes gapperi	1033	20	39	4		1	63	45		52	7	94	1358
Myodes rutilus	10,383	335	40	15	856	65	534	256	87	905	792	422	14,690
Neotoma cinerea									4	2			6
Ondatra zibethicus	494	6	1	9	2		9	38	42	98		7	706
Peromyscus keeni	5420	12	72	14	47	8	396	315		372	11	102	6769
Peromyscus maniculatus	32												32
Phenacomys intermedius	4												4
Synaptomys borealis	599	34		2	5		6	17	3	44		12	722
Muridae													
Mus musculus	11	35	18		2		5	1		33		5	110
Rattus norvegicus	57	1			13		1	3		23	9	1	108
Erethizontidae													
Erethizon dorsatum	41	7	6	6	7		7	78	10	61	1	21	245
LAGOMORPHA													
Ochotonidae													
Ochotona collaris	149	12		3	2	2	28	16	9	118	1	6	346
Leporidae													
Lepus americanus	760	29	2	18	4	4	22	54	43	236	5	13	1190
Lepus othus	182		1			3	5		1	63	1	4	260
Oryctolagus cuniculus	9									1			10
SORICOMORPHA													
Soricidae													
Sorex alaskanus	1									2			3
Sorex cinereus	8814	271	40	27	100	20	190	341	10	561	339	171	10,884
Sorex hoyi	362	5	1				6	1		12	47	3	437
Sorex jacksoni	49				1		3	16	13	3		3	88
Sorex monticolus	3608	215	49	13	160	11	168	334	15	505	22	215	5315
Sorex palustris	27		1		4					1		4	37
Sorex pribilofensis	32	3			10	4	2	4		139	4	22	220

Appendix 4 (continued)

SPECIES	UAM	AMNH	CAS (SU)	FMNH	KU	MCZ	MSB	MVZ	PSM	USNM	UWBM	Other*	TOTAL
Sorex tundrensis	625	6	2	8	35	24	8	75	8	168	8	27	994
Sorex ugyunak	115		1		33		3			4		6	162
Sorex yukonicus	34										3	1	38

CHIROPTERA

Vespertilionidae

	UAM	AMNH	CAS (SU)	FMNH	KU	MCZ	MSB	MVZ	PSM	USNM	UWBM	Other*	TOTAL
Eptesicus fuscus												1	1
Lasionycteris noctivagans	3	1											4
Myotis californicus	11						2						13
Myotis keenii	2									1			3
Myotis lucifugus	261		9	17	2		38	29		28		13	397
Myotis volans	3							1					4

CARNIVORA

Felidae

	UAM	AMNH	CAS (SU)	FMNH	KU	MCZ	MSB	MVZ	PSM	USNM	UWBM	Other*	TOTAL
Lynx canadensis	5111	88	26	57	1		3	80	83	106	2	46	5603
Puma concolor	2												2

Canidae

	UAM	AMNH	CAS (SU)	FMNH	KU	MCZ	MSB	MVZ	PSM	USNM	UWBM	Other*	TOTAL
Canis latrans	140		2	3			6	3	23	49		7	233
Canis lupus	3531	11	44	42	32	32	137	55	45	161	17	64	4171
Vulpes lagopus	991	7	42	16	27	28	27	306	998	586	18	77	3123
Vulpes vulpes	174	12	17	98	12	11	4	74	281	417	8	75	1183

Ursidae

	UAM	AMNH	CAS (SU)	FMNH	KU	MCZ	MSB	MVZ	PSM	USNM	UWBM	Other*	TOTAL
Ursus americanus	416	48	8	16	4	4	53	77	78	407	4	103	1218
Ursus arctos	905	121	22	49	19	33	22	121	70	1204	11	123	2700
Ursus maritimus	432	388	8	14	5	5	6	13	10	88	9	61	1039

Otariidae

	UAM	AMNH	CAS (SU)	FMNH	KU	MCZ	MSB	MVZ	PSM	USNM	UWBM	Other*	TOTAL
Callorhinus ursinus	37	38	111	51	7	28	1	16	10	1052	11	46	1408
Eumetopias jubatus	230	7	7	3		14	1	159	10	71	12	9	523
Zalophus californianus	1							1		1			3

Odobenidae

	UAM	AMNH	CAS (SU)	FMNH	KU	MCZ	MSB	MVZ	PSM	USNM	UWBM	Other*	TOTAL
Odobenus rosmarus	2099	24	5	5		13	4	13		53	17	95	2328

Phocidae

	UAM	AMNH	CAS (SU)	FMNH	KU	MCZ	MSB	MVZ	PSM	USNM	UWBM	Other*	TOTAL
Cystophora cristata	1												1
Erignathus barbatus	1350	4	8	4			1	31	12	40	3	19	1472
Histriophoca fasciata	240	5	2	4		4		3	9	9	1	15	292
Mirounga angustirostris	2									1	1		4
Pagophilus groenlandicus													0
Phoca largha	624								1	1	3		629
Phoca vitulina	1232	20	2	3	1	3		5	1	105	4	15	1391
Pusa hispida	2769	5	4		2	2	4	25	140	188	5	28	3172

Mustelidae

	UAM	AMNH	CAS (SU)	FMNH	KU	MCZ	MSB	MVZ	PSM	USNM	UWBM	Other*	TOTAL
Enhydra lutris	1661	14	21	17	21	4	3	24	17	627	90	209	2708
Gulo gulo	1381	9	28	31	9	8	24	24	39	66	9	53	1681
Lontra canadensis	208	2	6	13	1	1	52	24	16	68	3	19	413
Martes americana	3463	2	21	569		2	32	37	17	190	55	82	4470
Martes caurina	126							1					127
Martes pennanti	3												3
Mustela erminea	452	12	16	11	29	8	76	159	87	358	4	57	1269
Mustela nivalis	49	11	7		2	2	6	1	5	24	1	48	156
Neovison vison	1653	1	22	5	2	7	22	91	22	455	1	54	2335

Procyonidae

	UAM	AMNH	CAS (SU)	FMNH	KU	MCZ	MSB	MVZ	PSM	USNM	UWBM	Other*	TOTAL
Procyon lotor												1	1

ARTIODACTYLA

Cervidae

	UAM	AMNH	CAS (SU)	FMNH	KU	MCZ	MSB	MVZ	PSM	USNM	UWBM	Other*	TOTAL
Alces americanus	110	27	2	13	7	1	9	9	3	26	3	39	249
Cervus canadensis	2												2

Appendix 4 (continued)

SPECIES	UAM	AMNH	CAS (SU)	FMNH	KU	MCZ	MSB	MVZ	PSM	USNM	UWBM	Other*	TOTAL
Odocoileus hemionus	85	19	2	2		1	1	36	1	41	1	103	292
Rangifer tarandus	811	65	4	14	4		34	41	13	266	6	61	1319
Bovidae													
Bison bison	41									2			43
Oreamnos americanus	73	6	4	8				5	3	9	1	21	130
Ovibos moschatus	40	4	1	3			3	3	1	8	2	8	73
Ovis dalli	221	40	1	18	46		15	28	3	100	6	47	525

CETACEA

SPECIES	UAM	AMNH	CAS (SU)	FMNH	KU	MCZ	MSB	MVZ	PSM	USNM	UWBM	Other*	TOTAL
Balaenidae													
Balaena mysticetus	220	1	2						1	25	8	77	334
Eubalaena japonica										2	1		3
Balaenopteridae													
Balaenoptera acutorostrata	3	1								4			8
Balaenoptera borealis													0
Balaenoptera musculus										3	2		5
Balaenoptera physalus		3								10	2		15
Megaptera novaeangliae	3	1	2			2				15			23
Eschrichtiidae													
Eschrichtius robustus		1					1	1			1	2	6
Delphinidae													
Globicephala macrorhynchus		?	1							?			1
Grampus griseus													0
Lagenorhynchus obliquidens		1								2			3
Lissodelphis borealis													0
Orcinus orca	2	2								5	2		11
Pseudorca crassidens													0
Stenella attenuata										1			1
Monodontidae													
Delphinapterus leucas	318	3	2			1		2	2	13	2	3	346
Monodon monoceros	1								1		1		3
Phocoenidae													
Phocoena phocoena	27		1				4	1		30	2		65
Phocoenoides dalli	3	2			1			1		19		2	28
Physeteridae													
Kogia breviceps	1												1
Physeter catodon										1	2	2	5
Ziphiidae													
Berardius bairdii	2									4		1	7
Mesoplodon stejnegeri	6	1								15			22
Ziphius cavirostris	5									14		1	20
TOTAL	80,876	2638	858	1348	4315	848	4242	7338	2576	14,211	1901	4289	125,440

*Other institutions include ANSP, CMNH, CU, DMNS, LACM, LSUMZ, MMNH, MSU, MSUM, ROM, SDNHM, TTU, UBC, UCLA, UMNH, and YPM.

Appendix 5. Habitats of Alaska Mammals as Modified from Key Habitats Used in ADFG's Comprehensive Wildlife Conservation Strategy, www.sf.adfg.state.ak.us/statewide/ngplan/ NG_outline.cfm (accessed 12 August 2007).

SPECIES	FOREST Boreal	FOREST Coastal Temperate	SHRUB-LAND Tall	SHRUB-LAND Low	TUNDRA Alpine	TUNDRA Arctic	TUNDRA Maritime	GRASSLAND/WETLAND Grass	GRASSLAND/WETLAND Sedge	GRASSLAND/WETLAND Bog	GRASSLAND/WETLAND Marsh	FRESHWATER AQUATIC Lakes & Ponds	FRESHWATER AQUATIC Rivers & Streams	FRESHWATER AQUATIC Riparian Zone	MARINE & COASTLINE Nearshore	MARINE & COASTLINE Shelf	MARINE & COASTLINE Oceanic	MARINE & COASTLINE Rocky Islands, Beaches, and Cliffs	SEA ICE Fast	SEA ICE Pack	OTHER Rocks/Caves	OTHER Disturbed/Sparsely Vegetated	OTHER Artificial structures
RODENTIA																							
Sciuridae																							
Glaucomys sabrinus	X	X																					
Marmota broweri					X																X		
Marmota caligata					X																X		
Marmota monax	X																						
Spermophilus parryii					X	X		X													X	X	
Tamiasciurus hudsonicus	X	X																					X
Castoridae																							
Castor canadensis												X	X	X									
Dipodidae																							
Zapus hudsonius	X		X	X	X			X	X		X			X									
Zapus princeps		X	X	X				X	X	X	X			X									
Cricetidae																							
Dicrostonyx groenlandicus				X	X	X	X	X	X					X							X		
Lemmus trimucronatus				X	X	X	X	X	X		X			X							X		
Microtus abbreviatus				X				X	X	X											X		
Microtus longicaudus				X	X	X		X	X					X							X		
Microtus miurus				X	X	X	X	X			X			X							X		
Microtus oeconomus			X	X		X	X	X	X	X	X			X									
Microtus pennsylvanicus								X	X	X				X									
Microtus xanthognathus	X													X									
Myodes gapperi	X	X	X	X										X									
Myodes rutilus	X	X	X	X	X									X									
Neotoma cinerea			X	X	X																X	X	
Ondatra zibethicus											X	X	X	X									
Peromyscus keeni		X	X	X			X	X													X	X	X
Peromyscus maniculatus*				X	X																X	X	X
Phenacomys intermedius				X	X																		
Synaptomys borealis	X	X	X	X				X		X				X							X		

Appendix 5 (continued)

SPECIES	FOREST: Boreal	FOREST: Coastal Temperate	SHRUB-LAND: Tall	SHRUB-LAND: Low	TUNDRA: Alpine	TUNDRA: Arctic	TUNDRA: Maritime	GRASSLAND/WETLAND: Grass	GRASSLAND/WETLAND: Sedge	GRASSLAND/WETLAND: Bog	GRASSLAND/WETLAND: Marsh	FRESHWATER AQUATIC: Lakes & Ponds	FRESHWATER AQUATIC: Rivers & Streams	FRESHWATER AQUATIC: Riparian Zone	MARINE & COASTLINE: Nearshore	MARINE & COASTLINE: Shelf	MARINE & COASTLINE: Oceanic	MARINE & COASTLINE: Rocky I Islands, Beaches, and Cliffs	SEA ICE: Fast	SEA ICE: Pack	OTHER: Rocks/Caves	OTHER: Disturbed/Sparsely Vegetated	OTHER: Artificial structures
Muridae																							
*Mus musculus**																						X	X
*Rattus norvegicus**			X	X			X											X				X	X
*Rattus rattus**		X	X	X			X											X				X	X
Erethizontidae																							
Erethizon dorsatum	X		X		X			X			X			X							X	X	X
LAGOMORPHA																							
Ochotonidae																							
Ochotona collaris					X																X	X	
Leporidae																							
Lepus americanus	X	X	X	X										X									
Lepus othus			X	X	X	X	X							X									
*Oryctolagus cuniculus**		X	X	X			X																
SORICOMORPHA																							
Soricidae																							
Sorex alaskanus	X		X	X				X			X	X	X	X									
Sorex cinereus	X	X	X	X	X			X	X	X	X			X							X		
Sorex hoyi	X							X	X	X			X	X									
Sorex jacksoni							X	X													X		
Sorex monticolus	X	X	X	X	X			X	X		X	X	X	X							X		
Sorex palustris	X	X	X						X			X	X	X									
Sorex pribilofensis							X														X		
Sorex tundrensis	X		X	X	X	X	X	X			X			X									
Sorex ugyunak					X	X	X	X													X		
Sorex yukonicus	X		X		X									X							X		

SPECIES	FOREST: Boreal	FOREST: Coastal Temperate	SHRUB-LAND: Tall	SHRUB-LAND: Low	TUNDRA: Alpine	TUNDRA: Arctic	TUNDRA: Maritime	GRASSLAND/WETLAND: Grass	Sedge	Bog	Marsh	FRESHWATER: Lakes & Ponds	Rivers & Streams	Riparian Zone	MARINE: Nearshore	Shelf	Oceanic	Rocky Islands, Beaches, and Cliffs	SEA ICE: Fast	Pack	OTHER: Rocks/Caves	Disturbed/Sparsely Vegetated	Artificial structures
CHIROPTERA																							
Vespertilionidae																							
Lasionycteris noctivagans		X									X	X	X	X							X		X
Myotis californicus		X									X	X	X	X							X		X
Myotis keenii		X												X							X		
Myotis lucifugus	X	X												X							X		X
Myotis volans		X												X							X		X
CARNIVORA																							
Felidae																							
Lynx canadensis	X													X									
Puma concolor		X	X																				
Canidae																							
Canis latrans	X	X	X	X	X			X						X									
Canis lupus	X	X	X	X	X	X	X	X						X							X	X	
Vulpes lagopus					X	X	X								X				X	X			
Vulpes vulpes	X	X	X	X	X		X	X						X								X	
Ursidae																							
Ursus americanus	X	X	X	X	X			X						X							X		
Ursus arctos	X	X	X	X	X	X		X	X					X							X		
Ursus maritimus						X	X												X	X			
Otariidae																							
Callorhinus ursinus																	X	X					
Eumetopias jubatus															X	X		X					
Zalophus californianus															X	X		X					
Odobenidae																							
Odobenus rosmarus															X	X		X		X			

Appendix 5 (continued)

SPECIES	FOREST		SHRUB-LAND		TUNDRA			GRASSLAND/WETLAND				FRESHWATER AQUATIC			MARINE & COASTLINE				SEA ICE		OTHER		
	Boreal	Coastal Temperate	Tall	Low	Alpine	Arctic	Maritime	Grass	Sedge	Bog	Marsh	Lakes & Ponds	Rivers & Streams	Riparian Zone	Nearshore	Shelf	Oceanic	Rocky Islands, Beaches, and Cliffs	Fast	Pack	Rocks/Caves	Disturbed/Sparsely Vegetated	Artificial structures
Phocidae																							
Cystophora cristata																X				X			
Erignathus barbatus																				X			
Histriophoca fasciata																	X			X			
Mirounga angustirostris																	X	X					
Pagophilus groenlandicus																	X			X			
Phoca largha															X	X				X			
Phoca vitulina															X			X					
Pusa hispida																			X				
Mustelidae																							
Enhydra lutris															X	X							
Gulo gulo	X	X	X	X	X	X																	
Lontra canadensis	X	X	X	X								X	X	X	X								
Martes americana	X	X																					
Martes caurina		X																					
Martes pennanti	X	X																					
Mustela erminea	X	X	X	X	X	X						X	X	X									
Mustela nivalis	X		X	X	X	X															X		
Neovison vison	X	X	X	X				X			X	X	X	X	X			X					
Procyonidae																							
Procyon lotor*		X												X	X			X					

ARTIODACTYLA

SPECIES	Boreal	Coastal Temperate	Tall	Low	Alpine	Arctic	Maritime	Grass	Sedge	Bog	Marsh	Lakes & Ponds	Rivers & Streams	Riparian Zone	Nearshore	Shelf	Oceanic	Rocky Islands, Beaches, and Cliffs	Fast	Pack	Rocks/Caves	Disturbed/Sparsely Vegetated	Artificial structures
Cervidae																							
Alces americanus	X	X	X		X									X									
Cervus canadensis*		X	X					X						X									
Odocoileus hemionus		X	X					X						X									
Rangifer tarandus	X			X	X	X	X																
Bovidae																							
Bison bison	X							X						X									

Appendix 5 (continued)

SPECIES	FOREST		SHRUB-LAND		TUNDRA			GRASSLAND/WETLAND				FRESHWATER AQUATIC			MARINE & COASTLINE				SEA ICE		OTHER		
	Boreal	Coastal Temperate	Tall	Low	Alpine	Arctic	Maritime	Grass	Sedge	Bog	Marsh	Lakes & Ponds	Rivers & Streams	Riparian Zone	Nearshore	Shelf	Oceanic	Rocky Islands, Beaches, and Cliffs	Fast	Pack	Rocks/Caves	Disturbed/Sparsely Vegetated	Artificial structures
Oreamnos americanus					X																		
Ovibos moschatus						X	X																
Ovis dalli					X																X		
CETACEA																							
Balaenidae																							
Balaena mysticetus																X				X			
Eubalaena japonica																X	X						
Balaenopteridae																							
Balaenoptera acutorostrata															X	X	X			X			
Balaenoptera borealis																	X						
Balaenoptera musculus																	X						
Balaenoptera physalus															X	X	X						
Megaptera novaeangliae																X	X						
Eschrichtiidae																							
Eschrichtius robustus															X	X	X						
Delphinidae																							
Globicephala macrorhynchus																	X						
Grampus griseus																	X						
Lagenorhynchus obliquidens																X	X						
Lissodelphis borealis																	X						
Orcinus orca															X	X	X						
Pseudorca crassidens																	X						
Stenella attenuata																	X						
Monodontidae																							
Delphinapterus leucas															X					X			
Monodon monoceros																X				X			
Phocoenidae																							
Phocoena phocoena															X	X	X						
Phocoenoides dalli																X	X						

Appendix 5 (continued)

SPECIES	FOREST Boreal	FOREST Coastal Temperate	SHRUB-LAND Tall	SHRUB-LAND Low	TUNDRA Alpine	TUNDRA Arctic	TUNDRA Maritime	GRASSLAND/WETLAND Grass	GRASSLAND/WETLAND Sedge	GRASSLAND/WETLAND Bog	GRASSLAND/WETLAND Marsh	FRESHWATER AQUATIC Lakes & Ponds	FRESHWATER AQUATIC Rivers & Streams	FRESHWATER AQUATIC Riparian Zone	MARINE & COASTLINE Nearshore	MARINE & COASTLINE Shelf	MARINE & COASTLINE Oceanic	MARINE & COASTLINE Rocky / Islands, Beaches, and Cliffs	SEA ICE Fast	SEA ICE Pack	OTHER Rocks/Caves	OTHER Disturbed/Sparsely Vegetated	OTHER Artificial structures
Physeteridae																							
Kogia breviceps																	X						
Physeter catodon																	X						
Ziphiidae																							
Berardius bairdii																	X						
Mesoplodon stejnegeri																	X						
Ziphius cavirostris																	X						

* indicates an introduced species.

Appendix 6. Conservation Status of Alaska Mammals.

Taxon	Distributional Status	HERITAGE		ADFG	ESA	IUCN	CITES	COSEWIC
		Alaska	Global					
RODENTIA								
Castor canadensis	{Island exotic}	SNR	G5			LC		
C. c. phaeus	Island endemic	S3				DD		
Dicrostonyx groenlandicus		S3	G5			LC		
D. g. exsul	Island endemic	S4				DD		
D. g. rubricatus		S4						
D. g. unalascensis	Island endemic	SNR	G3			DD		
Erethizon dorsatum		S5	G5			LC		
Glaucomys sabrinus		S4	G5		PS	LC		
G. s. griseifrons	Island endemic	S2?				EN		
Lemmus trimucronatus		SNR	G5			LC		
L. t. harroldi	Island endemic	S4						
L. t. nigripes	Island endemic	S3						
Marmota broweri	Endemic	S4	G4			LC		
Marmota caligata		SNR	G5			LC		
M. c. sheldoni	Island endemic	S2S3				DD		
M. c. vigilis	Endemic	S3?				DD		
Marmota monax		SNR	G5			LC		
Microtus abbreviatus	Endemic	S3	G3					
M. a. abbreviatus	Island endemic	S2				DD		
M. a. fisheri	Island endemic	S3				DD		
Microtus longicaudus		SNR	G5			LC		
M. l. coronarius	Endemic	S3				DD		
Microtus miurus		*S4*	*G4*			*LC*		
Microtus oeconomus		*SNR*	*G5*			*LC*		
M. o. amakensis	Island endemic	S2				DD		
M. o. elymocetes	Island endemic	S2				DD		
M. o. innuitus	Island endemic	S3				DD		
M. o. popofensis	Island endemic	S3				DD		
M. o. punukensis	Island endemic	S1				DD		
M. o. sitkensis	Island endemic	S3				DD		
M. o. unalascensis	Island endemic	S3						
M. o. yakutatensis		S4						
Microtus pennsylvanicus		SNR	G5		PS	LC		
M. p. admiraltiae	Island endemic	S3				NT		
Microtus xanthognathus		SNR	G5			LC		
Mus musculus	Exotic	SNR	G5			LC		
Myodes gapperi		SNR	G5			LC		
M. g. solus	Island endemic	S3				DD		
M. g. stikinensis		SNR						
M. g. wrangeli	Island endemic	S2S3						
Myodes rutilus		SNR	G5			LC		
M. r. albiventer	Island endemic	S3						
M. r. glacialis	Endemic	S3						
M. r. insularis	Island endemic	S3						
M. r. orca	Endemic	S3						
Neotoma cinerea		SNR	G5			LC		
Ondatra zibethicus	{Island exotic}	SNR	G5			LC		
Peromyscus keeni		S3	G5			LC		
Peromyscus maniculatus	Island exotic	SNR	G5			LC		
Phenacomys intermedius		SNR	G5			LC		
Rattus norvegicus	Exotic	SNR	G5			LC		
Rattus rattus	Island exotic	SNR	G5			LC		
Synaptomys borealis		S4	G4			LC		
Spermophilus parryii	{Island exotic}	SNR	G5			LC		
S. p. kodiacensis	Island endemic	S3				DD		
S. p. lyratus	Island endemic	S3				DD		
S. p. nebulicola	Island endemic	S3				DD		
S. p. osgoodi	Endemic	S3?						
Tamiasciurus hudsonicus	{Island exotic}	S5	G5		PS	LC		
T. h. picatus	Endemic	S3?						
Zapus hudsonicus		S5?	G5		PS	LC		
Zapus princeps		SNR	G5			LC		

Taxon	Distributional Status	HERITAGE Alaska	HERITAGE Global	ADFG	ESA	IUCN	CITES	COSEWIC
LAGOMORPHA								
Lepus americanus	{Island exotic}	SNR	G5			LC		
Lepus othus		SNR	G3G4			LC		
Ochotona collaris		S5	G5			LC		
Oryctolagus cuniculus	Island exotic	SNR	G5			LC		
SORICOMORPHA								
Sorex alaskanus	Endemic	SH	GHQ			LC		
Sorex cinereus		S5	G5			LC		
S. c. streatori		SNR				LC		
Sorex hoyi		SNR	G5			LC		
Sorex jacksoni	Island endemic	S3	G3			EN		
Sorex monticolus		SNR	G5			LC		
S. m. malitiosus	Island endemic	S3						
Sorex palustris		SNR	G5			LC		
Sorex pribilofensis	Island endemic	S2	G3			EN		
Sorex tundrensis		SNR	G5			LC		
Sorex ugyunak		S5	G5			LC		
Sorex yukonicus		SNR	GU					
CHIROPTERA								
Eptesicus fuscus	Vagrant	S2?	G5			LC		
Lasionycteris noctivagans		S1S3B	G5			LC		
Myotis californicus		S1S3B	G5			LC		
Myotis keenii		S1S3	G2G3			LC		DD
Myotis lucifugus		S3S4	G5			LC		
Myotis volans		S2?B	G5			LC		
CARNIVORA								
Callorhinus ursinus		S3	G3			VU		T
Canis latrans		S5	G5			LC		
Canis lupus		S4	G4	PS		LC	A2	NAR
C. l. ligoni	Endemic	S2S3	G4				A2	
Cystophora cristata	Vagrant	SNR	G4G5			LC		NAR
Enhydra lutris	{Reintroduced}	S4	G4	PS		LC	A2	T
E. l. kenyoni		S2S3	G4T4	LT	EN	A2		
Erignathus barbatus		SNR	G4G5			LC		NAR
Eumetopias jubatus		S2	G3			EN		SC
Population east of 144° W		S2		SSCC	LT			
Population west of 144° W		SNR		SSCC	LE			
Gulo gulo		SNR	G4			VU		SC
G. g. katschemakensis	Endemic	S3?						
Histriophoca fasciata		SNR	G5			LC		
Lontra canadensis		SNR	G5			LC	A2	
L. c. kodiacensis	Island endemic	S4					A2	
L. c. mira		S3S4					A2	
Lynx canadensis		S4?	G5	PS:LT		LC	A2	NAR
Martes americana	{Island exotic}	SNR	G5			LC		PS
Martes caurina	Island endemic	SNR	GNR					
Martes pennanti		SNR	G5	PS:C		LC		
Mirounga angustirostris		SNR	G5			LC		NAR
Mustela erminea		SNR	G5			LC		PS
M. e. celenda	Island endemic	S4?						
M. e. haidarum	Island endemic	SNR						
M. e. kadiacensis	Island endemic	S4?						
M. e. seclusa	Island endemic	S2?						
Mustela nivalis		SNR	G5			LC		
Neovison vison		SNR	G5			LC		PS
Odobenus rosmarus		S4	G4			LC	A2	SC
Pagophilus groenlandicus	Vagrant	SNR	G5			LC		
Phoca largha		SNR	G4G5			LC		
Phoca vitulina		S3	G5			LC		NAR
P. v. richardsi		S4S5		SSOC		LC		NAR

Taxon	Distributional Status	HERITAGE Alaska	HERITAGE Global	ADFG	ESA	IUCN	CITES	COSEWIC
Procyon lotor	Island exotic	SNR	G5			LC		
Puma concolor		SNR	G5		PS	NT	A2	PS
Pusa hispida		SNR	G5			LC		NAR
Ursus americanus		SNR	G5		PS	LC	A2	NAR
U. a. emmonsii		S3?					A2	
U. a. perniger	Endemic	S4					A2	
Ursus arctos		SNR	G4		PS	LC	A2	SC
Ursus maritimus		S3	G3G4		LE	VU	A2	SC
Vulpes lagopus	{Island exotic}	SNR	G5			LC		
V. l. pribilofensis	Island endemic	S3S4						
Vulpes vulpes	{Island exotic}	S5	G5			LC		
Zalophus californianus		SNR	G5			LC		NAR

ARTIODACTYLA

Taxon	Distributional Status	HERITAGE Alaska	HERITAGE Global	ADFG	ESA	IUCN	CITES	COSEWIC
Alces americanus		SNR	G5			LC		
Bison bison bison	Exotic	SNR	G4		PS	CD		PS
Cervus elaphus	Island exotic	SNR	G5		PS			
Odocoileus hemionus	{Exotic}	SNR	G5		PS	LC		
Oreamnos americanus	{Island exotic}	SNR	G5			LC		
Ovibos moschatus	Reintroduced	S4S5	G4			LC		
Ovis dalli		SNR	G5			LC		
O. d. kenaiensis	Endemic	S3S4						
Rangifer tarandus	{Island exotic}	SNR	G5		PS	LC		SC

CETACEA

Taxon	Distributional Status	HERITAGE Alaska	HERITAGE Global	ADFG	ESA	IUCN	CITES	COSEWIC
Balaena mysticetus		S2	G3		LE	CD	A1	T
Bering-Chukchi-Beaufort pop.				SSOC		CD		SC
Balaenoptera acutorostrata		SNR	G5			NT	A1	NAR
Balaenoptera borealis		S2B	G3		LE	EN	A1	E
Pacific population								E, DD
Balaenoptera musculus		S2B	G2	E	LE	EN	A1	E
Balaenoptera physalus		S2B	G3G4		LE	EN	A1	T
Berardius bairdii		SNR	G4			CD	A1	NAR
Delphinapterus leucas		SNR	G4		PS	VU	A2	PS
Cook Inlet population		S1			LE	CR		
Eschrichtius robustus		S3B	G4		PS:LE	CD	A1	SC
Eubalaena japonica		S1	G1	SSOC	E	EN	A1	E
Globicephala macrorhynchus		SNR	G5			CD	A2	NAR
Grampus griseus		SNR	G5			DD	A2	NAR
Kogia breviceps	Vagrant	SNR	G4			LC	A2	NAR
Lagenorhynchus obliquidens		SNR	G5			LC	A2	NAR
Lissodelphis borealis		SNR	G4			LC	A2	NAR
Megaptera novaeangliae		S2B	G3	E	LE	VU	A1	T
Mesoplodon stejnegeri		SNR	G3			DD	A2	NAR
Orcinus orca		SNR	G4G5			CD	A2	PS
NE Pacific resident population		SNR						T
NE Pacific offshore population		SNR						SC
NE Pacific transient population		SNR						T
Phocoena phocoena		S2S3	G4G5			VU	A2	SC
Phocoenoides dalli		SNR	G4G5			CD	A2	NAR
Physeter catodon		S2S3	G3G4		LE	VU	A1	NAR
Pseudorca crassidens	Vagrant	SNR	G4			LC	A2	NAR
Stenella attenuata	Vagrant	SNR	G5			CD	A2	
Ziphius cavirostris		SNR	G4			DD	A2	NAR

Key

Distribution Status

{ } = subpopulation or evolutionarily significant unit (ESU) of concern within the state

Exotic = not native; introduced by human agency

Heritage. National Heritage Network and The Nature Conservancy (http://www.natureserve.org/ explorer)

G = global (status throughout its range)

Q = taxonomic status questionable

S = subnational (status in Alaska)

1 = critically imperiled; 2 = imperiled; 3 = rare or uncommon; 4 = not rare, long-term concern; 5 = widespread, abundant, secure; ? = insufficient data; R = reported to occur; NR = not ranked; B = breeding; SH = occurred historically

ADFG. Alaska Department of Fish & Game (http://www.wildlife.alaska.gov)

E = endangered

SSOC = state species of concern

ESA. U.S. Endangered Species Act of 1973, as amended by the U.S. Fish and Wildlife Service and the U.S. National Marine Fisheries Service (http://fws.gov/endangered)

C = candidate

LE = listed endangered

LT = listed threatened

PS = partial status (applies only to portion of species' range)

SC = special concern

IUCN. International Union for Conservation of Nature and Natural Resources (http://www.redlist.org)

CD = conservation dependent

CR = critically endangered

DD = data deficient

EN = endangered

LC = least concern

LR = lower risk

NT = near threatened

VU = vulnerable

CITES. Convention on International Trade in Endangered Species of Wild Fauna and Flora (http://www. cites.org)

A1 = Appendix I (most critically endangered)

A2 = Appendix II (species not necessarily now threatened with extinction but may become so unless trade is closely controlled)

COSEWIC. Committee on the Status of Endangered Wildlife in Canada (http://www.cosewic.gc.ca)

DD = data deficient

E = endangered and facing imminent extirpation or extinction

NAR = not at risk

PS = partial status (applies only to portion of species' range)

SC = particularly sensitive to human activities or natural events

T = threatened and likely to become endangered if limiting factors are not reversed

XT = extirpated

Appendix 7. Introductions and Translocations of Mammals In Alaska.

SPECIES	REGION	RELEASE SITE	SOURCE	DATE	PRESENT STATUS	REMARKS
RODENTS						
northern flying squirrel *Glaucomys sabrinus*	SE	A number of small islands in Sea Otter Sound	El Capitan Island	<1999	Unknown	Repeated releases of "nuisance" animals by private individual
hoary marmot *Marmota caligata*	SE	Prince of Wales Island	Juneau area	1930, 1931	No animals remaining	(Bailey 1993; Elkins and Nelson 1954)
	SW	Sud Island (Barren Islands)	Unknown	ca. 1930	Numerous in 1974 (Bailey 1976)	
arctic ground squirrel *Spermophilus parryii*	SW	As delineated in Table 2 (p. 71), ground squirrels presently occur on Aghiyuk, Aliksemit, Amaknak, Anowik, Atkins, Bendel, Bird, Chankliut, Chernabura, Cherni, Chirikof, Chowiet, Heredeen, Holiday, Kak, Karpa, Kateekuk, Kavalga, Kiliktagik, Kodiak, Koniuji (Big and Little), Marmot, Nagai, Nakchamik, Near, Simeonof, Spectacle, Tugidak, Turner, Umnak, Unalaska, Unavikshak, Unimak, Ushagat, and Woody islands. The origin of most of these populations remains clouded.				
red squirrel *Tamiasciurus hudsonicus*	SE	Admiralty Island	Unknown	late 1940s or early 1950s	Occur throughout island	(MacDonald and Cook 2007)
		Baranof and Chichagof islands	Juneau area	1930, 1931	Occur throughout, and now documented on Inian, Kruzof, Moser, Partofshikof, and Yakobi islands, and on islets in Sitka Sound	
		Prince of Wales Island	Unknown	Unknown	No animals remaining	(Fay and Sease 1985)
	SW	Afognak and Kodiak islands	Anchorage area	1952	Viable populations	
American beaver *Castor canadensis*	SE	Baranof Island	Prince of Wales Island	1927	Viable population	Restock of depleted population
	SW	Kodiak Island	Cordova area	1925	Viable population	Now established on Afognak and several other islands in the archipelago
		Raspberry Island	Cordova area	1929	Viable population	
common muskrat *Ondatra zibethicus*	SE	Prince of Wales Island	Haines area	1929	No animals remaining	
	SW	Kodiak, Afognak, Whale, and Spruce islands	Copper River	1925	Viable populations	
		Long Island	Kodiak Island	1929	Viable population	
		Unalaska Island	Unknown	<1920	No animals remaining	(Murie 1959)
	W	St. George Island	Nushagak area	1913	Died out first winter	(Preble 1923)

358   RECENT MAMMALS OF ALASKA

SPECIES	REGION	RELEASE SITE	SOURCE	DATE	PRESENT STATUS	REMARKS
North American deermouse *Peromyscus maniculatus*	SC	Copper River Basin, Tiekel River near Tiekel	Unknown	Unknown	Unknown status; probably introduced	(MacDonald et al. in press)
	SW	Shemya Island	Unknown	≤1978	Extant and possibly expanding range	
house mouse *Mus musculus*	SE, SC, SW, C, W	Preserved specimens and recent reports are from the communities of Anchorage, Chugiak, Eagle River, Fairbanks, Palmer, Sitka, and Wrangell, and from the islands of Hog, Kiska, Kodiak, St. Paul, and Unalaska.		As early as 1872 (St. Paul Island)	Status and distribution poorly understood anywhere in the state	
brown rat *Rattus norvegicus*	SE, SC, SW, C, W	Reported from the communities of College, Cordova, Craig, Douglas, Fairbanks, Homer, Juneau, Kenai, Ketchikan, King Cove, Kotzebue, Nome, Petersburg, Sitka, St. Michael, Tanana, Valdez, Wasilla, and Wrangell. Island records and reports include Adak, Akutan, Amaknak, Amchitka, Atka, Attu, Baranof, Bat, Bird Rock, Bolshoei Islets, Douglas, Great Sitkin, Kagalaska, Kiska, Kodiak, Little Kiska, Makarius, Mitkof, Ogangen, Prince of Wales, Rat, Revillagigedo, Sanak, Seal Rocks, Sedanka, Shemya, Unalaska, and Wrangell islands.		First known record prior to 1780 from Rat Island	Status and distribution poorly understood anywhere in the state	(Bailey 1993; Brechbill 1977; Fritts 2006)
roof rat *Rattus rattus*	SW	Shemya Island	Unknown	Remains reported in 1995; trapped on island every year since 2004	Established population	(Taylor and Brooks 1995; S. Ebbert, pers. comm.)

SPECIES	REGION	RELEASE SITE	SOURCE	DATE	PRESENT STATUS	REMARKS
LAGOMORPHS						
snowshoe hare *Lepus americanus*	SE	Juneau area and Douglas Island	Haines area	<1920	Viable populations	(Bailey 1920; Wenrich 1922)
		Admiralty, Otstoia, and Smeaton islands	Washington State	1923–1924	No animals remaining	(MacDonald and Cook 2007)
	SE	Cape, Prince of Wales, and Village islands	Anchorage area	1924	No animals remaining	
	SW	Kodiak and Afognak islands	Anchorage area	1934	Viable populations	(Burris and McKnight 1973)
		Long and Woody islands	Kodiak Island	1952	Extant	
		Popof Island	Kodiak Island	1955	Extant	
		Marmot Island	Unknown	Unknown	Considered common by Chumbley et al. (1997)	
		Raspberry Island	Unknown	Unknown	Unknown (Manville and Young 1965)	
Alaska hare *Lepus othus*	SW	Chirikof Island	Unknown	1891	No animals remaining	(Bailey 1993)
feral European (domestic) rabbit *Oryctolagus cuniculus*	SE	Betton Island	Unknown	<1995	Unknown	(MacDonald and Cook 2007)
	SC	Middleton Island	Domestic	1954	Viable population (1975)	(O'Farrell 1965)
	SW	Hog, Kanaga, Rabbit, and Umnak islands		ca. 1917–1940	Unknown	(Bailey 1993; Burris and McKnight 1973)
		Anangula, Ananiuliak, Poa, and Tangik islands			Extant	(AMNWF 2006)

SPECIES	REGION	RELEASE SITE	SOURCE	DATE	PRESENT STATUS	REMARKS
CARNIVORES						
arctic fox *Vulpes lagopus*	SE	Animals released on at least 187 islands	Stocks for fur farming ventures along coastal Alaska originally from native populations on the Commander (Russia) and Pribilof islands	1899–1929	No animals remaining	(Bailey 1993; MacDonald and Cook 2007)
	SC	Introduced to at least 62 coastal islands		1890–1929	By 1993, no animals remaining, except possibly on Glacier, East Chugach, Elizabeth, and Nuka islands	(Bailey 1993)
	SW	Starting on Attu Island, at least 166 islands in this region were once stocked with arctic foxes (Bailey 1993).		1750–1931	By 1993, foxes still remaining on 22 islands in the Aleutians, 11 islands off the Alaska Peninsula, and 1 (Chirikof) south of Kodiak (Bailey 1993)	Active eradication efforts by AMNWR staff began in 1949, and by 2002 foxes had been removed from 39 islands (Ebbert and Byrd 2002).
red fox *Vulpes vulpes*	SE	Cleft, Dry, Kupreanof, Passage, and Sokoi islands	Various sources for commercial fur farming	1894–1929	No animals remaining	(Bailey 1993)
	SC	At least 9 islands stocked		1891–1929	No animals remaining	(Bailey 1993)
	SW	At least 46 islands stocked	Usually from Fox Islands, but originally from Siberia on Kanaga, Adak, and perhaps other central Aleutian Islands (Bailey 1993)	1840–1929	Still present on Chuginadak and Tigalda islands (AMNWF 2006)	

SPECIES	REGION	RELEASE SITE	SOURCE	DATE	PRESENT STATUS	REMARKS
wolf *Canis lupus*	SE	Coronation Island		1960, 1963	None remaining by early 1970s (Burris and McKnight 1973)	Experimental population
feral dog *Canis lupus familiaris*	SW	Kodiak Island	Domestic	Reported 2001, 2004	Extant on southwest end of island	Occasionally forming packs and hunting deer (Van Daele 2004)
	C	Interior Alaska	Domestic	Studied 1979–1981	Unknown	(Gipson 1983)
sea otter *Enhydra lutris*	SE	Yakutat Bay and Chichagof, Yakobi, Biorka, Barrier, and Heceta islands	Amchitka Island	1965–1969	Viable population	Restocking of depleted populations
	SW	Attu Island	Amchitka Island	1956	Unsuccessful, but reestablished in mid-1960s through natural immigration	(Franzmann 1988)
	W	Otter Island	Amchitka Island	1955	Small extant population in the Pribilofs a result of translocations or possibly immigration of otters from the Alaska Peninsula	Restocking of depleted population
		St. Paul Island	Amchitka Island	1959		
		St. George Island	Amchitka Island	1968		
American marten *Martes americana*	SE	Prince of Wales Island	Behm Canal	1934	Viable populations	(Burris and McKnight 1973)
		Baranof Island	Thomas Bay	1934	Viable populations	Now also documented on Kruzof, Otstoia, Catherine, Partofschikof, and Yakobi islands (MacDonald and Cook 2007)
		Chichagof Island	Baranof, Mitkof, Revillagigedo, and Wrangell islands; Stikine River drainage; Anchorage area	1949, 1951, 1952	Viable populations	
	SC	Kayak and Patterson islands	Unknown	early 1940s	Present on Kayak Island (ADFG, 1978) and considered "common" there (M. E. Isleib, pers. comm., 1978)	(Elkins and Nelson 1954)
	SW	Afognak Island	Lake Minchumina area	1952	Viable populations	

SPECIES	REGION	RELEASE SITE	SOURCE	DATE	PRESENT STATUS	REMARKS
American mink *Neovison vison*	SE	Strait Island	Petersburg Fur Experimental Farm	1956	Unknown	(Burris and McKnight 1973)
	SC	Montague Island	Petersburg Fur Experimental Farm	1951	Present	Status on island prior to translocation unknown
	SW	Kodiak Island	Petersburg Fur Experimental Farm	1952	No animals remaining	
feral domestic ferret *Mustela putorius furo*	SE	Revillagigedo and Grant islands	Domestic	1980s	No animals reported in more recent years	(MacDonald and Cook 2007)
raccoon *Procyon lotor*	SE	Singa Island	Indiana	1941	Present on nearby El Capitan Island in 1999	All melanistic (Scheffer 1947)
		Japonski Island	Unknown	1950	No animals remaining by early 1970s	Some also seen on nearby Baranof Island during that time period
	SW	Long Island	Several Midwestern states	<1936	None reported since 1990 and presumed extirpated	

UNGULATES

SPECIES	REGION	RELEASE SITE	SOURCE	DATE	PRESENT STATUS	REMARKS
feral horse *Equus caballus*	SW	Unalaska and Umnak islands	Domestic	1984	Extant	(AMNWF 2006)
wild boar *Sus scrofa*	SW	Marmot Island	Unknown	1984	Surveys in 2006 found none present (van Daele, pers. comm.)	(Lloyd et al. 1987)
moose *Alces americanus*	SE	Berners Bay	Susitna and Matanuska valleys	1958, 1960	Viable population	(Burris and McKnight 1973)
		Chickamin River	Anchorage area	1963, 1964	None seen in recent years (Porter 2004a)	
	SC	Copper River Delta	Kenai Peninsula, Susitna River drainage, Matanuska Valley, Anchorage area	1949–1958	Viable populations	Expanded eastward to Cape Yakutaga by mid-1970s
		Kalgin Island	Anchorage area	1957, 1958, 1959	Viable population	
	SW	Kodiak Island	Southcentral mainland	1966, 1967	No animals remaining	

Appendix 7 (continued)

SPECIES	REGION	RELEASE SITE	SOURCE	DATE	PRESENT STATUS	REMARKS
wapiti *Cervus canadensis*	SE	Kruzof Island	Washington State	1926, 1927	No animals remaining	(Burris and McKnight 1973)
		Revillagigedo Island	Washington State, Afognak Island	1937, 1963, 1964	No animals remaining	
		Gravina Island	Afognak and Raspberry islands	1962	No animals remaining	
		Annette Island	Afognak and Raspberry islands	1963	No animals remaining	
		Etolin Island	Oregon	1987	Viable, expanding population	
	SW	Afognak Island	Washington State	1929	Viable, expanding populations	(MacDonald and Cook 2007)
Sitka black-tailed deer *Odocoileus hemionus sitkensis*	SE	Yakutat Bay	Rocky Pass (between Kuiu and Kupreanof islands)	1934	Small but viable population	
		Taiya Valley	Southeast Alaska	1951, 1952, 1956	No animals remaining	
		Sullivan Island	Southeast Alaska	1951–1954	Viable population	
		Kupreanof Island	Southeast Alaska	1979	Unsuccessful (Franzmann 1988)	Done to supplement deer numbers on island
	SC	Hinchinbrook and Hawkins islands	Sitka area	1916–1923	Viable, expanding populations	
		Homer, Kenai Peninsula	Sitka area	1923	No animals remaining	(Franzmann 1988)
	SW	Long Island	Sitka area, Prince of Wales Island	1924, 1930	Viable, expanding population	
		Kodiak Island	Rocky Pass (between Kuiu and Kupreanof islands)	1934	Viable, expanding population	Evidence presented by van Daele (2005) suggests deer may have been introduced to the Kodiak Archipelago as early as the turn of the last century.

SPECIES	REGION	RELEASE SITE	SOURCE	DATE	PRESENT STATUS	REMARKS
caribou/reindeer *Rangifer tarandus*	SC	Kenai Peninsula	Nelchina Basin	1965, 1966, 1985, 1986	Viable population	
	SW	Adak Island Atka, Umnak, and Unalaska islands	Nelchina Basin	1958, 1959	Viable population Extant (AMNWF 2006)	Reindeer removed from Hagemeister Island in early 1990s (Ebbert and Byrd 2002)
		Nushagak Peninsula	146 animals from Alaska Peninsula near Becharof Lake	1988	556 animals found during January 2008 survey; down from a peak of around 1,400 in 1997	(Tom Paul, pers. comm. 2009)
	W	Reindeer pastures currently include the entire Baldwin and Seward peninsulas, as well as permitted areas near Shaktoolik, Stebbins, and on St. Lawrence Island. There are also small herds in Palmer, Delta Junction, and the Kenai Peninsula. Reindeer herds currently occur on the islands of Nunivak, St. Paul, and St. George.		1892–present	An estimated 17,650 reindeer currently reside in Alaska, with about 10,000 on the Seward Peninsula.	(Jernsletten and Klokov 2002)
American bison *Bison bison*	SC	Copper River	Delta area	1950	Currently about 110 animals (Tobey 2004a)	
		Chitina	Delta area	1962	About 50 animals (Tobey 2004b)	
	SW	Popof Island	Montana	1954	About 100 animals (Fall et al. 1993)	
	C	Delta	Montana	1928, 1930	Currently about 360 animals (DuBois 2004)	
		Farewell	Delta area	1965, 1968	About 350 animals (Boudreau 2004)	
feral domestic cattle *Bos taurus*	SW	Chirikof Island	Believed first brought to the island for food by an American whaling company	≥ 1888	Last animals removed in 2003	

Free-ranging cattle reportedly still present on Akun, Long, Sanak, Umnak, Unalaska, and Wosnesenski islands. Feral animals on Simeonof Island were removed in the 1980s.

SPECIES	REGION	RELEASE SITE	SOURCE	DATE	PRESENT STATUS	REMARKS
mountain goat *Oreamnos americanus*	SE	Baranof Island	Tracy Arm	1923	Viable, expanding population	(Mooney 2004)
		Chichagof Island	Southeast Alaska	1953–1955	No animals remaining	(MacDonald and Cook 2007)
		Revillagigedo Island	Adjacent mainland	1983, 1991	Viable, expanding population	(Porter 2004b)
	SC	Cecil Rhode Mountain, Kenai Peninsula	Kenai Peninsula	1983	Unknown	Restocking of a depleted population (Smith and Nichols 1984)
	SW	Kodiak Island	Kenai Peninsula, Anchorage area	1952, 1953	Viable, expanding population	(van Daele and Crye 2004b)
muskox *Ovibos moschatus*	W	Nunivak Island	Greenland	1935, 1936	Viable population	(Burris and McKnight 1973; Franzmann 1988; Klein 2000)
		Nelson Island	Nunivak Island	1967, 1968	Viable population	Emigrating animals subsequently established a small population on the Yukon-Kuskokwim Delta
		Seward Peninsula	Nunivak Island	1970, 1981	Viable population	
		Cape Thompson	Nunivak Island	1970	Viable population	
	N	Camden Bay, Kavik River	Nunivak Island	1969, 1970	Viable population	(Jingfors and Klein 1982; Reynolds and Ross 1984)
feral domestic sheep *Ovis aries*	SW	Unalaska and Umnak islands	Domestic		Extant (AMNWF 2006)	
Dall's sheep *Ovis dalli*	SW	Kodiak Island	Kenai Peninsula	1964–1967	No animals remaining	(Franzmann 1988; L. van Daele, pers. comm., 2007)

Appendix 8. Mammal Occurrence on Islands in Southeast Alaska Based on Archived Specimens (● = species present and vouchered, ■ = species introduced and vouchered)

The following matrix lists species (rows, grouped by order) against islands of the Southeast Region (columns). ● = species present and vouchered; ■ = species introduced and vouchered.

Species \ Island	[Mainland]	Admiralty	Anguilla	Annette	Back	Baker	Baranof	"S" Barrier	Beardslee	Bell	Betton	Black	Bluff	Brothers	Buck	Bushy	Cap	Castle	Cat	Chichagof	Conclusion	Cone	Coronation	Couverden	Dall	Deer	Dog	Douglas	Duck	Duke
LAGOMORPHA																														
Lepus americanus	●																													
Ochotona collaris	●																													
RODENTIA																														
Erethizon dorsatum	●																											●		
Rattus norvegicus	■						■																					■		
Mus musculus							■																							
Synaptomys borealis	●					●					●																			
Phenacomys intermedius	●																													
Peromyscus keeni	●	●	●	●		●	●		●					●			●	●		●			●	●	●	●	●	●	●	●
Ondatra zibethicus	●	●																												
Neotoma cinerea	●																													
M. pennsylvanicus	●	●																												
M. oeconomus	●						●													●										
Microtus longicaudus	●		●	●																●					●			●	●	
Lemmus trimucronatus	●																													
M. rutilus	●																											●		
Myodes gapperi	●									●		●															●			
Z. princeps	●																													
Zapus hudsonius	●																													
Castor canadensis	●																													
Tamiasciurus hudsonicus	■						■													■								●		
Spermophilus parryii	●																													
Marmota caligata	●																											●		
Glaucomys sabrinus	●							●																	●					
ARTIODACTYLA																														
Oreamnos americanus	●																													
Odocoileus hemionus	●	●					●							●						●			●					●		
Cervus elaphus	●																													
Alces americanus	●																													
CARNIVORA																														
U. arctos	●	●					●							●						●										
Ursus americanus	●																		●				●	●						
Neovison vison	●	●	●				●							●			●			●						●				
Mustela erminea	●	●		●			●													●					●			●		
M. pennanti	●																													
Martes americana/caurina	●	●					■													■										
Lontra canadensis	●	●																												
Gulo gulo	●																													
Puma concolor	●																													
Lynx canadensis	●																													
Vulpes vulpes	●																													
C. lupus	●		●	●																●			●					●		
Canis latrans	●																													
CHIROPTERA																														
M. volans		●																												
M. lucifugus	●	●					●													●				●						
M. keenii																														
Myotis californicus																														
Lasionycteris noctivagans	●																													
SORICOMORPHA																														
S. palustris	●																													
S. monticolus	●	●	●	●		●	●		●	●	●	●					●			●				●	●	●	●	●		●
S. cinereus	●						●													●							●	●		
Sorex alaskanus	●																													

Presence of mammal species on islands of the Southeast Region (● = present; ■ = special record).

ISLAND	Lepus americanus	Ochotona collaris	Erethizon dorsatum	Rattus norvegicus	Mus musculus	Synaptomys borealis	Phenacomys intermedius	Peromyscus keeni	Ondatra zibethicus	Neotoma cinerea	M. pennsylvanicus	M. oeconomus	Microtus longicaudus	Lemmus trimucronatus	M. rutilus	Myodes gapperi	Z. princeps	Zapus hudsonius	Castor canadensis	Tamiasciurus hudsonicus	Spermophilus parryii	Marmota caligata	Glaucomys sabrinus	Oreamnos americanus	Odocoileus hemionus	Cervus elaphus	Alces americanus	U. arctos	Ursus americanus	Neovison vison	Mustela erminea	M. pennanti	Martes americana/caurina	Lontra canadensis	Gulo gulo	Puma concolor	Lynx canadensis	Vulpes vulpes	C. lupus	Canis latrans	M. volans	M. lucifugus	M. keenii	Myotis californicus	Lasionycteris noctivagans	S. palustris	S. monticolus	S. cinereus	Sorex alaskanus
Eagle																														●																	●		
Echo																							●								●																		
El Capitan								●					●																																		●	●	
Emmons								●																																									
Esquibel								●					●																																		●		
Etolin																●				●			●		●	■			●	●	●		●		●				●										
Fair								●					●																																				
Forrester																																																	
near Garden								●																						●	●																	●	
Gavanski								●																																									
Gedney																																		●													●		
Goat						●														●			●																			●					●	●	
Grant								●												●								●																			●	●	
Gravina								●												●																											●	●	
Grief																																							●									●	
Halleck																																							●									●	
Hassler																																																	
Heceta																●			●						●									●					●								●	●	
Herbert Graves																																																	
Hill								●					●																																				
Hoot																				■																											●		
Horseshoe								●												●							●																				●		
Hotspur																				■			●							●																	●		
Inian												●													●														●									●	
Kadin								●			●																												●										
Keene								●																				●					●														●		
Kosciusko								●					●							■								●	●				■	●													●		
Krestof																				●														●														●	
Kruzof			●					●												●					●				●	●	●		●	●					●								●	●	
Kuiu						●		●					●												●					●	●																	●	
Kupreanof			●			●		●					●							●					●				●	●	●		●	●		●			●								●	●	

RECENT MAMMALS OF ALASKA

Appendix 8 (continued)

The following matrix lists mammal species (grouped by order) against islands of the Southeast Region. A filled circle (●) or filled square (■) indicates a recorded occurrence.

SOUTHEAST REGION — ISLAND	SORICOMORPHA				CHIROPTERA					CARNIVORA													ARTIODACTYLA				RODENTIA																					LAGOMORPHA	
	Sorex alaskanus	S. cinereus	S. monticolus	S. palustris	Lasionycteris noctivagans	Myotis californicus	M. keenii	M. lucifugus	M. volans	Canis latrans	C. lupus	Vulpes vulpes	Lynx canadensis	Puma concolor	Gulo gulo	Lontra canadensis	Martes americana/caurina	M. pennanti	Mustela erminea	Neovison vison	Ursus americanus	U. arctos	Alces americanus	Cervus elaphus	Odocoileus hemionus	Oreamnos americanus	Glaucomys sabrinus	Marmota caligata	Spermophilus parryii	Tamiasciurus hudsonicus	Castor canadensis	Zapus hudsonius	Z. princeps	Myodes gapperi	M. rutilus	Lemmus trimucronatus	Microtus longicaudus	M. oeconomus	M. pennsylvanicus	Neotoma cinerea	Ondatra zibethicus	Peromyscus keeni	Phenacomys intermedius	Synaptomys borealis	Mus musculus	Rattus norvegicus	Erethizon dorsatum	Ochotona collaris	Lepus americanus
Lemesurier		●	●																																		●	●											
Lester		●	●																																														
Lincoln																			●	●					●																	●							
Long			●																																							●							
Lowrie			●			●																																				●							
Lulu			●																																							●							
Marble			●																																		●					●							
Mary			●																																		●					●							
Misery																																		●															
Mitkof		●	●		●			●			●				●		●		●	●	●		●		●					●■							●		●			●				■			
Moser		●																																			●					●							
North																●																																	
Noyes			●																																		●												
Onslow																									●												●												
Orr																											●										●												
Otstoia			●														■																																
Owl																														■																			
Partofshikof																	■																																
Peratrovitch																				●																													
Percy			●																																														
Pleasant			●																						●																								
Pow			●														■?																									●							
Prince of Wales		●							●		●					●			●	●	●				●		●			●	●						●					●							
Rapids		●																		●										●							●					●							
Read																				●																													
Revillagigedo		●	●				●	●			●				●	●	●		●	●	●				●		●			●	●	●		●			●				●	●		●		■			
Ring								●																																									
San			●																																							●							
San Fernando			●																																							●							
San Juan			●																																							●							
Bautista																																																	

Order	Species	Sangao	Santa Rita	Sergief	Shelikof	Shelter	Shrubby	Spanish	St. Ignace	Stevenson	Stone	Suemez	Sukkwan	Sullivan	Swan	Tatoosh	Thorne	Tuxekan	Vank	Warren	Woewodski	Woronkofski	Wrangell	Yakobi	Young	Zarembo
LAGO-MORPHA	Lepus americanus																									
	Ochotona collaris																									
RODENTIA	Erethizon dorsatum																						●			
	Rattus norvegicus																									
	Mus musculus																						■			
	Synaptomys borealis																						●			
	Phenacomys intermedius																									
	Peromyscus keeni	●	●		●	●	●	●				●	●		●		●	●	●	●	●	●	●			●
	Ondatra zibethicus			●																						
	Neotoma cinerea																									
	M. pennsylvanicus			●															●				●			
	M. oeconomus																							●		
	Microtus longicaudus	●		●				●				●	●	●			●	●		●	●		●			●
	Lemmus trimucronatus																									
	M. rutilus																								●	
	Myodes gapperi																						●			
	Z. princeps																									
	Zapus hudsonius																									
	Castor canadensis																									●
	Tamiasciurus hudsonicus											●			●				●			●	■			
	Spermophilus parryii																									
	Marmota caligata																									
	Glaucomys sabrinus											●		●	●								●			
ARTIODACTYLA	Oreamnos americanus																									
	Odocoileus hemionus											●								●		●	●			●
	Cervus elaphus																									
	Alces americanus																									
CARNIVORA	U. arctos																									
	Ursus americanus																						●			
	Neovison vison						●					●									●		●			●
	Mustela erminea											●											●			●
	M. pennanti																									
	Martes americana/caurina																				●		●			
	Lontra canadensis						●											●			●	●	●			
	Gulo gulo																									
	Puma concolor																						●			
	Lynx canadensis																									
	Vulpes vulpes																									
	C. lupus											●			●								●			●
	Canis latrans																									
CHIROPTERA	M. volans																						●			
	M. lucifugus																						●			
	M. keenii																						●			
	Myotis californicus																						●			
	Lasionycteris noctivagans																						●			
SORICO-MORPHA	S. palustris																									
	S. monticolus	●	●	●	●	●	●	●				●	●		●			●		●	●		●		●	●
	S. cinereus																						●	●		
	Sorex alaskanus																									

Appendix 9. Mammal Occurrence on Islands in Southcentral Alaska Based on Archived Specimens (● = species present and vouchered, ■ = species introduced and vouchered)

Order	Species	[Mainland]	Augustine	Bligh	Chenega	Chisik	Crafton	Disc	Egg	Eleanor	Elrington	Esther	Evans	Green	Hawkins	Hesketh	Hinchinbrook	Ingot	Kalgin	Kayak	Knight	Latouche	Middleton	Montague	Naked	Peak	Squire
LAGOMORPHA	Oryctolagus cuniculus																						■				
	Lepus americanus	●																									
	Ochotona collaris	●																									
RODENTIA	Erethizon dorsatum	●																									
	Synaptomys borealis	●															●										
	Ondatra zibethicus	●																									
	M. pennsylvanicus	●																									
	M. oeconomus	●			●	●	●	●	●	●			●				●		●	●	●	●		●	●		
	M. miurus	●																									
	Microtus longicaudus	●																									
	Lemmus trimucronatus	●																									
	Dicrostonyx groenlandicus	●																									
	Myodes rutilus	●				●							●	●		●					●						
	Zapus hudsonius	●																									
	Castor canadensis	●																		●							
	Tamiasciurus hudsonicus	●																									
	Spermophilus parryii	●	●																								
	Marmota caligata	●															●								●		
	Glaucomys sabrinus	●																									
ARTIODACTYLA	Ovis dalli	●																									
	Oreamnos americanus	●																									
	Bison bison	■																									
	Rangifer tarandus	●																									
	Odocoileus hemionus		■										■	■	■		■	■			■			■	■		■
	Alces americanus	●																									
CARNIVORA	U. arctos	●			●								●				●				●			●			
	Ursus americanus	●																									
	Neovison vison	●						●			●																
	M. nivalis	●																									
	Mustela erminea	●															●										
	Martes americana	●																									
	Lontra canadensis	●								●							●				●			●	●	●	
	Gulo gulo	●																									
	Lynx canadensis	●																									
	Vulpes vulpes	●																									
	C. lupus	●																									
	Canis latrans	●						●																			
BATS	Myotis lucifugus	●																									
SORICOMORPHA	S. yukonicus	●																									
	S. tundrensis	●																									
	S. palustris	●																									
	S. monticolus	●				●		●	●	●	●	●					●		●	●				●			
	S. hoyi	●																									
	Sorex cinereus	●			●		●		●		●		●	●			●		●		●			●			

See Bailey (1993) for a complete listing of island introductions of arctic and red foxes for commercial fur farming.

Appendix 10. Mammal Occurrence on Southwest Alaska Islands

Key: ● = occurrence (circle); ■ = occurrence (square); * and †, ‡ = footnote markers as printed in species names.

Order	Species	Mainland	Adak	Afognak	Agattu	Akutan	Alaid	Amak	Amchitka	Amlia	Anangula	Ananiuliak	Atka	Atkins	Attu	Avatanak	Big Konuji	Bird	Carlisle	Chernabura	Chirikof	Chowiet	Chuginadak	Deer	Expedition	Glen	Great Sitkin	Green	Herendeen	Hog	Holiday	Kagalaska	Kateekuk	Kavalga	Kiliktagik	Kiska	Kodiak	Little Kiska	Little Koniuji	Little Tanaga	Nagai	Poa	Popof
SORICOMORPHA	Sorex cinereus	●																																									
	Sorex hoyi	●																																									
	Sorex monticolus	●														●													●							●						●	●
	Sorex tundrensis	●																																									
	Sorex yukonicus	●																																									
BATS	Myotis lucifugus	●																																			●						
CARNIVORA	Canis latrans	●																																									
	Canis lupus	●																																									
	Vulpes lagopus *	●	■		■		■		■	■			■		■	■	■			■																■	■	■			■		
	Vulpes vulpes *	●	●																				■													●							
	Lynx canadensis	●																																									
	Gulo gulo	●																																									
	Lontra canadensis	●	●																																	●			●				
	Martes americana	●																																									
	Mustela erminea	●																																			●						
	Mustela nivalis	●																																									
	Neovison. vison	●																						●																			
	Ursus americanus	●																																									
	Ursus. arctos	●	●																																		●						
ARTIODACTYLA	Alces americanus	●																																									
	Cervus canadensis	■	■																																								
	Odocoileus hemionus	■	■																								■		■								■						
	Rangifer tarandus	■	■										■																														
	Ovis dalli	●																																									
RODENTIA	Marmota caligata	●																																									
	Spermophilus parryii †	●														■	■	■	■	■	■						■				●	■	●	■			■		■	■	■		
	Tamiasciurus hudsonicus	●																																			■						
	Castor canadensis	●																																									
	Zapus hudsonius	●																																									
	Dicrostonyx groenlandicus	●																						●																			
	Lemmus trimucronatus	●																																									
	Microtus oeconomus	●	●		●				●				●										●			●										●							●
	Microtus pennsylvanicus	●																																									
	Mus musculus																																				■						
	Myodes rutilus	●																																									
	Ondatra zibethicus	●																																									
	Peromyscus maniculatus	●																																									
	Rattus norvegicus	■		■		■			■				■														■				■												
	Synaptomys borealis	●																																									
	Erethizon dorsatum	●																																									
LAGOMORPHA	Ochotona collaris	●																																									
	Lepus americanus	●																																		■							●
	Lepus othus	●																																									
	Oryctolagus cuniculus ‡	●																																									

Appendix 10 (continued)

RECENT MAMMALS OF ALASKA

SOUTHWESTERN REGION — ISLAND

Data matrix of species occurrence by island (● and ■ = recorded occurrence). Taxonomic groups: LAGOMORPHA, RODENTIA, ARTIODACTYLA, CARNIVORA, BATS, SORICOMORPHA.

Species	Islands (with marker)
Rattus norvegicus	Round ■, Sedanka ■, Unalaska ■
Peromyscus maniculatus	Semisopochnoi ■
Mus musculus	Unalaska ■
Microtus oeconomus	Round ●, Sanak ●, Tigalda ●, Tugidak ●, Ugamak ●, Uganik ●, Umnak ●, Unalaska ●, Unalga ●, Unga ●, Unimak ●
Dicrostonyx groenlandicus	Umnak ●, Unalaska ●, Unga ●
Spermophilus parryii †	Seguam ●, Semidi ■, Umnak ■, Unalaska ●, Unalga ●, Woody ■
Rangifer tarandus	Umnak ■, Unalaska ■, Unimak ●
Odocoileus hemionus	Raspberry ■, Spruce ■, Spectacle ■, Ukolnoi ■, Whale ■
Ursus arctos	Simeonof ●, Sitkalidak ●, Unimak ●, Village ●
Mustela erminea	Unimak ●
Mustela nivalis	Unimak ●
Lontra canadensis	Spruce ●
Vulpes vulpes *	Tigalda ●, Tugidak ■, Umnak ●, Unalaska ●, Unimak ●, Village ●
Vulpes lagopus *	Sedanka ■, Seguam ■, Semidi ■, Ugamak ■, Uganik ■, Ulak ■, Unimak ●
Sorex monticolus	Round ●, Sanak ●, Tugidak ●, Unalaska ●, Unimak ●
Sorex cinereus	Unimak ●

Islands listed (rows): Raspberry, Rat, Round, Sanak, Sedanka, Seguam, Semidi, Semisopochnoi, Shemya, Simeonof, Sitkalidak, Shuyak, Spectacle, Spruce, Tangik, Tigalda, Tugidak, Ugaiushak, Ugamak, Uganik, Ukolnoi, Ulak, Uliaga, Umnak, Unalaska, Unalga, Unga, Unimak, Ushagat, Village, Walrus, Whale, Woody

* See Appendix 7 and Bailey (1993) for a listing of arctic and red fox introductions; † see Table 2 (p. 71) for more detailed information on arctic ground squirrel introductions; ‡ see Species Account of European rabbit for details.

Appendix 11.

Mammal Occurrence on Islands in Western Alaska Based on Archived Specimens (● = species present and vouchered, ■ = species introduced and vouchered)

Order	Species	Mainland	Hall	King	Little Diomede	Nelson	Nunivak	Otter	Punuk	Sledge	St. George	St. Lawrence	St. Matthew	St. Paul	Walrus
LAGOMORPHA	L. othus	●			●										
LAGOMORPHA	Lepus americanus	●													
RODENTIA	Erethizon dorsatum	●													
RODENTIA	Ondatra zibethicus	●													
RODENTIA	M. xanthognathus	●													
RODENTIA	M. oeconomus	●							●			●			
RODENTIA	M. miurus	●													
RODENTIA	Microtus abbreviatus		●										●		
RODENTIA	Lemmus trimucronatus	●					●					●			
RODENTIA	Dicrostonyx groenlandicus	●									●	●			
RODENTIA	Myodes rutilus	●									●				
RODENTIA	Zapus hudsonius	●													
RODENTIA	Castor canadensis	●													
RODENTIA	Tamiasciurus hudsonicus	●													
RODENTIA	Spermophilus parryii	●										●			
RODENTIA	M. caligata	●													
RODENTIA	Marmota broweri	●													
ARTIODACTYLA	Ovis dalli	●													
ARTIODACTYLA	Ovibos moschatus	■				■	■								
ARTIODACTYLA	Rangifer tarandus	●													
ARTIODACTYLA	Alces americanus	●													
CARNIVORA	U. maritimus	●			●					●		●	●	●	
CARNIVORA	U. arctos	●										●			
CARNIVORA	Ursus americanus	●													
CARNIVORA	Neovison vison	●					●								
CARNIVORA	M. nivalis	●													
CARNIVORA	Mustela erminea	●													
CARNIVORA	Martes americana	●													
CARNIVORA	Lontra canadensis	●													
CARNIVORA	Gulo gulo	●													
CARNIVORA	Lynx canadensis	●													
CARNIVORA	V. vulpes	●					●					●			
CARNIVORA	Vulpes lagopus	●	●	●			●	●			●	●	●	●	●
CARNIVORA	Canis lupus	●										●			
SORICOMORPHA	S. yukonicus	●													
SORICOMORPHA	S. ugyunak	●													
SORICOMORPHA	S. tundrensis	●													
SORICOMORPHA	S. pribilofensis	●												●	
SORICOMORPHA	S. monticolus	●													
SORICOMORPHA	S. jacksoni	●										●			
SORICOMORPHA	S. hoyi	●													
SORICOMORPHA	Sorex cinereus	●					●								

WESTERN REGION — ISLAND

RECENT MAMMALS OF ALASKA

Appendix 12. Pleistocene Mammals of Alaska and Yukon Territory

ORDER	COMMON NAME	SCIENTIFIC NAME	PLEISTOCENE		RECENT		REFERENCES
			Early and Middle Pleistocene	Late Pleistocene	Early and Middle Holocene	Late Holocene	
PROBOSCIDEA—elephants	woolly mammoth	Mammuthus primigenius		X	X		Guthrie 2004; Solow et al. 2006; Veltre et al. 2008
	mammoth	Mammuthus sp.	X				Harington 1978; Morlan 2003; Storer 2004a
	American mastodon	Mammut americanum		X			Harington 1990; Kurtén and Anderson 1980
PILOSA—sloths	Jefferson's ground sloth	Megalonyx jeffersonii		X			Kurtén and Anderson 1980; McDonald et al. 2000
SIRENIA—manatees, dugong, and sea cow	Steller's sea cow	Hydrodamalis gigas		X	X	X†	Domning 1978; Gard et al. 1972; Harington 1978; Savinetsky et al. 2004; Whitmore and Gard 1977
PRIMATES—primates	Human	Homo sapiens		X	X	X	Cinq-Mars and Morlan 1999; Heaton and Grady 2003; Porter 1988; Yesner 2001
RODENTIA—rodents	hoary marmot	Marmota caligata		X	X	↑	Georgina 2001; Heaton and Grady 2003; Yesner 2001
	woodchuck	Marmota monax		?	X	↑	Harington 1989; Morlan 1983
	marmot	Marmota sp.	X				Guthrie and Matthews 1971; Harington 1978
	arctic ground squirrel	Spermophilus parryii	X	X	X	X→	Bowers 1978; Georgina 2001; Guthrie 1968a, 1990a; Harington 2003; Hill 1942; Jopling et al. 1981; Kurtén and Anderson 1980; Yesner 2001; Zazula et al. 2007
	chipmunk	Tamias (Eutamias/Neotamias)	X			↑	Storer 2004a (as cf. minimus)
	American beaver	Castor canadensis	X	X	X	X→	Ackerman et al. 1997; Georgina 2001; Harington 1978, 2003; Heaton and Grady 2003; Holmes 1996; Jopling et al. 1981; Kurtén and Anderson 1980; Taber 1943
	giant beaver	Castoroides ohioensis	X	X			Harington 1978; Jopling et al. 1981
	collared lemming	Dicrostonyx sp.	X	X	X	X→	Georgina 2001; Jopling et al. 1981; Repenning et al. 1964
	Hopkins's lemming	Predicrostonyx hopkinsi	X				Guthrie and Matthews 1971

Appendix 12 (continued)

ORDER	COMMON NAME	SCIENTIFIC NAME	PLEISTOCENE		RECENT		REFERENCES
			Early and Middle Pleistocene	Late Pleistocene	Early and Middle Holocene	Late Holocene	
RODENTIA—rodents, continued	brown lemming	Lemmus trimucronatus	X	X	X	X→	Georgina 2001; Guthrie 1968a; Guthrie and Matthews 1971; Heaton and Grady 2003; Jopling et al. 1981; Repenning et al. 1964
	steppe lemming	Guildayomys matthewsi	X				Storer 2004a
	Cape Deceit vole	Microtus (Lasiopodomys) deceitensis	X				Bell et al. 2004; Guthrie and Matthews 1971; Storer 2003a
	singing vole	Microtus miurus	X	X	X	X→	Georgina 2001; Guthrie 1968a; Jopling et al. 1981; Repenning et al. 1964
	Morlan's vole	Microtus morlani	X				Storer 2004a
	root (tundra) vole	Microtus oeconomus	X?	X	X	X→	Georgina 2001; Harington 1990; Jopling et al. 1981; Morlan 1984; Storer 2003b
	meadow vole	Microtus pennsylvanicus		X	X	X→	Georgina 2001; Heaton 2001; Jopling et al. 1981; Morlan 1989
	taiga vole	Microtus xanthognathus	X	X	X	X→	Georgina 2001; Jopling et al. 1981; Storer 2003b
	Mimomys vole	Mimomys cf. virginianus	X				Repenning 2003
	red-backed vole	Myodes (Clethrionomys) sp.		X	X	X→	Georgina 2001; Harington 1978, 1990; Heaton and Grady 2003; Repenning et al. 1964
	common muskrat	Ondatra zibethicus	X	X	X	X→	Georgina 2001; Harington 1978; Jopling et al. 1981
	northwestern deermouse	Peromyscus keeni		?	?	X→	Heaton and Grady 2003
	deermouse	Peromyscus sp.	X				Storer 2003b
	heather vole	Phenacomys sp. (incl. deeringensis, intermedius)	X	X		(→)	Guthrie and Matthews 1971; Heaton and Grady 2003; Jopling et al. 1981; Martin et al. 1986; Repenning and Brouwers 1992; Repenning et al. 1987; Storer 2003a
	bog lemming	Synaptomys (Mictomys) sp.	X	X	X	X(→)	Georgina 2001; Harington 1978: Heaton and Grady 2003; Irving et al. 1989; Repenning et al. 1964; Storer 2003a
	N.A. porcupine	Erethizon dorsatum	X	X	X	X→	Ackerman 1996; Georgina 2001; Harington 1978; Heaton 2001; Kurtén and Anderson 1980

RECENT MAMMALS OF ALASKA

ORDER	COMMON NAME	SCIENTIFIC NAME	PLEISTOCENE		RECENT		REFERENCES
			Early and Middle Pleistocene	Late Pleistocene	Early and Middle Holocene	Late Holocene	
LAGOMORPHA—pikas and hares	collared/American pika	Ochotona collaris/princeps	X	X	X	↑	Guthrie 1973; Harington 1978; Jopling et al. 1981; Mead 1987; Yesner 2001
	giant pika	Ochotona whartoni	X	X			Guthrie and Matthews 1971; Jopling et al. 1981; Mead 1987; Mead and Grady 1996
	snowshoe hare	Lepus americanus	X	X	X	X↑	Harington 1978; Holmes 1996; Jopling et al. 1981; Yesner 2001
	Alaskan/arctic hare	Lepus othus/arcticus	X	X		(↑)	Best and Henry 1994; Jopling et al. 1981; Kurtén and Anderson 1980; Weber et al. 1981; Yesner et al. 2000
SORICOMORPHA—shrews	cinereus shrew	Sorex cf. cinereus	X			↑	Storer 2004a
	dusky shrew	Sorex monticolus		X?		X↑	Heaton and Grady 2003
	tundra shrew	Sorex tundrensis (arcticus)		X	X	↑	George 1988; Hopkins 1967; Yesner 2001
	shrew	Sorex sp. (incl. Microsorex)	X	X	X	(↑)	Georgina 2001; Guthrie and Matthews 1971; Harington 1980b; Storer 2003a, 2004a
CHIROPTERA—bats	bat	Unknown species	X				Storer 2004b
CARNIVORA—carnivores	American scimitar cat	Homotherium serum		X			Guthrie 1990a; Kurtén and Anderson 1980; Youngman 1993
	Canadian lynx	Lynx canadensis		X		↑	Kurtén and Anderson 1980; Youngman 1993
	American (Beringian) lion	Panthera leo atrox (?vereshchagini/spelaea)		X			Beebe and Hulland 1988; Burger et al. 2004; Grayson 1991; Guthrie 1990a; Harington 1996a; Kurtén 1985; Porter 1988
	cougar	Puma concolor		X		↑	Harington 1977; Jung and Merchant 2005; MacDonald and Cook 2007; Youngman 1993
	wolf	Canis lupus	X	X	X	↑	Georgina 2001; Guthrie 1990b; Harington 1990; Jopling et al. 1981; Nowak 1973; Savinetsky et al. 2004
	dhole	Cuon cf. alpinus	?	X			Harington 1978; Kurtén and Anderson 1980; Youngman 1993

ORDER	COMMON NAME	SCIENTIFIC NAME	PLEISTOCENE		RECENT		REFERENCES
			Early and Middle Pleistocene	Late Pleistocene	Early and Middle Holocene	Late Holocene	
CARNIVORA—carnivores, continued	arctic fox	*Vulpes (Alopex) lagopus*	X	X	X	↑	Georgina 2001; Heaton and Grady 2003; Jopling et al. 1981; Repenning 1983; Savinetsky et al. 2004; Yesner 2001; Youngman 1993
	red fox	*Vulpes vulpes*	X	X	X	↑	Georgina 2001; Heaton and Grady 2003; Kurtén and Anderson 1980; Péwé and Hopkins 1967; Sattler 1997; Yesner 2001; Youngman 1993
	giant short-faced bear	*Arctodus simus*		X			Harington 1980a, 1996b; Kurtén and Anderson 1980; Matheus 1997
	American black bear	*Ursus americanus*		X	X	X→	Dixon et al. 1979; Heaton and Grady 2003; Kurtén and Anderson 1980; Sattler 1997
	brown bear	*Ursus arctos*		X	X	X→	Barnes et al. 2002; Heaton and Grady 2003; Savinetsky et al. 2004
	polar bear	*Ursus maritimus*			X	X→	Davis 2001; Giddings 1964; Veltre et al. 2008
	northern fur seal	*Callorhinus ursinus*		X		X→	Harington 1978; Kurtén and Anderson 1980; Savinetsky et al. 2004
	Steller's sea lion	*Eumetopias jubatus*		X	X	X→	Domning 1978; Harington 1978; Heaton and Grady 2003; Kurtén and Anderson 1980; Savinetsky et al. 2004; Whitmore and Gard 1977
	walrus	*Odobenus rosmarus*	X	X		X→	Péwé and Hopkins 1967; Whitmore and Gard 1977; Domning 1978; Harington 1978; Kurtén and Anderson 1980; Dyke et al. 1999; Davis 2001
	bearded seal	*Erignathus barbatus*	X	X		X→	Kurtén and Anderson 1980; Savinetsky et al. 2004
	ribbon seal	*Histriophoca fasciata*		X		X→	Repenning 1983
	harbor/spotted seal	*Phoca vitulina/largha*		X		X→	Davis 2001; Harington 1978; Heaton and Grady 2003; Kurtén and Anderson 1980; Repenning 1983; Savinetsky et al. 2004
	ringed seal	*Pusa hispida*	X	X		X→	Davis 2001; Harington 1978; Heaton and Grady 2003; Kurtén and Anderson 1980; Repenning 1983; Savinetsky et al. 2004

RECENT MAMMALS OF ALASKA

ORDER	COMMON NAME	SCIENTIFIC NAME	PLEISTOCENE		RECENT		REFERENCES
			Early and Middle Pleistocene	Late Pleistocene	Early and Middle Holocene	Late Holocene	
CARNIVORA—carnivores, continued	sea otter	Enhydra lutris	X?		X	X→	Kurtén and Anderson 1980; Repenning 1983; Savinetsky et al. 2004; Youngman 1993
	wolverine	Gulo gulo	X	X	X	↑	Harington 1990; Jopling et al. 1981; Kurtén and Anderson 1980; Porter 1988; Youngman 1993
	N.A. river otter	Lontra canadensis		X	X	X→	Georgina 2001; Harington 1977; Heaton and Grady 2003; Yesner 2001; Youngman 1993
	American marten	Martes americana		X	X	X→	Ackerman et al. 1997; Georgina 2001; Youngman 1993
	noble marten	Martes (caurina) nobilis		X	X		Anderson 1970; Grayson 1984, 1991; Kurtén and Anderson 1980; Youngman 1993; Youngman and Schueler 1991
	fisher	Martes pennanti	X	X		↑	Harington 1977, 1990; Jopling et al. 1981; Youngman 1993
	ermine	Mustela erminea	X	X	X	X→	Georgina 2001; Heaton and Grady 2003; Irving et al. 1989; Kurtén and Anderson 1980; Youngman 1993
	steppe ferret	Mustela eversmanni		X			Anderson 1973, 1977; Youngman 1993, 1994
	black-footed ferret	Mustela nigripes		X			Kurtén and Anderson 1980; Youngman 1993, 1994
	least weasel	Mustela nivalis		X		↑	Harington 1989; Youngman 1993
	Jackson's weasel	Mustela jacksoni	X				Storer 2004b
	American mink	Neovison vison		X		↑	Anderson 1977; Heaton and Grady 2003; Youngman 1993
	American badger	Taxidea taxus		X			Anderson 1977; Flint 1957; Kurtén and Anderson 1980; Youngman 1993
	short-faced skunk	Brachyprotoma cf. obtusata	?	X			Grayson 1991; Harington 1990; Kurtén and Anderson 1980; Youngman 1986, 1993
PERISSODACTYLA—odd-toed ungulates	caballoid horse(s)	Equus spp. (caballus/ferus, lambei; others)	X	X			Burke and Cinq-Mars 1998; Guthrie 2001b, 2003; Harington 1979, 1990; Jopling et al. 1981; Porter 1988; Solow et al. 2006; Weinstock et al. 2005

ORDER	COMMON NAME	SCIENTIFIC NAME	PLEISTOCENE		RECENT		REFERENCES
			Early and Middle Pleistocene	Late Pleistocene	Early and Middle Holocene	Late Holocene	
PERISSODACTYLA—odd-toed ungulates, continued	stilt-legged (hemionid) horse(s)	*Equus* spp.	?	X			Guthrie 2001b, 2003; Harington 1978; Weinstock et al. 2005
ARTIODACTYLA—even-toed ungulates	flat-headed peccary	*Platygonus compressus*		X			Beebe 1980; Kurtén and Anderson 1980
	moose	*Alces americanus (alces)*		X	X	X↑	Guthrie 2006; Porter 1988
	broad-fronted moose	*Alces (Cervalces) latifrons*	X	X			Grayson 1991; Guthrie 1968b; Kurtén and Anderson 1980; Péwé and Hopkins 1967
	western camel	*Camelops hesternus*		X			Weber et al. 1981; Youngman 1994
	wapiti or E. red deer	*Cervus canadensis*	X	X	X	X† reintroduced	Guthrie 1966, 1983, 2006; Guthrie and Matthews 1971; Harington 2003; Kurtén and Anderson 1980; Péwé and Hopkins 1967; Porter 1988
	mule deer	*Odocoileus hemionus*			X	X↑	Heaton and Grady 2003
	caribou	*Rangifer tarandus*	X	X	X	X↑	Guthrie and Matthews 1971; Heaton and Grady 2003; Jopling et al. 1981; Péwé and Hopkins 1967; Porter 1988; Savinetsky et al. 2004
	bison	*Bison sp. (priscus/bison; alaskensis, others)*	X	X	X	X† reintroduced	Guthrie 1970, 1990a, 2006; Harington 1980a; McDonald 1981; Porter 1988; Shapiro et al. 2004; Stephenson et al. 2001
	helmeted (Harlan's) muskox	*Bootherium bombifrons*	X	X			McDonald and Ray 1989; Péwé and Hopkins 1967; Porter 1988
	tundra muskox	*Ovibos moschatus*	X	X	X	X† reintroduced	Harington 1980a; Péwé 1975; Porter 1988
	Staudinger's muskox	*Praeovibos priscus*	X				Kurtén and Anderson 1980; Péwé and Hopkins 1967
	saiga	*Saiga tatarica*	X	X			Guthrie et al. 2001; Harington and Cinq-Mars 1995; Porter 1988; Youngman 1994
	Soergel's muskox	*Soergalia mayfieldi*	X				Jopling et al. 1981

RECENT MAMMALS OF ALASKA

ORDER	COMMON NAME	SCIENTIFIC NAME	PLEISTOCENE		RECENT		REFERENCES
			Early and Middle Pleistocene	Late Pleistocene	Early and Middle Holocene	Late Holocene	
ARTIODACTYLA—even-toed ungulates, continued	Dall's sheep	*Ovis dalli*	X	X	X	X→	Bowers 1978; Guthrie 1968b, 1983; Kurtén and Anderson 1980; Péwé and Hopkins 1967; Porter 1988; Savinetsky et al. 2004
CETACEA—whales	bowhead	*Balaena mysticetus*				X→	Savinetsky et al. 2004
	beluga	*Delphinapterus leucas*		X		X→	Giddings 1964; Repenning 1983; Savinetsky et al. 2004
	gray whale	*Megaptera novaeangliae*		X		X→	Repenning 1983; Savinetsky et al. 2004

(*Early and Middle Pleistocene* = through Illinoian glacial; *Late Pleistocene* = Sangamonian interglacial through Wisconsinan glacial; *Early and Middle Holocene* = 10,000 to 5,000 years before present; *Late Holocene* = 5,000 years ago to present; X = fossil remains; (→) = species still present today; → = at least one species in genus still extant; † = extirpated/extinct by Late Holocene)

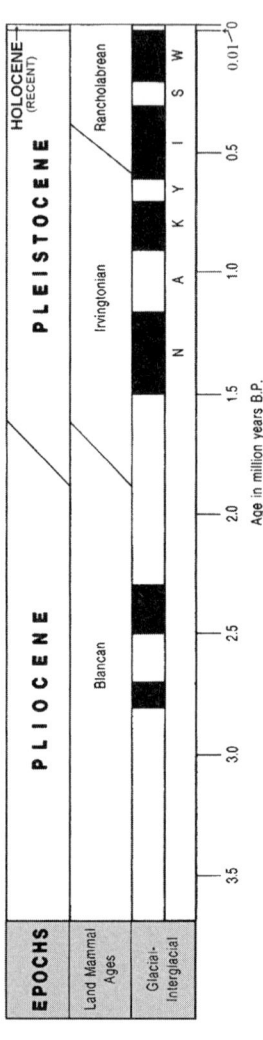

Late Pliocene and Pleistocene chronology (after Kurtén and Anderson 1980)

Index

Mammalian Scientific and Common Names in Text

Note: *Italicized* page numbers indicate illustrations. **Boldface** page numbers denote the beginning of species accounts.

A

abbreviatus, Microtus, **89,** 91
abbreviatus, Microtus abbreviatus, 89
abietorum, Vulpes alascensis, 162
abietorum, Vulpes vulpes, 162
ablusus, Citellus plesius, 70
ablusus, Spermophilus parryii, 70
actuosa, Martes americana, 201
actuosa, Mustela americana, 201
acutorostrata, Balaenoptera, **240**–241
admiraltiae, Microtus, 96
admiraltiae, Microtus pennsylvanicus, 96, 97
alascensis
 Evotomys, 100
 Lemmus, 87
 Lemmus trimucronatus, 87
 Mustela erminea, 206
 Myotis lucifugus, 148
 Putorius richardsoni, 206
 Sorex monticolus, 132
 Sorex obscurus, 132
 Ursus, 168
 Ursus arctos, 168
 Vulpes, 163
 Vulpes vulpes, 162, 163
 Zapus hudsonius, 81
Alaska arctic hare, 123–125
Alaska fur seal, 176
Alaska hare, **123**–125
Alaska haymouse, 91–93
Alaska marmot, **65**–66
alaskanus, Sorex, 134, **135**
alaskanus, Sorex navigator, 135

Alaska tiny shrew, 127, **139**–142
Alaska tundra hare, 123–125
Alaska vole, 91
albiventer, Clethrionomys, 100
albiventer, Myodes rutilus, 100
Alces, 214, 215–217
alces, Alces, 215
alces, Canis lupus, 159
alcorni, Microtus pennsylvanicus, 96
Alexandrine rat, 114
algidus, Peromyscus keeni, 104
algidus, Peromyscus maniculatus, 104, 106
alpina, Ochotona, 119
amakensis, Microtus, 93
amakensis, Microtus oeconomus, 93, 95
americana, Martes, 194, **200**–203
American beaver, 77, **78**–79
American bison, **226**–227
American black bear, **166**–168
American buffalo, 226
American marten, 194, **200**–203
American mink, **210**–211
American moose, 215–217
American pine marten, 200
americanum, Mammut, 53
americanus
 Alces, 214, **215**–217
 Lepus, 121, **122**–123, 125
 Oreamnos, **227**–229
 Ursus, **166**–168
 Ursus americanus, 166

American water shrew, *127,* **134**–135
andersoni, Alces americana, 215
andersoni, Alces americanus, 215
angliae, Balaena novae, 245
angustirostris, Macrorhinus, 188
angustirostris, Mirounga, **188,** *189*
aniakensis, Mustela vison, 210
aniakensis, Neovison vison, 210
arctica, Mustela erminea, 206, 208
arctic fox, **161**–162, 173
arctic ground squirrel, *62,* **69**–74
arctic marmot, 65
arctic right whale, 235
arcticus
 Lepus, 123, 124, 125
 Putorius, 206, 207
 Sorex, 137, 138
 Sorex personatus, 128
Arctocephalus, 175–176, 187
arctos, Ursus, 165, **168**–171, 172
arnuxii, Berardius, 272
Asiatic wild dog, 158
asiaticus, Rangifer tarandus, 222
athabascae, Bison bison, 226
attenuata, Stenella, **257**–259
attenuata, Stenella attenuata, 257
attenuatus, Steno, 258
australis, Balaena glacialis, 235, 237

B

bairdii, Berardius, 271, **272**–273

Baird's beaked whale, *271,* **272**–273

Balaena, 234, 235–237, *239,* 243, 245

Balaenidae family, 234–238

Balaenoptera, 240–244, 248

Balaenopteridae family, 239–246

banded seal, 187

barbatus, Erignathus, 172, **185**–187

barn rat, 113

barren ground shrew, **138**–139

barrowensis, Spermophilus, 70

bearded seal, 172, **185**–187

beluga, 259, **260**–261

belugae, Castor canadensis, 78

Berardius, 271, 272–273

beringensis, Spermophilus, 70

beringiana, Ursus arctos, 168

beringiana, Vulpes vulpes, 162

beringianus, Ursus arctos, 168–169

Bering Sea beaked whale, 273

big brown bat, **144**

Bison, 226–227

bison, Bison, **226**–227

bison, Bison bison, 226

black-backed shrew, 137

black bear, 166–167, 171

blackfish *(Globicephala macrorhynchus),* 251

blackfish *(Orcinus orca),* 255

black-footed lemming, 87

black fox, 162

black rat, 114

black right whale, 237

black whale, 237

bladdernose seal, 185

blue fox, 161

blue rorqual, 242

blue whale, 239, **242**–243

bonaerensis, Balaenoptera, 240

bonasus, Bison, 226

borealis
    Balaenoptera, **241**–242
    Lissodelphis, **254**–255
    Peromyscus maniculatus, 106
    Synaptomys, **108**–110

boreal red-backed vole, 98

Bovidae family, 225–233

bowhead, 234, **235**–236, 237, *238*

breviceps, Kogia, 267, **268**

breviceps, Physeter, 268

brevirostris, Orcaella, 261

Brooks Range marmot, 65

broweri, Marmota, **65**–66

broweri, Marmota caligata, 66, 67

brown bear, 161, *165,* **168**–171

brown lemming, 86, **87**–89

brown rat, *111,* **113**–114

bushy-tailed woodrat, **101**–102

C

California bat, 145–146

California gray whale, 248–249

California myotis, **145**–146

californiana, Otaria, 179

Californian myotis, 145–146

californianus, Zalophus, **179**–180

californianus, Zalophus californianus, 179

California sea lion, **179**–180

California sealion, 179–180

californicus, Myotis, **145**–146

caligata, Marmota, **67**–68

caligata, Marmota caligata, 67

caligatus, Arctomys, 67

camtschatica, Sorex, 138

canadensis
    Castor, 77, **78**–79
    Cervus, 214, **217**–219
    Lontra, **199**–200
    Lynx, 151, **152**–153
    Lynx canadensis, 152
    Marmota monax, 68
    Ovis, 231

Canadian beaver, 78–79

Canadian Lynx, 151, **152**–153

Canadian otter, 199–200

Canidae family, 156–164

Canis, 157–161

cantator, Microtus, 91

cantator, Microtus miurus, 91

capenus, Lepus, 126

caribou, *vi,* 159, 160, 161, 173, 214, **221**–224

caribou, Rangifer tarandus, 221–222

Castor, 77, 78–79

Castoridae family, 77–79

catodon, Physeter, 267, **269**–270

caurina, Martes, 194, 200–201, **203**–204

caurinus, Myotis californicus, 145

cavirostris, Ziphius, **274**

celenda, Mustela erminea, 207

Cervidae family, 214–224

Cervus, 214, 217–219

chestnut-cheeked vole, 97

chrotorrhinus, Microtus, 97

chukchenis, Rangifer tarandus, 222

cinerea, Neotoma, **101**–102

cinereus
    Sorex, **128**–129, 130–131, 138
    Sorex arcticus, 128
    Sorex cinereus, 128

Cinereus shrew, **128**–129

coalfish whale, 241

collared lemming, **85**–87

collared pika, *118,* **119**–120

collaris, Lagomys, 119

collaris, Ochotona, 118, **119**–120

columbiae, Oreamnos americanus, 228

common minke whale, **240**–241

common muskrat, **102**–104

common porpoise, 264

common rorqual, 243

common seal, 191

common shrew, 128

concolor, Puma, **153**–155

coronarius, Microtus, 90

coronarius, Microtus longicaudus, 90

Coronation Island vole, 90

cougar, 151, **153**–155

coyote, **157**–158, 163

crassidens, Phocaena, 256

crassidens, Pseudorca, **256**–257

crested seal, 185

Cricetidae family, 84–110

cristata, Cystophora, **185**

cristata, Phoca, 185

cross fox, 162

cuniculus, Oryctolagus, **125**–126

cuon, 158

Cuvier's beaked whale, **274**

Cystophora, 185

D

dalli
    Lepus americanus, 122
    Ovis, 225, **231**–233
    Ovis dalli, 232
    Ovis montana, 231
    Phocaena, 265
    Phocoenoides, 263, **265**–266
    Synaptomys borealis, 108–109
    Ursus, 169
    Ursus arctos, 169

Dall's porpoise, **265**–266

Dall's sheep, 225, **231**–233

davidsoni, Balaenoptera acutorostrata, 240

dawsoni, Evotomys, 100

dawsoni, Myodes rutilus, 100

Dawson red-backed mouse, 100

deer mouse, 104–105

Delphinapterus, 259, 260–261

Delphinidae family, 250–258

dhole, 158

Dicrostonyx, 85–87

Dipodidae family, 80–83

divergens, Odobenus rosmarus, 182

divuliana, Martes, 206

domestic rabbit, 125, 126

domesticus, Mus, 112

domesticus, Mus musculus, 112
dorsatum, Erethizon, 115, **116**–117
drummondii, Microtus pennsylvanicus, 96
dugong and sea cow, 56–58
Dugongidae family, 56–58
dusky shrew, **131**–134

E

eastern red deer, 217–218
elaphus, Cervus, 217–218
elassodon, Sorex longicauda, 132
elassodon, Sorex monticolus, 132
Elephantidae family, 53–55
elk (*Alces alces*), 215
elk (*Cervus canadensis*), 214, 217–219
elymocetes, Microtus, 93
elymocetes, Microtus oeconomus, 93
emmonsii, Ursus americanus, 166
empetra, Spermophilus, 70
endoecus, Microtus operarius, 93
energumenos, Neovison vison, 210
energumenos, Putorius vison, 210
Enhydra, 195–197
Eptesicus, 144
Erethizon, 115, 116–117
Erethizontidae family, 115–117
Erignathus, 172, 185–187
ermine, *194,* **206**–208
erminea, Mustela, 194, **206**–208
Eschrichtiidae family, 247–249
Eschrichtius, 247, 248–249
eskimo, Mustela nivalis, 208
eskimo, Putorius rixosus, 208
Eubalaena, 237–238
Eumetopias, 177–179, *180*
European rabbit, **125**–126
eximius, Sorex hoyi, 129
exsul, Dicrostonyx, 85, 86
exsul, Dicrostonyx groenlandicus, 85

F

false killer whale, **256**–257
familiaris, Canis, 159
familiaris, Canis lupus, 158, 159
Fannin's sheep, 231
fasciata, Histriophoca, **187**–188
fasciata, Phoca, 187, 190
Felidae family, 151–155
fiber, Castor, 77, 78
finback whale, 243
fin whale, *239,* **243**–244
fisher, **205**–206
fisher cat, 205
fisheri, Microtus abbreviatus, 89
foina, Martes, 204
forest deer mouse, 104–105
fuscus, Eptesicus, **144**

G

gapperi, Myodes, **98**–100
Gapper's red-backed vole, 98
giant sperm whale, 269
gibbosus, Eschrichtius, 248
gigas
    Alces, 215
    Alces americanus, 215
    Hydrodamalis, 56, **57**–58
    Manati, 57
gilmorei, Microtus oeconomus, 94
glacialis
    Balaena, 235
    Balaena glacialis, 235, 237
    Clethrionomys dawsoni, 100
    Myodes rutilus, 100
    Orcinus, 255
    Sorex, 132
Glacier Bay water shrew, **135**
glacier bear, 166
Glaucomys, 63–65
Globicephala, 251–252
goosebeak whale, 274
graffmani, Stenella attenuata, 257
granti, Rangifer, 222
granti, Rangifer tarandus, 221, 222
gray whale, *247,* **248**–249
gray wolf, 158
great northern rorqual, 242
great northern sea-cow, 57
great polar whale, 235
Greenland right whale, 235
Greenland seal, 188
griseifrons, Glaucomys sabrinus, 63, 65
griseus, Delphinus, 252
griseus, Grampus, 252, **252**
grizzly bear, 168
groenlandica, Phoca, 189
groenlandicus
    Dicrostonyx, **85**–87
    Pagophilus, **188**–189
    Rangifer tarandus, 221
groundhog, 68
Guadalupe fur seal, **175**–176
Guadalupe fur-seal, 175
Gulo, 197–199, 205
gulo, Gulo, **197**–199, 205
gyas, Ursus arctos, 169
gyas, Ursus dalli, 169

H

haidarum, Mustela erminea, 206, 208
hair seal, 191
hairy-winged myotis, 149
hallensis, Vulpes, 161
hallensis, Vulpes lagopus, 161
harbor porpoise, *263,* **264**–265

harbor seal, **191**–192
harp seal, **188**–189
harrimani, Vulpes, 163
harrimani, Vulpes vulpes, 163
harringtoni, Oreamnos, 229
harroldi, Lemmus, 87
harroldi, Lemmus trimucronatus, 87
heather vole, 107
helvolus, Arvicola, 87
helvolus, Lemmus trimucronatus, 87
hemionus
    Cervus, 219
    Odocoileus, **219**–220, 221
    Odocoileus hemionus, 219, 220
herring whale, 243
hispida
    Phoca, 171, 192
    Pusa, 184, **192**–193
    Pusa hispida, 192
Histriophoca, 187–188
hoary marmot, **67**–68
hollisteri, Sorex cinereus, 128
Hominidae family, 59–61
Homo sapiens, 60–61
hooded seal, **185**
hoofed lemming, 85
house mouse, *111,* **112**
hoyi, Microsorex (see hoyi, Sorex)
hoyi, Sorex, **129**–130
hudsonicus, Sciurus, 74
hudsonicus, Tamiasciurus, **74**–76
hudsonius
    Dipus, 81
    Zapus, 80, **81**–82
    Zapus hudsonius, 81
human, **60**–61
human being, 60
humpback whale, **244**–246
hump whale, 244
hunchbacked whale, 244
hydrochaeris, Hydrochoerus, 77
Hydrodamalis, 56, 57–58
hydromus, Sorex, 135, 136
hylaeus
    Gulo, 197
    Peromyscus, 104
    Peromyscus keeni, 104
hyperborea, Ochotona, 119

I

ice bear, 171
incolatus, Canis latrans, 157
ingens, Lutreola vison, 210
ingens, Neovison vison, 210
initis, Mustela erminea, 207
innuitus, Microtus, 93
innuitus, Microtus oeconomus, 93
insularis, Evotomys dawsoni, 100
insularis, Myodes rutilus, 100

insular vole, **89**
intermedius, Phenacomys, **107**–108
intermedius, Phenacomys interme-
dius, 107

J

jacksoni, Sorex, 128, **130**–131,
138
Japan finner, 241
japonica, Eubalaena, **237**–238
jubata, Phoca, 177
jubatus, Eumetopias, **177**–179, *180*

K

kadiacensis
   Microtus, 94
   Mustela erminea, 207
   Putorius arcticus, 207
kamtschaticus, Microtus, 93
katschemakensis, Gulo, 197
katschemakensis, Gulo gulo, 197
Keen bat, 146
keeni, Peromyscus, **104**–105, 106
keenii, Myotis, **146**–148
keenii, Vespertilio subulatus, 147
Keen's long-eared bat, 146
Keen's long-eared myotis, 146
Keen's mouse, 104–105
Keen's myotis, **146**–148
kenaiensis
   Martes americana, 201
   Mustela americana, 201
   Ovis dalli, 232
   Tamiasciurus hudsonicus, 74
   Ursus americanus, 166
   Vulpes, 163
   Vulpes vulpes, 163
kennedyi, Oreamnos, 228
kennedyi, Oreamnos americanus,
228
kennicottii, Arctomys, 70
kennicottii, Spermophilus parryii,
70
kenyoni, Enhydra lutris, 195
killer whale, **255**–256
kodiacensis
   Lontra canadensis, 199
   Lutra canadensis, 199
   Spermophilus parryii, 70
Kogia, 267, 268
kolymensis, Ursus arctos, 169

L

Lagenorhynchus, 253–254
lagopus
   Canis, **161**–162
   Vulpes, 161
   Vulpes lagopus, 161
land otter, 199

laptevi, Odobenus rosmarus, 182
Larga seal, 189
largha, Phoca, **189**–190
Lasionycteris, 144–145
latifrons, Alces, 217
latrans, Canis, **157**–158
least weasel, **208**–209
lemming mouse, 108
Lemmus, 86, 87–89
lenensis, Rangifer tarandus, 222
Leporidae family, 121–126
Lepus, 121, 122–126
lesser rorqual, 240
leucas, Delphinapterus, 259,
260–261
leucas, Delphinus, 260
leucogaster, Sorex, 131, 138
ligoni, Canis lupus, 159
Lissodelphis, 254–255
little brown bat, *143,* 148–149
little brown myotis, 143,
148–149
little piked whale, 240
littoralis
   Microtus longicaudus, 90–91
   Microtus mordax, 90
   Microtus oeconomus, 94
longicauda, Sorex monticolus, 132
longicauda, Sorex obscurus, 132
longicaudus, Microtus, 84, **90**–91
longicrus, Myotis volans, 150
longicrus, Vespertilio, 150
long-legged bat, 149
long-legged myotis, **149**–150
long-tailed suslik, 69
long-tailed vole, *84,* **90**–91
Lontra, 199–200
lotor, Procyon, 212, **213**
lotor, Procyon lotor, 213
lucifugus, Myotis, 143, **148**–149
lupus, Canis, 157, **158**–161
luscus, Gulo gulo, 197
lutris, Enhydra, **195**–197
lynx, 152–153
Lynx, 151, 152–153
lynx, Lynx, 152
lyratus, Citellus, 70
lyratus, Spermophilus parryii, 70

M

macfarlani
   Lepus americanus, 122
   Microtus, 93
   Microtus oeconomus, 93–94
macrocephalus, Physeter, 269
macrorhinus, Peromyscus keeni,
104–105
macrorhinus, Sitomys, 104

macrorhynchus, Globicephala,
251–252
malitiosus, Sorex monticolus, 132
malitiosus, Sorex obscurus, 132
Mammuthus, 53, 54–55
maniculatus, Peromyscus, 104,
106–107
marginata, Caperea, 248
maritimus, Ursus, 168, **171**–173
Marmota, 65–68
Martes, 194, 200–206
martes, Martes, 200, 204
masked shrew, 128
meadow jumping mouse, *80,*
81–82
meadow vole, **95**–97
Megaptera, 244–246
melaena, Globicephala, 251
melampeplus, Neovison vison, 210
melampeplus, Putorius vison, 210
melampus, Martes, 200
melas, Globicephala, 251, 252
Mesoplodon, 273–274
Microsorex (see hoyi, Sorex)
Microtus, 84, 89–97, *91*
middendorffi, Ursus, 169
middendorffi, Ursus arctos, 169
mink, 194, 210–211
minusculus, Lemmus, 87
minusculus, Lemmus trimucrona-
tus, 87
minutissimus, Sorex, 139
mira, Lontra canadensis, 199
mira, Lutra canadensis, 199
Mirounga, 188, *189*
missoulensis, Felis concolor, 153
missoulensis, Puma concolor, 153
miurus, Microtus, 89, **91**–93, 97
miurus, Microtus miurus, 92
modern man, 60
mollipilosus, Lynx canadensis, 152
monax, Marmota, **68**–69
monoceros, Monodon, **261**–262
Monodon, 261–262
Monodontidae family, 259–262
montane shrew, 131, 132
monticolus, Sorex, **131**–134
moose, *214,* **215**–217
moschatus
   Bos, 229
   Ovibos, **229**–231
   Ovibos moschatus, 229
mountain goat, **227**–229
mountain lion, 153
mouse-weasel, 208
mule deer, **219**–220
Muridae family, 111–114
muriei, Microtus, 92
muriei, Microtus miurus, 92

Mus, 111, 112
musculus
   Balaenoptera, **242**–243
   Balaenoptera musculus, 242
   Mus, 111, **112**
muskox, **229**–231
muskrat, 84, 102–104, *103, 104*
Mustela, 194, 201, 205–210
Mustelidae family, 194–211
Myodes, 98–101
myops, Erethizon dorsatum, 116
myops, Erethizon epixanthus, 116
Myotis, 143, 145–150
mysticetus, Balaena, 234,
   235–236, *239*

N
nanuk or nanook, 171
narwhal, 259, **261**–262
narwhale, 261
nauticus, Erignathus barbatus, 186
navigator
   Neosorex, 134
   Sorex, 134
   Sorex palustris, 134
navigator shrew, 134
nearctic brown lemming, 87
nearctic collared lemming, 85
nebulicola, Citellus, 70
nebulicola, Spermophilus parryii, 70
nelsoni
   Cervus canadensis, 218
   Dicrostonyx, 85
   Dicrostonyx groenlandicus, 85
   Neotoma, 101–102
   Neovison, 210–211
nesolestes, Lutreola vison, 210
nesolestes, Neovison vison, 210
nesophila, Martes caurina, 203
nigrescens, Erethizon dorsatum, 116
nigripes, Lemmus trimucronatus,
   87, 88
nigripes, Myodes, 87
nivalis, Mustela, **208**–209
nivicola, Ovis, 231
nobilis, Martes, 204
noctivagans, Lasionycteris, **144**–145
noctivagans, Vespertilio, 144
nodosa, Megaptera, 245
North American beaver, 78
North American deermouse,
   106–107
North American porcupine, *115,*
   116–117
North American river otter,
   199–200
northern bog lemming, **108**–110
northern collared lemming, 85
northern elephant seal, **188**
northern flying squirrel, **63**–65

northern four-toothed whale,
   272
northern fur seal, *174,* **176**–177
northern giant bottlenose whale,
   272
northern minke whale, 240
northern red-backed mouse, 100
northern red-backed vole,
   100–101
northern right whale, 237
northern right whale dolphin,
   254–255
northern right whale porpoise,
   254
northern river otter, 199
northern sea lion, 177
northern vole, 93
northern water shrew, 134
North Pacific bottlenose whale,
   272
North Pacific fur seal, 176
North Pacific right whale,
   237–238
northwestern deermouse,
   104–105
norvegicus, Rattus, 111, **113**–114
Norway rat, 113
novaeangliae, Megaptera, **244**–246

O
obesus, Trichechus, 182
obliquidens, Lagenorhynchus,
   253–254
obscurus
   Lagenorhynchus, 253
   Sorex, 131–132
   Sorex monticolus, 132
occidentalis, Neotoma, 101
occidentalis, Neotoma cinerea,
   101–102
oceanicus, Peromyscus keeni, 105
oceanicus, Peromyscus sitkensis, 105
Ochotona, 118, 119–120
Ochotonidae family, 118–120
ochracea, Marmota, 68
ochracea, Marmota monax, 68
Odobenidae family, 181–183
Odobenus, 181, 182–183
Odocoileus, 214, 219–221
oeconomus, Microtus, 92, **93**–95
ohioensis, Castoroides, 79
Old World rabbit, 125
Ondatra, 102–104
operarius
   Arvicola, 94
   Microtus, 93
   Microtus oeconomus, 94
optiva, Lutra canadensis, 199
orca, 255

orca
   Delphinus, 255
   Evotomys, 100
   Myodes rutilus, 100
   Orcinus, 250, **255**–256
Orcinus, 250, 255–256
Oreamnos, 227–229
oreas, Microtus miurus, 92
oregonensis, Felix, 154
oregonensis, Puma concolor, 153,
   154
Oryctolagus, 125–126
osgoodi, Spermophilus, 70
osgoodi, Spermophilus parryii, 70,
   73
Otariidae family, 174–180
othus, Lepus, **123**–125
othus, Lepus othus, 124
Ovibos, 229–231
Ovis, 225, 231–233

P
pacifica
   Lontra canadensis, 199
   Lutra hudsonica, 199
   Martes pennanti, 205
   Mustela canadensis, 205
Pacific marten, **203**–204
Pacific pilot whale, 251
Pacific right whale *(Eubalaena*
   japonica), 237
Pacific right whale *(Lissodelphis*
   borealis), 254
Pacific walrus, 182–183
Pacific white-sided dolphin,
   253–254
Pacific white-striped dolphin,
   253
packrat, 101
Pagophilus, 188–189
pallasi, Ochotona, 119
pallidus, Eptesicus, 144
pallidus, Eptesicus fuscus, 144
palustris, Sorex, 127, **134**–135
pambasileus, Canis, 159
pambasileus, Canis lupus, 159
paneaki, Microtus miurus, 92
pantropical spotted dolphin,
   257–259
pardinus, Lynx, 152
Parry ground squirrel, 69
parryii, Spermophilus, 62, **69**–74
parvus, Morenocetus, 238
pekan, 205
peninsulae, Dicrostonyx groenlandi-
   cus, 85
peninsulae, Dicrostonyx unalascen-
   sis, 85
pennanti, Martes, **205**–206
Pennant's marten, 205

pennsylvanicus, Microtus, **95**–97
perniger, Ursus americanus, 166
pernox, Myotis, 148
pernox, Myotis lucifugus, 148
Peromyscus, 104–107
peronii, Lissodelphis, 254
petrensis, Marmota monax, 68
petulans, Sciurus hudsonicus, 74
petulans, Tamiasciurus hudsonicus, 74
phaeus
   Castor canadensis, 78, 79
   Evotomys, 98
   Myodes gapperi, 98
Phenacomys, 107–108
philippii, Arctocephalus, 187
Phoca, 171, 176, 177, 185, 187, 189–192
Phocidae family, 184–193
Phocoena, 263, 264–265
phocoena, Phocoena, 263, **264**–265
Phocoenidae family, 263–266
Phocoenoides, 263, 265–266
physalus
   Balaena, 243
   Balaenoptera, **243**–244
   Balaenoptera physalus, 243–244
Physeter, 267, 268, 269–270
Physeteridae family, 267–270
picatus, Sciurus hudsonicus, 74
picatus, Tamiasciurus hudsonicus, 74
pika, 120
piked whale, 240
pine marten, 200
pine squirrel, 74
plains bison, 226
plesius, Spermophilus empetra, 70
plesius, Spermophilus parryii, 70
poadromus, Lepus, 124
poadromus, Lepus othus, 124
polar bear, **171**–173
polar fox, 161
pollack whale, 241
popofensis, Microtus oeconomus, 94, 95
popofensis, Microtus unalascensis, 94
porcupine, 116
portenkoi, Sorex, 131, 138
pot whale, 269
preblei, Tamiasciurus hudsonicus, 75
pribilofensis, Sorex, 128, 129, 131, **135**–137, 138
pribilofensis, Vulpes lagopus, 161
Pribilof fur seal, 176
Pribilof Island shrew, **135**–137
primigenius, Elephas, 54
primigenius, Mammuthus, 53, **54**–55

princeps
   Ochotona, 118, 119
   Zapus, **82**–83
    Zapus princeps, 83
priscus, Bison, 227
Procyon, 212, 213
Procyonidae family, 212–213
Pseudorca, 256–257
pugnax, Ursus americanus, 166
puma, 153
Puma, 153–155
punukensis, Microtus innuitus, 94
punukensis, Microtus oeconomus, 94, 95
Pusa, 184, 192–193
pusilla, Ochotona, 118
pygmy shrew, **129**–130
pygmy sperm whale, **268**

Q
Queen Charlotte marten, 203

R
raccoon, **213**
Rangifer, vi, 221–224
ratticeps, Microtus, 93
Rattus, 111, 113–114
rattus, Rattus, 113, **114**
razorback whale, 243
red fox, *156,* **162**–164
red squirrel, **74**–76
reindeer, 221, 222, 223, 224
ribbon seal, **187**–188
richardii, Halicyon, 191
richardii, Phoca vitulina, 191
richardsi, Phoca, 191
richardsonii, Mustela erminea, 206
right whale, 237
ringed seal, *184,* **192**–193
Risso's dolphin, **252**
river otter, 200
rixosa, Mustela, 208
rixosa, Mustela nivalis, 208
robusta, Balaenoptera, 248
robustus, Eschrichtius, 247, **248**–249
rocky mountain goat, 227
roof rat, **114**
roosevelti, Cervus, 218
roosevelti, Cervus canadensis, 218
root vole, **93**–95
rosmarus, Odobenus, 181, **182**–183
rubidus, Microtus pennsylanicus, 96
rubricatus, Arvicola, 85
rubricatus, Dicrostonyx groenlandicus, 85
Rudolphi's rorqual, 241
rufus, Lynx, 152
rutilus, Myodes, 98, **100**–101

S
sabre-toothed beaked whale, 273
sabrinus, Glaucomys, **63**–65
saddle-backed seal, 188
saltator, Zapus, 82
saltator, Zapus princeps, 82–83
salva, Mustela erminea, 207
sapiens, Homo, **60**–61
sardine whale, 241
saturatus, Evotomys gapperi, 99
saturatus, Myodes gapperi, 99
saxamans, Neotoma, 101
scammoni, Balaenoptera acutorostrata, 240
scammoni, Globicephala, 251
Sciuridae family, 62–76
sea otter, **195**–197
seclusa, Mustela erminea, 207
sei whale, **241**–242
septentrionalis, Myotis keenii, 146
sharpheaded finner, 240
sheldoni, Marmota caligata, 67, 68
ship rat, 114
short-finned blackfish, 251
short-finned pilot whale, **251**–252
short-tailed weasel, 206
shumaginensis, Sorex alascensis, 132
shumaginensis, Sorex monticolus, 132
Sibbald's rorqual, 242
Siberian lemming, 87
sibiricus, Lemmus, 87
sibiricus, Rangifer tarandus, 221, 222
sik-sik, 69
silver fox, 162
silver-haired bat, **144**–145
sima, Kogia, 268
similis, Sorex vagrans, 132
singing vole, **91**–93
Sitka black-tailed deer, 214, 219, 220, 221
Sitka deer, 219
Sitka deer mouse, 104–105
Sitka mouse, 104–105
Sitka white-footed mouse, 104–105
sitkensis
   Microtus, 94
   Microtus oeconomus, 94, 95
   Odocoileus columbianus, 219
   Odocoileus hemionus, 214, 219, 220
   Peromyscus, 104, 105
   Peromyscus keeni, 105
   Ursus, 169
   Ursus arctos, 169
snowshoe hare, *121,* **122**–123
snowshoe rabbit, 122

solus, *Clethrionomys gapperi*, 99
solus, *Myodes gapperi*, 99
Sorex, 128–142
Soricidae family, 127–142
southern red-backed mouse, 98
southern red-backed vole,
  98–100
spatulatus, Fiber, 103
spatulatus, Ondatra zibethicus, 103
Spermophilus, 62, 69–74
sperm whale, *267*, **269**–270
spotted seal, **189**–190
spray porpoise, 265
St. Lawrence Island shrew,
  130–131
St. Matthew Island vole, 89
stejnegeri, Mesoplodon, **273**–274
Stejneger's beaked whale,
  273–274
Steller sea lion, 177
Steller's sea cow, *56*, **57**–58
Steller's sea lion, **177**–179, *180*
Stenella, 257–259
*stevensoni, Dicrostonyx groenlandi-
  cus*, 85
stevensoni, Dicrostonyx unalascensis,
  85, 86
stikeenensis, Ursus, 169
stikeenensis, Ursus arctos, 169
stikinensis, Clethrionomys gapperi,
  99
stikinensis, Myodes gapperi, 99
stoat, 206
stonei
  Citellus, 70
  Ovis dalli, 231
  Rangifer, 221, 222
Stone's sheep, 231
streatori, Sorex cinereus, 128
streatori, Sorex personatus, 128
subarcticus, Lemmus trimucronatus,
  87
sulphur-bottom, 242
Synaptomys, 108–110

T

taiga vole, **97**–98
taimyrensis, Rangifer tarandus, 222
Tamiasciurus, 74–76
*tananaensis, Microtus pennsylvani-
  cus*, 96
tarandus, Rangifer, vi, **221**–224
thinhorn sheep, 231
timidus, Lepus, 123, 124, 125
Toklat vole, 91
torquatus, Dicrostonyx, 85

townsendi, Arctocephalus, **175**–176
trimucronatus, Lemmus, 86, **87**–89
truei, Synaptomys borealis, 109
tundra hare, 123–125
tundra mammoth, 54
tundra redback vole, 100
tundrarum, Canis, 159
tundrarum, Canis lupus, 159
tundra saddle-backed shrew, 137
tundra shrew, **137**–138
tundra vole, 93
tundra wolf, 158
tundrensis, Sorex, **137**–138
tundrensis, Sorex tundrensis, 137
tyrannus, Ursus maritimus, 173

U

ugyunak, Sorex, 128, 129, 131,
  138–139
ugyunak, Sorex cinereus, 138
unalascensis
  Dicrostonyx groenlandicus, 85
  Microtus, 94
  Microtus oeconomus, 94
ungava, Phenacomys, 107
unicorn whale, 261
Ursidae family, 165–173
ursina, Phoca, 176
ursinus, Callorhinus, 174, **176**–177
Ursus, 165, 166–173

V

Vancouver Island marten, 203
varying hare, 122
varying lemming, 122
vellerosus, Microtus, 91
vellerosus, Microtus longicaudus, 91
Vespertilionidae family,
  143–150
vigilis, Marmota, 67
vigilis, Marmota caligata, 67, 68
vison, Mustela, 210
vison, Neovison, **210**–211
vitulina, Phoca, 189, 190,
  191–192
volans, Myotis, **149**–150
vomerina, Phocaena, 264
vomerina, Phocoena phocoena, 264
Vulpes, 156, 161–164
vulpes, Vulpes, 156, 161, **162**–164

W

walrus, *181*, **182**–183
wapiti, 214, **217**–219
water shrew, 134–135
watsoni, Clethrionomys dawsoni,
  100

watsoni, Myodes rutilus, 100
weasel, 208
western heather vole, **107**–108
western jumping mouse, **82**–83
whistler, 67
white bear, 171
whitefin porpoise, 265
whiteflank porpoise, 265
white-footed mouse (*Peromyscus
  keeni*), 104
white-footed mouse (*Peromyscus
  maniculatus*), 106
whitesided porpoise, 265
white whale, 260
wolf, **158**–161
wolverine, **197**–199
wood bison, 226, 227
woodchuck, **68**–69
woolly mammoth, **54**–55
wrangeli
  Evotomys, 99
  Myodes gapperi, 99
  Synaptomys, 109

X

xanthognatha, Arvicola, 97
xanthognathus, Microtus, **97**–98

Y

yakutatensis
  Microtus, 94
  Microtus oeconomus, 94
  Rangifer tarandus, 222
yellow-cheeked vole, 97
yellow-nosed vole, 97
yukonensis
  Glaucomys sabrinus, 63, 64
  Lemmus, 87
  Lemmus trimucronatus, 87
  Lutra canadensis, 199
  Sciuropterus, 63
yukonicus, Sorex, 127, **139**–142

Z

Zalophus, 179–180
zalophus, Fiber zibethicus, 103
zalophus, Ondatra zibethicus, 103
zaphaeus, Glaucomys sabrinus, 63
zaphaeus, Sciuropterus alpinus, 63
Zapus, 80, 81–82
zibellina, Martes, 200, 203
zibethicus, Ondatra, **102**–104
Ziphiidae family, 271–274
Ziphius, 274